Activation and Functionalization of C–H Bonds

ACS SYMPOSIUM SERIES **885**

Activation and Functionalization of C–H Bonds

Karen I. Goldberg, Editor
University of Washington

Alan S. Goldman, Editor
Rutgers, The State University of New Jersey

Sponsored by the
ACS Division of Inorganic Chemistry, Inc.

American Chemical Society, Washington, DC

Library of Congress Cataloging-in-Publication Data

Activation and functionalization of C–H Bonds / Karen I. Goldberg, editor, Alan S. Goldman, editor; sponsored by the ACS Division of Inorganic Chemistry, Inc.

p. cm.—(ACS symposium series ; 885)

Includes bibliographical references and index.

ISBN 0–8412–3849–9 (alk. paper)

1. Hydrocarbons—Congresses. 2. Activation (Chemistry)—Congresses. 3. Organometallic compounds—Congresses. 4. Catalysis—Congresses.

I. Goldberg, Karen I., 1962- II. Goldman, Alan S., 1958- III. American Chemical Society. Division of Inorganic Chemistry, Inc. IV. American Chemical Society. Meeting (223rd : 2002 : Orlando, Fla.) V. Series.

QD305.H5A25   2004
547'.01—dc22   2004041123

The paper used in this publication meets the minimum requirements of American National Standard for Information Sciences—Permanence of Paper for Printed Library Materials, ANSI Z39.48–1984.

Copyright © 2004 American Chemical Society

Cover design by Nadine Summa and Alan Goldman.

Distributed by Oxford University Press

All Rights Reserved. Reprographic copying beyond that permitted by Sections 107 or 108 of the U.S. Copyright Act is allowed for internal use only, provided that a per-chapter fee of $27.25 plus $0.75 per page is paid to the Copyright Clearance Center, Inc., 222 Rosewood Drive, Danvers, MA 01923, USA. Republication or reproduction for sale of pages in this book is permitted only under license from ACS. Direct these and other permission requests to ACS Copyright Office, Publications Division, 1155 16th St., N.W., Washington, DC 20036.

The citation of trade names and/or names of manufacturers in this publication is not to be construed as an endorsement or as approval by ACS of the commercial products or services referenced herein; nor should the mere reference herein to any drawing, specification, chemical process, or other data be regarded as a license or as a conveyance of any right or permission to the holder, reader, or any other person or corporation, to manufacture, reproduce, use, or sell any patented invention or copyrighted work that may in any way be related thereto. Registered names, trademarks, etc., used in this publication, even without specific indication thereof, are not to be considered unprotected by law.

PRINTED IN THE UNITED STATES OF AMERICA

Foreword

The ACS Symposium Series was first published in 1974 to provide a mechanism for publishing symposia quickly in book form. The purpose of the series is to publish timely, comprehensive books developed from ACS sponsored symposia based on current scientific research. Occasionally, books are developed from symposia sponsored by other organizations when the topic is of keen interest to the chemistry audience.

Before agreeing to publish a book, the proposed table of contents is reviewed for appropriate and comprehensive coverage and for interest to the audience. Some papers may be excluded to better focus the book; others may be added to provide comprehensiveness. When appropriate, overview or introductory chapters are added. Drafts of chapters are peer-reviewed prior to final acceptance or rejection, and manuscripts are prepared in camera-ready format.

As a rule, only original research papers and original review papers are included in the volumes. Verbatim reproductions of previously published papers are not accepted.

ACS Books Department

Dedication

To
Kent, Hannah, Maya, and Bernice
Danka, Olivia, Leo, and Rhoda

Contents

Preface xiii

1. **Organometallic C–H Bond Activation: An Introduction** 1
   Alan S. Goldman and Karen I. Goldberg

Oxidative Addition Reactions

2. **C–H Bond Activation by Iridium and Rhodium Complexes: Catalytic Hydrogen–Deuterium Exchange and C–C Bond-Forming Reactions** 46
   Steven R. Klei, Kian L. Tan, Jeffrey T. Golden, Cathleen M. Yung, Reema K. Thalji, Kateri A. Ahrendt, Jonathan A. Ellman, T. Don Tilley, and Robert G. Bergman

3. **Alkane Complexes as Intermediates in C–H Bond Activation Reactions** 56
   William D. Jones, Andrew J. Vetter, Douglas D. Wick, and Todd O. Northcutt

4. **C–C versus C–H Bond Oxidative Addition in PCX (X=P,N,O) Ligand Systems: Facility, Mechanism, and Control** 70
   Boris Rybtchinski and David Milstein

5. **Kinetic and Equilibrium Deuterium Isotope Effects for C–H Bond Reductive Elimination and Oxidative Addition Reactions Involving the *Ansa*–Tungstenocene Methyl–Hydride Complex [Me$_2$Si(C$_5$Me$_4$)$_2$]W(Me)H** 86
   Kevin E. Janak, David G. Churchill, and Gerard Parkin

6. **Alkane C–H Bond Activation by O-Donor Ir Complexes** 105
   Gaurav Bhalla, Xiang Yang Liu, Antek Wong-Foy, C. J. Jones, and Roy A. Periana

7. Lanthanide Complexes: Electronic Structure and H–H, C–H, and Si–H Bond Activation from a DFT Perspective 116
   Lionel Perrin, Laurent Maron, and Odile Eisenstein

C–H Bond Activations in Organic Synthesis, Alkane Dehydrogenation, and Related Chemistry

8. Catalytic, Thermal, Regioselective Functionalization of Alkanes and Arenes with Borane Reagents 136
   John F. Hartwig

9. C–H Bond Functionalization in Complex Organic Synthesis 155
   Dalibor Sames

10. Pincer and Chelate N-Heterocyclic Carbene Complexes of Rh, Ir, and Pd: Synthetic Routes, Dynamics, Catalysis, Abnormal Binding, and Counterion Effects 169
    Anthony R. Chianese and Robert H. Crabtree

11. Sequential Hydrocarbon C–H Bond Activations by 16-Electron Organometallic Complexes of Molybdenum and Tungsten 184
    Peter Legzdins and Craig B. Pamplin

12. Alkane Transfer-Dehydrogenation Catalyzed by a Pincer-Ligated Iridium Complex 198
    Alan S. Goldman, Kenton B. Renkema, Margaret Czerw, and Karsten Krogh-Jespersen

13. DFT Calculations on the Mechanism of (PCP)Ir-Catalyzed Acceptorless Dehydrogenation of Alkanes: Realistic Computational Models and Free Energy Considerations.. 216
    Karsten Krogh-Jespersen, Margaret Czerw, and Alan S. Goldman

14. Double Cyclometalation: Implications for C–H Oxidative Addition with PCP Pincer Compounds of Iridium 234
    Hani A. Y. Mohammad, Jost C. Grimm, Klaus Eichele, Hans-Georg Mack, Bernd Speiser, Filip Novak, William C. Kaska, and Hermann A. Mayer

Activation of C–H Bonds by Pt(II) (Shilov) Chemistry

15. **C–H Bond Activation at Pt(II): A Route to Selective Alkane Oxidation?** 250
    Alan F. Heyduk, H. Annita Zhong, Jay A. Labinger, and John E. Bercaw

16. **Competitive Trapping of Equilibrating Pt(IV) Hydridoalkyl and Pt(II) Alkane Complexes: A Valuable Tool for the Mechanistic Investigation of C–H Activation Reactions** 264
    Mats Tilset, Lars Johansson, Martin Lersch, and Bror J. Wik

17. **Mechanisms of Reactions Related to Selective Alkane Oxidation by Pt Complexes** 283
    Jennifer L. Look, Ulrich Fekl, and Karen I. Goldberg

18. **Hydrocarbon C–H Bond Activation with Tp'Pt Complexes** 303
    Cynthia M. Norris and Joseph L. Templeton

19. **C–H Bond Activation with Neutral Platinum Methyl Complexes** 319
    Carl N. Iverson, Charles A. G. Carter, John D. Scollard, Melanie A. Pribisko, Kevin D. John, Brian L. Scott, R. Tom Baker, John E. Bercaw, and Jay A. Labinger

20. **Issues Relevant to C–H Activation at Platinum(II): Comparative Studies between Cationic, Zwitterionic, and Neutral Platinum(II) Compounds in Benzene Solution** 334
    Jonas C. Peters, J. Christopher Thomas, Christine M. Thomas, and Theodore A. Betley

Oxidations

21. **Non-Organometallic Mechanisms for C–H Bond Oxidation: Hydrogen Atom versus Electron versus Hydride Transfer** 356
    James M. Mayer, Anna S. Larsen, Jasmine R. Bryant, Kun Wang, Mark Lockwood, Gordon Rice, and Tae-Jin Won

22. Oxidation of Organic Molecules with Molecular
    Oxygen Catalyzed by Heterometallic Complexes 370
    Patricia A. Shapley

23. Catalytic Hydrocarbon Oxidations
    Involving Coreductants 379
    Joseph E. Remias and Ayusman Sen

24. New Applications of Electrophilic Aromatic C–H
    Activation with Metal Trifluoroacetates 393
    Vladimir V. Grushin, David L. Thorn,
    William J. Marshall, and Viacheslav A. Petrov

25. Reengineering of Organic-Based Metal Active Sites
    for Oxidations and Oxygenations 407
    Sergiu M. Gorun

Indexes

Author Index 424

Subject Index 426

Preface

This volume is an outgrowth of a Symposium on *Activation and Functionalization of C–H Bonds* sponsored by the American Chemical Society (ACS) Division of Inorganic Chemistry, Inc. at the National Meeting in Orlando, Florida, April 7–9, 2002. This very exciting Symposium convinced us, as symposium organizers, of the great value of compiling the contributions of the invited speakers so that a wider audience would be able to appreciate the recent advances made in this important field.

The C–H bond is perhaps the most fundamental linkage in organic chemistry. It is certainly the most widespread. It is also notoriously resistant to selective chemical transformations. Therefore the development of the ability to selectively functionalize C–H bonds presents a tremendous intellectual challenge and one with almost incalculable potential rewards; the term "Holy Grail of chemistry" is often used (or perhaps abused) in this context. Potential applications range from the conversion of methane to methanol or higher alkanes, to the synthesis of the most complex organic molecules. This volume reveals current progress directly relevant to this entire range of applications. The approaches taken by the contributing authors also cover a broad spectrum, ranging from fundamental theoretical and experimental studies to the emergence of systems of immediate applicability.

This book does not pretend to be a comprehensive overview of the current work in C–H bond activation. Neither a symposium nor a monograph could achieve that. Had we been more conscientious about truth in advertising, a more precise title might have been "activation and functionalization of C–H bonds (particularly of aliphatic groups), with emphasis on homogeneous systems that involve the formation of metal-carbon bonds". We hope the reader forgives our less informative but also less unwieldy title. Even with the more circumscribed range, we can only cover a fraction (hopefully a representative one) of the exciting work in this field. Areas outside our targeted range include, among others, the vast fields of biological and heterogeneous C–H activation. Yet with all respect for the great importance of these fields, our view is that homogeneous transition-metal-based systems (or closely related immobilized and biphasic counterparts), offer the greatest potential for further developments in synthetic chemistry.

We thank the authors for their contributions, written with great dedication—even in some cases initiated with great reluctance. The ACS Division of Inorganic Chemistry, Inc. is thanked for financial support of this symposium. We greatly appreciate the support and advice given by Bob Hauserman and Stacy VanDerWal in acquisitions and Margaret Brown in editorial/production of the ACS Books Department throughout the publication process.

Alan S. Goldman
Department of Chemistry and Chemical Biology
Rutgers, The State University of New Jersey
New Brunswick, NJ 08903
732–445–5232 (telephone)
732–445–5312 (fax)
agoldman@rutchem.rutgers.edu

Karen I. Goldberg
Department of Chemistry
University of Washington
Seattle, WA 98195
206–616–2973 (telephone)
206–616–4209 (fax)
goldberg@chem.washington.edu

Activation and Functionalization of C–H Bonds

Chapter 1

Organometallic C–H Bond Activation: An Introduction

Alan S. Goldman[1] and Karen I. Goldberg[2]

[1]Department of Chemistry and Chemical Biology, Rutgers, The State University of New Jersey, New Brunswick, NJ 08903
[2]Department of Chemistry, University of Washington, Seattle, WA 98195

Abstract An introduction to the field of activation and functionalization of C-H bonds by solution-phase transition-metal-based systems is presented, with an emphasis on the activation of aliphatic C-H bonds. The focus of this chapter is on stoichiometric and catalytic reactions that operate via organometallic mechanisms, i.e., those in which a bond is formed between the metal center and the carbon undergoing reaction

The carbon-hydrogen bond is the un-functional group. Its unique position in organic chemistry is well illustrated by the standard representation of organic molecules: the presence of C-H bonds is indicated simply by the absence of any other bond. This "invisibility" of C-H bonds reflects both their ubiquitous nature and their lack of reactivity. With these characteristics in mind it is clear that if the ability to selectively functionalize C-H bonds were well developed, it could potentially constitute the most broadly applicable and powerful class of transformations in organic synthesis. Realization of such potential could revolutionize the synthesis of organic molecules ranging in complexity from methanol to the most elaborate natural or unnatural products.

The following chapters in this volume offer a view of many of the current leading research efforts in this field, in particular those focused on the activation of C-H bonds by "organometallic routes" (i.e., those involving the formation of a bond between carbon and a metal center). This is a very large, diverse, and highly active field; we cannot hope to provide a comprehensive or even a particularly representative sampling of the vast literature within this introductory chapter. Several excellent review articles have been published on various aspects of this field, some fairly recently (*1*). Our goal in this chapter is to convey enough background and insight to help the reader appreciate and put into context the remainder of this volume. Our emphasis on background comes at the expense of current work, but we are confident that the subsequent chapters of this volume provide ample coverage of the current state of this field.

Setting the Stage for C-H Activation (and the Opening Acts)

The activation of C-H bonds by transition metal-based systems has a long history. Bacteria have been practicing it for several billion years! On a more conscious level, in the context of "modern" chemistry, Grushin, Thorn and co-authors note (Chapter 24) that C-H activation by transition metals dates back to at least 1898. As Sames points out however (Chapter 9), the term "C-H activation" has acquired mechanistic connotations. Mechanistically, the 1898 Dimroth reaction, an electrophilic attack on an arene π-system followed by deprotonation of the resulting cation, does not qualify as "C-H activation". In this chapter we place strong emphasis on reactions of alkanes, with discussion of arene and other C-H bonds limited to examples relevant to alkanes.

The renaissance of inorganic/organometallic chemistry of the 1950s and early 1960s did not include obvious contributions to C-H activation. Nevertheless, a number of seminal observations suggested (to the most astute observers) possible directions for C-H activation (*2*). The bond of H_2, in terms of polarity and bond strength, is the closest relative to the C-H bond. In 1955, Halpern found that Cu^{2+} could heterolytically cleave H_2 (eq 1) (*3*); this reaction led to Cu-catalyzed reduction of various metal ions by H_2.

$$Cu^{2+} + H_2 \rightarrow [CuH]^+ + H^+ \tag{1}$$

Several other metals including mercury(II) and silver(I) were soon found to display similar behavior (*4*).

In 1962, Vaska reported oxidative addition of H_2 to $Ir(PPh_3)_2(CO)Cl$ (eq 2) (*5*).

$$\underset{\underset{PPh_3}{|}}{\overset{\overset{PPh_3}{|}}{Cl-Ir-CO}} + H_2 \longrightarrow \underset{\underset{PPh_3}{|}}{\overset{\overset{PPh_3}{|}}{Cl-\underset{OC}{Ir}-H}} \tag{2}$$

Vaska argued strongly that the reaction was genuinely oxidative (i.e. not just a formal change in oxidation state from I to III) (6); to a significant degree this view has endured to the present.

These two reactions, electrophilic and oxidative activation of dihydrogen, foreshadowed the two approaches to C-H bond activation by transition metal complexes that are currently most widespread. While it is often considered that the driving forces behind them are very different, even opposite, the relationship between them is actually far from opposite. The electrophilic activations were, even in the earliest reports, proposed to proceed via coordination of H_2 and subsequent loss of H^+ (4).

$$Cu^{2+} + H_2 \rightarrow [Cu(H_2)]^{2+} \rightarrow [Cu(H)]^+ + H^+ \qquad (3)$$

Likewise, oxidative addition of hydrogen is now understood to frequently proceed through an intermediate dihydrogen complex (eq 4) (7) (or, at the very least, a transition state closely resembling a dihydrogen complex).

$$M + H_2 \rightarrow M(H_2) \rightarrow M(H)_2 \qquad (4)$$

Thus these reaction pathways have at least as much in common as they have distinguishing them.

The first reported example of "C-H activation" by a transition metal complex is often attributed to Chatt. Ru(0)(dmpe)$_2$ was generated, leading to activation of a C-H bond of a ligand phosphinomethyl group or of a C-H bond of naphthalene (eq 5) (8).

These reactions are clearly examples of oxidative addition, in analogy with Vaska's result of eq 2.

Roughly at the same time as Chatt's discovery, Wilkinson's hydrogenation catalyst was reported (9). The mechanism of the catalytic cycle, which alternates between Rh(I) and Rh(III) intermediates, explicitly involves both H_2 addition and what is regarded (at least in today's language) as a reductive elimination of a C-H bond. Shortly thereafter Shilov extended olefin hydrogenation to platinum catalysts (10). With an understanding that a critical step in this reaction was C-H elimination from a Pt alkyl to give Pt(II), the reversibility of that step was investigated. As shown in eq 6, it was indeed found that Pt(II) could react with C-H bonds (11); H/D exchange between methane and D_2O was observed in the presence of Pt(II).

$$CH_4 + D_2O \xrightarrow[D_2O \,/\, CH_3CO_2D]{K_2PtCl_4} CH_3D + HDO \qquad (6)$$

The mechanism was suggested to proceed via an electrophilic substitution (eq 7).

$$Pt(II) + R\text{-}H \rightarrow Pt(II)\text{-}R + H^+ \qquad (7)$$

In analogy with the above-described modes of H_2 addition, the reactions of C-H bonds with the highly electron-rich (nucleophilic) Ru(dmpe)$_2$ and the electrophilic Pt(II) species might seem best described as driven by entirely different forces. Accordingly C-H bond activations are often classified as proceeding via either nucleophilic (oxidative addition) or electrophilic modes. But these two classes of C-H activation have much in common, perhaps even more so than the corresponding modes of H_2 activation. Both reactions appear to proceed in many cases, perhaps most, through a σ-bond complexed intermediate (7,12-14). Even more striking, putatively electrophilic activations, in some cases (perhaps most), proceed via complete oxidative addition (followed by deprotonation of the resulting metal hydride) (15). Conversely, one of the earliest examples of alkane activation was recognized by Crabtree as proceeding through "oxidative" addition - to a highly *electrophilic* cationic iridium complex (16).

Thus the relationship between electrophilic and oxidative activation is certainly not a dichotomy. It is therefore with reluctance that we perpetuate this distinction, for organizational purposes, in the sections below. The fact that this distinction works so well as an organizational tool may stem from the fact that workers in the field have focused in either one direction or the other, making it a somewhat self-propagating categorization. However, we hope that the reader will take away from the discussion that follows that this distinction is blurring rapidly and its usefulness is likely outlived.

"Electrophilic" systems

In 1972, Shilov published a dramatic advance on his earlier report of C-H activation with Pt(II) (eq 6) (*17*). The addition of Pt(IV) to the aqueous reaction of $PtCl_4^{2-}$ with methane lead to the production of the selectively oxidized species methanol and methyl chloride! This reaction, shown in eq 8, is catalytic in Pt(II) but is, unfortunately, stoichiometric in Pt(IV).

$$CH_4 + PtCl_6^{2-} + H_2O\ (Cl^-) \xrightarrow[\substack{H_2O \\ 120\ °C}]{PtCl_4^{2-}} CH_3OH\ (CH_3Cl) + PtCl_4^{2-} + 2\ HCl \quad (8)$$

Despite the impracticality of using platinum as a stoichiometric oxidant, this thirty-year old "Shilov system" remains to date one of relatively few catalytic systems that actually accomplish selective alkane functionalization under mild conditions. That distinction and the fact that this was the first example of such a transformation justify the prominent status of "Shilov chemistry" in the field of C-H activation. It is noteworthy that Shilov's exciting discovery was not immediately followed up on in the early 1970s by researchers outside of the Soviet Union. The significant time lag in wide recognition of this important advance can perhaps be attributed to the political climate at the time which limited communication and trust between Soviet and Western bloc scientists. It was not until the 1980s that Shilov's work was "rediscovered" by the by the West. However, in the past twenty years, research in the area of "Shilov chemistry" has been very active (*18*). Six chapters in this volume (Chapters 15-20) make up a section entitled "Shilov Chemistry" and describe some of the most recent research in the activation and functionalization of hydrocarbon C-H bonds using platinum(II) metal centers. In addition, in Chapter 9, Sames details the elegant use of platinum(II) for the selective activation of C-H bonds in complex organic molecules.

A significant fraction of research in the area of "Shilov Chemistry" has concentrated on determining the mechanism of the Shilov reaction (eq 8). A reasonable mechanistic scheme for the catalytic cycle (Figure 1) was proposed by Shilov not long after his initial report of the oxidation (*19,20*). In the first step of this cycle, a methylplatinum(II) intermediate is formed by reaction of Pt(II) with alkane. The product from this C-H activation step is then oxidized to a methylplatinum(IV) species in the second step. Either reductive elimination involving the Pt(IV) methyl group and coordinated water or chloride or, alternatively, nucleophilic attack at the carbon by an external nucleophile (H_2O or Cl^-) was proposed to generate the functionalized product and reduce the Pt center back to Pt(II) in the final step. Over the past two decades, convincing experimental support has been offered for this general mechanism.

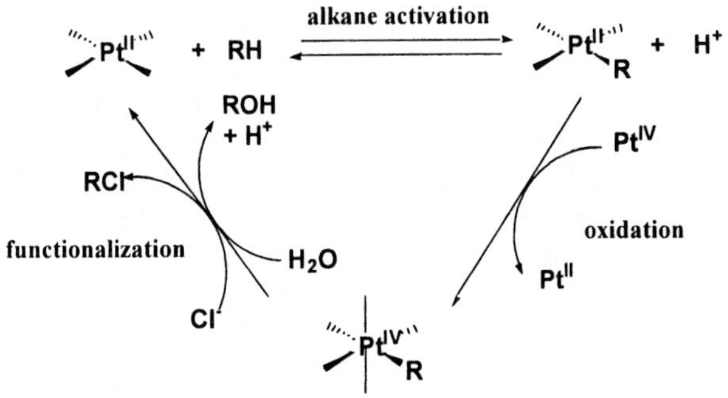

Figure 1. Proposed mechanism for Shilov's platinum catalyzed alkane oxidation

The cycle shown in Figure 1 is also thought to form the basis of other related Pt-based C-H functionalization systems *(21,22)*. However, the cycle as pictured is quite general and the intimate mechanisms of the individual steps (C-H activation, oxidation and functionalization) may be different in each particular system. For example, whether the activation of C-H bonds by platinum(II) proceeds via an electrophilic pathway or an oxidative addition followed by platinum deprotonation (eq 9) has been extensively debated. Recently, experimental evidence has been presented to support the oxidative addition pathway for model platinum(II) systems, in particular for those bearing nitrogen ligands *(15)*. Stable Pt(IV) alkyl hydrides have even been produced via the reaction of a Pt(II) species with alkanes *(23)*, demonstrating that oxidative addition of C-H bonds to Pt(II) is indeed a viable pathway for alkane activation. However, this evidence collected in model systems does not necessarily implicate oxidative addition in platinum-catalyzed alkane functionalization reactions, and the particular platinum(II) species involved in the actual Shilov system may in fact react via electrophilic activation. Calculational studies on the activation of methane by $Pt(H_2O)_2Cl_2$ (thought to be the active catalyst in the Shilov process) indicated that the barriers to both mechanistic pathways were very similar *(24)*.

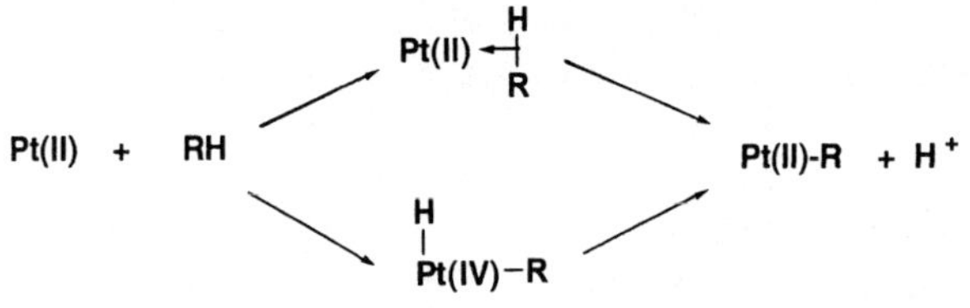

$$Pt(II) + RH \begin{matrix} \nearrow Pt(II) \leftarrow\!\!\!\!\overset{H}{\underset{R}{|}} \searrow \\ \searrow \underset{Pt(IV)-R}{\overset{H}{|}} \nearrow \end{matrix} Pt(II)\text{-}R + H^+ \quad (9)$$

The second two steps of the Shilov process, oxidation and functionalization, are also of great interest and considerable mechanistic work has been carried out on these reactions as well (*18*). Of particular significance, elegant labeling experiments by Bercaw and Labinger convincingly showed that Pt(IV) acts as an external oxidant (*25*). This important result opens the possibility of utilization of a more economical oxidant in the functionalization. Toward that end, researchers have in recent years reported the use of other oxidants including SO_3 (see below), Cl_2, H_2O_2 and O_2/heteropolyacids or O_2/Cu(II) to carry out this transformation (*21,22,26*). Overall, success has been modest and to date, no commercially viable system has been developed. The final step of the Shilov cycle, the carbon-heteroatom bond formation step leading to product, occurs via nucleophilic attack at the carbon. Most persuasive are studies demonstrating stereochemical inversion at carbon in model systems (*27*). Strong evidence that nucleophilic attack occurs at five-coordinate Pt(IV) rather than six-coordinate Pt(IV) in carbon-heteroatom bond formation has also been presented (*18,27,28*).

The most significant advance on the Shilov system appeared in 1998 (*21*). Researchers at Catalytica discovered that methane could be selectively converted to methyl bisulfate by a platinum(II) bipyrimidine catalyst in concentrated sulfuric acid (eq 10). An impressive 72% yield (89% conversion of methane, 81% selectivity) was observed in this reaction which formally uses SO_3 as the oxidant.

$$CH_4 + 2\ H_2SO_4 \xrightarrow{\text{(bpym)PtCl}_2} CH_3OSO_3H + 2\ H_2O + SO_2 \quad (10)$$

The mechanism of this reaction has been under investigation and a general scheme similar to that shown in Figure 1 for the Shilov reaction has been proposed. The intimate mechanism of the C-H activation step which forms the Pt(II)-methyl intermediate was recently examined using computational methods and it was found that an electrophilic mechanism for C-H activation was favored over oxidative addition (*29*). In contrast, for the closely related complex $(NH_3)_2PtCl_2$ (which was discovered to be a more active but less stable catalyst for methane oxidation under similar conditions), oxidative addition was favored over the electrophilic pathway. Other calculations, which differ in which ligands are replaced by methane and whether cationic platinum species or complete ion pairs are considered, reach opposite conclusions (*30*). The intimate mechanism of the C-H activation step in platinum-catalyzed alkane oxidation reactions should thus be affected by the solvent medium, a factor which is difficult to treat computationally. Clearly, this mechanistic issue is not yet resolved for Pt(II)

C-H bond activation and comparison between the two pathways continues to be an active topic of discussion in the field.

Prior to Catalytica's report of the Pt(II)/H_2SO_4 system, the same researchers reported a related Hg(II)/H_2SO_4 system for methane oxidation (*31*). Since an oxidative addition pathway is not accessible to Hg(II), the electrophilic route for C-H activation was favored. However, Sen and co-workers have suggested that an alternative pathway of electron transfer, wherein Hg(II) acts as a one-electron outer-sphere oxidant, is also consistent with the experimental data and should be considered for this reaction (*1a*).

Others have also followed up on Shilov's original observations, and numerous systems involving oxidation of alkanes by metal cations in acidic solution have been reported (*18*). While some work, most notably Shilov chemistry, has been carried out in aqueous solutions, strong acids (e.g. H_2SO_4, CF_3CO_2H) are particularly desirable as solvents for alkane oxidation reactions. In the sulfuric acid systems, as described above, the acid can act as both the solvent and the oxidant. Common to all strong acid reactions, however, is the fact that strong acids, and even their conjugate bases, are weak ligands (although highly polar); therefore the electrophilicity of the metal ion is enhanced in these solvents. In addition, the products of such reactions (including water, e.g., eq 10) are protonated by acidic solvent. Finally, the partially oxidized products formed are not the alcohols but rather the esters, which are better protected from overoxidation (*32*).

There have been several review articles and chapters reporting on alkane oxidation in acidic solution (*33-35*). Here we will only highlight a few additional examples to give a flavor of this approach. It will also be noted that detailed mechanistic understanding of these reactions has been severely hampered by the complexity of the systems.

Building on some earlier results of Pd(II) catalyzed electrophilic alkane functionalization reactions, Sen reported an interesting bimetallic electrophilic system for the selective oxidation of methane and lower alkanes (*36*). Methane was oxidized to methanol and methyl trifluoroacetate by Pd/C and $CuCl_2$ in a mixture of trifluoroacetic acid and water in the presence of dioxygen and carbon monoxide. The requirement of carbon monoxide as a coreductant in this reaction is reminiscent of monooxygenase systems (see section on Non-Organometallic C-H Activation later in this chapter) wherein only one of the two oxygens in dioxygen is utilized for substrate oxidation. Carbon monoxide was also required for a related $RhCl_3$ system which generates methanol or its ester methyl perfluorobutyrate and acetic acid from methane, carbon monoxide and dioxygen in the presence of several equivalents of Cl^- and I^- in a mixture of perfluorobutyric acid and water (*37*). Changes in the solvent composition significantly affect the ratio of methanol and methyl ester to acetic acid. The different products are proposed to result from a competition between attack of the nucleophile on the Rh-Me to form the methanol or methyl ester versus CO migratory insertion followed by nucleophilic attack. For ethane and higher alkanes, products derived from C-C cleavage (e.g. methanol from ethane) were also observed. While evidence was presented against the involvement of free alkyl radicals in both the Pd/Cu and the Rh based systems, the mechanisms of these complex reactions remain to be determined.

Researchers have also found that ligation of the metal center can have a significant effect on such reactions. For example, Strassner recently reported the catalytic conversion of methane to methyl trifluoroacetate using palladium(II) N-heterocyclic carbene complexes in a mixture of trifluoroacetic acid and trifluoroacetic acid anhydride with potassium peroxodisulfate as an oxidant (*38*). The same reaction carried out with $Pd(OAc)_2$ as a catalyst proceeded in a slightly lower yield. The authors also found that iodide complexes of the palladium N-heterocyclic carbene catalysts were inactive in contrast to their bromide analogs. This difference was attributed to a lower basicity of the iodide which inhibited the creation of a free coordination site by protonation of the halide ligand. The open coordination site is presumably needed for an electrophilic activation. Radical pathways must be also be considered, however, whenever strong oxidants such as peroxodisulfate are used. It is noteworthy in this context that Sen has reported that peroxodisulfate reacts with methane (via a radical path) to give alcohols, in the absence of metal, at 105-115 °C; at lower temperatures the role of metals could be to catalyze reactions of this nature, rather than the direct activation of a C-H bond (*39*)

Several years ago, Fujiwara reported the carboxylation of alkanes with CO, using $Pd(OAc)_2$ and/or $Cu(OAc)_2$ in trifluoroacetic acid with potassium peroxodisulfate as an oxidant (*40*). $Cu(OAc)_2$ was found to be the most effective catalyst for methane conversion to acetic acid. Just last year, Periana found another interesting methane carbonylation reaction. In this case, carbon monoxide is not added but rather two moles of methane undergo an oxidative condensation (*41*) (eq 11). Using $PdSO_4$ in concentrated sulfuric acid, methane was converted to a mixture of methyl bisulfate and acetic acid. The yield for this reaction is low (ca. 10%) but the selectivity for methyl bisulfate and acetic acid was > 90%.

$$2\ CH_4 + 4\ H_2SO_4 \xrightarrow{Pd(II)} CH_3CO_2H + 4\ SO_2 + 6\ H_2O \qquad (11)$$

While these metal ion/acid reactions are promising in that selectively functionalized products are formed (particularly for methane oxidation), the yields, selectivities and rates are still far from commercially viable levels. Further work in determining the optimal ligands for these reactions will no doubt assist in this regard. There are also significant costs associated with separation of methanol and water from acid solutions, and this issue will have to be addressed for large scale application of an electrophilic late metal/acid system.

Oxidative Addition of C-H Bonds: Early Examples

Concomitant with Shilov's original work on electrophilic systems, observations from other laboratories seemed to point in an entirely opposite direction as the most promising route to C-H activation *(8,42,43)*. Led by Chatt's 1965 discovery of the reactions of the extremely electron-rich species Ru(dmpe)$_2$ (eq 5), several other examples were discovered in which apparently electron-rich metal centers were found capable of oxidatively adding C-H bonds.

These early examples all involved addition of aryl C-H bonds (as in the Chatt system of eq 5) or intramolecular addition of ligand C-H bonds (either aryl or alkyl). For example, Green reported that photoelimination of dihydrogen from Cp$_2$WH$_2$ resulted in benzene C-H bond addition (eq 12) *(43)*.

$$Cp_2WH_2 + C_6H_6 \xrightarrow[-H_2]{h\nu} Cp_2W(H)(Ph) \tag{12}$$

Cyclometalation of ligand aryl groups (eq 13) was discovered early *(44)* and ultimately found to be quite common. Whitesides *(42)* reported that an aliphatic, remote (γ) C-H bond underwent addition (eq 14).

$$Ph_3P-Ir(PPh_3)_2-Cl \longrightarrow (C_6H_4\text{-}PPh_2)Ir(H)(PPh_3)_2Cl + PPh_3 \tag{13}$$

$$(Et_3P)_2Pt(CH_2CMe_3)_2 \longrightarrow (Et_3P)_2Pt(\text{metallacyclobutane}) + CMe_4 \tag{14}$$

By no means did all early examples of C-H addition involve metal centers that were obviously very electron-rich (note the W(II) and Pt(II) centers of eqs. 12 and 14); nevertheless, the nature of oxidative addition (or at least the nomenclature) suggested that increased electron richness at the metal center was favorable. And while Chatt's Ru(dmpe)$_2$ example was something of an extreme case, it probably had an undue influence simply by virtue of being the first such example. Ironically, it was later found by Roddick (45) that Ru[(C$_2$F$_5$)$_2$PCH$_2$CH$_2$P(C$_2$F$_5$)$_2$)]$_2$, a much less electron-rich analogue of Ru(dmpe)$_2$, was apparently *more* reactive toward alkane and arene C-H bonds.

The very favorable intramolecular activation of aryl C-H bonds was exploited by Shaw to design remarkably stable ligands via cyclometalation (46).

$$\text{(15)}$$

(^{R}PCP)IrHCl

The propensity of the bisphosphine shown in eq 15 to undergo cyclometalation yielding the corresponding "^{R}PCP" tridentate ligated complexes (an early example of what might now be called self-assembly) was immediately demonstrated with a wide range of simple metal halide salts including Rh, Ir, Ni, Pd, Pt (46). This example and other "pincer" ligands would eventually play another important, though supporting, role in C-H activation (for example, the work described in Chapters 4, 12, 13, 14 in this volume) (47). In addition, reaction of analogous diphosphines methylated at the 2-position led to the first example in which a C-C bond was added in a manner analogous to the C-H activations described in this section (48).

In a sense, the culmination of these observations of intramolecular and intermolecular aryl C-H oxidative addition appeared in 1982 when Bergman and then Graham reported the first examples of intermolecular addition of alkane C-H bonds to give stable alkyl iridium hydrides (49,50). Clearly the first example of such reactivity must constitute a major milestone and this breakthrough has justly achieved celebrity.

$$\text{(16)}$$

Nevertheless, what may be the most important aspect of this work is perhaps still under-appreciated. Bergman found a remarkable selectivity among C-H bonds. The Cp*Ir(PMe$_3$) (Cp* = η^5-C$_5$Me$_5$) metal center showed a striking preference for cleavage of *stronger* C-H bonds (e.g. aryl > 1° > 2° >> 3°) (*51*). This result was soon supported and elaborated upon in work by Graham (*50*), Jones (*52*), Flood (*53*), Field (*54*), and others. It soon became apparent that both kinetics *and* thermodynamics of C-H addition to transition metal complexes (at least of the late metals) were more favorable for less-substituted alkyl groups. The importance of this cannot be underestimated. First, it allows transition metals to show selectivity that is opposite (and therefore complementary) to most other reagents capable of cleaving C-H bonds. Such selectivity is desirable in many cases. For example, terminal functionalization of *n*-alkanes is a much-pursued goal. Second, and perhaps even more important, this surprising selectivity for stronger C-H bonds can lead to selectivity for activation of the alkane versus activation of the derivatized product. In virtually all cases, a derivatized product has weaker C-H bonds than the parent alkane; therefore most species capable of activating C-H bonds show selectivity for the products, resulting in further, typically unwanted, transformations (e.g. the over-oxidation of methane to CO_2).

Stoichiometric Oxidative Addition Chemistry

Cp and Tp Complexes

The (cyclopentadienyl)iridium complex of eq 16 and various related species, including Cp and Tp (trispyrazolylborate) complexes of rhodium and iridium and to a lesser extent rhenium, have been the subject of a wide range of studies that have afforded detailed pictures of the mechanism and thermodynamics of oxidative addition of C-H bonds. (We will use abbreviations "Cp" and "Tp" to refer to the parent ligands, e.g. C$_5$H$_5$, as well as derivatives, e.g. C$_5$Me$_5$ or C$_5$MeH$_4$). In particular, such complexes have been used to develop a detailed understanding of the critical issue of selectivity and the question of intermediates formed along the C-H activation pathways.

The Cp*Ir(PMe$_3$) system was found to give selectivity for 1° vs. 2° C-H bonds of *n*–alkanes on the order of 5-10:1 (per C-H bond). Impressive as this selectivity is, Bergman found that the Cp*Rh complexes appeared to be much more selective; reaction with *n*-alkanes gave *exclusively* the primary alkyl rhodium hydrides (>50:1). Cyclohexane, however, possessing only 2° C-H bonds, showed reactivity comparable to *n*-alkanes. It was further shown in competition experiments that reaction rates of different n-alkanes with Cp*Rh(PMe₃) were nearly proportional to the number of carbons (e.g. the ratio of n-octane:n-propane reactivity was 2.9 ± 0.5 or ~ 8:3); this seems inconsistent with the observation that only the terminal C-H groups appeared to be reactive. This apparent contradiction was resolved with labeling experiments that demonstrated that the metal fragment could undergo C-H reductive elimination, migrate to an adjacent carbon, and add a C-H bond at the new site (Figure 2) (*55*).

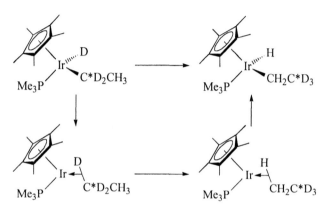

Figure 2. Double-labeling experiment demonstrating intramolecular "chain-walking" (55)

The migration was demonstrated to be intramolecular and it was proposed to proceed via intermediate σ-bonded alkane complexes as shown in Figure 2. Regardless of the nature of any such intermediate alkane complex, the ability to "walk" down an alkane chain was clearly implied by this result, which thus explained the selectivity for longer n-alkanes. The rate-determining step was apparently addition of the alkane (either σ-bond coordination or C-H addition) with little, if any, selectivity for 1° C-H bonds. Chain-walk then followed and either the 1° product was the thermodynamic sink or, alternatively, a metal center migrating between σ-bonds underwent oxidative addition much more rapidly with 1° C-H bonds.

Arenes undergo addition more rapidly than alkanes to Cp*ML ("L" in this chapter refers to monodentate charge-neutral ligands in general, but most commonly trialkylphosphines). A series of extremely elegant experiments by Jones demonstrated that the rate-determining step of arene C-H addition was coordination to give η^2-C-C bound complexes (*52*). Thus the kinetic isotope effect (KIE) for reaction with C_6H_6/C_6D_6 was near unity (1.05). Reaction with 1,3,5-$C_6H_3D_3$, however, gave C-D and C-H addition products in a ratio of 1.4:1. Thus the product-determining step, subsequent to rate-determining η^2-C-C coordination, is C-H(D) addition with a significant KIE. Confirming these conclusions, π-complexed arene intermediates were later directly observed (56-58).

In a striking reversal of the above-noted results for arene addition to Cp*Rh(PMe$_3$), Bergman found that ethylene C-H addition to Cp*Ir(PMe$_3$) does *not* proceed via π-coordination (59). Formation of the C-H addition product was found to be *faster* than π-coordination. In fact, it was found that the C-H addition product reacts to give the π-coordinated product (i.e. not the other way around). (Analogous results, with an iron complex, were reported by Field shortly thereafter (60).)

Isotope effects have been exploited with particular effectiveness in the context of CpM complexes (14). A key result by Jones was the 1985 discovery that C-H elimination from Cp*Rh aryl hydrides can have an inverse KIE (14,52,61); soon thereafter Bergman obtained inverse KIE's with analogous iridium alkyl hydrides (62). This can be explained in terms of a two-step mechanism involving a pre-equilibrium with a σ-complex (or π-complex) intermediate; the kinetic isotope effect (EIE) for such a species is expected to be inverse due to the much greater force constant of the C-H bond versus the M-H bond. If the KIE for loss of coordinated hydrocarbon is then less than or near unity (a reasonable assumption as the C-H bond is already formed in the alkane complex intermediate) then the overall KIE is inverse.

$$L_nM\genfrac{}{}{0pt}{}{R}{H} \underset{}{\overset{K_H/K_D < 1}{\rightleftharpoons}} L_nM \leftarrow |\genfrac{}{}{0pt}{}{R}{H} \xrightarrow{k_H/k_D \approx 1} L_nM + RH \qquad (17)$$
$$KIE = (K_H/K_D)(k_H/k_D)$$

However, a single step mechanism in which the transition state (TS) has a nearly fully formed C-H(D) bond, as is the case in the TS for the 2nd step in eq 17, will likewise have an inverse KIE.

Reductive eliminations of alkane from a number of other metal alkyl hydride complexes have also been characterized by inverse KIE values (14,63). The observation of H/D scrambling reactions between the hydride position and the C-H bonds of the alkyl group appears to correlate with those systems exhibiting inverse KIEs (15). Such behavior would be consistent with the mechanism shown in eq 17. For a more complete discussion of KIEs and σ-complexes in C-H reductive elimination and oxidative addition reactions, the reader is referred to Chapters 3 and 5 in this volume and recent reports on the topic (*14,63*).

Alkane σ-bond complexes have been spectroscopically observed by a number of workers, as early as 1972 (*64*). Of particular relevance to C-H addition are of course those systems in which the observed alkane complex is demonstrated to be an intermediate on the reaction path. Low-temperature work by Bergman and co-workers on the femtosecond to microsecond time scale has elucidated in depth the process of C-H activation by CpIr- and TpIr-containing fragments. The many important aspects of this work include (a) competition between alkanes and noble gas atoms for coordination at the vacant coordination site, (b) rates of alkane coordination and oxidative cleavage, (c) selectivity, indicating greater reactivity of large alkanes but lower rates of oxidative cleavage, (d) surprisingly large isotope effects for both coordination and oxidative cleavage, and (e) the κ^3-κ^2 ring-opening of Tp complexes prior to oxidative cleavage (*65*).

In the context of this large body of work on C-H addition, it is easy to overlook a critical point: the addition of C-H bonds to transition metals is generally a thermodynamically unfavorable process (*66*). Reductive elimination is the rule while addition is the exception. Even addition to coordinatively unsaturated complexes is generally unfavorable – and it is much more so after factoring in the energetic cost of dissociating a ligand to open a coordination site. Generally, those examples where C-H addition is thermodynamically favorable (many of the aforementioned examples are driven uphill photochemically) involve considerable effort and energy (literally and figuratively) to generate highly reactive, coordinatively unsaturated, species. At the risk of over-simplification, it would probably be fair to say that the "difficulty" associated with C-H activation is typically more related to thermodynamics than to kinetics; in those cases where the thermodynamics are favorable, the kinetics are often quite facile. The classic illustration of this was offered by Graham and Rest who found that methane addition to photogenerated CpM(CO) occurred at temperatures as low as 12 K(!)(*67*).

Seminal work by Hoffmann (extended Hückel calculations) offered a theoretical framework to understand the propensity of CpRhL and CpIrL complexes to activate C-H bonds (*68*). The 16-electron CpML d^8 species were compared with 16-electron square-planar ML_4 d^8 complexes. When the square-planar species was "bent back", at a significant energetic cost, it gave a fragment possessing a high-energy filled orbital of π-symmetry (d_{yz}; Figure 3) which mixes with the C-H σ* orbital, and a low energy empty σ-symmetry hybrid orbital. This fragment was shown to be isolobal to CpML (as well as to CH_2), and both fragments were calculated to smoothly add C-H bonds (Figure 3).

Figure 3. C-H bond approaching isolobal (d^8) fragments, "bent-back" square-planar, and CpML, showing the key (filled) d_{yz} orbital. An empty (σ-acceptor) hybrid orbital, pointing along the z axis, is not shown.

The nature of the Cp ring precluded formation of a relaxed state, analogous to the square-planar complex, and thus the TS for addition was that much more accessible. While the "strain" present in the CpML species was thereby used to interpret the facile kinetics of C-H addition, it may serve equally well to explain the favorable thermodynamics of addition. This may be applicable to other Cp-bearing species, in contrast with monodentate-ligated complexes, which might reach a lower-energy arrangement of ligands rather than undergoing C-H addition. Accordingly, it has been shown that addition of alkane C-H bonds to CpReL$_2$ is favorable (*69*), and that the thermodynamics of addition (of C-H and other X-Y bonds) to CpIrL and Cp$_2$W are similarly very favorable as compared with addition to Ir(PR$_3$)$_2$(CO)Cl (*70*). Much more surprising than these other "electron-rich" examples, however, was the 1993 report by Bergman (*71,72*) of C-H activation by cationic Ir(III) complexes [CpIrLR]$^+$. Oxidative addition was not directly observed but alkyl metathesis reactions were reported (eq 18).

$$[\text{Cp(Me}_3\text{P})\text{Ir(CH}_3\text{)L}]^+ + \text{R-H} \longrightarrow [\text{Cp(Me}_3\text{P})\text{Ir(R)L}]^+ + \text{CH}_4 + \text{L} \quad (18)$$
$$(\text{L} = \text{N}_2, \text{CH}_2\text{Cl}_2)$$

By all rights, these Ir(III) complexes should bear no relation to low-valent CpM(I)L complexes; they certainly cannot be considered particularly "electron-rich". Thus although the cation of eq 18 obviously has d-electrons, there was good reason to consider the possibility that the reaction proceeded via σ-bond metathesis (a reaction discussed in more detail later in this Chapter), rather than oxidative addition (to give Ir(V)) followed by reductive elimination. The reaction was calculated (DFT) by Hall, however, to proceed via oxidative addition (*73,74*). This conclusion later received strong experimental support from Tilley and Bergman (*75*) who directly observed oxidative addition of C-H bonds of the silyl groups of Cp*Ir(PMe$_3$)(SiR$_3$) to give the corresponding cyclometalated Ir(V) complexes. It might be inferred therefore that the frequent involvement of iridium (and by extension platinum), in C-H activation, has little to do with the degree of "electron-richness" or electrophilicity of its complexes, but simply results from a tendency of these metals to form strong bonds to carbon.

Concluding this section we note that our understanding of C-H addition is based in large part on studies such as those discussed above with Cp and Tp complexes. The reactivity of the C-H addition products formed in such studies is dominated by reversible C-H elimination. This limited reactivity has allowed for extensive and informative mechanistic investigations; unfortunately, however, it has also precluded the development of such systems for catalytic hydrocarbon functionalization.

Non-Cp-Bearing Species; 3-coordinate d^8 and Other Configurations

A survey of the C-H activation chemistry of non-Cp-bearing species reveals a dominant role of 3-coordinate d^8 metal centers. Indeed even addition to TpML fragments, despite the apparently close relationship to CpML, can proceed via a κ^3-κ^2 dechelation to give the κ^2 three-coordinate d^8 intermediate (65). The importance of the 3-coordinate d^8 configuration will be further underscored in the section following this one, on catalytic reactions. Nevertheless, even among those metals most likely to add C-H bonds via intermediates of this configuration (e.g. Fe, Ru, Os, Ir, Pt), alternative pathways are well documented.

The addition of various C-H bonds to osmium was reported by Flood beginning in 1986 (76) (eq 19).

$$(PMe_3)_4Os\begin{smallmatrix}{}^tBu\\H\end{smallmatrix} + R\text{-}H \longrightarrow (PMe_3)_4Os\begin{smallmatrix}R\\H\end{smallmatrix} + CMe_4 \qquad (19)$$

Elegant kinetic and mechanistic studies demonstrated the intermediacy of three-coordinate d^8 Os(PMe$_3$)$_3$ for methane addition in cyclohexane solvent. This might appear to support the importance of both low oxidation state centers and the role of the 3-coordinate d^8 configuration. However, this work had been preceded by equally rigorous studies showing that Os(II) intermediates, [Os(PMe$_3$)$_3$(R)(H)], were capable of directly adding the C-H bonds of benzene, SiMe$_4$, and PMe$_3$ (53). In some respect this prefigured the dual Ir(I)/Ir(III) pathways discussed above.

In parallel with Flood's osmium work, Field reported that Fe(depe)$_2$ underwent addition of sp^2 and sp^3 C-H bonds, including the selective addition of *n*-pentane 1° C-H bonds (54). Thus, early on, alkylphosphino complexes of all three members of the Group 8 triad were found to effect C-H addition (8).

$$\text{(20)}$$

It has not been determined if the mechanism of eq 20 involves dechelation (ring-opening) of the depe ligand.

Werner reported that addition of P^{i}Pr$_3$ to a benzene solution of [(cyclooctene)$_2$IrCl]$_2$ resulted in benzene C-H bond addition to give (P^{i}Pr$_3$)$_2$Ir(Cl)(Ph)(H) (77). Kaska had previously reported (^{tBu}PCP)RhH$_2$ along with spectroscopic indications of analogous (^{tBu}PCP)Rh-containing C-H addition products (78). Whereas most examples of addition to three-coordinate d^8 species were inferred in catalytic cycles, or were followed by ligand addition to give 6-coordinate complexes, these examples gave stable 5-coordinate d^6 addition products. Crystallography revealed a remarkable structure for (P^{i}Pr$_3$)$_2$Ir(Cl)(Ph)(H): a severely distorted trigonal bipyramid (P atoms apical) with a C-Ir-H angle of 78 °, far less than the ideal 120 ° of a regular trigonal bipyramid (77).

Eisenstein and co-workers used computational methods to investigate the nature of 5-coordinate d^6 complexes, particularly L$_2$IrHXY (79,80). In this seminal work two energetic minima were calculated: the distorted trigonal bipyramid and the square pyramid with apical H. Particularly for complexes L$_2$IrHRX (R = hydrocarbyl or H) π-donation from X was found to play a critical role in determining structure, and subsequent studies have also demonstrated the highly favorable role of π-donation on reaction thermodynamics and kinetics (81-84).

The importance of 3-coordinate d^8 metal centers extends beyond the range of classical oxidative addition chemistry of low-valent metal centers. This configuration has played perhaps an even more central role in "electrophilic" C-H activation. In 1988, the first example of intermolecular C-H bond activation by a Pt(II) complex was reported by Whitesides (85). (PMe$_3$)$_2$Pt(neopentyl)(OTf) was thermolyzed in benzene-d$_6$ to give neopentane-d$_1$ and (PMe$_3$)$_2$Pt(C$_6$D$_5$)(OTf). Intermolecular C-H activation of alkanes by Pt(II) was not discovered until 1997 when Bercaw and Labinger reported the reaction of [(tmeda)PtMe(NC$_5$F$_5$)]$^+$ (tmeda = Me$_2$NC$_2$H$_4$NMe$_2$) ^{13}CH$_4$ to yield [(tmeda)Pt(^{13}CH$_3$)(NC$_5$F$_5$)]$^+$ (86). The activation of C-H bonds by [(diimine)PtMe(OH$_2$)]$^+$ quickly followed (87). All of these Pt(II) complexes (and several others that followed) that were found to activate C-H bonds share in common one labile ligand. It has been generally accepted that a hydrocarbon replaces this ligand on the path to C-H activation; thus, it is again a T-shaped d^8-ML$_3$ species (isolobal to CH$_2$) that activates the C-H bond. Since no Pt(IV) species and only the Pt(II) hydrocarbyl product is observed in these reactions, it was not possible to determine whether the reactions proceeded via a σ-bond metathesis mechanism or through C-H oxidative addition to form a five-coordinate Pt(IV) species followed by C-H reductive elimination (eq 21).

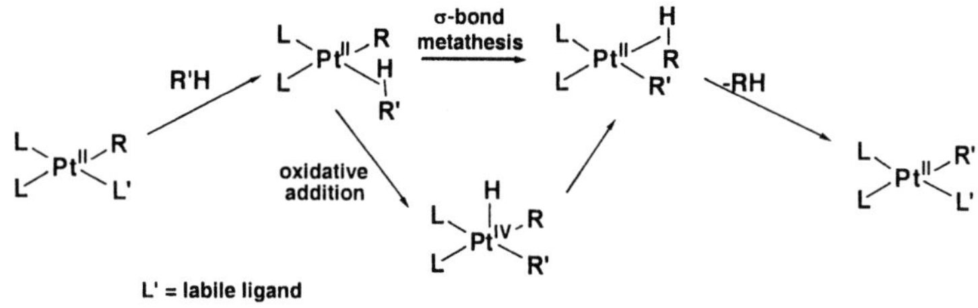

(21)

L' = labile ligand

Wick and Goldberg reported in 1999 that the reaction of KTp^{Me_2}PtMe$_2$ with B(C$_6$F$_5$)$_3$ in hydrocarbon solvent yielded the Pt(IV) hydrocarbyl hydrides (Figure 4) (23). In this case, the facially coordinating ligand Tp was able to trap the reactive five-coordinate Pt(IV) alkyl hydride and stabilize it as a six-coordinate complex (88). In the proposed mechanism (Figure 3), the role of the B(C$_6$F$_5$)$_3$ is to remove a methide and open the coordination site at Pt(II) to allow interaction of the hydrocarbon with a d^8-ML$_3$ fragment.

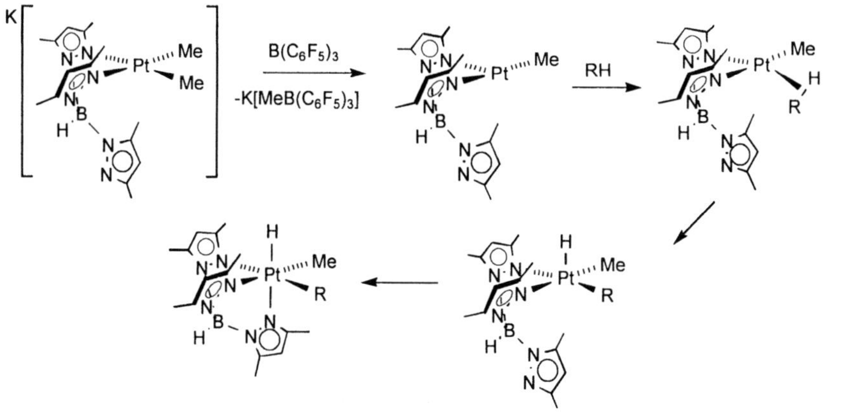

Figure 4. Proposed mechanism for the reaction of B(C$_6$F$_5$)$_3$ with KTp^{Me_2}PtMe$_2$ in hydrocarbon solvent to yield Pt(IV) hydrocarbyl hydrides.

In contrast to Ir(III), five-coordinate Pt(IV) hydrocarbyl hydride species have not been isolated or even unambiguously observed. However, five-coordinate Pt(IV) alkyl complexes and five-coordinate Pt(IV) silyl hydride complexes have been isolated and fully characterized (*89*). Square-pyramidal type geometry around platinum has been documented.

Considerable insight into the mechanism of C-H activation by Pt(II) species has also been garnered from studies of protonation of Pt(II) alkyls and from studies of C-H reductive elimination from Pt(IV) complexes (*15,18,88,90-92*). An extensive study of protonation of Pt(II) methyl complexes provided strong evidence for the involvement of both Pt(IV) methyl hydrides and Pt(II) σ-methane complexes as intermediates in the rapid protonolysis reactions of Pt(II) methyls (*15,91*). The principle of microscopic reversibility requires that these species would also be involved in C-H activation reactions at these Pt(II) centers. Similarly mechanistic studies of reductive elimination from isolable six-coordinate Pt(IV) complexes like (bpma)PtMe$_2$H (bpma = bis(pyridylmethyl)amine) and Tp^{Me2}PtMe$_2$H support the involvement of Pt(II) σ-methane complexes on the reaction pathway (*92b,e*). Isotopic scrambling of deuterium between the hydride and the methyl positions is observed along with inverse kinetic isotope effects. C-H reductive elimination from Tp^{Me2}PtMe$_2$H generates an intermediate which activates solvent C-H bonds, for example forming Tp^{Me2}PtPh$_2$H in benzene (*92e*).

Platinum(0) complexes are also known to oxidatively add C-H bonds to form Pt(II) alkyl hydrides (*93*). To promote this reactivity, bidentate ligands were used so that a cis-d^{10}ML$_2$ (rather than trans) configuration was enforced. Such a fragment, like 3-coordinate d^8, CpML, and TpML, is also isolobal to CH$_2$ (*68*).

Catalytic Functionalization Involving Oxidative Addition of C-H Bonds.

Insertion of CO and Isonitriles into C-H Bonds

Of the many reactions in which transition metal alkyls are intermediates, carbonylations (hydroformylation and the Monsanto acetic acid process to name two) are certainly among the most important. Thus, soon after the initial spate of discoveries of stoichiometric alkane C-H oxidative addition to CpML and other fragments, attempts were made to incorporate such reactions into catalytic carbonylation cycles. To our knowledge, no such attempts were directly successful. However in 1983, Eisenberg discovered that Rh(PPh$_3$)$_2$(CO)Cl photochemically catalyzed the carbonylation of benzene (*94-96*).

$$\text{benzene} + CO\ (1\ atm) \xrightarrow[h\nu]{trans\text{-}Rh(PPh_3)_2(CO)Cl} \text{benzaldehyde} \quad (22)$$

As this reaction was assumed to proceed via photoextrusion of a ligand (either CO or PPh$_3$) the importance of three-coordinate d^8 fragments with respect to carbonylation was highlighted early. Accordingly, in fast timescale studies by Ford, C-H activation by Rh(PR$_3$)$_2$Cl was observed directly (97). Further support for the activity of the Rh(PR$_3$)$_2$Cl fragment was subsequently offered by additional fast-timescale studies (98), low-temperature matrix photolysis (99), and conventional kinetics (100)).

Tanaka, noting that Bergman's Cp*Ir(PR$_3$) system indicated the superiority of PMe$_3$ vs. PPh$_3$, investigated the trimethylphosphine analog of Eisenberg's catalyst (101). Rh(PMe$_3$)$_2$(CO)Cl was indeed found to be somewhat more effective for benzene carbonylation, but more importantly, it was found to carbonylate alkanes as well (eq 23). In a somewhat puzzling twist, it was found that with short-wavelength light, *n*-alkanes could be carbonylated with regioselectivity for the terminal position, but longer wavelengths gave primarily branched aldehydes (102).

$$\text{n-pentane} \xrightarrow[CO\ (1\ atm)\quad h\nu]{Rh(PMe_3)_2(CO)Cl} \text{linear aldehyde} + \text{branched aldehyde} \quad (23)$$

This reaction was proposed by Field and Goldman to proceed via two distinct mechanisms (100,103,104). The long-wavelength mechanism effected C-H activation via photoelimination of CO, followed by C-H addition. The short-wavelength mechanism was proposed to proceed via direct C-H oxidative addition to an intact, four coordinate, photoexcited species; a second photon was then required to induce alkyl-to-CO migration. Based on these findings, it was possible to select conditions that resulted in greatly increased turnover numbers, quantum yields and selectivity via the short-wavelength mechanism (100,104). The duality of mechanisms received strong support from TRIR studies by Ford (98) and Bitterwolf (99).

Carbonylation of hydrocarbons is an endoergic reaction (for alkanes much more so than arenes), and therefore must be driven by either light or coupling with a secondary reaction to obtain any significant catalytic yields. The closely related insertion of isocyanides, however, is exothermic. Cp*Rh(NCR) (105) and TpRh(NCR) (106,107) fragments are capable of activating C-H bonds (the latter has been exploited to yield a wealth of insight into selectivity and mechanism (*107,108*)), but these species do not afford catalytic functionalization. Jones has, however, developed several systems capable of catalyzing insertion to give aldimines (eq 24) (*109,110*) as well as intramolecular insertion to give indoles (*111,112*). These systems are based upon d^8 metal centers (Fe(0), Ru(0) and Rh(I)) bearing mono- and bidentate phosphines.

$$R-H \ + \ C\equiv N-R' \ \xrightarrow{\text{catalyst}} \ R-\overset{N-R'}{\underset{H}{\Big\langle}} \quad\quad (24)$$

In view of the rich C-H activation chemistry of Tp- and Cp-bearing complexes, discussed above, it is worth considering the scarcity of catalytic reactions in this context. Pathways for desirable catalytic reactions would inevitably involve the formation of several unsaturated intermediates. As noted above, the inability of CpM fragments to adopt stabilized unsaturated configurations (as compared with complexes of monodentate ligands) favors the thermodynamics of C-H addition. Given the importance of coordinatively unsaturated intermediates in catalysis, it seems reasonable to speculate that those same factors that favor stoichiometric reactions, by raising the energy of such intermediates, could disfavor catalytic reactions.

Insertion of unsaturated C-C bonds into C-H bonds

There are many transition-metal based systems (e.g. catalysts for olefin oligomerization or polymerization) that display a very low barrier to the insertion of olefins or other unsaturates into metal-carbon bonds. In that context, it is perhaps surprising that there are few examples in which such unsaturated species can be inserted into alkyl C-H bonds (i.e. effect alkyl C-H addition across a double or triple bond). However, in the past decade, great progress has been made with respect to insertion into aryl and vinyl C-H bonds (*113*). This disparity between sp^2 and sp^3 C-H bonds is striking and can probably be attributed only in part to the generally greater reactivity of the former. Other steps in the catalytic cycles, particularly C-C reductive elimination, are probably also favored by sp^2- versus sp^3- hybridized carbon atoms.

An early example of such a reaction was the hydrophenylation of diphenylketene (eq 25) (*114*).

$$\text{C}_6\text{H}_6 \ + \ Ph_2C=C=O \ \xrightarrow[\text{CO (30 atm) 200 °C}]{Rh_4(CO)_{12}} \ \underset{Ph}{\overset{Ph}{\Big\rangle}}\!\!=\!\!\underset{O}{\overset{Ph}{\Big\langle}} \quad (25)$$

The most significant breakthrough in this field was reported by Murai in 1993 (*115*); Ru(PPh$_3$)$_3$(CO)$_2$ and Ru(PPh$_3$)$_3$(CO)H$_2$ were found to catalyze the insertion of olefins into the ortho C-H bonds of aromatic ketones (eq 26).

$$\text{ArC(O)-H} + \text{CH}_2=\text{CHY} \xrightarrow[\text{or Ru(PPh}_3)_3(\text{CO})_2]{\text{Ru(PPh}_3)_3(\text{CO})\text{H}_2} \text{ArC(O)-CH}_2\text{CH}_2\text{Y} \quad (26)$$

Y = H, Si(OR)$_3$, alkyl

From this reaction the most extensive and well developed body of work in the area of unsaturate insertion into C-H bonds has emerged (*1f*). We will focus on this work as exemplary of this class of reactions although we would be very wary of generalizations extended to any other catalyst systems.

The scope of the system of eq 26 has been extended to include insertion of alkynes as well as olefins (*116*) and, more importantly, to a very broad range of substrates with a directing group (either N- or O-coordinating) *syn* to an sp^2 C-H bond (eq 27) (*1f*).

$$\text{(diene-E-H)} + \text{CH}_2=\text{CHY} \xrightarrow[\text{or Ru(PPh}_3)_3(\text{CO})_2]{\text{Ru(PPh}_3)_3(\text{CO})\text{H}_2} \text{(diene-E-CH}_2\text{CH}_2\text{Y)} \quad (27)$$

E = O, NR

Yields are frequently high, even quantitative, in both reagents. Accordingly, as Pfeffer has noted (*1g*), in terms of practical utility this is currently the singularly most valuable example of catalytic C-H bond functionalization.

The Ru-catalyzed insertions are very much favored by "directing" groups that are conjugated with the π-system of the C-H bond undergoing reaction (e.g. carbonyl or imine). It appears almost certain that this group plays a role more complex than simply bringing the C-H bond into the proximity of the metal center (i.e. minimizing the entropic cost of C-H addition). One possible role, indicated by DFT calculations by Morokuma (*117*), is to facilitate C-H addition via a transition state as indicated in eq 28. Note that although the net result is the same, eq 28 represents a pathway very different from that normally operative for simple C-H oxidative addition. We note in this context, however, that not all substrates transformed by this system possess the diene motif as shown in eq 27; limited examples have been reported in which even sp^3 C-H bonds undergo addition. (*118*)

$$\text{[diene-E-H···Ru]} \longrightarrow \text{[diene-E···Ru-H]} \longrightarrow \text{[diene-E-Ru-H]} \quad (28)$$

Subsequent to C-H addition, olefin insertion could be plausibly envisaged to proceed via either of two classes of mechanism involving: (a) initial insertion into the resulting Ru-C bond, or (b) insertion into the resulting Ru-H bond. In each case the appropriate reductive elimination (C-H or C-C respectively) would follow. Murai has presented strong evidence, some of which follows, for the hydrometalation path (b) and the overall mechanism of Figure 5. (Note that the dihydride precursor is believed to give rise to a zerovalent ruthenium species such as $Ru(CO)_n(PPh_3)_m$).

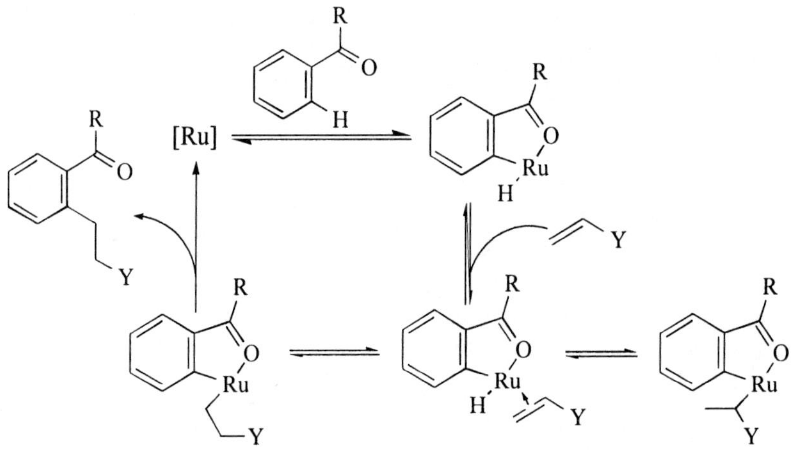

Figure 5. Mechanism proposed by Murai for Ru(0)-catalyzed "site-directed" addition of C-H bonds to olefins.

Methyl benzoate does *not* undergo coupling with olefins; however $C_6D_5CO_2CH_3$ undergoes full H/D exchange between the ortho-aryl deuterons and the vinylic protons of $CH_2=CHSi(OEt)_3$ indicating that C-D addition and olefin insertion into the resulting Ru-D bond both occur reversibly as in Figure 5. In accord with this result, coupling reactions showed a significant ^{13}C KIE (for the ortho carbon) implicating C-C coupling as the rate-determining step (*119*). In an unusual example of chemoselectivity in this system, only aryl esters with electron-withdrawing groups undergo coupling (*120*) (whereas aryl ketones tolerate electron-withdrawing and -donating groups). Murai has noted that this chemoselectivity indicates that the TS for the rate-determining C-C elimination is not a simple 3-centered σ-complex. Rather it suggests migration of the alkyl anion to the aryl ring; the anionic charge of the alkyl group can be delocalized onto the aryl group. This may further explain the importance of conjugation between the directing group and the C-H bond undergoing addition, in that the charge can also be delocalized on the directing group and back to the metal center. All of this suggests a TS for C-C elimination that is analogous to that calculated for C-H elimination as shown (for C-H addition) in eq 28.

Alkane Dehydrogenation

As a class of molecules, olefins are the most important feedstock in the chemical industry. The ability to dehydrogenate alkanes to olefins therefore has tremendous potential value. The high effectiveness of many organometallic hydrogenation catalysts suggested the possibility of related species catalyzing dehydrogenation. This point was appreciated by Crabtree who reported in 1979 that $[(Me_2CO)_2IrH_2(PPh_3)_2]^+$ dehydrogenated cyclopentane and cyclooctane to give the corresponding cycloalkadiene iridium complexes (16). The reaction required a hydrogen acceptor to offset the enthalpy of dehydrogenation (ΔH = ca. +30 kcal/mol for dehydrogenation of alkane to give free olefin and H_2). Mechanistically, the presumed role of the acceptor is to generate an active Ir(I) species. *t*-Butylethylene (TBE) was discovered to have the optimal balance of being strongly hydrogen-accepting and weakly coordinating (eq 29).

$$[IrH_2(solv)_2L_2]^+ + \bigcirc + 3 \text{ }^tBu\text{-}/ \longrightarrow \text{[Cp-Ir(H)(L)]} + 3 \text{ }^tBu\text{-}/ \quad (29)$$

It was proposed that the reaction proceeded via C-H "oxidative" addition although driven largely by the electrophilicity of the cation (16).

Baudry, Ephritikhine, and Felkin subsequently reported a similar cycloalkane reaction with L_2ReH_7, in which the dehydrogenated alkane remained bound to the metal center, also using TBE as hydrogen acceptor (121). This was extended to *n*-pentane which gave an η^4-trans-1,3-pentadiene complex. Rather remarkably, attempts to dislodge the coordinated diene with trimethyphosphite gave free 1-alkenes (α-olefins) (122). Selective conversion of *n*-alkane to 1-alkene was thus achieved, a reaction with great potential value if it could be effected catalytically in high yield. Cycloalkanes were soon reported to be dehydrogenated catalytically using L_2ReH_7 and L_2IrH_5 (123). Turnover numbers up to 70 were achieved, limited by catalyst decomposition.

The first well-characterized catalytic alkane dehydrogenation system was reported by Crabtree (124,125). $L_2IrH_2(\kappa^2\text{-}O_2CCF_3)$ was found to catalyze dehydrogenation of cycloalkanes and somewhat less effectively, *n*-alkanes. It was found that dehydrogenation could be catalyzed either with the use of a sacrificial hydrogen acceptor, or by driving the reaction photochemically. Catalyst deactivation was a critical issue in this and both prior and subsequent alkane dehydrogenation systems, and deactivation in this case was demonstrated to take place by P-C hydrogenolysis of the PR_3 ligand (124,125). Isotope effects and the isolation of the benzene C-H addition product $L_2Ir(Ph)(H)(O_2CCF_3)$ provided the first strong support for an oxidative addition pathway. Perhaps most significant, this report made explicit the importance of carboxylate dechelation, to yield a three-coordinate d^8 unit after transfer of the H atoms. Notably, virtually all other systems in this section are also based upon three-coordinate d^8 fragments, and in particular the ML_2X unit where M = Rh or Ir and X is a halide or anionic group.

Saito found that Wilkinson's Catalyst and related rhodium complexes could catalyze dehydrogenation in the absence of acceptor, thermochemically, when H_2 is driven off by reflux, albeit with low turnover numbers (126). Shortly thereafter Aoki and Crabtree generalized this reaction; they reported, inter alia, that L_2WH_6 catalyzed acceptorless dehydrogenation, the first example of homogeneous alkane dehydrogenation catalyzed (with or without acceptor) by a non-platinum-group metal (127).

Soon after Tanaka's report of carbonylation catalyzed by $Rh(PMe_3)_2(CO)Cl$ (101), it was discovered that the same complex catalyzed photodehydrogenation of alkanes to give alkenes in the absence of CO (e.g. eq 30) (128,129). Quantum yields and turnovers were much higher than were obtained for carbonylation. This reaction might be considered the first example of highly efficient alkane functionalization catalyzed via an organometallic route.

$$\text{cyclohexane} \xrightarrow[h\nu]{\textit{trans-}Rh(PMe_3)_2(CO)Cl} \text{cyclohexene} + H_2 \uparrow \qquad (30)$$

Mechanistic studies of this system led Maguire and Goldman to the conclusion that the role of the photon was expulsion of CO (and not photoelimination of H_2 from an intermediate). The resulting $Rh(PMe_3)_2Cl$ fragment then reacts *thermochemically* with alkane to give $H_2Rh(PMe_3)_2Cl$; CO then displaces H_2 to complete the catalytic cycle (129). This implied that this robust fragment offered great potential as a catalyst for thermochemical transfer-dehydrogenation. Attempts to generate thermochemical precursors of $Rh(PMe_3)_2Cl$ in the absence of a strongly binding ligand like CO led only to formation of an inactive dimer. It was found, however, that dihydrogen could react with either monomeric or dimeric species to enter the catalytic cycle via $H_2Rh(PMe_3)_2Cl$ (130). The result, surprisingly, is transfer dehydrogenation co-catalyzed by dihydrogen (e.g., eq 31).

$$\text{cyclohexane} + \text{CH}_2\text{=CH}^t\text{Bu} \xrightarrow[H_2 \text{ atm}]{Rh(PMe_3)_2ClL \text{ or } [Rh(PMe_3)_2Cl]_2} \text{cyclohexene} + \text{CH}_3\text{-CH}_2^t\text{Bu} \qquad (31)$$

While the system shown in eq 31 gave excellent rates and turnover numbers, the presence of hydrogen atmosphere led to the hydrogenation of more than one mol acceptor per mol of dehydrogenated product. In an effort to circumvent the need for hydrogen, related complexes that would not dimerize were investigated, including pincer complexes containing the unit (PCP)Rh (*131*). These were found to be very poor catalysts; independently, however, Jensen and Kaska found that the iridium analogues gave excellent results for cycloalkane dehydrogenation, particularly at elevated temperatures (eq 32)(*132*).

$$\text{cyclooctane} + {}^t\text{Bu-CH=CH}_2 \xrightarrow[150-200\ °C]{(^R\text{PCP})\text{IrH}_2;\ R = {}^t\text{Bu}} \text{cyclooctene} + {}^t\text{Bu-CH}_2\text{-CH}_3 \quad (32)$$

Thus, precursors of (^{R}PCP)Ir and Rh(PR$_3$)$_2$Cl gave excellent catalysts, in contrast to the "converse" pair, (^{R}PCP)Rh and Ir(PR$_3$)$_2$Cl. Calculations (DFT) by Krogh-Jespersen (*133*) indicated that this surprising relationship was related to the respective M-H bond strengths of the fragments. Addition of H-H to the model fragments Rh(PH$_3$)$_2$Cl and Ir(PH$_3$)$_2$Ph were calculated to be comparably exothermic, by ca. 28 kcal/mol; this value is consistent with the alkane-to-metal/metal-to-alkene hydrogen transfers being approximately thermoneutral. In contrast, the enthalpies of addition to the "converse" pair, Ir(PH$_3$)$_2$Cl and Rh(PH$_3$)$_2$Ph, were calculated to be much greater and much smaller respectively.

The stability at high temperature exhibited by (^{tBu}PCP)IrH$_n$ was exploited to give rates and turnover numbers for acceptorless dehydrogenation greater than those achieved with any previous catalysts (*133*).

$$\text{cyclodecane} \xrightarrow[\text{reflux}]{(^R\text{PCP})\text{IrH}_2;\ R = {}^t\text{Bu},\ {}^i\text{Pr}} \text{cyclodecene} + \text{H}_2 \uparrow \quad (33)$$

The ^{iPr}PCP derivative gave rates and TONs that were still higher (*134*); more importantly, this complex was found to catalyze the selective dehydrogenation of *n*-alkanes to give α-olefins (*135*).

$$\text{CH}_3(\text{CH}_2)_n\text{CH}_2\text{CH}_3 \xrightarrow[\underset{\text{AH}_2}{A}]{(^R\text{PCP})\text{IrH}_n} \text{CH}_3(\text{CH}_2)_n\text{CH=CH}_2 \quad (34)$$

Although the ^{iPr}PCP complex appeared to afford greater regioselectivity than the ^{tBu}PCP complex, it was found that both complexes in fact gave very high kinetic selectivity for α-olefins (*135*). Due to more rapid isomerization relative to dehydrogenation, however, the build-up of α-olefin tended to be lower with the ^{tBu}PCP complex although this was found to be highly dependent on reaction conditions.

Borylation

Perhaps the most notable exception to the observation that few Cp-containing complexes have offered catalytic activity is the hydroborylation of hydrocarbons. Cp*-bearing complexes (of Rh, Ir and Re) have been found to catalyze the reaction of eq 35 for alkanes (*136*) and arenes (*137*).

$$(R'O_2)B\text{-}B(OR')_2 + 2\ R\text{-}H \rightarrow 2\ R\text{-}B(OR')_2 + H_2 \tag{35}$$

The systems display the high selectivity characteristic of other transition-metal-based systems, particularly for the primary position of *n*-alkanes (*136*). The reaction has significant potential value owing to the synthetic versatility of alkyl and aryl boranes in cross-couplings and other reactions.

Eq 35 and related chemistry are also of great interest from a fundamental perspective. Initially one might envisage cycles similar to those invoked for H-D exchange reactions involving addition/elimination; in this case the boryl group, rather than a D atom, would end up bound to R. But the boryl group is not merely acting as a "trap" for a C-H bond after the task of activation is accomplished by the transition metal center. Rather, the empty p-orbital of a metalated boryl group plays an intimate role in promoting the activation process. Indeed, the development of this catalysis was based upon the observation that CpM(CO)$_n$(boryl) complexes showed stoichiometric reactivity towards alkanes, unlike their counterparts CpM(CO)$_n$R (R = alkyl, aryl or H) (*138,139*). (See Chapter 8 for a thorough discussion of this chemistry.)

Organometallic C-H Activation Modes Other than Oxidative Addition and Electrophilic

The modes of C-H bond activation discussed in the present section have been the subject of extensive research, although not on the scale seen for oxidative addition and electrophilic addition. Perhaps this is due to a scarcity – thus far – of such examples resulting in catalytic functionalization. In none of these cases however is there any obvious reason to believe that these modes are intrinsically incompatible with catalysis.

Sigma-bond metathesis

Almost concomitant with the development of C-H bond oxidative addition chemistry, a number of examples were reported in which d^0 metal alkyls were found to undergo σ-bond metathesis reactions with unactivated alkanes (*140*). A four-centered transition state, as shown in eq 36 is generally accepted for these reactions (see Chapter 7).

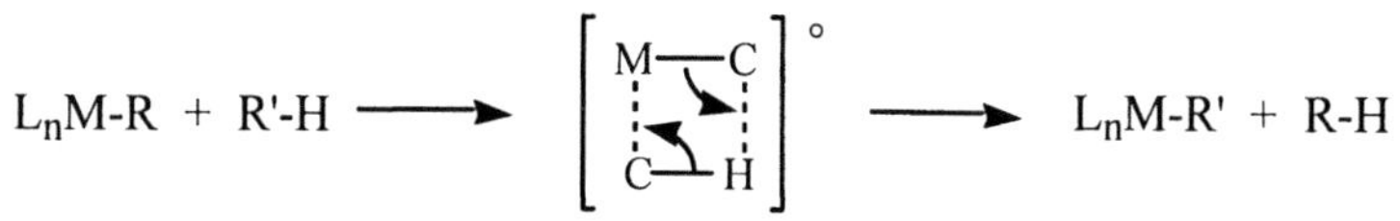

$$L_nM\text{-}R + R'\text{-}H \longrightarrow \left[\begin{array}{c} M\text{---}C \\ \vdots \ \ \vdots \\ C\text{---}H \end{array} \right]^\circ \longrightarrow L_nM\text{-}R' + R\text{-}H \quad (36)$$

Watson first reported in 1983 that Cp*$_2$LuMe underwent exchange with $^{13}CH_4$, in cyclohexane solvent, indicating that the preference for activation of less substituted hydrocarbons extended beyond simple electrophilic and oxidative additions (*141,142*). Certainly this selectivity could be viewed simply as another manifestation of the relatively greater metal-R bond strengths of less-substituted alkyls. It is tempting however to draw a close parallel to electrophilic activation in particular; the σ-bond metathesis can be viewed as an electrophilic addition in which the initially coordinated R group act as the base that receives the proton of the adding R'-H molecule. Conversely (and remarkably), Siegbahn and Crabtree have calculated a pathway for the prototype "electrophilic" system of Shilov that is apparently best described as a σ-bond metathesis (Figure 6); a chloride ligand accepts the proton of the σ-complexed alkane before the proton is transferred to solvent (*24*). (An oxidative addition path was also investigated, and calculated to be very close in energy (*24*).)

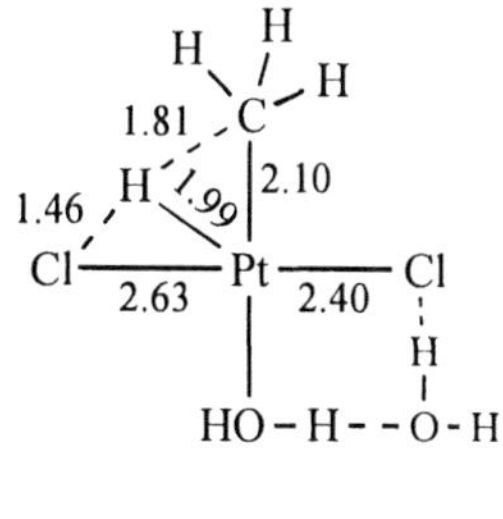

Figure 6. Transition state calculated for reaction of PtCl$_2$(H$_2$O)$_2$ with methane (including outer-sphere water). Key distances shown in Å. From reference (24).

Since metal centers that undergo σ-bond metathesis tend to be very closely related to those that catalyze Ziegler-Natta-type polymerization (i.e. undergo olefin insertion into M-R bonds), such complexes would seem to offer promise with respect to catalytic insertion of olefins into C-H bonds. An early indication of such promise was the report by Jordan of the insertion of propene into the C-H bond of 2-picoline (eq 37) (*143*).

$$\text{2-picoline} + \text{propene} \xrightarrow[H_2]{Cp_2Zr^+ \text{ intermediate}} \text{2-isopropylpyridine} \quad (37)$$

Very recently, Tilley has found that $Cp^*_2ScCH_3$ catalyzes (albeit with low turnovers) the addition of methane across the propene C-H bond to give isobutane (*144*).

σ-Bond metatheses of R'-H with M-R always result in formation of R-H, not R'-R bonds, when R and R' are alkyl groups. This would preclude the possibility of simple alkane-coupling reactions. Notably, this restriction does not apply to silanes and thus formation of silicon-silicon bonds is permissible, from cleavage of Si-H or even Si-C bonds (*145-147*); the origin of this difference is explained in Chapter 7.

1,2-addition

Wolczanski (*148*) and Bergman (*149*) independently reported in 1988 that Zr(IV) amido alkyl complexes could undergo 1,2 elimination of alkane to generate the corresponding transient imido complexes which then underwent addition of C-H bonds, including alkanes (eq 38) (*148*), across the Zr-N linkage.

$$(^tBuHN)_2Zr(R')(NH^tBu) \xrightarrow{-R'H} (^tBuHN)_2Zr=N^tBu \xrightarrow{+RH} (^tBuHN)_2Zr(R)(NH^tBu) \quad (38)$$

Legzdins has developed a rich chemistry involving the formally analogous addition of C-H bonds of hydrocarbons, including alkanes, across M=C (M = Mo or W) bonds (*150*) (see Chapter 11).

An even closer relationship with electrophilic addition can be considered for 1,2-addition than for a classic σ-bond metathesis (see above); in this case, the M-E π-bonding electrons (which presumably have some lone pair character on E) act as the base, accepting the proton of the adding C-H bond.

Activation of C-H bonds by metalloradicals

One of the most remarkable classes of "organometallic" C-H activation systems is based upon porphyrin complexes, mostly of rhodium(II) but also iridium(II), discovered and developed by Wayland. These radical-like paramagnetic systems are unreactive toward aryl C-H bonds, but they successfully cleave benzylic bonds, a marked contrast with systems operating via oxidative addition or electrophilic substitution. Such a preference would typically suggest a radical mechanism; however even higher selectivity is found for the much stronger C-H bond of methane than for that of toluene (*151*).

Kinetic studies, with the support of spectroscopic measurements and the effects of varying sterics, all led to resolution of this unusual selectivity issue. A termolecular transition state is operative, with one metal center forming a bond with the carbon while a second forms a bond with hydrogen (eq 39).

$$2\text{ (por)Rh(II)} + \text{R-H} \longrightarrow \left[(\text{por})\text{Rh}\cdots\text{H}\cdots\overset{\diagdown\;\;\diagup}{\underset{|}{\text{C}}}\cdots\text{Rh(por)} \right] \longrightarrow (\text{por})\text{Rh-H} + \text{R-Rh(por)}$$

(39)

Wayland subsequently synthesized complexes in which two metalloporphyrin units are tethered, held at distances consistent with the transition state of eq 39; these dimers have revealed high reactivity (*152*). In combination with a rich reaction chemistry of the metalloporphyrins with carbon monoxide and other small molecules (*153*), and a high selectivity for C-H activation of methane vs. methanol, such systems may offer great promise for catalysis.

The Mercat system

A unique photochemical system has been developed by Crabtree (*154*) which, at first glance, appears to fall outside the realm of organometallic chemistry. It is, however, closely tied to the field of organometallic C-H activation on both historical and molecular levels. During studies of photodehydrogenation catalyzed by $(CF_3CO_2)IrL_2H_2$, it was noticed that the presence of mercury (added to test for activity of colloidal iridium) resulted in formation of heavy hydrocarbons (e.g. eq 40). It was ultimately determined that a reaction takes place in the gas phase above the solution, where photo-excited mercury atoms react with alkanes to give alkyl radicals and free H atoms; the iridium complex was found to play no role in this reaction.

$$\text{R-H} \xrightarrow{\text{Hg}^*} \text{R}\bullet + \text{H}\bullet \longrightarrow 1/2\,\text{R-R} + 1/2\,\text{H}_2\uparrow \qquad (40)$$

As expected of a radical reaction, the system displays selectivity for the weakest C-H bond in the molecule. But several factors independent of the C-H activation step lead to a remarkable – and remarkably useful – selectivity, for the formation of homodimeric (as in eq 40) or heterodimeric (R-R′) products, that is not found in other radical based systems. The most critical of such factors is that dimerization of radicals generally affords products with C-H bonds weaker than those of the parent hydrocarbons (e.g. two secondary radicals couple to give a product with two tertiary C-H bonds); these therefore undergo relatively rapid secondary reactions in the presence of radicals. But in the gas-above-solution Mercat system, the dimeric products naturally condense and remain safely in solution after their formation. The system is applicable to alkanes but has been extended to include C-H activation of most major classes of heteroatom-containing organic molecules (alcohols, ketones, amines, etc.).

Calculations on the C-H activation step support insertion of the photo-excited mercury atoms into C-H bonds. This is a surprising result in view of the observed selectivity, which is characteristic of radical reactions (*155*).

$$\text{Hg}^*(^3P_1) + \text{R-H} \longrightarrow {}^3[\text{R-Hg-H}] \longrightarrow \text{R}\bullet + \text{H}\bullet + \text{Hg}$$

Non-organometallic C-H activation

The focus of this chapter is on "organometallic" C-H activation; i.e. C–H activation pathways involving the formation of metal-carbon bonds. Science, however, has a way of disrespecting such tidy definitions, and we feel it would be a disservice to many readers if we were to not draw attention to the vast field of metal-based "non-organometallic C-H activation" systems. Selected classes within this category are discussed in the present section.

Monooxygenases

Perhaps the most actively studied of all C-H functionalization systems are monooxygenases; these enzymes are nature's most widely used tool for C-H activation (*156*). Cytochrome P-450 and methane monooxygenase (MMO) (*157*) are particularly important examples. This very large and diverse class of enzymes effects the net insertion of an oxygen atom into C-H bonds, as well as oxidations of other substrates. The name monooxygenases derives from the fact that only one atom of each dioxygen consumed is used for the substrate oxidation. In vivo, a sacrificial reductant (NADH or another source of "H$^-$") is used to activate the O$_2$ molecule. The resulting formation of water from one of the two O atoms can be viewed as the thermodynamic price, paid to break the O-O bond, ultimately yielding a high-energy high-oxidation-state metal oxo complex willing to donate this O atom to the hydrocarbon.

$$\text{"H}^-\text{"} + \text{H}^+ + \text{O}_2 + \text{L}_n\text{M} \xrightarrow{\text{H}_2\text{O}} \text{L}_n\text{M=O} \xrightarrow{\text{R-H}} \text{L}_n\text{M} + \text{ROH} \qquad (41)$$

The mechanism of this second (deceptively simple-looking) step is the subject of tremendous interest. Numerous pathways have been proposed, and it appears that no single one of them is operative for all substrates. For many years it was confidently believed that these reactions generally proceeded via a so-called "rebound mechanism" elucidated by Groves (*158*). An H atom is transferred to the oxo ligand and the resulting carbon-based radical then abstracts the resulting OH group from the metal center (eq 42).

$$\text{L}_n\text{M=O} + \text{H-R} \longrightarrow \text{L}_n\text{M-OH} + \cdot\text{R} \longrightarrow \text{L}_n\text{M} + \text{HOR} \qquad (42)$$

Even this seemingly simple mechanism is in fact anything but simple: the "transfer of an H atom" is a useful accounting concept but not very meaningful mechanistically. Obviously nothing resembling a true (free) H atom is involved in reaction 42, which is of course actually the transfer of a proton and an electron – stepwise, or otherwise. An important breakthrough in this context was Mayer's discovery that the kinetics of H-transfer are best predicted based on correlation with thermodynamics and not with the radical character of the metal complex (or ligand) in question (*159*). More information on these types of reactions can be found in Chapter 21.

At least in the particularly important case of cytochrome P-450, compelling evidence has been presented for the very widespread occurrence of pathways which have some radical character, but in which there are no free radical intermediates (*160*). Complicating matters further is evidence for concomitant OH$^+$ insertion pathways, in which a peroxo-iron complex acts as OH$^+$ donor (*160*).

Fenton-type Chemistry

Fenton reported in the 1890s that iron(II) catalyzed the reactions of H_2O_2 with alkanes and other substrates. Related chemistry of other metals, particularly cobalt, has since been developed in depth (*161*). These systems are of great industrial importance and biological interest. The initial step in the reaction mechanism presumably involves reduction of peroxide by the metal (eq 43).

$$M(II) + H_2O_2 \longrightarrow M(III)OH + \cdot OH \qquad (43)$$

It has been widely believed that Fenton chemistry is merely comprised of metal-catalyzed peroxide decomposition and the reactions of the resulting oxy radicals. The oxidized metal is presumed to be regenerated by oxidation of peroxide. However, this has been called into question and the fate of the oxidized metal center and the other half of the peroxide unit (represented above as M(III)OH) is unclear. The presumed hydroxide intermediate shown in eq 43 could react with C-H bonds, acting either as a protonated high oxidation state metal-oxo species or as a simple oxo ligand. Recently, the intermediacy of active Fe(IV) oxo species has been proposed (*162,163*), based upon high-level calculational studies. Thus the distinction between the chemistry in this section and the preceding section has become blurred (*156*).

Barton's "Gif" systems, comprised of iron salts, added ligands, peroxides and pyridine/acetic acid solvent, present a situation somewhat converse to that described above for Fenton chemistry. Although unusual selectivity displayed by the Gif systems (e.g. conversion of alkanes to ketones) led to the assumption that they involved the direct reaction of iron oxo species with C-H bonds, strong evidence has been presented that the chemistry is actually that of oxygen- and carbon-centered radicals (*164*).

Vitamin B_{12}

Organometallic chemistry is surprisingly rare in Nature. The most well-studied organometallic system that occurs naturally (in species including humans and micro-organisms) is that of co-enzyme B_{12}. Ironically, its function is to catalyze C-H activation reactions via a non-organometallic mechanism (*165*). The corrin-ring-bearing complex contains a cobalt-deoxyadenosyl (metal-carbon) bond, which is intrinsically weak (and weakened further by the enzyme environment and the steric requirements of the corrin ring) and which readily undergoes homolytic cleavage (eq b1). The resulting deoxyadenosyl radical can then abstract a hydrogen atom from a wide range of biologically important substrates to give radicals that undergo spontaneous 1,2-functional group shifts. The rearranged radical then retrieves the adenosyl-bound H-atom to complete the 1,2-rearrangement.

$$L_nCo\text{-}CH_2Ad \rightleftharpoons L_nCo\bullet + \bullet CH_2Ad \quad (CH_2Ad = \text{deoxyadensyl})$$

Summary

The past 25 years have seen tremendous strides made toward the goal of selective catalytic functionalization of C-H bonds by organometallic systems. Studies of stoichiometric reactions have contributed greatly to this progress; it is primarily through such studies that we learn about the actual process of C-H activation. Studies of catalytic systems have led directly to improved systems; such studies have also elucidated the problems faced in the development of such systems, revealing many issues that are quite distinct from the challenge of stoichiometric C-H activation. Systems of significant practical utility have only just begun to emerge from this field. Extrapolating the current rate of progress, however, leads us to believe that the next 25 years will see the appearance of a diverse array of valuable systems involving substrates ranging from methane to the most complex targets of organic synthesis.

Acknowledgement

We would like to thank the many colleagues who have helped us to understand the field of C-H activation and organometallic chemistry more generally. This includes, among others, the authors of the following chapters in this volume. The authors of the reviews in reference 1 are thanked in particular, both for these very valuable reviews and in many cases for insights conveyed through both formal and informal channels. We are especially indebted to Prof. Alexander Shilov for his contributions to this field, to this chapter, and to the Symposium upon which this book is based. We also thank the National Science Foundation and Department of Energy for support.

References

1. Some recent reviews of C-H bond activation by organometallic complexes: (a) Sen, A. *Acc. Chem. Res.* **1998**, *31*, 550. (b) Guari, Y.; Sabo-Etiennne, S.; Chaudret, B. *Eur. J. Inorg. Chem.* **1999**, 1047. (c) Jones, W. D. *Science* **2000**, *287*, 1942. (d) Crabtree, R. H. *J. Chem. Soc., Dalton Trans.* **2001**, *17*, 2437. (e) Labinger, J. A.; Bercaw, J. E. *Nature* **2002**, *417*, 507. (f)

Kakiuchi, F.; Murai, S. *Acc. Chem. Res.* **2002**, *35*, 826. (g) Ritleng, V.; Sirlin, C.; Pfeffer, M. *Chem. Rev.* **2002**, *102*, 1731.
2. Halpern, J. *Discussions of the Faraday Society* **1968**, No. *46*, 7.
3. (a) Halpern, J.; Peters, E. *Journal of Chemical Physics* **1955**, *23*, 605. (b) Peters, E.; Halpern, J. *Journal of Physical Chemistry* **1955**, *59*, 793.
4. (a) Webster, A. H.; Halpern, J. *Journal of Physical Chemistry* **1956**, *60*, 280. (b) Halpern, J. *Journal of Physical Chemistry* **1959**, *63*, 398.
5. Vaska, L.; DiLuzio, J. W. *J. Am. Chem. Soc.* **1962**, *84*, 679.
6. (a) Vaska, L. *Acc. Chem. Res.* **1968**, *1*, 335. (b) Vaska, L.; Werneke, M. F. *Ann. N. Y. Acad. Sci.* **1971**, *172*, 546.
7. (a) Kubas, G. J.; Ryan, R. R.; Swanson, B. I.; Vergamini, P. J.; Wasserman, H. J. *J. Am. Chem. Soc.* **1984**, *106*, 451. (b) Kubas, G., J. *Acc. Chem. Res.* **1988**, *21*, 120.
8. Chatt, J.; Davidson, J. M. *J. Chem. Soc.* **1965**, 843.
9. Young, J. F.; Osborn, J. A.; Jardine, F. H.; Wilkinson, G. *Chem. Comm.* **1965**, 131.
10. Khrushch, A. P.; Tokina, L. A.; Shilov, A. E. *Kinetika i Kataliz* **1966**, *7*, 901.
11. Gol'dshleger, N. F.; Tyabin, M. B.; Shilov, A. E.; Shteinman, A. A. *Zhurnal Fizicheskoi Khimii* **1969**, *43*, 2174.
12. Kubas, G. J. *Metal dihydrogen and σ-bond complexes*; Kluwer Academic/Plenum Publishers: New York, 2001.
13. (a) Brookhart, M.; Green, M. L. H. *J. Organomet. Chem.* **1983**, *250*, 395. (b) Brookhart, M.; Green, M. L. H.; Wong, L. L. *Prog. Inorg. Chem.* **1988**, *36*, 1.
14. Jones, W. D. *Acc. Chem. Res.* **2003**, *36*, 140.
15. Stahl, S. S.; Labinger, J. A.; Bercaw, J. E. *J. Am. Chem. Soc.* **1996**, *118*, 5961.
16. (a) Crabtree, R. H.; Mihelcic, J. M.; Quirk, J. M. *J. Am. Chem. Soc.* **1979**, *101*, 7738. (b) Crabtree, R. H.; Quirk, J. M. *J. Organomet. Chem.* **1980**, *199*, 99. (c) Crabtree, R. H.; Mellea, M. F.; Mihelcic, J. M.; Quirk, J. M. *J. Am. Chem. Soc.* **1982**, *104*, 107.
17. Gol'dshleger, N. F.; Es'kova, V. V.; Shilov, A. E.; Shteinman, A. A. *Zhurnal Fizicheskoi Khimii* **1972**, *46*, 1353.
18. (a) Shilov, A. E.; Shul'pin, G. B. *Chem. Rev.* **1997**, *97*, 2879. (b) Stahl, S. S.; Labinger, J. A.; Bercaw, J. E. *Angew. Chem. Int. Ed.* **1998**, *37*, 2181. (c) Shilov, A. E.; Shul'pin, G. B. *Activation and Catalytic Reactions of Saturated Hydrocarbons in the Presence of Metal Complexes;* Kluwer: Boston, 2000. (d) Fekl, U.; Goldberg, K. I. *Adv. Inorg. Chem.* **2003**, *54*, 259.
19. Shilov, A. E.; Shteinman, A. A. *Coord. Chem. Rev.* **1977**, *24*, 97.
20. Kushch, L. A.; Lavrushko, V. V.; Misharin, Y. S.; Moravskii, A. P.; Shilov, A. E. *Nouveau Journal de Chimie* **1983**, *7*, 729.
21. Periana, R. A.; Taube, D. J.; Gamble, S.; Taube, H.; Satoh, T.; Fujii, H. *Science* **1998**, *280*, 560.

22. (a) Horváth, I. T.; Cook, R. A.; Millar, J. M.; Kiss, G. *Organometallics* **1993**, *12*, 8. (b) DeVries, N.; Roe, D. C.; Thorn, D. L. *J. Mol. Catal. A* **2002**, *189*, 17.
23. Wick, D. D.; Goldberg, K. I. *J. Am. Chem. Soc.* **1997**, *119*, 10235.
24. Siegbahn, P. E. M.; Crabtree, R. H. *J. Am. Chem. Soc.* **1996**, *118*, 4442.
25. (a) Luinstra, G. A.; Wang, L.; Stahl, S. S.; Labinger, J. A.; Bercaw, J. E. *Organometallics* **1994**, *13*, 755. (b) Wang, L.; Stahl, S. S.; Labinger, J. A.; Bercaw, J. E. *J. Mol. Catal. A* **1997**, *116*, 269.
26. (a) Geletii, Y. V.; Shilov, A. E. *Kinet. Catal.* **1983**, *24*, 413. (b) Lin, M.; Shen, C.; Garcia-Zayas, E. A.; Sen, A. *J. Am. Chem. Soc.* **2001**, 123, 1000.
27. (a) Luinstra, G. A.; Labinger, J. A.; Bercaw, J. E. *J. Am. Chem. Soc.* **1993**, *115*, 3004. (b) Luinstra, G. A.; Wang, L.; Stahl, S. S.; Labinger, J. A.; Bercaw, J. E. *J. Organomet. Chem.* **1995**, *504*, 75.
28. Williams, B. S.; Goldberg, K. I. *J. Am. Chem. Soc.* **2001**, *123*, 2576.
29. Kua, J.; Xu, X.; Periana, R. A.; Goddard, W. A., III *Organometallics* **2002**, *21*, 511.
30. (a) Mylvaganam, K.; Bacskay, G. B.; Hush, N. S. *J. Am. Chem. Soc.* **2000**, *122*, 2041. (b) Gilbert, T. M.; Hristov, I.; Ziegler, T. *Organometallics* **2001**, *20*, 1183
31. Periana, R. A.; Taube, D. J.; Evitt, E. R.; Loffler, D. G.; Wentrcek, P. R.; Voss, G.; Masuda, T. *Science* **1993**, *259*, 340.
32. Sen, A. *Acc. Chem. Res.* **1998**, *31*, 550.
33. Shilov, A. E. *Activation of Saturated Hydrocarbons by Transition Metal Complexes*; Reidel: Dordrecht, Netherlands, 1984, pp 212 pp.
34. Crabtree, R. H. *Chem. Rev.* **1995**, *4*, 987.
35. Sen, A. *Topics in Organometallic Chemistry* **1999**, *3*, 81.
36. Lin, M.; Hogan, T.; Sen, A. *J. Am. Chem. Soc.* **1997**, *119*, 6048.
37. Lin, M.; Hogan, T. E.; Sen, A. *J. Am. Chem. Soc.* **1996**, *118*, 4574.
38. Muehlhofer, M.; Strassner, T.; Herrmann, W. A. *Angew. Chem., Int. Ed.* **2002**, *41*, 1745.
39. Lin, M.; Sen, A. *J. Chem. Soc., Chem. Comm.* **1992**, 892.
40. (a) Fujiwara, Y.; Takaki, K.; Watanabe, J.; Uchida, Y,; Taniguchi, H. *Chem. Lett.* **1989**, 1687. (b) Nakata, K.; Yamaoka, Y.; Miyata, T.; Taniguchi, Y.; Takaki, K.; Fujiwara, Y. *J. Organomet. Chem.* **1994**, *473*, 329. (c) Jia, C.; Kitamura, T.; Fujiwara, Y. *Acc. Chem. Res.* **2001**, *34*, 633.
41. Periana, R. A.; Mironov, O.; Taube, D.; Bhalla, G.; Jones, C. J. *Science* **2003**, *301*, 814.
42. Foley, P.; Whitesides, G. M. *J. Am. Chem. Soc.* **1979**, *101*, 2732.
43. Green, M. L. H.; Knowles, P. J. *J. Chem. Soc., Chem. Comm.* **1970**, 1677.
44. Bennett, M. A.; Milner, D. L. *Chem. Comm.* **1967**, 581.
45. Koola, J. D.; Roddick, D. M. *J. Am. Chem. Soc.* **1991**, *113*, 1450.
46. Moulton, C. J.; Shaw, B. L. *J. Chem. Soc., Dalton Trans.* **1976**, 1020.
47. (a) van der Boom, M. E.; Milstein, D. *Chem. Rev.* **2003**, *103*, 1759. (b) Singleton, J. T. *Tetrahedron* **2003**, *59*, 1837.

48. Gozin, M.; Weisman, A.; Bendavid, Y.; Milstein, D. *Nature* **1993**, *364*, 699.
49. Janowicz, A. H.; Bergman, R. G. *J. Am. Chem. Soc.* **1982**, *104*, 352.
50. Hoyano, J. K.; Graham, W. A. G. *J. Am. Chem. Soc.* **1982**, *104*, 3723.
51. Bergman, R. G. *Science* **1984**, *223*, 902.
52. (a) Jones, W. D.; Feher, F. J. *J. Am. Chem. Soc.* **1982**, *104*, 4240. (b) Jones, W. D.; Feher, F. J. *J. Am. Chem. Soc.* **1984**, *106*, 1650. (c) Jones, W. D.; Feher, F. J. *Acc. Chem. Res.* **1989**, *22*, 91.
53. (a) Desrosiers, P. J.; Shinomoto, R. S.; Flood, T. C. *J. Am. Chem. Soc.* **1986**, *108*, 1346. (b) Desrosiers, P. J.; Shinomoto, R. S.; Flood, T. C. *J. Am. Chem. Soc.* **1986**, *108*, 7964. (c) Shinomoto, R. S.; Desrosiers, P. J.; Harper, T. G. P.; Flood, T. C. *J. Am. Chem. Soc.* **1990**, *112*, 704.
54. (a) Baker, M. V.; Field, L. D. *Organometallics* **1986**, 5, 821. (b) Baker, M. V.; Field, L. D. *J. Am. Chem. Soc.* **1986**, 108, 7433. (c) Baker, M. V.; Field, L. D. *J. Am. Chem. Soc.* **1987**, 109, 2825. (d) Field, L. D.; George, A. V.; Messerle, B. A. *Chem. Comm.* **1991**, 1339.
55. Periana, R. A.; Bergman, R. G. *J. Am. Chem. Soc.* **1986**, *108*, 7332.
56. Jones, W. D.; Dong, L. *J. Am. Chem. Soc.* **1989**, *111*, 8722.
57. Belt, S. T.; Dong, L.; Duckett, S. B.; Jones, W. D.; Partridge, M. G.; Perutz, R. N. *J. Chem. Soc., Chem. Comm.* **1991**, 266.
58. Jones, W. D.; Partridge, M. G.; Perutz, R. N. *J. Chem. Soc., Chem. Comm.* **1991**, 264.
59. Stoutland, P. O.; Bergman, R. G. *J. Am. Chem. Soc.* **1985**, *107*, 4581.
60. Baker, M. V.; Field, L. D. *J. Am. Chem. Soc.* **1986**, *108*, 7436.
61. Jones, W. D.; Feher, F. J. *J. Am. Chem. Soc.* **1985**, *107*, 620.
62. Buchanan, J. M.; Stryker, J. M.; Bergman, R. G. *J. Am. Chem. Soc.* **1986**, *108*, 1537.
63. (a) Churchill, D. G.; Janak, K. E.; Wittenberg, J. S.; Parkin, G. *J. Am. Chem. Soc.* **2003**, *125*, 1403. (b) Janak, K. E.; Parkin, G. *J. Am. Chem. Soc.* **2003**, *125*, 6889.
64. (a) Graham, M. A.; Perutz, R. N.; Poliakoff, M.; Turner, J. J. *J. Organomet. Chem.* **1972**, *34*, C34. (b) Hall, C.; Perutz, R. N. *Chem. Rev.* **1996**, *96*, 3125.
65. (a) Wasserman, E. P.; Moore, C. B.; Bergman, R. G. *Science* **1992**, *255*, 315. (b) Bromberg, S. E.; Yang, H.; Asplund, M. C.; Lian, T.; McNamara, B. K.; Kotz, K. T.; Yeston, J. S.; Wilkens, M.; Frei, H.; Bergman, R. G.; Harris, C. B. *Science* **1997**, *278*, 260. (c) McNamara, B. K.; Yeston, J. S.; Bergman, R. G.; Moore, C. B. *J. Am. Chem. Soc.* **1999**, *121*, 6437. (d) Asplund, M. C.; Snee, P. T.; Yeston, J. S.; Wilkens, M. J.; Payne, C. K.; Yang, H.; Kotz, K. T.; Frei, H.; Bergman, R. G.; Harris, C. B. *J. Am. Chem. Soc.* **2002**, *124*, 10605.
66. Halpern, J. *Inorg. Chim. Acta* **1985**, *100*, 41.
67. Rest, A. J.; Whitwell, I.; Graham, W. A. G.; Hoyano, J. K.; McMaster, A. D. *J. Chem. Soc., Chem. Commun.* **1984**, 624.
68. Saillard, J.; Hoffmann, R. *J. Am. Chem. Soc.* **1984**, *106*, 2006.

69. Wenzel, T. T.; Bergman, R. G. *J. Am. Chem. Soc.* **1986**, *108*, 4856.
70. Nolan, S. P.; Hoff, C. D.; Stoutland, P. O.; Newman, L. J.; Buchanan, J. M.; Bergman, R. G.; Yang, G. K.; Peters, K. S. *J. Am. Chem. Soc.* **1987**, *109*, 3143.
71. Burger, P.; Bergman, R. G. *J. Am. Chem. Soc.* **1993**, *115*, 10462.
72. Arndtsen, B. A.; Bergman, R. G. *Science* **1995**, *270*, 1970.
73. Strout, D. L.; Zaric, S.; Niu, S.; Hall, M. B. *J. Am. Chem. Soc.* **1996**, *118*, 6068.
74. Niu, S.; Hall, M. B. *J. Am. Chem. Soc.* **1998**, *120*, 6169.
75. Klei, S. R.; Tilley, T. D.; Bergman, R. G. *J. Am. Chem. Soc.* **2000**, *122*, 1816.
76. Harper, T. G. P.; Shinomoto, R. S.; Deming, M. A.; Flood, T. C. *J. Am. Chem. Soc.* **1988**, *110*, 7915.
77. Werner, H.; Höhn, A.; Dziallas, M. *Angew. Chem. Int. Ed. Engl.* **1986**, *25*, 1090.
78. Nemeh, S.; Jensen, C.; Binamira-Soriage, E.; Kaska, W. C. *Organometallics* **1983**, *2*, 1442.
79. Jean, Y.; Eisenstein, O. *Polyhedron* **1988**, *7*, 405.
80. Riehl, J. F.; Jean, Y.; Eisenstein, O.; Pelissier, M. *Organometallics* **1992**, *11*, 729.
81. Cundari, T. R. *J. Am. Chem. Soc.* **1994**, *116*, 340.
82. Krogh-Jespersen, K.; Goldman, A. S. in *ACS Symposium Series 721, Transition State Modeling for Catalysis*; D. G. Truhlar and K. Morokuma, Eds.; American Chemical Society: Washington DC, 1999; 151.
83. Clot, E.; Eisenstein, O. *J. Phys. Chem. A* **1998**, *102*, 3592.
84. Krogh-Jespersen, K.; Czerw, M.; Zhu, K.; Singh, B.; Kanzelberger, M.; Darji, N.; Achord, P. D.; Renkema, K. B.; Goldman, A. S. *J. Am. Chem. Soc.* **2002**, *124*, 10797.
85. Brainard, R. L.; Nutt, W. R.; Lee, T. R.; Whitesides, G. M. *Organometallics* **1988**, *7*, 2379.
86. Holtcamp, M. W.; Labinger, J. A.; Bercaw, J. E. *J. Am. Chem. Soc.* **1997**, *119*, 848.
87. Johansson, L.; Tilset, M.; Labinger, J. A.; Bercaw, J. E. *J. Am. Chem. Soc.* **2000**, *122*, 10846.
88. Examples of stable six coordinate Pt(IV) alkyl hydrides: (a) Puddephatt, R. J. *Coord. Chem. Rev.* **2001**, *219-221*, 157. (b) O'Reilly, S.; White, P. S.; Templeton, J. L. *J. Am. Chem. Soc.* **1996**, *118*, 5684. (c) Canty, A.J.; Dedieu, A.; Jin, H.; Milet, A.; Richmond, M. K. *Organometallics* **1996**, *15*, 2845. (d) Hill, G. S.; Vittal, J. J.; Puddephatt, R. J. *Organometallics* **1997**, *16*, 1209. (e) Prokopchuk, E. M.; Jenkins, H. A.; Puddephatt, R. J. *Organometallics* **1999**, *18*, 2861. (f) Haskel, A.; Keinan, E. *Organometallics* **1999**, *21*, 247 . (g) Reinartz, S.; Brookhart, M.; Templeton, J. L. *Organometallics* **2002**, *18*, 4677 (h) Iron, M. A.; Lo, H. C.; Martin, J. M. L.; Keinan, E. *J. Am. Chem. Soc.* **2002**, *124*, 7041. (i)

Crumpton-Bregel, D. M.; Goldberg, K. I. *J. Am. Chem. Soc.* **2003**, *125*, 9442.
89. (a) Fekl, U.; Kaminsky, W.; Goldberg, K. I. *J. Am. Chem. Soc.* **2001**, *123*, 6423. (b) Reinartz, S.; White, P. S.; Brookhart, M.; Templeton, J. L. *J. Am. Chem. Soc.* **2001**, *123*, 6425. (c) Fekl, U.; Goldberg, K. I. *J. Am. Chem. Soc.* **2002**, *124*, 6804.
90. (a) Hill, G. S.; Rendina, L. M.; Puddephatt, R. J. *Organometallics* **1995**, *14*, 4966. (b) Stahl, S. S.; Labinger, J. A.; Bercaw, J. E. *J. Am. Chem. Soc.* **1995**, *117*, 9371.
91. Wik, B. J.; Lersch, M.; Tilset, M. *J. Am. Chem. Soc.* **2002**, *124*, 12116.
92. (a) Jenkins, H. A.; Yap, G. P. A.; Puddephatt, R. J. *Organometallics* **1997**, *16*, 1946. (b) Fekl, U.; Zahl, A.; van Eldik, R. *Organometallics* **1999**, *18*, 4156. (c) Prokopchuk, E. M.; Puddephatt, R. J. *Organometallics* **2003**, *22*, 563. (d) Prokopchuk, E. M.; Puddephatt, R. J. *Organometallics* **2003**, *22*, 787. (e) Jensen, M. P.; Wick, D. D.; Reinartz, S.; Templeton, J. L.; Goldberg, K. I. *J. Am. Chem. Soc.* **2003**, *125*, 8614
93. Hackett, M.; Ibers, J. A.; Jernakoff, P.; Whitesides, G. M. *J. Am. Chem. Soc.* **1986**, *108*, 8094.
94. Fisher, B. J.; Eisenberg, R. *Organometallics* **1983**, *2*, 764.
95. Kunin, A. J.; Eisenberg, R. *J. Am. Chem. Soc.* **1986**, *108*, 535.
96. Kunin, A. J.; Eisenberg, R. *Organometallics* **1988**, *7*, 2124.
97. Spillett, C. T.; Ford, P. C. *J. Am. Chem. Soc.* **1989**, *111*, 1932.
98. Bridgewater, J. S.; Lee, B.; Bernhard, S.; Schoonover, J. R.; Ford, P. C. *Organometallics* **1997**, *16*, 5592.
99. Bitterwolf, T. E.; Scallorn, W. B.; Bays, J. T.; Weiss, C. A.; Linehan, J. C.; Franz, J.; Poli, R. *J. Organomet. Chem.* **2002**, *652*, 95.
100. Rosini, G. P.; Boese, W. T.; Goldman, A. S. *J. Am. Chem. Soc.* **1994**, *116*, 9498.
101. Sakakura, T.; Tanaka, M. *Chem. Lett.* **1987**, 249.
102. Sakakura, T.; Tanaka, M. *J. Chem. Soc., Chem. Commun.* **1987**, 758.
103. Boyd, S. E.; Field, L. D.; Partridge, M. G. *J. Am. Chem. Soc.* **1994**, *116*, 9492.
104. Rosini, G. P.; Zhu, K.; Goldman, A. S. *J. Organomet. Chem.* **1995**, *504*, 115.
105. Jones, W. D.; Feher, F. J. *Organometallics* **1983**, *2*, 686.
106. Jones, W. D.; Hessell, E. T. *J. Am. Chem. Soc.* **1992**, *114*, 6087.
107. Northcutt, T. O.; Wick, D. D.; Vetter, A. J.; Jones, W. D. *J. Am. Chem. Soc.* **2001**, *123*, 7257.
108. Wick, D. D.; Reynolds, K. A.; Jones, W. D. *J. Am. Chem. Soc.* **1999**, *121*, 3974.
109. Jones, W. D.; Foster, G. P.; Putinas, J. *J. Am. Chem. Soc.* **1987**, *109*, 5047.
110. Jones, W. D.; Hessell, E. T. *Organometallics* **1990**, *9*, 718.
111. Jones, W. D.; Kosar, W. P. *J. Am. Chem. Soc.* **1986**, *108*, 5640.
112. Hsu, G. C.; Kosar, W. P.; Jones, W. D. *Organometallics* **1994**, *13*, 385.

113. (a) Lenges, C. P.; Brookhart, M. *J. Am. Chem. Soc.* **1999**, *121*, 6616. (b) Tan, K. L.; Bergman, R. G.; Ellman, J. A. *J. Am. Chem. Soc.* **2001**, *123*, 2685. (c) Lail, M.; Arrowood, B. N.; Gunnoe, T. B. *J. Am. Chem. Soc.* **2003**, *125*, 7506.
114. Hong, P.; Yamazaki, H.; Sonogashira, K.; Hagihara, N. *Chem. Lett.* **1978**, 535.
115. Murai, S.; Kakiuchi, F.; Sekine, S.; Tanaka, Y.; Kamatani, A.; Sonoda, M.; Chatani, N. *Nature* **1993**, *366*, 529.
116. Kakiuchi, F.; Yamamoto, Y.; Chatani, N.; Murai, S. *Chem. Lett.* **1995**, 681.
117. Matsubara, T.; Koga, N.; Musaev, D. G.; Morokuma, K. *J. Am. Chem. Soc.* **1998**, *120*, 12692.
118. (a) Chatani, N.; Ishii, Y.; Ie, Y.; Kakiuchi, F.; Murai, S. *J. Org. Chem.* **1998**, *63*, 5129. (b) Chatani, N.; Asaumi, T.; Yorimitsu, S.; Ikeda, T.; Kakiuchi, F.; Murai, S. *J. Am. Chem. Soc.* **2001**, *123*, 10935.
119. Kakiuchi, F.; Ohtaki, H.; Sonoda, M.; Chatani, N.; Murai, S. *Chem. Lett.* **2001**, 918.
120. Sonoda, M.; Kakiuchi, F.; Kamatani, A.; Chatani, N.; Murai, S. *Chem. Lett.* **1996**, 109.
121. Baudry, D.; Ephritikhine, M.; Felkin, H. *J. Chem. Soc., Chem. Comm.* **1980**, 1243.
122. Baudry, D.; Ephritikhine, M.; Felkin, H.; Zakrzewski, J. *J. Chem. Soc., Chem. Comm.* **1982**, 1235.
123. (a) Baudry, D.; Ephritikhine, M.; Felkin, H.; Holmes-Smith, R. *Chem. Comm.* **1983**, 788. (b) Felkin, H.; Fillebeen-Khan, T.; Gault, Y.; Holmes-Smith, R.; Zakrzewski, J. *Tetrahedron Lett.* **1984**, *25*, 1279. (c) Felkin, H.; Fillebeen-Khan, T.; Holmes-Smith, R.; Lin, Y. *Tetrahedron Lett.* **1985**, *26*, 1999.
124. Burk, M. J.; Crabtree, R. H.; Parnell, C. P.; Uriarte, R. J. *Organometallics* **1984**, *3*, 816.
125. Burk, M. J.; Crabtree, R. H. *J. Am. Chem. Soc.* **1987**, *109*, 8025.
126. Fujii, T.; Saito, Y. *J. Chem. Soc., Chem. Commun.* **1990**, 757.
127. Aoki, T.; Crabtree, R. H. *Organometallics* **1993**, *12*, 294.
128. (a) Nomura, K.; Saito, Y. *Chem. Comm.* **1988**, 161. (b) Sakakura, T.; Sodeyama, T.; Tokunaga, M.; Tanaka, M. *Chem. Lett.* **1988**, 263.
129. Maguire, J. A.; Boese, W. T.; Goldman, A. S. *J. Am. Chem. Soc.* **1989**, *111*, 7088.
130. (a) Maguire, J. A.; Goldman, A. S. *J. Am. Chem. Soc.* **1991**, *113*, 6706. (b) Maguire, J. A.; Petrillo, A.; Goldman, A. S. *J. Am. Chem. Soc.* **1992**, *114*, 9492.
131. Wang, K.; Goldman, M. E.; Emge, T. J.; Goldman, A. S. *J. Organomet. Chem.* **1996**, *518*, 55.
132. (a) Gupta, M.; Hagen, C.; Flesher, R. J.; Kaska, W. C.; Jensen, C. M. *Chem. Commun.* **1996**, 2083. (b) Gupta, M.; Hagen, C.; Kaska, W. C.; Cramer, R. E.; Jensen, C. M. *J. Am. Chem. Soc.* **1997**, *119*, 840.

133. Xu, W.; Rosini, G. P.; Gupta, M.; Jensen, C. M.; Kaska, W. C.; Krogh-Jespersen, K.; Goldman, A. S. *Chem. Comm.* **1997**, 2273.
134. Liu, F.; Goldman, A. S. *Chem. Comm.* **1999**, 655.
135. Liu, F.; Pak, E. B.; Singh, B.; Jensen, C. M.; Goldman, A. S. *J. Am. Chem. Soc.* **1999**, *121*, 4086.
136. (a) Chen, H.; Hartwig, J. F. *Angew. Chem., Intl. Ed.* **1999**, *38*, 3391. (b) Chen, H.; Schlecht, S.; Semple, T. C.; Hartwig, J. F. *Science* **2000**, *287*, 1995.
137. Iverson, C. N.; Smith, M. R., III *J. Am. Chem. Soc.* **1999**, *121*, 7696.
138. Waltz, K. M.; Hartwig, J. F. *Science* **1997**, *277*, 211.
139. Webster, C. E.; Fan, Y.; Hall, M. B.; Kunz, D.; Hartwig, J. F. *J. Am. Chem. Soc.* **2003**, *125*, 858.
140. Thompson, M. E.; Baxter, S. M.; Bulls, A. R.; Burger, B. J.; Nolan, M. C.; Santarsiero, B. D.; Schaefer, W. P.; Bercaw, J. E. *J. Am. Chem. Soc.* **1987**, *109*, 203.
141. Watson, P. L. *J. Am. Chem. Soc.* **1983**, *105*, 6491.
142. Watson, P. L.; Parshall, G. W. *Acc. Chem. Res.* **1985**, *18*, 51.
143. (a) Jordan, R. F.; Taylor, D. F. *J. Am. Chem. Soc.* **1989**, *111*, 778. (b) Rodewald, S.; Jordan, R. F. *J. Am. Chem. Soc.* **1994**, *116*, 4491.
144. Sadow, A. D.; Tilley, T. D. *J. Am. Chem. Soc.* **2003**, *125*, 7971.
145. Castillo, I.; Tilley, T. D. *Organometallics* **2001**, *20*, 5598.
146. Sadow, A. D.; Tilley, T. D. *J. Am. Chem. Soc.* **2003**, *125*, 9462.
147. Woo, H. G.; Tilley, T. D. *J. Am. Chem. Soc.* **1989**, *111*, 8043.
148. Cummins, C. C.; Baxter, S. M.; Wolczanski, P. T. *J. Am. Chem. Soc.* **1988**, *110*, 8731.
149. Walsh, P. J.; Hollander, F. J.; Bergman, R. G. *J. Am. Chem. Soc.* **1988**, *110*, 8729.
150. Wada, K.; Pamplin, C. B.; Legzdins, P.; Patrick, B. O.; Tsyba, I.; Bau, R. *J. Am. Chem. Soc.* **2003**, *125*, 7035.
151. (a) Del Rossi, K. J.; Wayland, B. B. *J. Am. Chem. Soc.* **1985**, *107*, 7941. (b) Del Rossi, K. J.; Wayland, B. B. *Chem. Comm.* **1986**, 1653. (c) Sherry, A. E.; Wayland, B. B. *J. Am. Chem. Soc.* **1990**, *112*, 1259. (d) Wayland, B. B.; Ba, S.; Sherry, A. E. *J. Am. Chem. Soc.* **1991**, *113*, 5305.
152. (a) Zhang, X.-X.; Wayland, B. B. *J. Am. Chem. Soc.* **1994**, *116*, 7897. (b) Cui, W.; Zhang, X. P.; Wayland, B. B. *J. Am. Chem. Soc.* **2003**, *125*, 4994.
153. Wayland, B. B.; Coffin, V. L.; Sherry, A. E.; Brennen, W. R. *ACS Symposium Series; Bonding and Energetics of Organometallic Compounds* **1990**, *428*, 148.
154. (a) Brown, S. H.; Crabtree, R. H. *Chem. Comm.* **1987**, 970. (b) Brown, S. H.; Crabtree, R. H. *J. Am. Chem. Soc.* **1989**, *111*, 2935. (c) Brown, S. H.; Crabtree, R. H. *J. Am. Chem. Soc.* **1989**, *111*, 2946. (d) Crabtree, R. H.; Brown, S. H.; Muedas, C. A.; Boojamra, C.; Ferguson, R. R. *Chemtech* **1991**, *21*, 634.
155. Siegbahn, P. E. M.; Svensson, M.; Crabtree, R. H. *J. Am. Chem. Soc.* **1995**, *117*, 6758.

156. *Biomimetic Oxidations Catalyzed by Transition Metal Complexes*; Meunier, B., Ed.; Imperial College Press: River Edge, NJ, 2000.
157. Baik, M.-H.; Newcomb, M.; Friesner, R. A.; Lippard, S. J. *Chem. Rev.* **2003**, *103*, 2385.
158. McLain, J. L.; Lee, J.; Groves, J. T. in *Biomimetic Oxidations Catalyzed by Transition Metal Complexes*; B. Meunier, Ed. 2000; 91.
159. Mayer, J. M. in *Biomimetic Oxidations Catalyzed by Transition Metal Complexes*; B. Meunier, Ed. 2000; 1.
160. Newcomb, M.; Toy, P. H. *Acc. Chem. Res.* **2000**, *33*, 449.
161. For a good introduction to this topic see: Tolman, C. A.; Druliner, J. D.; Nappa, M. J.; Herron, N. In *Activation and Functionalization of Alkanes*; Hill, C. L., Ed.; John Wiley & Sons: New York, 1989, pp 303-360.
162. Buda, F.; Ensing, B.; Gribnau, M. C. M.; Baerends, E. J. *Chemistry--A European Journal* **2001**, *7*, 2775.
163. Ensing, B.; Buda, F.; Baerends, E. J. *J. Phys. Chem. A* **2003**, *107*, 5722.
164. (a) Barton, D. H. R.; Doller, D. *Pure and Applied Chemistry* **1991**, *63*, 1567. (b) Barton, D. H. R.; Hu, B. *Pure and Applied Chemistry* **1997**, *69*, 1941. (c) Stavropoulos, P.; Celenligil-Cetin, R.; Tapper, A. E. *Acc. Chem. Res.* **2001**, *34*, 745.
165. (a) Halpern, J. *Science* **1985**, *227*, 869. (b) Geno, M. K.; Halpern, J. *J. Am. Chem. Soc.* **1987**, *109*, 1238.

Oxidative Addition Reactions

Chapter 2

C–H Bond Activation by Iridium and Rhodium Complexes: Catalytic Hydrogen–Deuterium Exchange and C–C Bond-Forming Reactions

Steven R. Klei, Kian L. Tan, Jeffrey T. Golden, Cathleen M. Yung, Reema K. Thalji, Kateri A. Ahrendt, Jonathan A. Ellman, T. Don Tilley, and Robert G. Bergman

Chemical Sciences Division, Lawrence Berkeley National Laboratory, and Department of Chemistry and Center for New Directions in Organic Synthesis, University of California, Berkeley, CA 94720

> The metal-mediated activation of carbon-hydrogen bonds has the potential of leading to important fundamental insights, as well as to advances in large-scale industrial and synthetic organic chemical processes. Our initial finding of C-H oxidative addition reactions in low-valent iridium (I) complexes was followed by a more recent discovery and exploration of C-H activation processes at Ir (III) centers. This paper reviews recent studies of the Ir (III) reaction that include work directed toward elucidation of its reaction mechanism and the development of catalytic hydrogen/deuterium exchange reactions. The article also discusses collaborative studies of rhodium (I)-catalyzed reactions that provide a method for adding complexity to a range of carbo- and heterocyclic arene compounds with structures analogous to those found in physiologically active organic compounds.

Introduction

The activation of carbon-hydrogen bonds in alkanes (and alkyl groups remote from other functional groups in organic molecules) has important implications for the development of new industrial processes, new methods for the synthesis of fine chemicals, and the efficient use of energy. However, such "unactivated" C-H bonds are normally so inert toward conventional reagents that until recently it has not been possible to activate and replace the hydrogen atoms with more synthetically reactive functional groups (*1-4*) This article reviews experiments carried out in our group, and in collaboration with other groups at Berkeley, aimed at discovering and understanding the mechanisms of metal-mediated C-H activation processes.

Carbon-Hydrogen Bond Activation at Methyliridium(III) Centers

Several years ago we (and simultaneously, Hoyano and Graham) discovered one of the first "C-H activation" reactions that converts alkanes into isolable metal complexes (eq 1) (*5-7*). The important step in this process involves the generation of an exceedingly reactive iridium, rhodium (*8,9*) or rhenium (*10*) fragment, in formal oxidation state +1, which undergoes an oxidative addition reaction leading to a metal (+3) product containing a new metal-carbon and metal-hydrogen bond.

More recently, we have prepared a series of higher-valent, methyliridium(III) complexes that also are capable of hydrocarbon C-H bond activation (*11-13*) These Ir(III)-based reactions differ from those observed with Ir(I) in that replacement of the methyl group with the alkyl group of the activated hydrocarbon is observed (eq. 2). The first of these reactions that we uncovered involved reaction of the iridium complex Cp*(PMe$_3$)Ir(CH$_3$)(OTf),

which bears a reactive trifluoromethanesulfonate ("triflate") ligand, with the C-H bonds of benzene (eq. 2, R = Ph). However, the reactivity of this material extends beyond aromatic C-H bonds, and includes primary alkanes and methane (established by carbon labeling), although the Ir(III) complexes do not attack as many different kinds of C-H bonds as their Ir(I) relatives. The corresponding cationic complexes (associated with fluorinated polyarylborate (BAr$_f^-$) counterions), bearing neutral and more loosely bound ligands such as dichloromethane, attack the same range of C-H bonds, but these complexes react much more rapidly than the triflates. The BAr$_f^-$ salts are typically made by treating Cp*(PMe$_3$)Ir(CH$_3$)(OTf) with NaBAr$_f$, but they can be generated under particularly mild conditions (temperatures as low as -80 °C) by abstracting a CH$_3^-$ group from the charge-neutral dimethyl complex [Cp*(PMe$_3$)Ir(CH$_3$)$_2$] with B(C$_6$F$_5$)$_3$. Finally, many of the C-H activation reactions observed with Ir(III) complexes lead to products formed by rearrangement of the first-formed intermediates. A summary of some of the C-H activation/rearrangement processes that we have observed is presented in Scheme 1.

Scheme 1

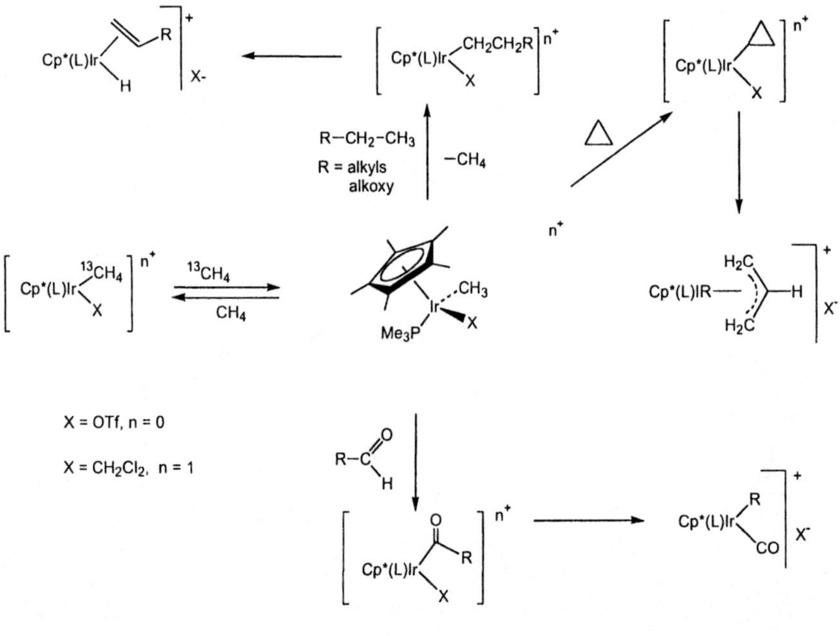

After exploring the generality of the Ir(III) C-H activation reactions, we began to address the question of their mechanism. A central problem concerned whether these transformations proceed by initial C-H oxidative addition to give very high oxidation state iridium(V) intermediates (path (a) in Scheme 2), or by concerted metal alkyl/C-H σ-bond metathesis (path (b) in Scheme 2) (*14*).

A number of lines of investigation have provided evidence that path (a) is correct.

Scheme 2

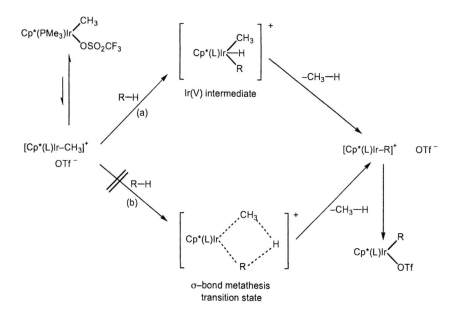

Probably the most convincing results were obtained in studies that were initially designed to compare the C-H activation reactions of this system with analogous processes that might occur at Si-H bonds. An unexpected, but very informative, result from these experiments is illustrated in Scheme 3. These show that initial Si-H activation occurs, but this is followed in the first-formed intermediate by intramolecular C-H activation, leading to a cyclometallated complex that exhibits a stable Ir(V) hydridoaryl center. Furthermore, treatment of this complex with acetonitrile causes C-H reductive elimination and reconversion of the metal center to the Ir(III) state. This result demonstrates that Ir(III)/Ir(V) interconversion can occur readily in this system, providing strong circumstantial evidence for path (a) in Scheme 2. DFT calculations carried out on the oxidative addition reaction of methane to [Cp*(PMe$_3$)IrCH$_3$]$^+$ support this conclusion (*15,16*).

Scheme 3

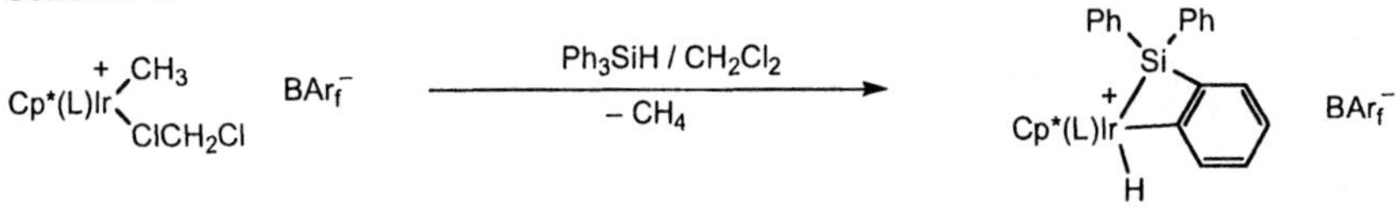

[Cp*(L)Ir(CH₃)(ClCH₂Cl)]⁺ BAr_f⁻ →(Ph₃SiH / CH₂Cl₂, −CH₄)→ [Cp*(L)Ir(C₆H₄-SiPh₂)(H)]⁺ BAr_f⁻

Carbon-Hydrogen Bond Activation at Hydridoiridium(III) Centers

The C-H activation chemistry observed with [Cp*(PMe₃)IrCH₃]⁺ raised the possibility that analogous reactivity could be seen with the corresponding cationic Ir(III) hydride, [Cp*(PMe₃)IrH]⁺. We were also interested in preparing this complex because of its potential use as a catalyst for alkane dehydrogenation. We judged that the mildest method for generating the cation would involve the use of B(C₆F₅)₃ to abstract a hydride anion from Cp*(PMe₃)IrH₂. However, this reaction was unsuccessful, leading instead to a mixture of trihydrides. We were finally able to prepare this complex by hydrogenolysis of [Cp*(PMe₃)IrCH₃]⁺ at very low temperature, as shown in Scheme 4. The mononuclear, cationic hydride is very sensitive, decomposing to polyhydrides at temperatures above -25 °C. However, at lower temperatures it is stable enough for spectroscopic characterization and chemical investigation (*17*).

Scheme 4

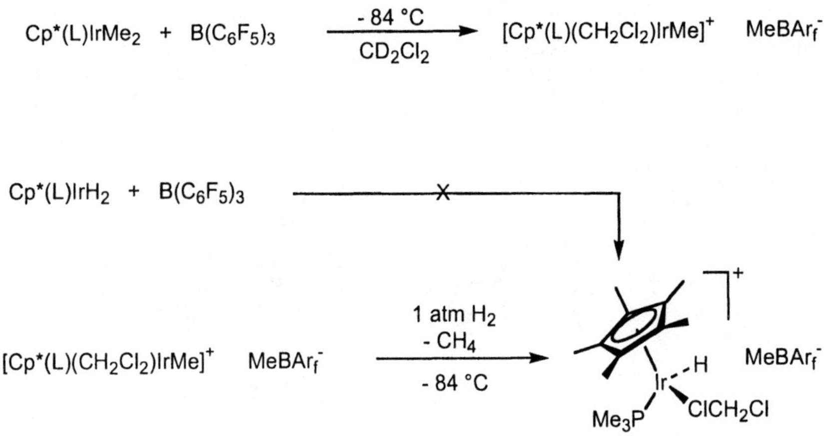

Cp*(L)IrMe₂ + B(C₆F₅)₃ →(−84 °C, CD₂Cl₂)→ [Cp*(L)(CH₂Cl₂)IrMe]⁺ MeBAr_f⁻

Cp*(L)IrH₂ + B(C₆F₅)₃ ——X——

[Cp*(L)(CH₂Cl₂)IrMe]⁺ MeBAr_f⁻ →(1 atm H₂, −CH₄, −84 °C)→ [Cp*(Me₃P)Ir(H)(ClCH₂Cl)]⁺ MeBAr_f⁻

Surprisingly, except in a few cases, BArf salts of [Cp*(PMe$_3$)IrH]$^+$ did not undergo the type of overall σ–bond metathesis reactions observed with [Cp*(PMe$_3$)IrCH$_3$]$^+$. Instead, this material undergoes hydrogen/deuterium exchange with a wide range of organic molecules (Scheme 5). Many of the

Scheme 5

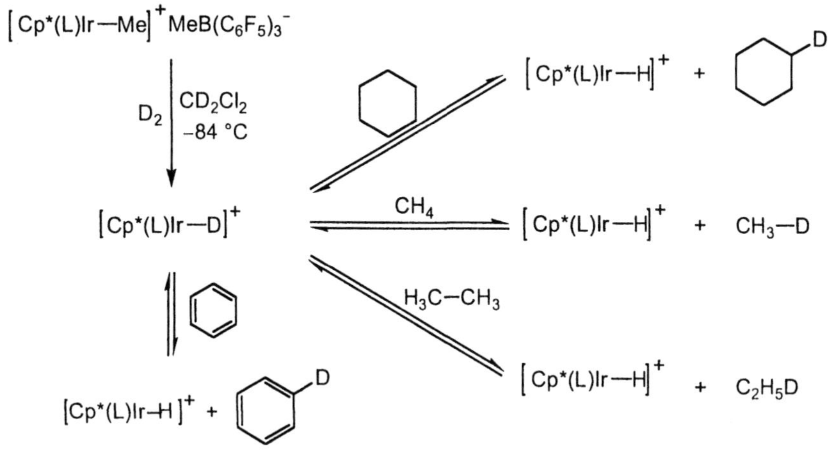

exchange reactions are surprisingly rapid, proceeding at a convenient rate even at -50 °C. Secondary C-H bonds, which are normally inert toward reaction with the methyl complex [Cp*(PMe$_3$)IrCH$_3$]$^+$, are activated in these reactions. Extrapolating from the reactions of [Cp*(PMe$_3$)IrCH$_3$]$^+$, we expect that the exchange processes mediated by [Cp*(PMe$_3$)IrH]$^+$ also proceed by initial C-H oxidative addition. However, in the latter case, the Ir(V) intermediates that are initially formed undergo R-H reductive elimination much more rapidly than H-H reductive elimination. The reason for this is probably thermodynamic, resulting from the fact that M-H bonds are generally stronger than M-R bonds.

The fact that these exchange reactions are reversible raised the possibility that the hydride complex could be used to catalyze H/D exchange from readily available deuterated organic molecules, such as d$_6$-benzene, into less easily deuterated molecules. This is successful; some of the compounds that have been deuterated in this way are illustrated in Scheme 6. The related polyhydride [Cp*(PMe$_3$)IrH$_3$]$^+$ BArf also serves as an H/D exchange catalyst precursor between d$_6$-benzene and organic molecules, but the reactions are much slower, proceeding at detectable rates only between 45 and 135 °C, depending

Scheme 6

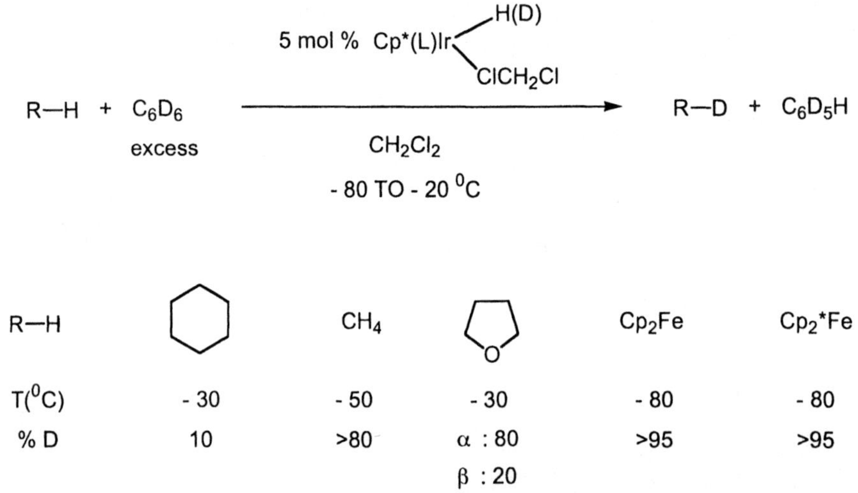

R—H + C$_6$D$_6$ (excess) $\xrightarrow[\text{CH}_2\text{Cl}_2]{\text{5 mol \% Cp*(L)Ir(H(D))(ClCH}_2\text{Cl)}}$ R—D + C$_6$D$_5$H

−80 TO −20 °C

R—H	C$_6$H$_{12}$	CH$_4$	THF	Cp$_2$Fe	Cp$_2$*Fe
T(°C)	−30	−50	−30	−80	−80
% D	10	>80	α : 80 β : 20	>95	>95

on the substrate. This is presumably because these reactions take place by rate-determining dissociation of H$_2$ from the trihydride, giving low concentrations of the active monohydride catalyst. In all of these reactions, complete H/D exchange has been observed with the fastest-reacting substrates, such as methane, ferrocene and decamethylferrocene, but high levels of deuteration have not yet been achieved with slower substrates (e.g., tetrahydrofuran, cyclohexane) due to competitive catalyst decomposition.

The most economical source of deuterium is D$_2$O. Therefore, since it is water-soluble, we have explored the possibility of using the more stable cationic iridium trihydride [Cp*(PMe$_3$)IrH$_3$]$^+$ as an H/D exchange catalyst precursor in aqueous solution. This is successful, and deuterium has been exchanged into a number of water-soluble organic molecules using this system. We have most recently found that other pentamethylcyclopentadienyliridium complexes also serve as catalyst precursors for these exchange reactions; the most efficient is [Cp*IrCl$_2$]$_2$ (*18*). The mechanisms of these reactions, and the identity of the actual operating catalyst that is present under operating conditions, are still under investigation.

The Use of C-H Bond-Activation in Carbon-Carbon Bond-Forming Reactions

Catalytic carbon-carbon bond forming reactions between aromatic compounds and other organic carbon fragments now play a central role in many areas of organic synthesis. Typically these reactions require the initial synthesis of a functionalized aromatic compound, such as an aryl halide, triflate, or boron compound, before it can be used in the carbon-carbon coupling reaction. A major step forward in efficiency and atom economy would be achieved if direct coupling between an organic substrate and a carbon atom at a C-H position could be achieved in a selective manner.

Significant progress toward this goal has been accomplished by Murai and his coworkers, who have developed carbon-carbon coupling reactions between aromatic compounds, CO and alkenes using ruthenium catalysts at elevated temperatures (19,20). Several other groups have also made contributions to this field. We have recently applied this type of chemistry to the development of C-H activated cyclization methods that involve the formation of new carbon-carbon bonds.

Our initial goal was the development of a catalyst that would achieve the cyclization of alkenyl-substituted aromatic imines of the type shown in eq. 3. After surveying a number of potential catalysts, we have found several organorhodium complexes that catalyze this transformation with an unusual

level of generality (21,22). Five- and six-membered rings can be formed, depending on the substitution at the alkenyl group, including those containing new stereocenters. In addition, both di- and tri-substituted alkenes undergo efficient cyclization without interference from double bond isomerization along the chain, which is normally a troublesome side reaction in these processes.

Most recently we have extended the rhodium-catalyzed coupling reactions to heteroaromatic systems. Because of the prominent role they play in the structures of important pharmaceuticals, the most extensively studied have been alkenylated benzimidazoles. These compounds also undergo cyclization with satisfying generality. In addition to exploring the scope of the cyclizations, we have also investigated their mechanism (23). Through the use of a combination of kinetic, isotope labeling and density functional theory methods, we have concluded that the benzimidazole reactions do not proceed via a conventional oxidative addition/alkene insertion/reductive elimination process. Instead, the mechanism that fits our data most effectively is that shown in Scheme 7, in which the N-heterocyclic carbene complex illustrated at the bottom of the cycle is formed as a critical intermediate, and exists as the resting state of the catalyst. Preliminary results have been obtained indicating that analogous intermolecular coupling reactions can be carried out successfully (and without double bond isomerization) using these catalysts (23).

Scheme 7

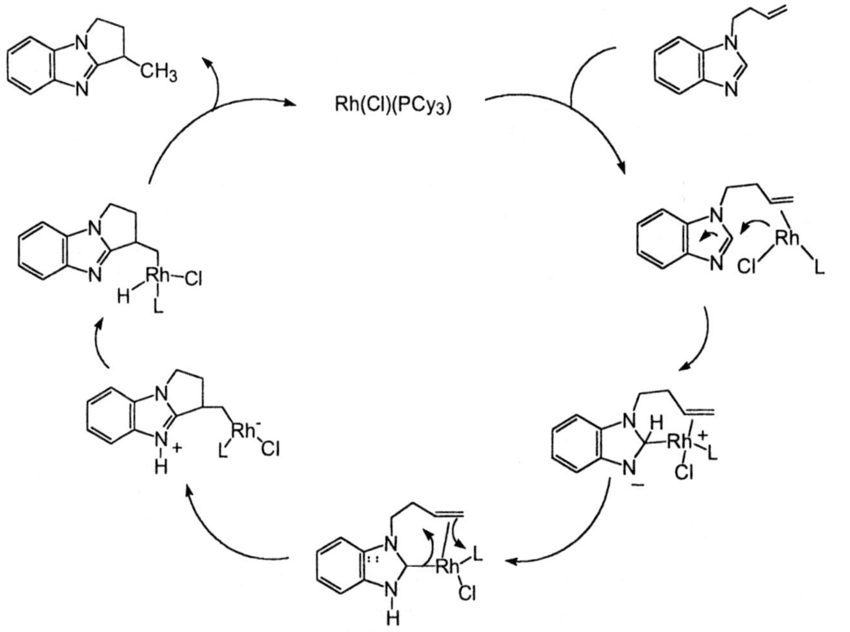

We also have early results which suggest that some of these reactions can be carried out enantioselectively (*24*).

Acknowledgments

This work benefited immensely from the experimental and intellectual talents of the graduate students and postdoctoral associates who participated in it and whose names are listed in the references. It also benefited from experimental assistance by the staff of the spectroscopic and X-ray diffraction services at Berkeley, and by important collaborations with other faculty members in the Department of Chemistry. The part of this research that was conducted in the R. G. Bergman laboratory was supported financially by the Director, Office of Energy Research, Office of Basic Energy Sciences, Chemical Sciences Division, U.S. Department of Energy, under Contract No. DE-AC03-76SF00098.

References and Notes

1. Bergman, R. G. *Science* **1984**, *223*, 902.
2. Shilov, A. E.; Shulpin, G. B. *Chem. Rev.* **1997**, *97*
3. Arndtsen, B. A.; Bergman, R. G.; Mobley, T. A.; Peterson, T. H. *Acc. Chem. Res.* **1995**, *28*, 154.
4. Labinger, J. A.; Bercaw, J. E. *Nature* **2002**, *417*, 507.
5. Janowicz, A. H.; Bergman, R. G. *J. Am. Chem. Soc.* **1982**, *104*, 352.
6. Janowicz, A. H.; Bergman, R. G. *J. Am. Chem. Soc.* **1983**, *105*, 3929.
7. Hoyano, J. K.; Graham, W. A. G. *J. Am. Chem. Soc.* **1982**, *104*, 3723.
8. Jones, W. D.; Feher, F. J. *Organometallics* **1983**, *2*, 562.
9. Periana, R. A.; Bergman, R. G. *Organometallics* **1984**, *3*, 508.
10. Bergman, R. G.; Seidler, P. F.; Wenzel, T. T. *J. Am. Chem. Soc.* **1985**, *107*, 4358.
11. Burger, P.; Bergman, R. G. *J. Am. Chem. Soc.* **1993**, *115*, 10462.
12. Arndtsen, B. A.; Bergman, R. G.; Mobley, T. A.; Peterson, T. H. *Acc. Chem. Res.* **1995**, *28*, 154.
13. Arndtsen, B. A.; Bergman, R. G. *Science* **1995**, *270*, 1970.
14. Klei, S. R.; Tilley, T. D.; Bergman, R. G. *J. Am. Chem. Soc.* **2000**, *122*, 1816.
15. Niu, S.; Hall, M. B. *J. Am. Chem. Soc.* **1998**, *120*, 6169.
16. Su, M.-D.; Chu, S.-Y. *J. Am. Chem. Soc.* **1997**, *119*, 5373.
17. Golden, J. T.; Andersen, R. A.; Bergman, R. G. *J. Am. Chem. Soc.* **2001**, *123*, 5837.
18. Klei, S. R.; Golden, J. T.; Tilley, T. D.; Bergman, R. G. *J. Am. Chem. Soc.* **2002**, *124*, 2092.
19. Murai, S.; Kakiuchi, F.; Sekine, S.; Tanaka, Y.; Kamatani, A.; Sonoda, M.; Chatani, N. *Nature* **1993**, *366*, 529.
20. Kakiuchi, F.; Murai, S. **1999**, 47.
21. Tan, K. L.; Bergman, R. G.; Ellman, J. A. *J. Am. Chem. Soc.* **2001**, *123*, 2685.
22. Thalji, R. K.; Ahrendt, K. A.; Bergman, R. G.; Ellman, J. A. *J. Am. Chem. Soc.* **2001**, *123*, 9692.
23. Tan, K. L.; Bergman, R. G.; Ellman, J. A. *J. Am. Chem. Soc.* **2002**, *124*, 3202.
24. Bergman, R. G.; Ellman, J. A.; Thalji, R. K. *unpublished results*.

Chapter 3

Alkane Complexes as Intermediates in C–H Bond Activation Reactions

William D. Jones, Andrew J. Vetter, Douglas D. Wick, and Todd O. Northcutt

Department of Chemistry, University of Rochester, Rochester, NY 14627

The rearrangements of alkyl deuteride complexes are monitored and a kinetic model used to extract the rate constants for the fundamental processes involving the unseen alkane σ-complexes. Isotope effects for C-H and C-D bond formation and cleavage have been measured to permit these determinations. The relative rates of the processes available to an alkane σ-complex (C-H oxidative cleavage, C-D oxidative cleavage, dissociation, migration to an adjacent C-H bond) have been determined for methane, ethane, propane, butane, pentane, and hexane.

Introduction

It has now been generally established that alkanes are activated by homogeneous transition metal complexes by way of initial complexation of the C-H bond to the metal, followed by oxidative cleavage of the C-H bond to produce an alkyl hydride product (eq 1) (1). Evidence for this pathway has come from a variety of observations, including scrambling of a metal-deuteride into the α-position of the alkyl group (2-10), temperature-dependent linewidth variations (11), and direct observation via NMR or IR spectroscopy (12,13). Despite these studies, little is known about the behavior of these alkane complexes due to their lability and the inability to determine the structure (i.e. binding site) in these species. In this article we present an overview of experiments that use deuterium scrambling to monitor the rearrangements of alkyl hydride complexes. Through rate measurements and kinetic modeling, it has proven possible to extract the relative rates of the various processes open to the alkane complex intermediates.

$$L_{n+1}[M] \xrightarrow{-L} [L_nM] \xrightarrow{R-H} [L_nM(R-H)] \xrightarrow{K_{eq}} L_nM\overset{R}{\underset{H}{\diagdown}} \qquad (1)$$

The direct observation of alkane complexes has been limited to a few experiments involving low temperature spectroscopy or transient absorption techniques. One of the more interesting of these is the report by Ball in which photolysis of CpRe(CO)$_3$ in cyclopentane at 180 K leads to the observation of a σ-cyclopentane complex (12). A resonance is observed at δ -2.32 for *two* hydrogens interacting with the metal center, suggesting either an η^2-H,H or a fluxional η^2-C,H structure as shown below (eq 2). Time resolved IR experiments using CpRe(CO)$_3$ in heptane solvent at 298 K have also shown evidence for a transient intermediate assigned as the σ-heptane complex. The species has an approximate lifetime of 10's of milliseconds before back reacting with CO in a bimolecular reaction (13).

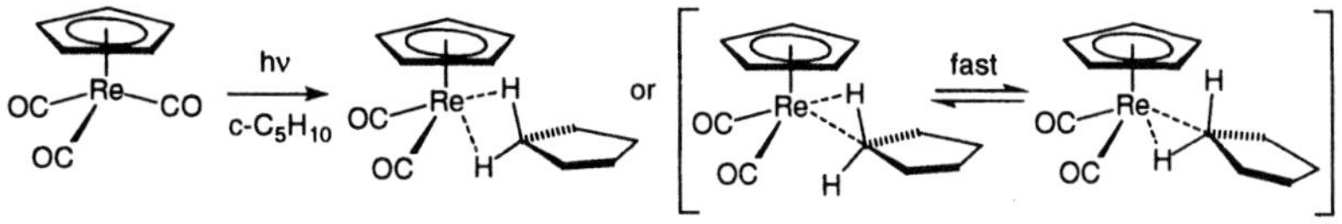

$$\text{(2)}$$

Another interesting example where a σ-alkane complex has been observed by transient IR spectroscopy is in the photolysis of Tp*Rh(CO)$_2$ in cyclohexane solution (Tp* = tris(3,5-dimethylpyrazolyl)borate) (14,15). A species assigned as the alkane complex is observed at 1990 cm^{-1} that converts to the alkyl hydride complex with a half-life of 230 ns. Conversion of the η^3-Tp* ligand to an η^2-Tp* ligand is seen to occur on the 200 ps timescale (eq 3). As with the IR study above, however, little can be ascertained regarding the structure of the alkane complex. Fast IR studies of Cp*Rh(CO)$_2$ and alkanes in liquid xenon and krypton solvents provide evidence for the intermediacy of alkane complexes of the type Cp*Rh(CO)(alkane), also characterized through the carbonyl IR absorptions (16-18).

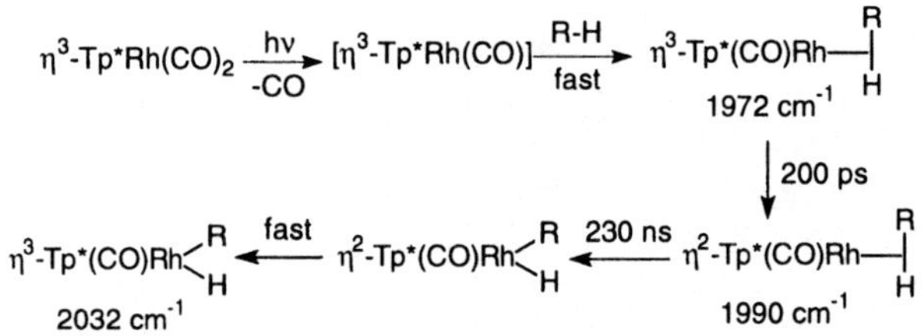

$$\eta^3\text{-Tp*Rh(CO)}_2 \xrightarrow[-CO]{h\nu} [\eta^3\text{-Tp*Rh(CO)}] \xrightarrow[\text{fast}]{R\text{-}H} \eta^3\text{-Tp*(CO)Rh}\begin{array}{c}R\\|\\H\end{array}$$
$$1972\ cm^{-1}$$

$$\downarrow 200\ ps$$

$$\eta^3\text{-Tp*(CO)Rh}\begin{array}{c}R\\<\\H\end{array} \xleftarrow{\text{fast}} \eta^2\text{-Tp*(CO)Rh}\begin{array}{c}R\\<\\H\end{array} \xleftarrow{230\ ns} \eta^2\text{-Tp*(CO)Rh}\begin{array}{c}R\\|\\H\end{array}$$
$$2032\ cm^{-1} \qquad\qquad\qquad\qquad 1990\ cm^{-1}$$

(3)

Our research group has been studying a related complex, [Tp*Rh(CNR)], for C-H bond activation reactions over the past decade (CNR = neopentyl isocyanide). The fragment can be conveniently prepared by irradiation of the carbodiimide adduct, Tp*Rh(CNR)(RN=C=NPh), which has nearly unit quantum efficiency for loss of the carbodiimide ligand (19). Early studies with benzene activation to produce the phenyl hydride complex Tp*Rh(CNR)PhH were proposed to occur by way of an unobserved η^2-benzene adduct, not unlike the mechanism proposed for alkane activation (20). Evidence for this adduct came from the observation that the phenyl deuteride complex was found to scramble the deuterium into the ortho, meta, and para positions of the phenyl ring much faster than the rate of benzene exchange. In addition, displacement of benzene from Tp*Rh(CNR)PhH by isocyanide was found to be an associative process with activation parameters consistent with attack of isocyanide on the metal complex. These experiments suggest a pre-equilibrium with the η^2-benzene complex prior to further reaction (eq 4). The associative kinetics were

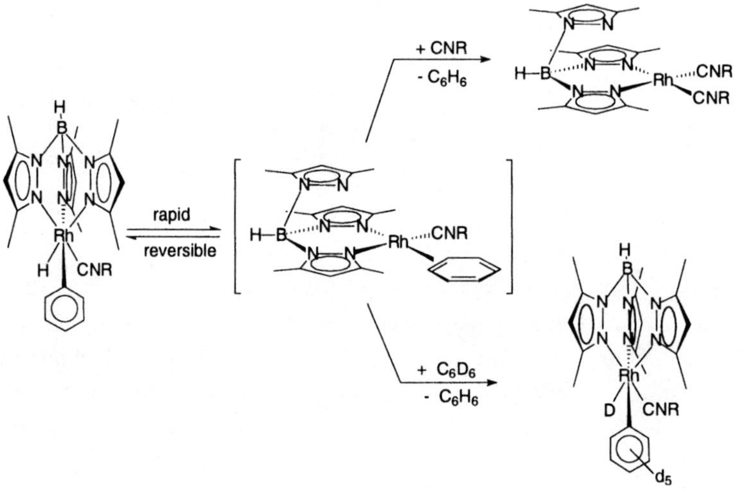

(4)

most consistent with the formulation of a d^8 square planar Rh(I) intermediate, and such rapid $\eta^3 - \eta^2$ interconversions are consistent with the earlier interpretations of the carbonyl derivative by Bergman and Harris (*14*).

The rhodium carbodiimide precursor to [Tp*Rh(CNR)] was also found to be valuable for the activation of alkanes, including methane, propane, pentane, cyclopentane, and cyclohexane (*21*). The complexes Tp*Rh(CNR)(R)H have stabilities toward alkane reductive elimination at ambient temperature in benzene solution on the order of minutes (for cyclohexane) to hours (for methane). In the case of the linear hydrocarbons, only *n*-alkyl products are formed, even at low temperatures. Yet the formation of cyclopentyl and cyclohexyl hydride adducts indicated that activation of secondary C-H bonds was possible. We set out to determine and understand the stabilities of the different isomers available in the alkane oxidative addition adducts, as well as the role that the σ-alkane complexes played in their formation and interconversions. The use of rhodium deuteride derivatives proved to be a valuable tool in these studies, as they are readily prepared by the metathesis reaction of the rhodium chloride complex (e.g. Tp*Rh(CNR)MeCl) with Cp_2ZrD_2, as reported in our full paper (*22*). In addition, the use of deuteride compounds to monitor the rearrangements introduced isotope effects on the rate of reaction, and these needed to be determined and taken into account in order to provide a meaningful interpretation of the observations. A full discussion of isotope effects in these reactions has recently appeared (*23*).

Isotope Effect Measurements

One of the key experiments that permitted our success in this project was the ability to prepare an isopropyl hydride complex from the corresponding isopropyl chloride complex. The metathesis reaction with Cp_2ZrH_2 was not immediate, but was fast enough to allow for complete formation of the isopropyl hydride complex in about 10 minutes. Furthermore, the isopropyl hydride complex was seen to convert into the *n*-propyl hydride complex competitively with loss of propane over the next hour or so. The unsaturated fragment reacted with the benzene solvent to give ultimately Tp*Rh(CNR)PhH as the thermodynamically most favored product (Figure 1). This experiment demonstrated that secondary alkyl hydride complexes have sufficient stability that they should have been observed during the photolysis of Tp*Rh(CNR)(RN=C=NPh) in propane solvent to generate Cp*Rh(CNR)(n-propyl)H. Taken together, *these experiments prove that activation of secondary C-H bonds does not occur in linear alkanes.*

A second related experiment was then undertaken using Cp_2ZrD_2 to prepare the isopropyl deuteride complex. A similar sequence of events is seen:

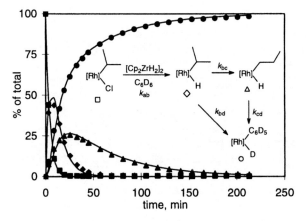

Figure 1. Rearrangement of isopropyl hydride complex

conversion of the isopropyl deuteride complex to the *n*-propyl hydride complex with competitive loss of propane-d_1 and formation of Tp*Rh(CNR)PhH. In this case, however, it is important to point out that no scrambling of deuterium is seen in the isopropyl deuteride complex. That is, one does not observe the formation of Tp*Rh(CNR)(CDMe$_2$)H as Tp*Rh(CNR)(CHMe$_2$)D goes away. It implies that the reductive coupling in the isopropyl deuteride (and hydride) complex is irreversible (eq 5). From this, *we conclude that reductive coupling of a secondary alkyl hydride complex is irreversible.* This observation is important, and is unique in the C-H activation literature. Furthermore, since the reaction is irreversible, the rate of disappearance of the isopropyl deuteride (or hydride) can be used to determine the fundamental rate of reductive coupling. From the ratio of these rates in the deuteride and hydride complexes, *we can therefore determine the isotope effect in just the reductive coupling step*, without complications from reversibility. From our measurements $k_{RC}^H/k_{RC}^D = 2.1$.

(5)

We have also been able to determine the isotope effect on the reverse reaction in this system, namely oxidative cleavage. For this purpose, the carbodiimide Tp*Rh(CNR)(RN=C=NPh) was irradiated in unreactive C_6F_6 solvent under 10 atm CH_2D_2 at 5 °C. The kinetic products formed were a 4.3:1 ratio of Tp*Rh(CNR)(CHD$_2$)H : Tp*Rh(CNR)(CH$_2$D)D, indicating a *kinetic isotope effect for oxidative cleavage of* $k_{OC}^H/k_{OC}^D = 4.3$. Upon standing at room temperature, this ratio adjusts to 1.7 : 1 before methane-d_2 is slowly lost over several hours. This experiment is interpreted in terms of the initial formation of a methane σ-complex that exchanges between the statistically equivalent C-H and C-D bonds before undergoing oxidative cleavage (eq 6).

$$\begin{array}{c}\text{Tp}^*\\ \text{Rh-CNR}\\ |\\ \text{RNC=NPh}\end{array} \xrightarrow[\substack{CH_2D_2\\ 10\ atm}]{h\nu} \left[\begin{array}{c}\text{Tp*LRh}-\overset{H}{\underset{D}{\overset{|}{C}}}-D\\ \Updownarrow\\ \text{Tp*LRh}-\overset{D}{\underset{H}{\overset{|}{C}}}-D\end{array}\right] \longrightarrow \underset{\substack{\downarrow\\ 1.7:1\ \text{ratio}\\ \text{at equilibrium}}}{\underset{4.3:1}{\text{Tp*LRh}\overset{CD_2H}{\underset{H}{\diagup}} + \text{Tp*LRh}\overset{CH_2D}{\underset{D}{\diagup}}}} \qquad (6)$$

These experiments lead to some interesting conclusions regarding isotope effects. First, both reductive coupling and oxidative cleavage have *normal kinetic isotope effects.* Second, the equilibrium isotope effect for Tp*Rh(CNR)(CHD$_2$)H ⇌ Tp*Rh(CNR)(CH$_2$D)D *is inverse*, since from the near-final product mixture in equation 6, $K_{H/D} = 1/1.7 = 0.59$. Third, the ratio of the two kinetic isotope effects can also be used to confirm the inverse equilibrium isotope effect, since $K_{H/D} = (k_{RC}^H/k_{RC}^D)/(k_{OC}^H/k_{OC}^D) = 2.1/4.3 = 0.49$. The agreement is quite good, considering that different alkanes were used in the measurements. The agreement suggests that kinetic isotope effects are likely to be of a similar magnitude for the other alkanes studied, and this assumption will be used in the discussions that follow.

Alkyl Hydride and Deuteride Reductive Elimination Studies

As mentioned above, alkane loss from Tp*Rh(CNR)(R)H is fairly slow at room temperature. Consequently, solutions of these complexes can be readily prepared by shaking a benzene solution of Tp*Rh(CNR)(R)Cl with sparingly soluble Cp$_2$ZrH$_2$ and then filtering to remove Cp$_2$ZrHCl. A comparison of the rate of alkane reductive elimination in the methyl, ethyl, n-propyl, n-butyl, n-pentyl, and n-hexyl complexes showed the methyl hydride to be the most stable ($t_{1/2}$ = 4.3 h), the ethyl hydride to be slightly less stable ($t_{1/2}$ = 1.1 h), and the longer n-alkyl hydrides to be the least stable ($t_{1/2}$ = 0.7 h). In benzene-d_6 solution, the loss of alkane is followed by irreversible formation of the phenyl hydride product Tp*Rh(CNR)(C_6D_5)D.

By using Cp_2ZrD_2 in the metathesis reaction, the alkyl deuterides can be conveniently prepared. The reaction is quick (<5 min) and the workup consists of simply filtering the insoluble zirconocenes. The first reaction examined was that of Tp*Rh(CNR)(CH_3)Cl to give Tp*Rh(CNR)(CH_3)D. The formation of Tp*Rh(CNR)(CH_2D)H was seen to occur to the extent of approximately 55% of the total (Figure 2). Loss of CH_3D occurs over several hours as Tp*Rh(CNR)(C_6D_5)D forms. Towards the end of the reaction, the ratio of Tp*Rh(CNR)(CH_3)D : Tp*Rh(CNR)(CH_2D)H can be seen to be 1 : ~6.3. Taking into account the 3-fold statistical preference for the latter species, the equilibrium isotope effect can be estimated to be ~0.48, in good agreement with the values seen above for CH_2D_2, again supporting the generality of the values of the isotope effects.

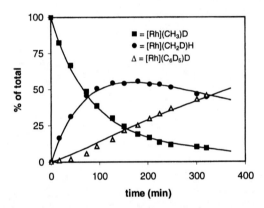

Figure 2. Distribution of Species for the Rearrangement of Tp*Rh(CNR)(CH$_3$)D in C$_6$D$_6$ at 25 °C. Symbols represent observed data. Solid lines represent simulated data.

With confidence that the values for the isotope effects for oxidative cleavage and reductive coupling were now known, a kinetic simulation of the scrambling was undertaken according to Scheme 1. In this simulation, it was found that the 3 rate constants for the reactions leading away from the σ-methane complex (k_{OC}^H, k_{OC}^D, k_d) were highly correlated (0.9999), which occurs because the σ-complex is not directly observed and therefore its absolute rate of disappearance cannot be determined. The *relative* rates for these three reactions, however, could be determined. This problem arises essentially because one does not know the absolute free energy of the methane σ-complex, but one can determine the difference in the barrier heights from the kinetic simulation

(Figure 3). In this picture, the value X corresponds to the stability of the alkane complex relative to the barrier for C-H bond oxidative cleavage, and the other barriers are then determined relative to it. From the values of the rate constants determined in this fit, one can extrapolate that *in a σ-methane complex of CH₄, oxidative cleavage would be 10.7 times faster than dissociation.*

Scheme 1:

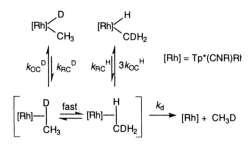

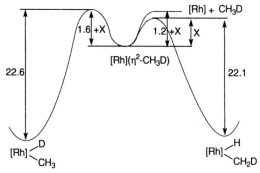

Figure 3. Free energy diagram for the activation of CH₃D by [Tp*Rh(CNR)]. Energies are in kcal/mol.

The next alkyl deuteride complex to be examined was the ethyl deuteride complex Tp*Rh(CNR)(CH₂CH₃)D, prepared from the chloro derivative and zirconium hydride. Upon standing in benzene solution, two scrambling processes are observed. First, scrambling of deuterium into the α-position is seen (~23%), followed by scrambling into the β-position (Figure 4).

The kinetic scheme for these processes in terms of alkane σ-complexes is shown in Scheme 2. The rate constants for the scheme can be fit using the values for the two kinetic isotope effects determined above, and with the same limitation that only the relative rates for the processes involving the σ-ethane complex can be determined. In this system, one new rate constant is introduced for the migration of the rhodium from one methyl group to the other, k_{m11}. From the values of these rate constants, the relative rates of oxidative cleavage : migration : dissociation with ethane are determined to be ~4:2:1. The fairly rapid end-to-end migration rate manifests itself in the observation of nearly equal amounts of α- and β-scrambled products prior to ethane-d_1 loss.

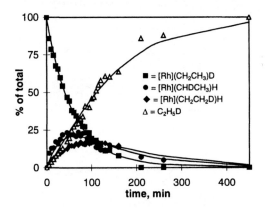

Figure 4. Distribution of Species for the Rearrangement of Tp*Rh(CNR)(CH$_2$CH$_3$)D in C$_6$D$_6$ at 25 °C. Symbols represent observed data. Solid lines represent simulated data.

Scheme 2:

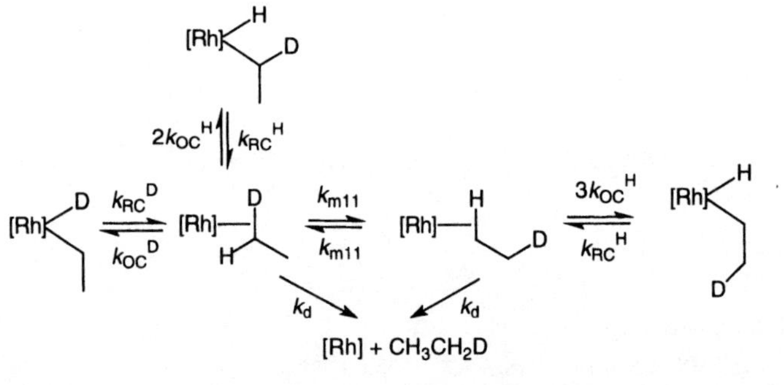

Similar deuterium migration experiments were conducted with the n-propyl, n-butyl, n-pentyl, and n-hexyl complexes. In each case, the chloro derivative Tp*Rh(CNR)(n-alkyl)Cl reacted quickly with Cp$_2$ZrD$_2$ to give Tp*Rh(CNR)(n-alkyl)D. The α-deutero product was seen to form to the same extent in all of these complexes (~20%), as in the ethyl case. This observation can be reconciled by noting that this scrambling sequence is quite similar in all of the n-alkyl derivatives. The scrambling of the deuterium to the far end of the alkyl chain was observed to occur to a diminishing extent as the alkyl chain length increased... it could not even be measured in the case of the n-hexyl complex. Figure 5 shows distribution of species plots for these four alkyl deuteride rearrangements, and the above features can be easily distinguished. The gradual change in terminal d-migration argues against a direct end-to-end migration path, which would be expected to be faster in the n-hexyl case.

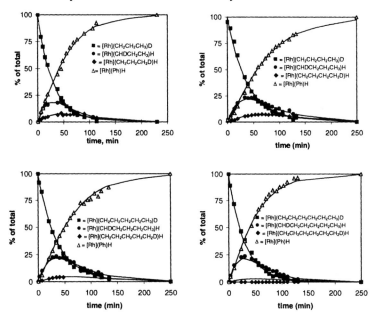

Figure 5. Distribution of Species for the Rearrangement of Tp*Rh(CNR)(n-alkyl)D in C_6D_6 at 25 °C. Symbols represent observed data. Solid lines represent simulated data.

Kinetic simulations of these reactions were undertaken, and the results are shown as the solid lines in the figures. The model used for the n-propyl rearrangement is shown in Scheme 3. As there are many independent rate constants in this model, it only proved possible to obtain a convergent fit if the data for the rearrangement of the isopropyl deuteride complex (see Figure 1) was also fit simultaneously, since this model contains many of the same rate constants.

The following key points could be ascertained from the fit: (1) the σ-propane complex with the terminal methyl group undergoes three processes, C-H oxidative cleavage (k_{OC}^H), migration to the methylene group (k_{m12}), or dissociation (k_{d1}), in a ~5:3:1 ratio; (2) the σ-propane complex with the

Scheme 3:

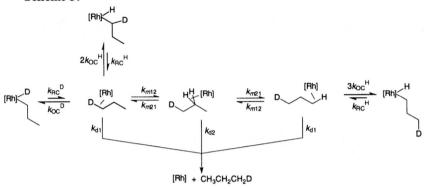

methylene group undergoes two processes, migration to the end (k_{m21}) or dissociation (k_{d2}), in a 1:1 ratio; (3) there is no observable oxidative cleavage of the secondary C-H bonds; (4) as with methane, the energies of the σ-complexes cannot be determined. Therefore, the absolute rates of these processes cannot be obtained, only their relative values. In addition, since there are *two* possible σ-complexes with propane, their relative energies cannot be determined, which means that the rates of the processes that interrelate these two species (k_{m12} and k_{m21}) cannot be determined either. This is shown in the rather complicated free energy diagram in Figure 6.

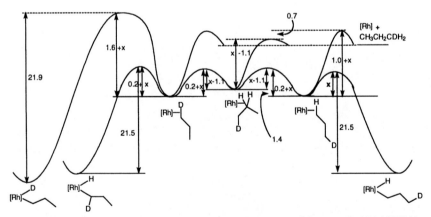

Figure 6. Free energy diagram for the deuterium scrambling in Tp*Rh(CNR)(n-propyl)D. Energies are in kcal/mol. The energies of the σ-alkane complexes are arbitrary.

In the simulation of the n-butyl complex, the same rate constants used to fit the n-propyl complex were used. In addition, one new rate constant was introduced for the rate of secondary to secondary σ-complex migration, k_{m22}. The fit shows that this migration up and down the interior of the chain is quite fast, with the ratio of migration (k_{m22}) vs. dissociation (k_{d2}) of 7 : 1. Simulation of the n-pentyl and n-hexyl derivatives required introduction of no new rate constants, and it can be seen in Figure 5 that the agreement of the observed and predicted behavior is quite good within the limits of measurement by ^{2}H NMR spectroscopy.

The relative rates of these processes can be summarized in side-by-side fashion as shown in Figure 7. It is important to remember that relative rates can only be compared for processes arising from a single σ-alkane complex. For example, one cannot compare the rate of dissociation of methane to that of ethane. In all cases, the same relative pattern for activation vs. migration vs. dissociation can be seen. In addition, 1°/2° migration patterns are similar. The rapid 2°-2° migration rate is striking, and indicates that if a metal coordinates to the middle of a linear alkane, it rapidly runs to the end of the chain. Once there, the rate of C-H oxidative bond cleavage exceeds all other processes. *The combination of these two factors is what leads to the observation of selective activation of the terminal C-H bonds in linear alkanes.*

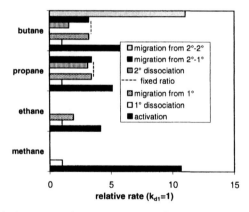

Figure 7. Relative rates of processes in σ-alkane complexes of linear alkanes.

One further question that can be asked is can the relative rates of coordination of an unsaturated metal complex to a linear alkane be determined (Scheme 4). Once the alkane is complexed, the above model for C-H bond migration, dissociation, and oxidative cleavage can then be applied to give oxidative addition product selectivities. Together, this information will lead to a comprehensive understanding of the details of C-H bond activation.

Scheme 4:

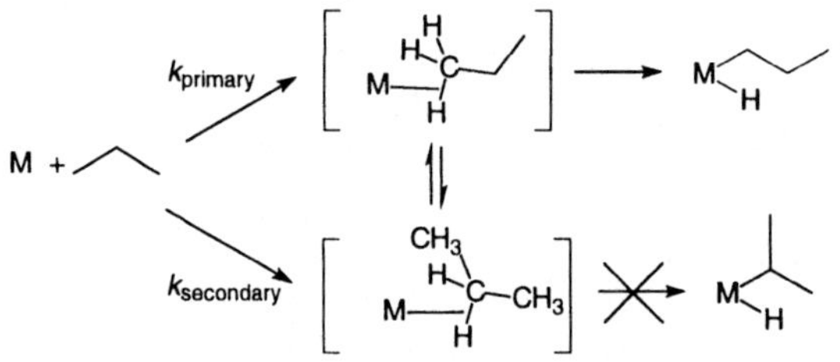

$$M + \overset{k_{primary}}{\underset{k_{secondary}}{\rightleftarrows}} \left[\begin{array}{c} \text{M--CH}_2\text{--CH}_2\text{R} \end{array} \right] \longrightarrow M\text{--propyl,H}$$

$$\left[\begin{array}{c} \text{M--CH(CH}_3)_2 \end{array} \right] \not\rightarrow M\text{--iPr,H}$$

Preliminary studies show that selectivities in alkane coordination can be measured. Irradiation of the carbodiimide complex Tp*Rh(CNR)(RN=C=NPh) in a 1:1 mixture of pentane and decane gives a 1:1.1 ratio of C-H activation products, respectively (eq 7). Note that if only methyl groups bind to an unsaturated metal, than a 1:1 ratio of products would be expected. The ever-so-slight excess of decane C-H activation product indicates that the predominate site for alkane binding is the methyl group, and that methylene coordination contributes only slightly. Simulations of these systems are underway to quantify the rates of coordination.

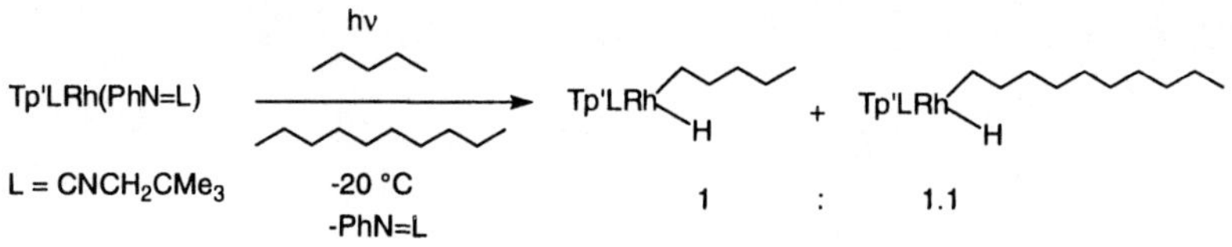

$$\text{Tp'LRh(PhN=L)} \xrightarrow[\substack{-20\ °C \\ -PhN=L}]{hv,\ \text{pentane/decane}} \text{Tp'LRh(pentyl)(H)} + \text{Tp'LRh(decyl)(H)} \quad (7)$$

L = CNCH$_2$CMe$_3$ 1 : 1.1

References

1. Hall, C.; and Perutz, R. *Chem. Rev.* **1996**, *96*, 3125.
2. Buchanan, J. M.; Stryker, J. M.; Bergman, R. G. *J. Am. Chem. Soc.* **1986**, *108*, 1537.
3. Periana, R. A.; Bergman, R. G. *J. Am. Chem. Soc.* **1986**, *108*, 7332.
4. Bullock, R. M.; Headford, C. E. L.; Hennessy, K. M.; Kegley, S. E.; Norton, J. R. *J. Am. Chem. Soc.* **1989**, *111*, 3897.
5. Parkin, G.; Bercaw, J. E. *Organometallics* **1989**, *8*, 1172.
6. Gould, G. L.; Heinekey, D. M. *J. Am. Chem. Soc.* **1989**, *111*, 5502.
7. Stahl, S. S.; Labinger, J. A.; Bercaw, J. E. *J. Am. Chem. Soc.* **1996**, *118*, 5961.
8. Wang, C; Ziller, J. W.; Flood, T. C. *J. Am. Chem. Soc.* **1995**, *117*, 1647.

9. Mobley, T. A.; Schade, C.; Bergman, R. G. *J. Am. Chem. Soc.* **1995**, *117*, 7822.
10. Chernaga, A.; Cook, J.; Green, M. L. H.; Labella, L.; Simpson, S. J.; Souter, J.; Stephens, A. H. H. *J. Chem. Soc., Dalton Trans.* **1997**, 3225.
11. Gross, C. L.; Girolami, G. S. *J. Am. Chem. Soc.* **1998**, *120*, 6605.
12 By NMR: Geftakis, S.; Ball, E. *J. Am. Chem. Soc.* **1998**, *120*, 9953.
13. By TRIR: Sun, X.-Z.; Grills, D. C.; Nikiforov, S. M.; Poliakoff, M.; George, M. W. *J. Am. Chem. Soc.* **1997**, *119*, 7521.
14. Lian, T.; Bromberg, S. E.; Yang, H.; Proulx, G.; Bergman, R. G.; Harris, C. B. *J. Am. Chem. Soc.* **1996**, *118*, 3769.
15. Bromberg, S.E.; Yang, H.; Asplund, M.C.; Lian, T.; McNamara, B. K.; Kotz, K. T.; Yeston, J. S.; Wilkens, M.; Frei, H.; Bergman, R. G.; Harris, C. B. *Science* **1997**, *278*, 260.
16. Weiller, B. H.; Wasserman, E. P.; Bergman, R. G.; Moore, C. B.; Pimentel, G. C. *J. Am. Chem. Soc.* **1989**, *111*, 8288.
17. Weiller, B. H.; Wasserman, E. P.; Moore, C. B.; Bergman, R. G. *J. Am. Chem. Soc.* **1993**, *115*, 4326.
18. Bengali, A. A.; Schultz, R. H.; Moore, C. B.; Bergman, R. G. *J. Am. Chem. Soc.* **1994**, *116*, 9585.
19. Jones, W. D.; Hessell, E. T. *Organometallics* **1992**, *11*, 1496.
20. Jones, W. D.; Hessell, E. T. *J. Am. Chem. Soc.* **1992**, *114*, 6087.
21. Jones, W. D.; Hessell, E. T. *J. Am. Chem. Soc.* **1993**, *115*, 554.
22. Northcutt, T. O.; Wick, D. D.; Vetter, A. J.; Jones, W. D. *J. Am. Chem. Soc.* **2001**, *123*, 7257.
23. Jones, W. D. *Acc. Chem. Res.* **2003**; *36*, 140.

Chapter 4

C–C versus C–H Bond Oxidative Addition in PCX (X=P,N,O) Ligand Systems: Facility, Mechanism, and Control

Boris Rybtchinski and David Milstein*

Department of Organic Chemistry, The Weizmann Institute of Science, Rehovot 76100, Israel

An overview of recent results regarding the activation of strong C-C and C-H bonds by Rh(I) and Ir(I) in PCX (X=P,N,O) type ligand systems is presented. Whereas both C-C and C-H oxidative addition involve non-polar 3-centered transition states and 14 electron intermediates, steric requirements differ markedly and chelation plays a much more important role in C-C than in C-H activation. Control over C-C vs C-H activation can be achieved by choice of ligand and solvent. Under optimal conditions, C-C activation can be thermodynamically and kinetically more favorable than C-H activation and proceed even at −70°C. Activation parameters for an apparent single-step metal insertion into a C-C bond were obtained. A combination of C-H and C-C activation in conjunction with oxidative addition of other bonds has led to unique methylene transfer chemistry.

Whereas C-H activation by late transition metals in homogeneous media is a topic of much research effort, resulting in mechanistic insight (*1*) and development of catalytic systems (*2*), the field of C-C bond activation is much less developed (*3*).

In general, thermodynamic and kinetic factors are expected to favor C-H over C-C activation. However, although M-H bonds in solution are in many cases significantly stronger than the M-C ones (*4*), appropriate design can make C-C bond activation in solution thermodynamically feasible (*3*). For example, M-C$_{Aryl}$ bonds are very strong in cases of rhodium and iridium (*5*), sometimes stronger than M-H bonds. Kinetic factors favoring C-H over C-C bond activation include (a) an easier approach of the metal center to C-H bonds (b) the statistical abundance of C-H bonds, and (c) a higher activation barrier for C-C vs C-H oxidative addition due to the more directional nature of the C-C bond (*3*).

C-C bond activation by soluble metal complexes in most of the reported systems is driven by strain relief, aromatization or the presence of a carbonyl group (*3*). Aiming at the activation of strong, unstrained C-C bonds, we chose to use bis-chelating PCP and PCX ligand systems (*6*) in order to bring the metal center to the proximity of the C-C bond and gain understanding as to whether insertion into a strong C-C bond is possible and what are the factors that might favor this process.

PCP: R=Ph, Me, ^{i}Pr, ^{t}Bu   PCX: X=NEt$_2$, OMe

C-C activation in these systems is expected to be irreversible and lead to a stable C-C activation product. The C$_{Me}$-C$_{Aryl}$ bonds of these ligands, targeted for metal insertion, are very strong (eg BDE(C$_6$H$_5$-CH$_3$) = 101.8±2 kcal mol^{-1}), stronger than the competing benzylic C-H bonds (compare BDE(C$_6$H$_5$CH$_2$-H) = 88±1 kcal mol^{-1}) (*7*). We demonstrated that C-C bond oxidative addition can take place in PCP ligand systems in the presence of hydrogen, the evolving methane being a significant thermodynamic driving force (*8*). We have also demonstrated that the process of metal insertion into a C-C bond can be direct (not requiring an additional driving force) and thermodynamically more favorable than C-H bond activation (*9*). Even C-C bonds as strong as those in aryl-CF$_3$ can be cleaved (*10*). When both sp^2-sp^3 and sp^3-sp^3 C-C bonds are available for activation, only activation of the former was observed (*11*). Catalytic C-C bond activation in the PCP system was also demonstrated (*12*). The scope of metal complexes capable of C-C activation in bis-chelating systems includes complexes of Ru(II) (*13*), Os(II) (*14*), Rh(I) (*3*), Ir(I) (*15*), Pt(II) (*13,16*) and Ni(II) (*17*).

Here we present an overview of our recent results describing C-H and C-C activation in PCX (X=P,N,O) systems. We have recently found that C-C bond activation can be very *facile*, and preferred over C-H activation both thermodynamically and kinetically. We have also found that a significant degree of *control* can be achieved regarding C-C vs C-H bond activation aptitudes. The obtained *mechanistic insight*, including the direct measurement of the activation parameters of the oxidative addition of a strong C-C bond as well as computational studies, reveals similarity in the electronic requirements for C-C and C-H oxidative addition but significantly different steric prerequisites. A combination of C-H and C-C bond activation, in conjunction with oxidative addition of other bonds has resulted in a unique *methylene transfer* reaction.

C-C Bond Activation: Facility

Bulky phosphines are advantageous ligands for the study of oxidative addition processes, since upon coordination to a metal center they generate a species having a sterically shielded vacant coordination site, favoring cyclometallation. Upon reaction of the tBu-PCP ligand with rhodium olefin dimers at *room temperature,* direct concurrent rhodium insertion into C-H and C-C bonds took place (Scheme 1) (*15*).

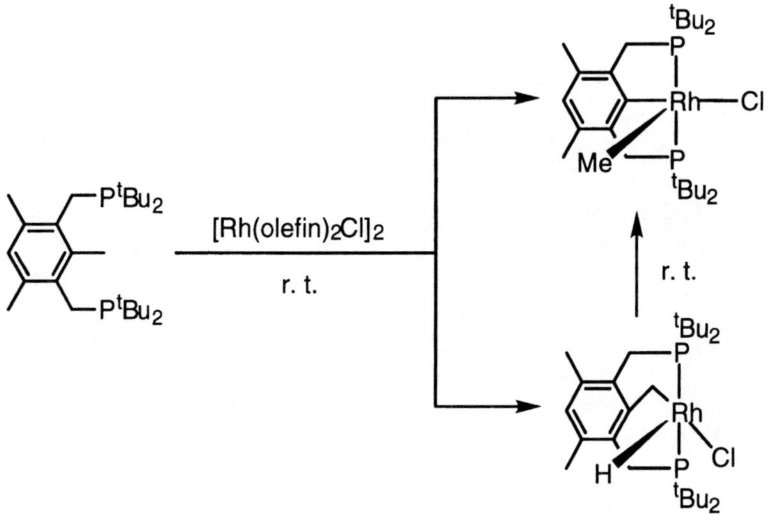

Scheme 1

Interestingly, it was found that associative displacement of the alkene by the phosphine takes place and that the *initial coordination of the diphosphine ligand to the rhodium olefin complex is the rate determining step for the entire process* rather than the C-C or C-H activation steps. In the case of Ir competitive formation of the C-H and C-C activation products also took place at *room temperature* (Scheme 2) but, contrary to the rhodium case, the product of C-H activation is stable under the reaction conditions (it converts to the C-C activation product at 100°C), facilitating mechanistic investigation of the process (vide infra). It is remarkable that in both tBu-PCP-Rh and tBu-PCP-Ir systems the insertion step into a very strong C-C bond in solution is not rate determining, although the overall process proceeds even at 25°C *(15)*.

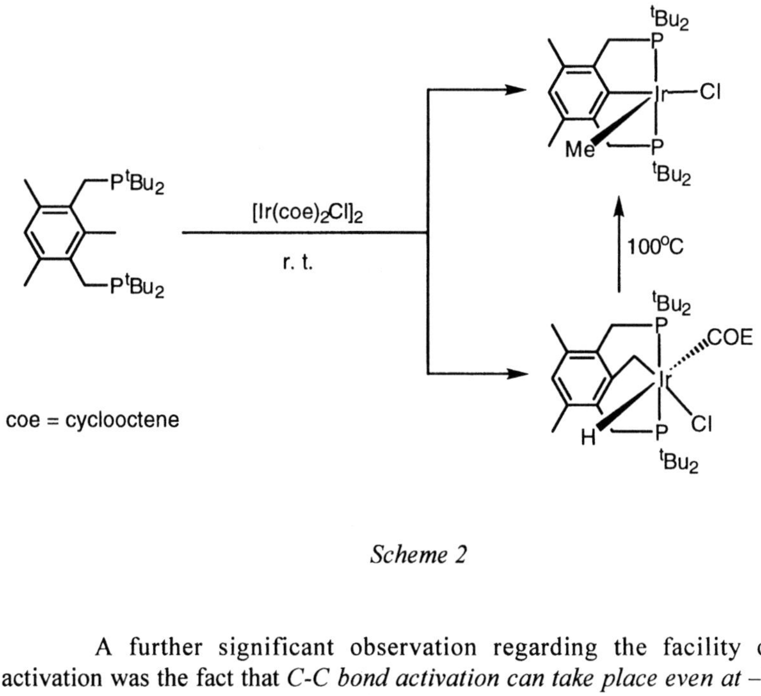

Scheme 2

A further significant observation regarding the facility of C-C activation was the fact that *C-C bond activation can take place even at −70 °C* in the case of the PCN ligand (here ligand coordination is not rate determining), with exclusive formation of the C-C bond activated product (Scheme 3). The reaction was shown to proceed through a 14e intermediate (see below) *(18)*.

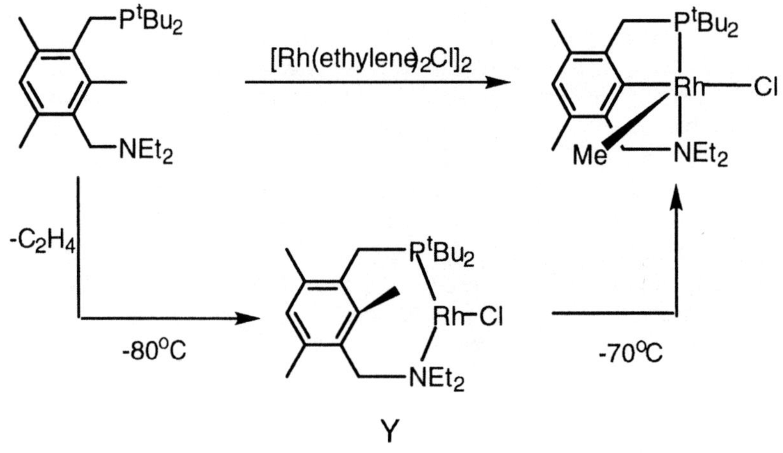

Scheme 3

It should be noted that formation of 14e intermediates takes place also in the case of the *Ph*-PCP-Rh system (Scheme 4) (*19*). C-C bond activation is very facile here as well: it is not a rate determining step and takes place at room temperature, indicating the general importance of 14e intermediates for facile C-C bond activation. The presence of the bulky iodide ligand results very likely in favorable formation of the unsaturated 14e intermediate in this case.

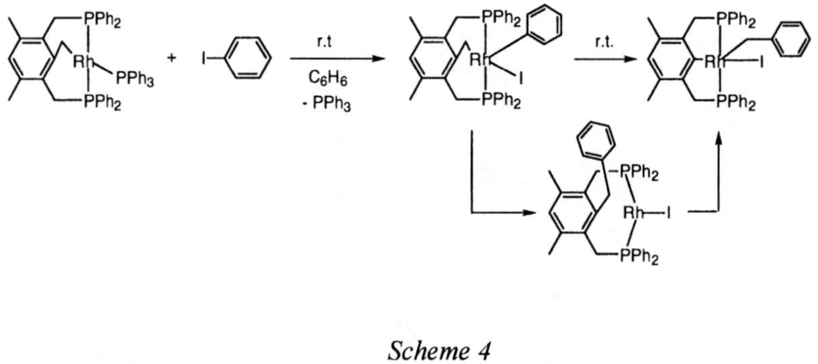

Scheme 4

C-H vs C-C Bond Activation in Neutral PCP and PCN-Rh Systems: Mechanistic Insight

The tBu-PCP-iridium and rhodium systems described above (Schemes 1,2) allowed a direct comparison of C-C and C-H activation processes (Fig. 1) (15). C-C activation was found to be thermodynamically and, importantly, kinetically preferred over C-H. The two processes proceed through similar transition states, most probably through 3-centered non-polar ones, as suggested by the small difference in the activation parameters between them, and by the fact that variations of the solvent and substituent in the aromatic ring of the ligand do not influence the $\Delta\Delta G^{\neq}$. This is in agreement with computational studies (20,21). However, direct measurement of the C-C insertion activation parameters was not possible in this system.

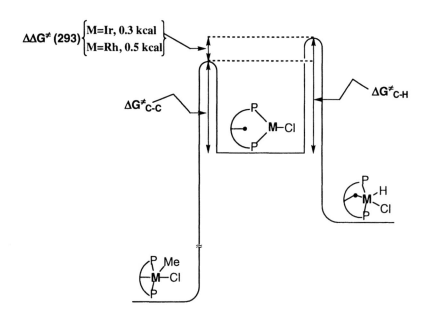

Figure 1. Reaction profile for C-C vs C-H oxidative addition in the tBu-PCP system
(Reproduced from reference 5. Copyright 1996 American Chemical Society.)

Further insight into the C-C bond activation mechanism was obtained by kinetic evaluation of a single step metal insertion into a carbon-carbon bond in solution in the PCN-Rh system (*18*). The 14e intermediate **Y** (Scheme 3), a frozen intermediate in the process, was formed and fully characterized at low temperature. It undergoes clean oxidative addition of the C-C bond, allowing the direct measurement of the activation parameters, that are as follows: $\Delta H^{\neq}$ = 15.0($\pm$0.4) kcal/mol, $\Delta S^{\neq}$ = -7.5($\pm$2.0) e. u., $\Delta G^{\neq}$(298) = 17.2($\pm$1.0) kcal/mol. As expected for a concerted oxidative addition process, the activation entropy is negative. The fact that it is only moderately negative indicates that the intermediate is already significantly ordered towards the insertion step. The PCN system was also studied computationally (*22*).

$$\left[\begin{array}{c}\text{structure}\end{array}\right]^{\ddagger}$$

TS$_{cc}$(PCN)

Thus, our kinetic study supports a 3-centered, non-polar transition state, TS$_{CC}$(PCN), for C$_{Ar}$-C oxidative addition to Rh(I), in agreement with our postulate in the case the tBu-PCP-Rh/Ir system described above. The obtained activation parameters are *the first data* for an apparent single step carbon-carbon bond activation by a metal complex.

PCN-Rh and Cationic PCP-Rh System: Control over C-C vs C-H Activation Aptitudes

In addition to the great facility of C-C oxidative addition, the PCN-Rh system demonstrates very high *selectivity* towards C-C vs C-H bond activation: no C-H activated products are observed in this system (Scheme 3) (*18,23*). Such high preference for C-C bond activation is most probably due to the optimal metal positioning for the insertion into the C-C bond. Thus, as both C-H and C-C activation are observed in the PCP systems mentioned above, changing of one phosphine arm for an amine results in preferential C-C activation, demonstrating that the C-C vs C-H reaction aptitudes can be controlled by appropriate ligand choice.

Having investigated the reactivity of neutral Rh and Ir complexes, we were interested in extending the scope to cationic metal centers. A cationic metal center was expected to exhibit different reactivity and selectivity due to lower electron density and possibly different interactions with the arene ring and the ligands. We found that in a *cationic* tBu-PCP rhodium system the reaction can be driven towards the exclusive activation of C-C or C-H bonds at room temperature by *solvent choice* (*24*). Thus, in acetonitrile exclusive C-H activation took place (Scheme 5), while in THF the C-C bond activation product was quantitatively formed (although C-H bond oxidative addition was also observed in THF at low temperature). This unique degree of control over metal insertion into strong C-H vs C-C bonds seems to be a consequence of solvent coordination. In THF an unsaturated intermediate, possessing a vacant coordination site capable of both C-H and C-C activation, may be formed. In the case of the better coordinating acetonitrile, solvent coordination blocks vacant coordination sites and generates an intermediate too bulky for insertion into the sterically more hindered C-C bond. Thus, unlike its neutral counterpart, which is almost insensitive to the solvent (due to lower tendency to bind solvent molecules), the cationic rhodium center appears to possess very good selectivity in C-C vs C-H activation, conveniently regulated by the reaction solvent.

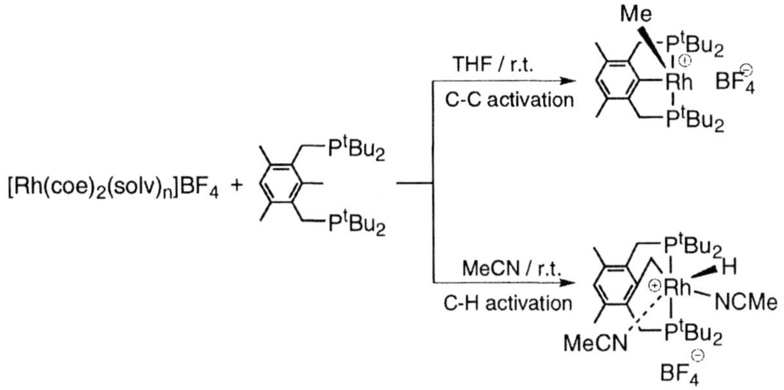

Scheme 5

C-H vs C-C Activation in a Cationic PCO-Rh System. Facility, Mechanism and Control

Our attempts to extend the scope of C-C bond activation to a mono-chelating PC system resulted in exclusive C-H bond activation (Scheme 6) (*25*). No C-C bond activation was observed using various conditions and reagents. Following this observation, a PCO-type ligand was designed in order to probe the role of the chelating effect (Scheme 7). One of its chelating moieties being an oxygen donor, this system mimics the electron density environment and general structure of the mono-chelating PC system, enabling both mono- and bis-chelating binding modes.

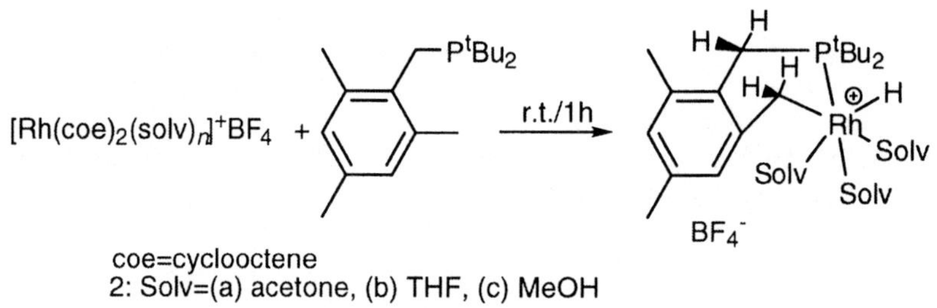
[Rh(coe)2(solv)n]+BF4 +
PtBu2
r.t./1h
H
H
H
H
PtBu2
Rh
Solv
Solv
Solv
BF4-
coe=cyclooctene
2: Solv=(a) acetone, (b) THF, (c) MeOH

Scheme 6

The PCO-Rh system undergoes C-H bond activation at room temperature, forming products with an open and closed methoxy arm (Scheme 7) (*26*). C-C bond activation takes place upon heating. Significantly, whereas C-H activation of the two methyl groups is observed, only the C-C bond between the chelating arms is activated, demonstrating that the chelating methoxy ligand is *both essential and sufficient* for C-C activation in this system. An attempt to isolate the C-H activation products led to an intriguing observation: removal of the solvent from a solution of the C-H activation complexes under vacuum at room temperature resulted in conversion of the C-H to the C-C activation product. Thus, C-H vs C-C bond activation can be controlled just by the presence or absence of the solvent. This effect appears to be due to the preferential coordination of the BF_4 counter anion to the metal center of the C-C activation product upon solvent evaporation.

The exclusive activation of the C-C bond situated between the phosphine and methoxy arm and the absence of C-C activation in the PC system suggest that *chelation is crucial for C-C bond activation.*

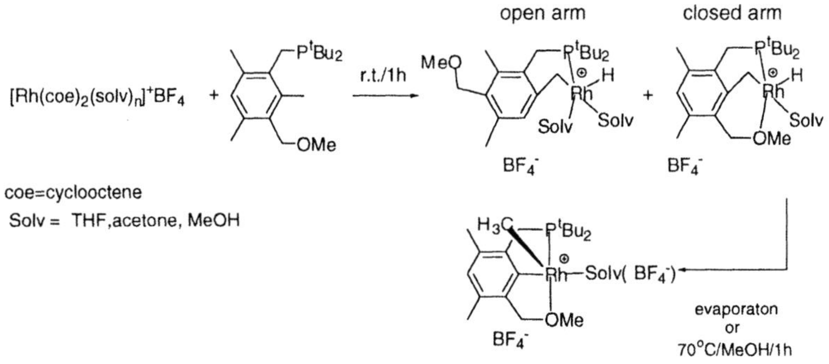

Scheme 7

Apparently, the chelate effect influences both the kinetics and thermodynamics of the process. An important kinetic factor is the bonding of the metal center by the two ligand arms, which brings the metal into close proximity of the C-C bond to be cleaved. The resulting C-C activation product is also stabilized by the presence of two five-membered chelate rings, contributing to the thermodynamic driving force of the reaction. Further insight regarding these factors was obtained from a computational study of the Rh-PCO system. The computational results are in good agreement with our experimental findings. The following important conclusions stem from our combined experimental and theoretical study.

Regarding the electronic requirements, 14e intermediates are active in both the "open" and "closed arm" systems (the lowest energy pathway proceeds through these intermediates in both C-C and C-H oxidative addition). The fact that the metal center coordinates solvent molecules in order to achieve a 14e configuration demonstrates the general importance of this electronic structure for C-H and C-C bond activation. Even relatively labile ligands such as methanol or the methoxy arm are capable of providing the electron density necessary for C-C and C-H activation. Thus, the electron density of the $[R_3PRh(MeOH)_2]^+$ and $[R_3PRh(MeOH)(Ligand-OMe)]^+$ fragments is adequate for C-H and C-C bond activation.

Comparison of the steric requirements for C-C and C-H bond activation reveal that there is a striking difference between the open and closed arm cases. Whereas in the closed arm system the transition states for both C-C and C-H activation have a similar geometry, in the open arm case their geometry differs significantly (Figure 2).

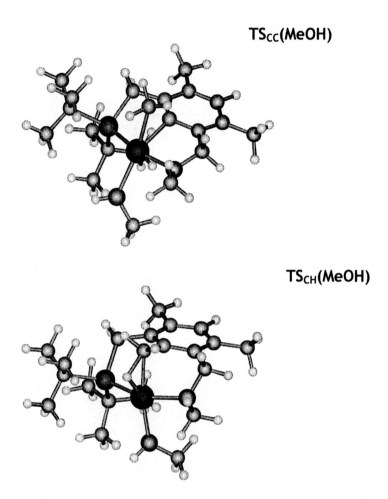

Closed arm: the metal center is situated in close proximity to both the C-H and C-C bonds in TS$_{CC}$ and TS$_{CH}$, resulting in similar barriers of activation.

Figure 2a. C-H and C-C activation transition states in the closed arm PCO system

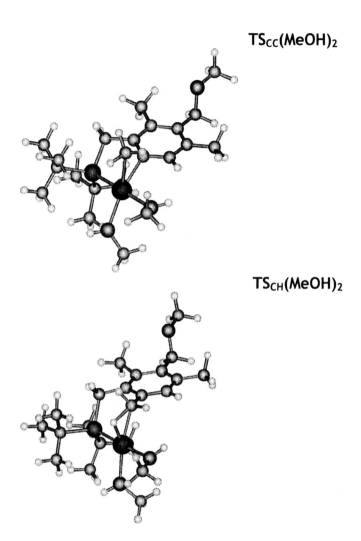

Open arm: C-H activation proceeds through TS$_{CH}$, in which the metal center is not situated in close proximity to the C-C bond. In the C-C activation transition state, TS$_{CC}$, the metal is in close vicinity to this bond.

Figure 2b. C-H and C-C activation transition states in the open arm PCO system

For insertion into the C-C bond, the metal must be in closer proximity to the bond than in the case of C-H insertion. *An additional activation energy cost is to be paid in order to bring the metal center closer to the C-C bond.* It seems to be a general factor, which is most probably due to the more directed character of the C-C bond in comparison to the C-H one. Thus, in order to cleave the C-C bond the metal center has not only to penetrate through the C-H bond "coat", but also to be in very close vicinity to the C-C bond because of its intrinsically higher directionality. Hence, steric requirements for C-C bond activation are far more restricting than for the C-H one and hence the importance of chelation which promotes the cleavage of the C-C bond by bringing the metal center close to it. On the other hand, the electronic requirements for both C-C and C-H are very similar and 14e intermediates play a key role in both activation processes.

Thus, in the PCO-Rh system C-C bond activation is a *facile process*: it can be achieved at moderate heating or at room temperature upon solvent evaporation. A remarkable *control* can be gained: the presence or absence of the solvent can drive the reaction towards C-H or C-C bond activation respectively. As far as the *mechanism* is concerned, steric limitations were found to play a critical role in C-H vs C-C bond activation aptitudes, electronic requirements being very similar.

C-H vs C-C Activation: Methylene Transfer

A combination of C-H and C-C oxidative addition led to the development of a unique reaction sequence: the methylene transfer reaction. In this reaction a kinetically favorable C-H activation process was utilized to form methylene-bridged complexes which in turn were used as methylene group donors (27). The reaction sequence involves oxidative addition of an appropriate substrate to the metal center in the methylene donor complex, subsequent reductive elimination, C-C cleavage and C-X (X=H, Si) reductive elimination leading to product formation. The CH_2 group was selectively inserted into Si-H, Si-Si and C_{Aryl}-H bonds (Scheme 8), utilizing a kinetically preferred C-H activation process to create a methylene donor complex and thermodynamically favored C-C activation processes to selectively insert the methylene group into a variety of strong, non-activated bonds.

The methylene-bridged Rh(I) complex can be regenerated by reaction of the "methylene depleted" Rh(I) complex with MeI and a base, the result being

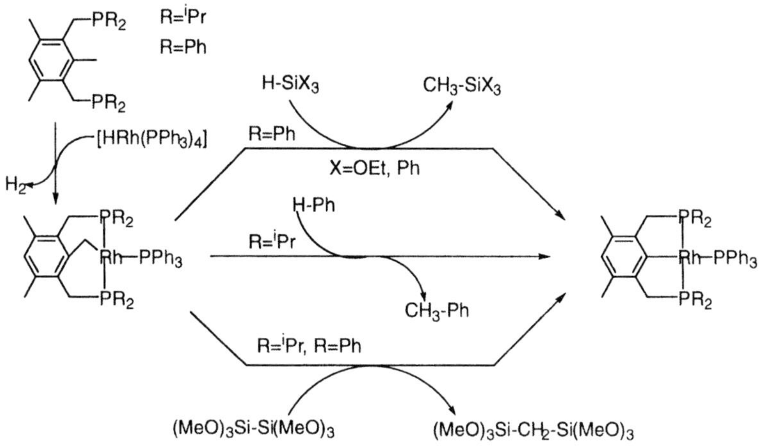

Scheme 8

the extraction of a methylene group from methyl iodide and its selective incorporation into other C-H, Si-H and Si-Si bonds. The "reverse" methylene transfer, which is essential for regeneration of the active methylene donating moiety, was also achieved with substituted methylene moieties which were successfully transferred to the bis-chelating ring, extending the potential scope of the reaction (19).

Conclusions

We have demonstrated that oxidative addition of strong C-C bonds can be facile and kinetically and thermodynamically more favorable than C-H activation. The mechanisms of C-C and C-H bond oxidative addition are similar as far as electronic factors are concerned and proceed through non-polar three-centered transition states, as supported by the direct measurement of the activation parameters for a single step C-C oxidative addition. On the other hand, C-C bond oxidative addition is much more influenced by sterics than the C-H one and, as a consequence, metal direction by chelation is more important for C-C than for C-H bond activation. However, under optimal metal positioning towards the C-C bond to be activated, the kinetic barrier for oxidative addition of a strong C-C bond is low (it can proceed at −70°C). A remarkable degree of control over C-C vs C-H activation can be achieved: (a) as far as the ligand is concerned, PCN is highly specific for C-H activation while PCP is less so, due to a directional ligand effect; (b) the selectivity of C-C vs C-H activation can be controlled by solvent choice and by counter anion coordination. The unique methylene transfer reaction utilizes kinetically preferred C-H bond activation and thermodynamic preferred C-C bond activation to selectively insert a CH_2 unit in a variety of strong bonds.

Acknowledgement

This work was supported by the Israel Science Foundation, the MINERVA foundation, Munich, Germany, the US-Israel Binational Science Foundation and the Helen and Martin Kimmel Center for Molecular Design. D.M. holds the Israel Matz Professorial Chair of Organic Chemistry.

References

1. Reviews: (a) Crabtree, R. H. *Chem. Rev.* **1985**, *85*, 245. (b) Jones, W. D.; Feher, F. J. *Acc. Chem. Res.* **1989**, 22, 91 (c) Arndtsen, B. A.; Bergman, R. G.; Mobley, T. A.; Peterson, T. H. *Acc. Chem. Res.* **1995**, *28*, 154. (d) J. A. Labinger, J. E. Bercaw *Nature* **2002**, *417*, 507.
2. For recent reviews on catalytic C-H activation, see: (a) Jensen, C. M. *Chem. Comm.* **1999**, 2443 (b) Jia, C. G.; Kitamura, T.; Fujiwara, Y. *Acc. Chem. Res.* **2001**, *34*, 633 (c) Ritleng, V.; Sirlin, C.; Pfeffer, M. *Chem. Rev.* **2002**, *102*, 1731 (d) Kakiuchi, F.; Murai, S. *Acc. Chem. Res.* **2002**, *35*, 826.
3. Reviews: (a) Rybtchinski, B.; Milstein, D. *Angew. Chem. Int. Ed.* **1999**, 38, 870 (b) Murakami, M.; Ito, Y. In *Topics in Organometallic Chemistry*, Murai, S., Ed.; Springer-Verlag Berlin Heidelberg, 1999; Vol. 3, p. 97.
4. Hourlet, R.; Halle, L. F.; Beauchamp, J. L. *Organometallics* **1983**, 2, 1818.
5. (a) Simões, A. M.; Beauchamp, J. L. *Chem. Rev.* **1990**, *90*, 629. (b) Nolan, S. P.; Hoff, C. D.; Stoutland, P. O.; Newman, L. J.; Buchanan, J. M.; Bergman, R. G.; Yang, G. K.; Peters, K. S. *J. Am. Chem. Soc.* **1987**, *109*, 3143 (c) Jones, W. D.; Feher, F. J. *J. Am. Chem. Soc.* **1984**, *106*, 1650.
6. Reviews on bis-chelating, pincer-type complexes: (a) Albrecht, M.; van Koten, G. . *Chem. Int. Ed.* **2001**, 40, 3750 (b) Vigalok, A.; Milstein, D; *Acc. Chem. Res.* **2001**, 34, 798 (c) van der Boom, M. E.; Milstein, D. *Chem. Rev.* **2003**, 103, 1759.

7. McMillen, D. F.; Golden, D. M. *Ann. Rev. Phys. Chem.* **1982**, *33*, 492.
8. Gozin, M.; Weisman, A.; Ben-David, Y.; Milstein, D. *Nature* **1993**, *364*, 699.
9. Liou, Sh.-Y.; Gozin, M.; Milstein, D. *J. Am. Chem. Soc.* **1995**, *117*, 9774.
10. (a) van der Boom, M. E.; Ben David, Y.; Milstein, D. *J. Chem. Soc. Chem. Comm.* **1998**, 917 (b) van der Boom, M. E.; Ben David, Y.; Milstein, D. *J. Am. Chem. Soc.* **1999**, *121*, 6652.
11. (a) Liou, Sh.-Y.; Gozin, M.; Milstein, D. *J. Chem. Soc. Chem. Comm.* **1995**, 1965 (b) van der Boom, M.E.; Liou, Sh.-Y.; Ben-David, Y.; Gozin, M.; Milstein, D. *J. Am. Chem. Soc.* **1998**, *120*, 13415.
12. Liou, Sh.-Y.; van der Boom, M. E.; Milstein, D. *J. Chem. Soc. Chem. Comm.* **1998**, 687.
13. van der Boom M. E.; Kraatz, H. B.; Hassner L.; Ben-David Y.; Milstein, D. *Organometallics* **1999**, *18*, 3873.
14. Gauvin R.M.; Rozenberg H.; Shimon L.J.W.; Milstein D. *Organometallics* **2001**, *20*, 1719.
15. Rybtchinski, B.; Vigalok, A.; Ben-David, Y.; Milstein, D. *J. Am. Chem. Soc.* **1996**, *118*, 12406.
16. van der Boom, M. E.; Kraatz, H.-B.; Ben-David, Y.; Milstein, D. *J. Chem. Soc. Chem. Comm.* **1996**, 2167.
17. van der Boom, M. E.; Uzan, O.; Milstein, D. unpublished results.
18. Gandelman, M.; Vigalok, A.; Konstantinovsky, L.; Milstein, D. *J. Am. Chem. Soc.* **2000**, *117*, 9774; Commentary: Crabtree, R. H. *Nature* **2000**, *408*, 415.
19. Cohen, R.; van der Boom, M. E.; Shimon, L. J. W.; Rozenberg H.; Milstein, D. *J. Am. Chem. Soc.* **2000**, *122*, 7723.
20. Sunderman, A.; Uzan, O.; Milstein, D.; Martin, J. M. L. *J. Am. Chem. Soc.* **2000**, *122*, 9848.
21. A computational study of C-C /C-H activation involving a *H*-PCP system suggests involvement of 4-coordinate η^1– arene intermediates: Cao, Z.; Hall, M.B. *Organometallics* **2000**, *19*, 3338.
22. Sunderman, A.; Uzan, O.; Martin, J. M. L. *Organometallics* **2001**, *20*, 1783.
23. Gandelman, M.; Vigalok, A.; Shimon, L. J. W.; Milstein, D. *Organometallics* **1997**, *16*, 3981.
24. Rybtchinski, B.; Milstein, D. *J. Am. Chem. Soc.* **1999**, *121*, 4528
25. Rybtchinski, B.; Konstantinovsky, L.; Shimon, L. J. W.; Vigalok, A.; Milstein, D. *Chem. Eur. J.* **2000**, *17*, 3287.
26. Rybtchinski, B.; Oevers, S.; Montag, M.; Vigalok, A.; Rozenberg, H.; Martin, J.M.L.; Milstein, D. *J. Am. Chem. Soc.* **2001**, *123*, 9064.
27. Gozin, M.; Aizenberg, M.;.Liou, Sh.-Y.; Weisman, A.; Ben-David, Y.; Milstein, D. *Nature* **1994**, *370*, 42.

Chapter 5

Kinetic and Equilibrium Deuterium Isotope Effects for C–H Bond Reductive Elimination and Oxidative Addition Reactions Involving the *Ansa*–Tungstenocene Methyl–Hydride Complex [Me$_2$Si(C$_5$Me$_4$)$_2$]W(Me)H

Kevin E. Janak, David G. Churchill, and Gerard Parkin*

Department of Chemistry, Columbia University, New York, NY 10027

The reductive elimination of methane from [Me$_2$Si(C$_5$Me$_4$)$_2$]W(CH$_3$)H and [Me$_2$Si(C$_5$Me$_4$)$_2$]W(CD$_3$)D is characterized by an inverse kinetic isotope effect (KIE). A kinetics analysis of the interconversion of [Me$_2$Si(C$_5$Me$_4$)$_2$]-W(CH$_3$)D and [Me$_2$Si(C$_5$Me$_4$)$_2$]W(CH$_2$D)H, accompanied by elimination of methane, provides evidence that the reductive coupling step in this system is characterized by a normal KIE and that the inverse KIE for overall reductive elimination is a result of an inverse *equilibrium* isotope effect (EIE), rather than being a result of an inverse KIE for a single step. Calculations on [H$_2$Si(C$_5$H$_4$)$_2$]W(Me)H support these results and further demonstrate that the interconversion between [H$_2$Si(C$_5$H$_4$)$_2$]W(Me)H and the σ-complex [H$_2$Si(C$_5$H$_4$)$_2$]-W(σ–HMe) is characterized by normal kinetic isotope effects for both reductive coupling and oxidative cleavage. Interestingly, the temperature dependencies of EIEs for coordination and oxidative addition of methane to the tungstenocene fragment {[H$_2$Si(C$_5$H$_4$)$_2$]W} are calculated to be very different, with the EIE for coordination approaching zero at 0K, while the EIE for oxidative addition approaches infinity.

Introduction

The oxidative addition and reductive elimination of C–H bonds at a transition metal center are reactions that are crucial to the functionalization of hydrocarbons (*1*). An important component of these transformations is that they are mediated by σ–complexes, [M](σ–HR), in which the hydrocarbon is coordinated to the metal by 3–center–2–electron M···H–C interactions (*2,3,4*). Evidence for the existence of these σ–complexes includes: (i) low temperature spectroscopic and room temperature flash kinetics studies (*3,5*), (ii) the observation of deuterium exchange between hydride and alkyl sites, *e.g.* [M](CH$_3$)D → [M](CH$_2$D)H, and (iii) the measurement of kinetic isotope effects (KIEs) (*6*). As a result of the existence of σ–complex intermediates, the terms "reductive elimination" (re) and "oxidative addition" (oa) do not correspond to elementary steps and additional terms are required to describe adequately the overall mechanism. Thus, reductive elimination consists of reductive coupling (rc) followed by dissociation (d), while the microscopic reverse, oxidative addition, consists of ligand association (a) followed by oxidative cleavage (oc), as illustrated in Scheme 1.

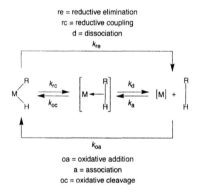

Scheme 1. Oxidative addition and reductive elimination mediated by σ–complex intermediates.

The present article describes experimental and computational studies designed to determine the kinetic and equilibrium isotope effects of the individual steps pertaining to oxidative addition and reductive elimination of methane involving the *ansa*–tungstenocene complex [Me$_2$Si(C$_5$Me$_4$)$_2$]W(Me)H.

Reductive Elimination of Methane from [Me$_2$Si(C$_5$Me$_4$)$_2$]W(Me)H

Previous studies have indicated that reductive elimination of methane from the tungstenocene methyl–hydride complexes Cp$_2$W(Me)H (*7*) and Cp*$_2$W(Me)H (*8*) is facile. The *ansa*-complex [Me$_2$Si(C$_5$Me$_4$)$_2$]W(Me)H likewise reductively eliminates methane; the tungstenocene intermediate so generated is trapped intramolecularly to give [Me$_2$Si(η^5–C$_5$Me$_4$)(η^6–C$_5$Me$_3$CH$_2$)]WH, or intermolecularly by benzene to give [Me$_2$Si(C$_5$Me$_4$)$_2$]W(Ph)H (Scheme 2) (*9*).

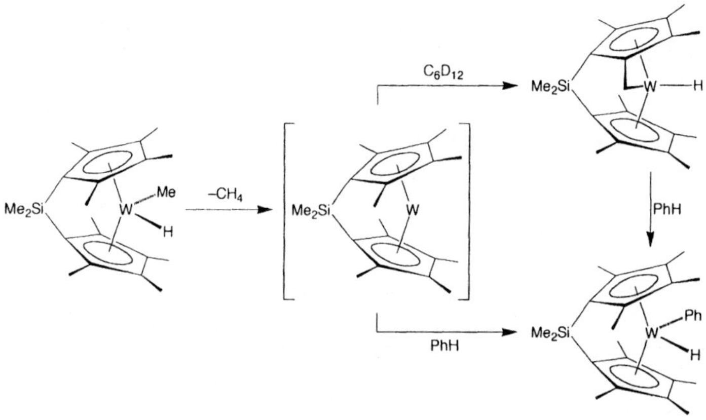

Scheme 2. Reductive elimination of methane from [Me$_2$Si(C$_5$Me$_4$)$_2$]W(Me)H.

By comparison with Cp*$_2$W(Me)H, two noteworthy aspects of the reductive elimination of methane from [Me$_2$Si(C$_5$Me$_4$)$_2$]W(Me)H are: (i) the *ansa* bridge substantially inhibits the reductive elimination of methane, with $k_{ansa}/k_{Cp*} = 0.03$ at 100°C (*10*); and (ii) the *ansa* bridge promotes intermolecular C–H bond activation, with {[Me$_2$Si(C$_5$Me$_4$)$_2$]W} being capable of being trapped by benzene to give the phenyl–hydride complex [Me$_2$Si(C$_5$Me$_4$)$_2$]W(Ph)H, whereas reductive elimination of methane from Cp*$_2$W(Me)H in benzene gives only the tuck-in complex Cp*(η^6–C$_5$Me$_4$CH$_2$)WH. Kinetics studies, however, indicate that although intermolecular oxidative addition of benzene is thermodynamically favored, intramolecular C–H bond cleavage within {[Me$_2$Si(C$_5$Me$_4$)$_2$]W} to give [Me$_2$Si(η^5–C$_5$Me$_4$)(η^6–C$_5$Me$_3$CH$_2$)]WH is actually kinetically favored.

Evidence for σ–Complex Intermediates: Kinetic Isotope Effects and Isotope Scrambling for [Me$_2$Si(C$_5$Me$_4$)$_2$]W(CH$_3$)H and its Isotopologues

Evidence that reductive elimination of methane from [Me$_2$Si(C$_5$Me$_4$)$_2$]W(Me)H proceeds via a σ–complex intermediate is provided by the observation of H/D exchange between the hydride and methyl sites of the isotopologue [Me$_2$Si(C$_5$Me$_4$)$_2$]W(CH$_3$)D resulting in the formation of [Me$_2$Si(C$_5$Me$_4$)$_2$]W(CH$_2$D)H (Scheme 3). Examples of such isotope exchange reactions are well known (*6*), and are postulated to occur by a sequence that involves: (i) reductive coupling to form a σ–complex intermediate, (ii) H/D exchange within the σ–complex, and (iii) oxidative cleavage to generate the isotopomeric methyl-hydride complex (Scheme 3).

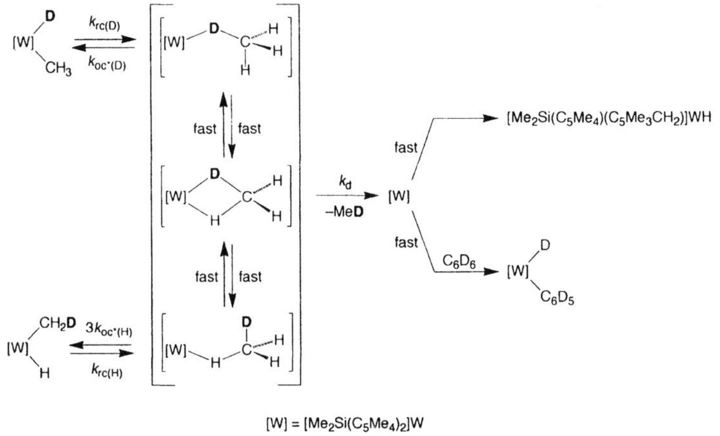

Scheme 3. H/D Exchange via a σ–complex intermediate.

Further evidence for the existence of a σ–complex intermediate in the reductive elimination of methane is obtained from the observation of an *inverse* (*i.e.* < 1) kinetic isotope effect of 0.45(3) for reductive elimination of CH_4 and CD_4 from $[Me_2Si(C_5Me_4)_2]W(CH_3)H$ and $[Me_2Si(C_5Me_4)_2]W(CD_3)D$ at 100°C. Specifically, the rate constant for reductive elimination is a composite of the rate constants for reductive coupling (k_{rc}), oxidative cleavage (k_{oc}), and dissociation (k_d), namely $k_{obs} = k_{rc}k_d/(k_{oc} + k_d)$. For a limiting situation in which dissociation is rate determining (*i.e.* $k_d \ll k_{oc}$), the expression simplifies to $k_{obs} = k_{rc}k_d/k_{oc} = K_\sigma k_d$, where K_σ is the equilibrium constant for the conversion of [M](R)H to [M](σ–RH). As such, the kinetic isotope effect for overall reductive elimination is $k_H/k_D = [K_{\sigma(H)}/K_{\sigma(D)}][k_{d(H)}/k_{d(D)}]$, where $K_{\sigma(H)}/K_{\sigma(D)}$ is the *equilibrium* isotope effect for the conversion of [M](R)H to [M](σ–RH) (Figure 1). If the isotope effect for dissociation of RH (*i.e.* $[k_{d(H)}/k_{d(D)}]$) is close to unity (since the C–H bond is close to being fully formed) (*6a*), the isotope effect on reductive elimination would then be dominated by the *equilibrium* isotope effect $K_{\sigma(H)}/K_{\sigma(D)}$ for formation of the σ–complex [M](σ–RH). The latter would be predicted to be inverse on the basis of the simple notion that deuterium prefers to be located in the stronger bond, *i.e* C–D *versus* M–D (*11*). Consequently, an *inverse* KIE would be predicted for the overall reductive elimination, without requiring an inverse effect for a single step (Figure 1) (*12*). Indeed, this explanation has been used to rationalize the inverse KIEs for a variety of alkyl

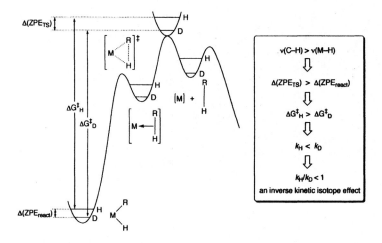

Figure 1. Origin of an inverse kinetic isotope effect for reductive elimination.

hydride complexes, including Cp*Ir(PMe$_3$)(C$_6$H$_{11}$)H (0.7) (*2*), Cp*Rh(PMe$_3$)(C$_2$H$_5$)H (0.5) (*12d*), Cp$_2$W(Me)H (0.75) (*7b*), Cp*$_2$W(Me)H (0.70) (*8*), [Cp$_2$Re(Me)H]$^+$ (0.8) (*13*), [(Me$_3$tacn)Rh(PMe$_3$)(Me)H]$^+$ (0.74) (*14*), (tmeda)Pt(Me)(H)(Cl) (0.29) (*15*), [Tp^{Me_2}]Pt(Me)$_2$H (0.81), and [Tp^{Me_2}]Pt(Me)(Ph)H ($\leq$ 0.78) (*16*).

It must be emphasized that whereas an inverse KIE is to be expected if the σ–complex is formed prior to the rate determining step, a normal KIE would be expected if the reductive coupling step is rate determining since reactions which involve X–H(D) cleavage in the rate determining step are typically characterized by k_H/k_D ratios greater than unity. Thus, for a limiting situation in which reductive coupling is rate determining (*i.e.* $k_d >> k_{oc}$), the rate constant for reductive elimination simplifies to $k_{obs} = k_{rc}$. Since the transition state for reductive coupling involves cleavage of the M–H bond, $k_{rc(H)}/k_{rc(D)}$ might be expected to be > 1 and so a normal KIE would be expected for such a situation. Examples of complexes that exhibit normal KIEs include (Ph$_3$P)$_2$Pt(Me)H (3.3) (*17a*), (Ph$_3$P)$_2$Pt(CH$_2$CF$_3$)H (2.2) (*17b*), and (Cy$_2$PCH$_2$CH$_2$PCy$_2$)Pt(CH$_2$Bu^t)H (1.5) (*17c*).

While the preequilibrium explanation (Figure 1) has found common acceptance for the rationalization of inverse kinetic isotope effects for reductive elimination of RH, it must be emphasized that there is actually very little direct kinetic evidence to support it because the kinetic isotope effects for the individual steps are generally unknown. Rather, the common acceptance is in large part due to the fact that inverse primary *kinetic* deuterium isotope effects for a single step reaction are not well-known, while inverse *equilibrium* isotope effects for reactions that involve the transfer of hydrogen from a metal to carbon are certainly precedented (*18*). The question, therefore, arises as to whether it is possible that the inverse KIE for reductive elimination could actually be a result of an inverse kinetic isotope effect on reductive coupling. While it is not possible to address this issue by studying the kinetics of reductive elimination of [Me$_2$Si(C$_5$Me$_4$)$_2$]W(CH$_3$)H and [Me$_2$Si(C$_5$Me$_4$)$_2$]W(CD$_3$)D, it is possible to address the issue by studying the elimination of CH$_3$D from [Me$_2$Si(C$_5$Me$_4$)$_2$]W(CH$_3$)D and [Me$_2$Si(C$_5$Me$_4$)$_2$]W(CH$_2$D)H (9). Specifically, [Me$_2$Si(C$_5$Me$_4$)$_2$]W(CH$_3$)D is observed to isomerize to [Me$_2$Si(C$_5$Me$_4$)$_2$]W(CH$_2$D)H on a time-scale that is comparable to the overall reductive elimination of CH$_3$D, and a kinetics analysis of the transformations illustrated in Scheme 3 permits the KIE for reductive coupling to be determined. However, it must be emphasized that not all rate constants can be determined uniquely, and only relative values may be derived for reactions pertaining to the σ–complex intermediates (oxidative cleavage or dissociation) since they are not spectroscopically detectable. Thus, for the purpose of the analysis, the value for $k_{oc*(D)}$ was arbitrarily set as unity and rapid interconversion between the various σ–complex intermediates was assumed such that they were modeled by a single species {[Me$_2$Si(C$_5$Me$_4$)$_2$]W(CH$_3$D)} with a single rate constant for the dissociation of methane (k_d). The simulation is illustrated in Figure 2, with the derived free energy surface presented in Figure 3. Significantly, a *normal* isotope effect of 1.4(2) is observed for $k_{rc(H)}/k_{rc(D)}$. Assuming that secondary effects do not play a dominant role in the reductive coupling of [Me$_2$Si(C$_5$Me$_4$)$_2$]W(CH$_2$D)H, the value of 1.4(2) provides an estimate of the primary KIE for reductive coupling of [Me$_2$Si(C$_5$Me$_4$)$_2$]W(Me)X (X = H, D) to form the σ–complex intermediate [Me$_2$Si(C$_5$Me$_4$)$_2$]W(σ–XMe).

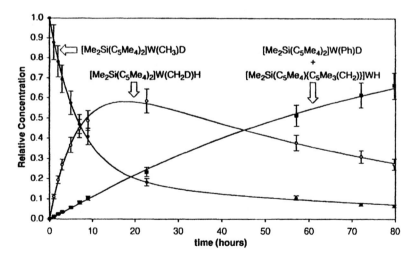

Figure 2. Kinetics simulation of isotope exchange within [Me$_2$Si(C$_5$Me$_4$)$_2$]W(CH$_3$)D and reductive elimination of methane.

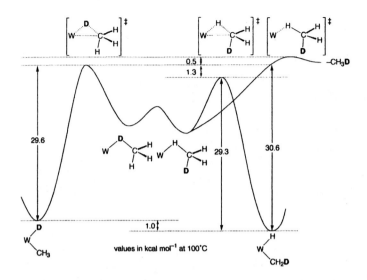

Figure 3. Free energy surface for interconversion of [Me₂Si(C₅Me₄)₂]W(CH₃)D and [Me₂Si(C₅Me₄)₂]W(CH₂D)H and elimination of methane at 100°C. Note that for each pair of isotopomers, it is the one with deuterium attached to the carbon in a terminal fashion that is the lower in energy.

The observation of a normal kinetic isotope effect for reductive coupling within [Me₂Si(C₅Me₄)₂]W(Me)H is significant because it supports the notion that the inverse nature of the KIE for the reductive elimination of methane is *not* a manifestation of an inverse KIE for a single step in the transformation, but is rather associated with an inverse *equilibrium* isotope effect. Of direct relevance to this issue, Jones, in the most definitive study performed to date, has recently demonstrated that the EIE for the interconversion of [Tp^{Me_2}]Rh(L)(Me)X and [Tp^{Me_2}]Rh(L)(σ–XMe) is *inverse* (0.5), even though the individual KIEs for oxidative cleavage (4.3) and reductive coupling (2.1) are *normal* (L = CNCH₂Bu^t; X = H, D) (*19*).

Although the notion that the reductive coupling of a methyl-hydride complex is characterized by a normal primary kinetic deuterium isotope effect is in line with the common understanding of KIEs (*6*), it has recently been proposed that the reductive coupling for [Tp]Pt(Me)H₂ is characterized by an *inverse* KIE of 0.76 (*20*). However, it has subsequently been recognized that the experiment performed is actually incapable of determining the KIE for reductive coupling unless the KIE for oxidative cleavage is known (*6a,9*). Furthermore, assigning the observed KIE to that for reductive coupling is only possible if the KIE for oxidative cleavage is unity. It is, therefore, evident that the experiment purported to determine an inverse kinetic isotope effect of 0.76 for the reductive coupling of [Tp]Pt(Me)X₂ (X = H, D) has been erroneously interpreted, and that the system does not provide the claimed unprecedented opportunity to study the initial step of reductive coupling in alkyl hydride compounds (*21*).

Computational Determination of Kinetic and Equilibrium Isotope Effects

In view of the experimental difficulty associated with determining the kinetic isotope effects for the individual steps comprising reductive elimination and oxidative addition, we have employed computational methods to determine these values for the reductive elimination of methane from [Me$_2$Si(C$_5$Me$_4$)$_2$]W(Me)H *(22,23)*. The mechanism for the reductive elimination reaction was determined by first performing a series of DFT (B3LYP) linear transit geometry optimizations that progressively couple the C$_{Me}$–H bond. The result of these calculations was the generation of the σ–complex intermediate [Me$_2$Si(C$_5$Me$_4$)$_2$]W(σ–HMe) *via* a {[Me$_2$Si(C$_5$Me$_4$)$_2$]W(σ–HMe)}‡ transition state (Figure 4).

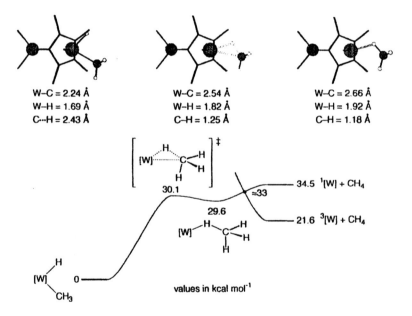

Figure 4. Calculated enthalpy surface for reductive elimination of CH$_4$ from [Me$_2$Si(C$_5$Me$_4$)$_2$]W(Me)H.

Subsequent dissociation of methane from the σ–complex [Me$_2$Si(C$_5$Me$_4$)$_2$]W(σ–HMe) generates the 16–electron tungstenocene intermediate, {[Me$_2$Si(C$_5$Me$_4$)$_2$]W}. However, an important consideration relevant to the dissociation of methane from the σ–complex intermediate [Me$_2$Si(C$_5$Me$_4$)$_2$]W(σ–HMe) is that the parent tungstenocene [Cp$_2$W] is known to be more stable as a triplet and thus dissociation of methane from singlet Cp$_2$W(σ–HMe) involves a spin crossover from the singlet to triplet manifold *(23a,b)*. Likewise, triplet {[Me$_2$Si(C$_5$Me$_4$)$_2$]W} is also calculated to be 12.9 kcal mol^{-1} more stable than the singlet. The geometry of the crossing point for the singlet–triplet interconversion during dissociation of methane from [Me$_2$Si(C$_5$Me$_4$)$_2$]W(σ–HMe) was estimated by using a procedure analogous to that used for [H$_2$C(C$_5$H$_4$)$_2$]W(σ–HMe) (23a). Specifically, a series of geometry optimizations were performed on singlet [Me$_2$Si(C$_5$Me$_4$)$_2$]W(σ–HMe) in which the W⋯C$_{Me}$ distance was progressively increased. At each point, the energy of the geometry optimized structure was determined in its triplet state, thereby allowing determination of the geometry for which the singlet and triplet states would be energetically degenerate (24). The derived crossing point for [Me$_2$Si(C$_5$Me$_4$)$_2$]W(σ–HMe) is observed to occur with a W⋯C$_{Me}$ distance of 3.3 Å, which is comparable to the value of 3.5 Å reported for [H$_2$C(C$_5$H$_4$)$_2$]W(σ–HMe) (23b). The computed enthalpy surface for the overall reductive elimination is illustrated in Figure 4.

The computation of isotope effects requires knowledge of the vibrational frequencies of the participating species. However, since frequency calculations are highly computationally intensive, it was necessary to perform such studies on a computationally simpler system in which the methyl groups of the [Me$_2$Si(C$_5$Me$_4$)$_2$] ligand are replaced by hydrogen atoms, i.e. [H$_2$Si(C$_5$H$_4$)$_2$]W(Me)H. This simplification considerably facilitates the calculation, while still retaining the critical features of the molecules of interest.

Kinetic isotope effects are conventionally determined by the expression KIE = k_H/k_D = SYM · MMI · EXC · ZPE or a modification that employs the Teller–Redlich product rule KIE = SYM · VP · EXC · ZPE (25). In these expressions, SYM is the symmetry factor (26,27), MMI is the mass-moment of inertia term, EXC is the excitation term, ZPE is the zero point energy term, and VP is the vibrational product, as defined in Scheme 4 (28,29).

$$SYM = \frac{\{(\sigma n)_H/(\sigma n)_D\}}{\{(\sigma n)^{\ddagger}_H/(\sigma n)^{\ddagger}_D\}} \quad MMI = \frac{(M^{\ddagger}_H/M_H)^{3/2}(I^{\ddagger}_H/I_H)^{1/2}}{(M^{\ddagger}_D/M_D)^{3/2}(I^{\ddagger}_D/I_D)^{1/2}} \quad VP = \Pi\,(v^{\ddagger}_{iH}/v^{\ddagger}_{iD})\,\Pi\,(v_{iD}/v_{iH})$$

$$EXC = \frac{\Pi\{[1-\exp(-u_{iH})]/[1-\exp(-u_{iD})]\}}{\Pi\{[1-\exp(-u^{\ddagger}_{iH})]/[1-\exp(-u^{\ddagger}_{iD})]\}} \quad ZPE = \frac{\exp\{\Sigma(u_{iH}-u_{iD})/2\}}{\exp\{\Sigma(u^{\ddagger}_{iH}-u^{\ddagger}_{iD})/2\}}$$

Scheme 4. Definitions of SYM, MMI, VP, and ZPE.[27,28]

The practical distinction between the two expressions is that the former requires the additional determination of the mass-moment of inertia term (MMI) for the structures of the molecules in question, while the latter requires determination of the vibrational product (VP) from the calculated frequencies. The two expressions should yield identical isotope effects given perfect data, but errors in computed frequencies may result in discrepancies (30). Therefore, we not only calculated the isotope effects by both of these methods, but also determined the isotope effects by using the thermodynamic values obtained directly from the DFT calculations (31). Significantly, the three methods yield very similar results, thereby providing an indication of the reliability of the calculations. In view of the similarity of the results obtained by the three methods, we present here only those derived from the expression, KIE = SYM · MMI · EXC · ZPE, since this is the one that is more commonly featured in the literature.

Calculated primary and secondary KIE values for the individual transformations pertaining to the overall reductive elimination of methane from $[H_2Si(C_5H_4)_2]W(Me)H$ are summarized in Table 1, illustrating several important points. Firstly, the primary KIE for reductive coupling of $[H_2Si(C_5H_4)_2]W(Me)X$ (X = H, D) to give the σ–complex $[H_2Si(C_5H_4)_2]W(\sigma-XMe)$ is small, but *normal* (1.05). Likewise, the microscopic reverse, *i.e.* oxidative cleavage of $[H_2Si(C_5H_4)_2]W(\sigma-XMe)$, is also *normal* (1.60). The equilibrium isotope effect (EIE) for the interconversion of $[H_2Si(C_5H_4)_2]W(Me)X$ and $[H_2Si(C_5H_4)_2]W(\sigma-XMe)$, however, is *inverse* (0.65), a consequence of the fact that the KIE for oxidative cleavage is greater than that for reductive coupling. Secondary isotope effects do not play a significant role, with values close to unity for the interconversion of $[H_2Si(C_5H_4)_2]W(CX_3)H$ and $[H_2Si(C_5H_4)_2]W(\sigma-HCX_3)$: $k_{rc(H)}/k_{rc(D)}$ = 1.02, $k_{oc(H)}/k_{oc(D)}$ = 1.09, and $K_{\sigma(H)}/K_{\sigma(D)}$ = 0.94. Analysis of the individual SYM, MMI, EXC and ZPE terms indicates that it is the zero point energy term that effectively determines the magnitude of the isotope effects for the interconversion of $[H_2Si(C_5H_4)_2]W(Me)H$ and $[H_2Si(C_5H_4)_2]W(\sigma-HMe)$ at 100 °C.

The KIE for dissociation of methane from a σ–complex has been postulated to be small (*7b*). Dissociation of methane from $[H_2Si(C_5H_4)_2]W(\sigma-HMe)$ would likewise be expected to exhibit a small KIE, especially since the C–H bond in the σ–complex is almost fully formed (d_{C-H} = 1.17 Å). Despite the complication that the transition state for dissociation occurs at the singlet–triplet crossing point (*32*), frequency calculations on singlet $[H_2Si(C_5H_4)_2]W(\sigma-HMe)$ with the geometry of the crossing point demonstrate that the KIEs for dissociation of methane are indeed close to unity (Table 1).

By predicting both a normal kinetic isotope effect for the reductive coupling step and an inverse kinetic isotope effect for the overall reductive elimination, the calculated isotope effects for reductive elimination of methane from $[H_2Si(C_5H_4)_2]W(Me)H$ are in accord with the experimental study on $[Me_2Si(C_5Me_4)_2]W(Me)H$. For example, the calculated inverse KIE for reductive elimination of methane from $[H_2Si(C_5H_4)_2]W(CH_3)H$ and $[H_2Si(C_5H_4)_2]W(CD_3)D$ (0.58) (*33*) compares favorably with the experimental value for $[Me_2Si(C_5Me_4)_2]W(CH_3)H$ and $[Me_2Si(C_5Me_4)_2]W(CD_3)D$ (0.45). Analysis of the isotope effects for the various steps provides conclusive evidence that the principal factor responsible for the inverse nature of the KIE for the overall reductive elimination is the inverse *equilibrium* isotope effect for the interconversion of $[H_2Si(C_5H_4)_2]W(Me)H$ and $[H_2Si(C_5H_4)_2]W(\sigma-HMe)$. The calculations therefore reinforce the notion that inverse primary kinetic isotope effects for reductive elimination of alkanes imply the existence of a σ–complex intermediate prior to rate determining loss of alkane. In addition to examining the kinetic isotope effects for loss of methane, it is instructive to evaluate the related equilibrium isotope effects. Interestingly, and in contrast to the neglible KIEs, the EIEs for dissociation of methane from the σ–complex [$K_{d(H)}/K_{d(D)}$] are large and inverse, as are those for complete reductive elimination [K_H/K_D] (Table 1).

Table 1. Primary (p) and secondary (s) isotope effects (IEs) pertaining to reductive elimination of methane from [$H_2Si(C_5H_4)_2$]W(Me)H at 100°C.[a]

		SYM	MMI	EXC	ZPE	IE
$k_{rc(H)}/k_{rc(D)}$	p	1	1.000	1.047	0.999	1.047
	s	1	0.996	0.984	1.040	1.019
	p & s	1	0.996	1.035	1.003	1.035
$k_{oc(H)}/k_{oc(D)}$	p	1	1.006	1.033	1.538	1.599
	s	1	0.997	1.052	1.039	1.090
	p & s	1	1.003	1.077	1.597	1.725
$K_{\sigma(H)}/K_{\sigma(D)}$	p	1	0.994	1.014	0.650	0.654
	s	1	0.999	0.935	1.001	0.936
	p & s	1	0.993	0.961	0.628	0.600
$k_{d(H)}/k_{d(D)}$	p	1	1.003	0.898	0.975	0.879
	s	1	0.980	0.924	1.229	1.114
	p & s	1	0.983	0.852	1.148	0.962
$k_{re(H)}/k_{re(D)}$	p	1	0.997	0.911	0.633	0.575
	s	1	0.979	0.864	1.231	1.042
	p & s	1	0.976	0.819	0.721	0.577
$K_{d(H)}/K_{d(D)}$	p	0.25	0.688	1.139	0.944	0.185
	s	0.25	0.364	1.802	1.433	0.235
	p & s	1	0.276	2.003	1.246	0.689
K_H/K_D	p	0.25	0.684	1.155	0.613	0.121
	s	0.25	0.364	1.685	1.435	0.220
	p & s	1	0.274	1.925	0.783	0.413

(a) Primary effects (p) correspond to reductive elimination of CH_3–H vs. CH_3–D; secondary effects (s) correspond to reductive elimination of CH_3–H vs. CD_3–H; primary and secondary effects (p&s) correspond to reductive elimination of CH_3–H vs. CD_3–D.

The inverse EIEs are a result of the SYM and MMI terms. Thus, the EIE for dissociation of methane from [$H_2Si(C_5H_4)_2$]W(σ–HCH_3) and [$H_2Si(C_5H_4)_2$]W(σ–DCD_3) is inverse due to the small value of the MMI term (0.28), a consequence of the fact that isotopic substitution has a substantial effect on the mass and moments of inertia of a molecule as small as methane. In addition to the MMI term, the SYM term also has a role in determining the EIE for dissociation of methane from $[H_2Si(C_5H_4)_2]W(\sigma-HCH_3)$ and $[H_2Si(C_5H_4)_2]W(\sigma-DCH_3)$ because the rotational symmetries of CH_3D and CH_4 are different.

An alternative perspective of the data presented in Table 1 is that the EIEs for both coordination (1.45) and oxidative addition (2.42) of methane to $\{[H_2Si(C_5H_4)_2]W\}$ are normal, such that the reactions of CH_4 are thermodynamically more favored than those of CD_4. To our knowledge, there are no experimental reports of the EIEs for coordination and oxidative addition of methane to a metal center (*34*), although there are several conflicting reports of EIE's for coordination of other alkanes. Specifically, Geftakis and Ball reported a normal EIE (1.33 at −93°C) for coordination of cyclopentane to [CpRe(CO)$_2$] (*35*), whereas on the basis of kinetics measurements, Bergman and Moore reported substantially inverse EIEs for the coordination of cyclohexane (≈ 0.1 at −100°C) and neopentane (≈ 0.07 at −108°C) to [Cp*Rh(CO)] (*36*).

In view of these differing results, Bullock and Bender have commented that the issue of whether coordination of an alkane would be characterized by a normal or inverse EIE is not trivial (*6b*). Furthermore, Bender has calculated that the EIE for coordination of CH_4 and CD_4 to $OsCl_2(PH_3)_2$ giving *trans*-(η^2-CH_4)$OsCl_2(PH_3)_2$ is almost unity at 27°C, but becomes inverse upon lowering the temperature, with a value of 0.66 at −108°C; Bender has also noted that this change is not intuitively obvious (*37*). Indeed, the inherent difficulty in predicting the EIE for coordination of an alkane is a consequence of the fact that isotopic substitution exerts different effects on the MMI, EXC, and ZPE terms, each of which have different temperature dependencies.

While Bender's calculations indicate that coordination of methane could be characterized by an inverse EIE, and thereby provide support for Bergman's studies on [Cp*Rh(CO)] (*36*), they do not address the issue of how coordination of cyclopentane to [CpRe(CO)$_2$] could be characterized by a normal EIE. Specifically, because the EIE is unity at infinite temperature, it is not clear how a normal EIE could ever arise if EIE's only become more inverse upon reducing the temperature. Since we calculated a normal EIE for coordination of methane to $\{[H_2Si(C_5H_4)_2]W\}$, we were intrigued to study its temperature dependence and determine whether it would become inverse upon lowering the temperature.

Most interestingly, rather than becoming inverse, the EIE at −100°C (1.57) actually *increased* slightly from the value at 100°C (1.45)! However, the relative insensitivity of the EIE with respect to temperature over this 200°C range is deceptive, as illustrated by the full temperature dependence illustrated in Figure 5. Thus, while the MMI term is temperature independent, the EXC and ZPE terms are strongly temperature dependent (*38*), but with dependencies that oppose each other so that the product of EXC and ZPE does not vary markedly in the range −100°C to 100°C. Indeed, the EIE does not become inverse in this system until very low temperatures (*ca.* −200°C) (*39*).

At *all* temperatures, the values of the ZPE and EXC terms are ≤ 1 and thereby favor an inverse EIE. However, these terms are mitigated by the large temperature independent MMI term (3.62) that dominates at temperatures ≥ *ca.*

−200°C. At low temperatures, the inverse EIE is a result of domination by the strongly inverse ZPE term. While the ZPE term is typically normal when the bond being broken is stronger than the one being formed (*11*), the effect may become inverse when the reaction results in the creation of a species with a greater number of isotope sensitive vibrations (*40*), as illustrated by the inverse EIEs associated with the coordination and oxidative addition of H_2 to a metal center (*41*). Since the ZPE term becomes zero at 0 K, the EIE likewise becomes zero.

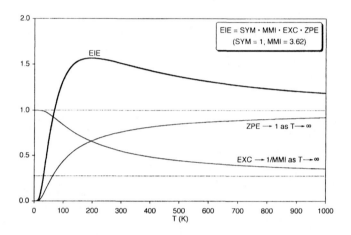

Figure 5. Calculated EIE as a function of temperature for coordination of methane to {[$H_2Si(C_5H_4)_2$]W} as determined by the individual MMI, EXC, and ZPE terms.

At high temperatures, it is the EXC rather than ZPE term that becomes dominant in reducing the value of the EIE for coordination of methane. However, since EXC does not become zero at infinite temperature, but rather becomes 1/MMI, the EXC term is incapable of causing the EIE to become inverse because the product MMI•EXC approaches unity (*42*).

It is also instructive to analyze the temperature dependence of the EIE in terms of the combined SYM•MMI•EXC term (which influences the entropy factor) and the ZPE term (which influences the enthalpy factor). Since EXC is unity at 0 K and is 1/MMI in the limit of infinite temperature, the product MMI•EXC varies from MMI to unity over this temperature range, as illustrated in Figure 6. Thus, at all temperatures the SYM•MMI•EXC entropy factor favors a normal EIE, while the ZPE entropy factor favors an inverse EIE. At high temperatures SYM•MMI•EXC dominates and the EIE is normal, while at low temperatures the ZPE term dominates and the EIE is inverse. If the form of the temperature dependence illustrated in Figures 5 and 6 applies to other systems, it is evident that it provides a means to rationalize *both* normal and inverse EIEs for alkane coordination. However, the precise form of the temperature dependence will depend critically on the structure of the σ–complex and its corresponding vibrational frequencies.

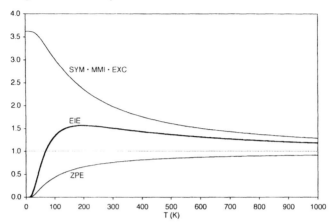

Figure 6. Calculated EIE for coordination of methane to {[H₂Si(C₅H₄)₂]W} analyzed as a product of the combined [SYM•MMI•EXC] and ZPE terms.

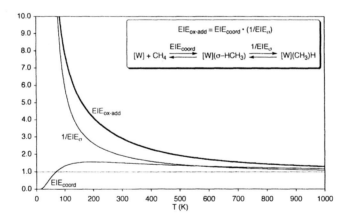

Figure 7. Markedly different temperature dependencies of the EIEs for coordination and oxidative addition of methane to {[H₂Si(C₅H₄)₂]W}.

Most interestingly, whereas the EIE for coordination of methane to {[H₂Si(C₅H₄)₂]W} approaches zero at low temperature, the EIE for oxidative addition is normal at all temperatures and actually approaches infinity at 0K (Figure 7) (*43*)! The dramatically different temperature dependencies of the

EIEs for methane coordination and oxidative addition is associated with the ZPE terms: the ZPE term for coordination of methane is inverse at all temperatures (and zero at 0K), while the ZPE term for oxidative addition is normal at all temperatures and approaches infinity at 0K. The ZPE term for coordination of methane is inverse because coordination results in the creation of additional isotope sensitive vibrations in the σ–complex $[H_2Si(C_5H_4)_2]W(\sigma-HMe)$ that, in combination, are sufficiently strong to counter those associated with C–H bond in methane. In contrast, the isotopically sensitive vibrations associated with the W–H bond of the methyl hydride complex $[H_2Si(C_5H_4)_2]W(Me)H$, namely a W–H stretch and two bends, are of sufficiently low energy that they do not counter those associated with the C–H bond that has that has been broken. As a result, the ZPE term for oxidative addition of the C–H bond is *normal*. An alternative way of analyzing the situation is to recognize that the EIE for oxidative addition is a product of the EIEs for (i) methane coordination and (ii) oxidative cleavage of $[H_2Si(C_5H_4)_2]W(\sigma-HMe)$ to $[H_2Si(C_5H_4)_2]W(Me)H$, thereby demonstrating that it is the latter EIE_{oc} (*i.e.* $EIE_{oc} = 1/EIE_{\sigma} = 1/[K_{\sigma(H)}/K_{\sigma(D)}]$ in Table 1) which is dominant in determining the EIE for the overall oxidative addition. As indicated above, the normal EIE for the oxidative cleavage of $[H_2Si(C_5H_4)_2]W(\sigma-HMe)$ and $[H_2Si(C_5H_4)_2]W(Me)H$ is dictated by the σ–complex exhibiting the greater ZPE stabilization because the hydrogen is attached to carbon, rather than only to tungsten in $[H_2Si(C_5H_4)_2]W(Me)H$.

Summary

In summary, the reductive elimination of methane from $[Me_2Si(C_5Me_4)_2]W(CH_3)H$ and $[Me_2Si(C_5Me_4)_2]W(CD_3)D$ is characterized by an inverse KIE and calculations on $[H_2Si(C_5H_4)_2]W(Me)H$ provide the first theoretical evidence that the origin of the inverse KIE is a manifestation of the existence of a σ–complex intermediate; thus, the inverse KIE is a consequence of an inverse equilibrium isotope effect for interconversion of $[H_2Si(C_5H_4)_2]W(Me)H$ and $[H_2Si(C_5H_4)_2]W(\sigma-HMe)$, with KIEs for both reductive coupling and oxidative cleavage being normal. Interestingly, the temperature dependencies of EIEs for coordination and oxidative addition of methane to the tungstenocene fragment $\{[H_2Si(C_5H_4)_2]W\}$ are calculated to be very different, with the EIE for coordination approaching zero at 0K, while the EIE for oxidative addition approaches infinity.

Acknowledgment

We thank the U. S. Department of Energy, Office of Basic Energy Sciences (DE-FG02-93ER14339) for support of this research.

References

1. For recent reviews, see: (a) Crabtree, R. H. *J. Chem. Soc., Dalton Trans.* **2001**, 2437-2450. (b) *Activation and Functionalization of Alkanes*; Hill, C. L., Ed.; John Wiley and Sons: New York, 1989. (c) Arndtsen, B. A.; Bergman, R. G.; Mobley, T. A.; Peterson, T. H. *Acc. Chem. Res.* **1995**, *28*, 154-162. (d) Ryabov, A. D. *Chem. Rev.* **1990**, *90*, 403-424. (e) Jones, W. D.; Feher, F. J. *Acc. Chem. Res.* **1989**, *22*, 91-100. (f) Special issue *J. Organomet. Chem.* **1996**, *504* (1-2). (g) Shilov, A. E.; Shul'pin, G. B. *Chem. Rev.* **1997**, *97*, 2879-2932. (h) Labinger, J. A.; Bercaw, J. E. *Nature* **2002**, *417*, 507-514.
2. Buchanan, J. M.; Stryker, J. M.; Bergman, R. G. *J. Am. Chem. Soc.* **1986**, *108*, 1537-1550 and reference 28 therein.
3. (a) Hall, C.; Perutz, R. N. *Chem. Rev.* **1996**, *96*, 3125-3146. (b) Crabtree, R. H. *Chem. Rev.* **1995**, *95*, 987-1007. (c) Crabtree, R. H. *Angew. Chem. Int. Edit. Engl.* **1993**, *32*, 789-805. (d) *Metal Dihydrogen and σ-Bond Complexes: Structure, Theory, and Reactivity*, Kubas, G. J.; Kluwer Academic/Plenum Publishers, New York, 2001.
4. Although hydrocarbon σ–complexes were originally formulated for an interaction of a transition metal with a single C–H bond, it has been noted that other coordination modes are possible, involving the simultaneous interaction with two or three C–H bonds. In this paper we use the term [M](σ–HR) to refer generally to σ–complexes without specifying the exact coordination mode, since this is often unknown.
5. For a recent example of a σ–complex that has been characterized by NMR spectroscopy, see: Geftakis, S.; Ball, G. E. *J. Am. Chem. Soc.* **1998**, *120*, 9953-9954.
6. For recent reviews, see: (a) Jones, W. D. *Acc. Chem. Res.* **2003**, *36*, 140-146. (b) Bullock, R. M.; Bender, B. R. *"Isotope Methods in Homogeneous Catalysis"* in *Encyclopedia of Catalysis*, I. T. Horváth Ed. (2002).
7. (a) Green, M. L. H. *Pure Appl. Chem.* **1984**, *56*, 47-58. (b) Bullock, R. M.; Headford, C. E. L.; Hennessy, K. M.; Kegley, S. E.; Norton, J. R. *J. Am. Chem. Soc.* **1989**, *111*, 3897-3908.
8. Parkin, G.; Bercaw, J. E. *Organometallics* **1989**, *8*, 1172-1179.
9. Churchill, D. G., Janak, K. E.; Wittenberg, J. S.; Parkin, G. *J. Am. Chem. Soc.* **2003**, *125*, 1403-1420.
10. Green has reported an even more dramatic effect, with [Me$_2$C(C$_5$H$_4$)$_2$]W(Me)H being stable to elimination of methane under the conditions studied. See: (a) Labella, L.; Chernega, A.; Green, M. L. H. *J. Chem. Soc., Dalton Trans.* **1995**, 395-402. (b) Chernega, A.; Cook, J.; Green, M. L. H.; Labella, L.; Simpson, S. J.; Souter, J.; Stephens, A. H. H. *J. Chem. Soc,. Dalton Trans.* **1997**, 3225-3243.
11. Wolfsberg, M. *Acc. Chem. Res.* **1972**, *5*, 225-233.

12. For the first reports of inverse KIE's for elimination of RH (R = H,[a] Ph,[b] Cy[c,d]), see: (a) Howarth, O. W.; McAteer, C. H.; Moore, P.; Morris, G. E. *J. Chem. Soc., Dalton Trans.* **1984**, 1171-1180. (b) Jones, W. D.; Feher, F. J. *J. Am. Chem. Soc.* **1985**, *107*, 620-630. (c) Stryker, J. M.; Bergman, R. G. *J. Am. Chem. Soc.* **1986**, *108*, 1537-1550. (d) Periana, R. A.; Bergman, R. G. *J. Am. Chem. Soc.* **1986**, *108*, 7332-7346.
13. Gould, G. L.; Heinekey, D. M. *J. Am. Chem. Soc.* **1989**, *111*, 5502-5504.
14. Wang, C. M.; Ziller, J. W.; Flood, T. C. *J. Am. Chem. Soc.* **1995**, *117*, 1647-1648.
15. Stahl, S. S.; Labinger, J. A.; Bercaw, J. E. *J. Am. Chem. Soc.* **1996**, *118*, 5961-5976.
16. Jensen, M. P.; Wick, D. D.; Reinartz, S.; White, P. S.; Templeton, J. L.; Goldberg, K. I. *J. Am. Chem. Soc.* **2003**, *125*, 8614-8624.
17. (a) Abis, L.; Sen, A.; Halpern, J. *J. Am. Chem. Soc.* **1978**, *100*, 2915-2916. (b) Michelin, R. A.; Faglia, S.; Uguagliati, P. *Inorg. Chem.* **1983**, *22*, 1831-1834. (c) Hackett, M.; Ibers, J. A.; Whitesides, G. M. *J. Am. Chem. Soc.* **1988**, *110*, 1436-1448.
18. See, for example, reference 10.
19. Northcutt, T. O.; Wick, D. D.; Vetter, A. J.; Jones, W. D. *J. Am. Chem. Soc.* **2001**, *123*, 7257-7270.
20. (a) Lo, H. C.; Haskel, A.; Kapont, M.; Keinan, E. *J. Am. Chem. Soc.* **2002**, *124*, 3226-3228. (b) Iron, M. A.; Lo, H. C.; Martin, J. M. L.; Keinan, E. *J. Am. Chem. Soc.* **2002**, *124*, 7041-7054.
21. Lo, H. C.; Haskel, A.; Kapont, M.; Keinan, E. *J. Am. Chem. Soc.* **2002**, *124*, 12626.
22. Janak, K. E.; Churchill, D. G.; Parkin, G. *Chem. Commun.* **2003**, 22-23.
23. For other calculations pertaining to $(Cp^R)_2M(Me)H$ (M = Mo,W) derivatives, see: (a) Green, J. C.; Jardine, C. N. *J. Chem. Soc., Dalton Trans.* **1998**, 1057-1061. (b) Green, J. C.; Harvey, J. N.; Poli, R. *J. Chem. Soc., Dalton Trans.* **2002**, 1861-1866. (c) Su, M.-D.; Chu, S.-Y. *J. Phys. Chem. A* **2001**, *105*, 3591-3597.
24. This approach is only capable of approximating the crossing point since the DFT method does not allow for interaction between the singlet and triplet surfaces.
25. (a) Wolfsberg, M.; Stern, M. J. *Pure Appl. Chem.* **1964**, *8*, 225-242. (b) Melander, L.; Saunders, W. H., Jr. *Reaction Rates of Isotopic Molecules*, Wiley–Interscience, New York (1980). (c) Carpenter, B. K. *Determination of Organic Reaction Mechanisms*, Wiley–Interscience, New York (1984).
26. Although a degree of confusion exists in the literature with respect to the application of symmetry numbers in transition state theory, it has been emphasized that symmetry numbers and not statistical factors, should be used. See: Pollak, E.; Pechukas, P. *J. Am. Chem. Soc.* **1978**, *100*, 2984-2991.

27. The symmetry factor includes both external (σ) and internal (n) symmetry numbers. See: Bailey, W. F.; Monahan, A. S. *J. Chem. Educ.* **1978**, *55*, 489-493.
28. $I = I_A I_B I_C$, the product of the moments of inertia about the three axes; $u_i = h\nu_i/k_B T$. Note that the EXC and ZPE terms do not include the imaginary frequencies associated with the reaction coordinate of the transition state, whereas they are included in VP.
29. For some important studies concerned with the computation of isotope effects in organometallic systems, see: (a) Slaughter, L. M.; Wolczanski, P. T.; Klinckman, T. R.; Cundari, T. R. *J. Am. Chem. Soc.* **2000**, *122*, 7953-7975. (b) Bender, B. R. *J. Am. Chem. Soc.* **1995**, *117*, 11239-11246. (c) Abu-Hasanayn, F.; Krogh-Jespersen, K.; Goldman, A. S. *J. Am. Chem. Soc.* **1993**, *115*, 8019-8023. (d) Bender, B. R.; Kubas, G. J.; Jones, L. H.; Swanson, B. I.; Eckert, J.; Capps, K. B.; Hoff, C. D. *J. Am. Chem. Soc.* **1997**, *119*, 9179-9190.
30. The existence of random errors in derived force constants and frequencies cause small discrepancies in the results derived by the two methods. For such situations, the expression employing the vibrational product has been suggested to be the more reliable of the two methods. See: Schaad, L. J.; Bytautas, L.; Houk, K. N. *Can. J. Chem.* **1999**, *77*, 875-878.
31. Internal symmetry numbers (n) are not included in the DFT entropy determination and so $KIE_{DFT} = [n_H/n_D]/[n^{\ddagger}_H/n^{\ddagger}_D] \cdot (k_H/k_D)_{DFT}$, where $(k_H/k_D)_{DFT}$ is calculated from $\Delta G^{\ddagger}_{DFT(H)}$ and $\Delta G^{\ddagger}_{DFT(D)}$ using the Eyring equation.
32. As such, the derived transition state is not a well defined stationary point on the enthalpy surface; nevertheless, we have also calculated frequencies at other points on the singlet dissociation surface and find similar values.
33. These calculations assume the preequilibrium approximation ($k_{oc} \gg k_d$) for reductive elimination (k_{re}), thereby corresponding to the most extreme inverse value for the system.
34. 1,2–addition of CH_4 and CD_4 across the Ti=N bond of $(Bu^t_3SiO)_2Ti=NSiBu^t_3$ is, nevertheless, characterized by a normal EIE (2.00 at 26.5 °C). See reference 34.
35. Geftakis, S.; Ball, G. E. *J. Am. Chem. Soc.* **1998**, *120*, 9953-9954.
36. (a) Schultz, R. H.; Bengali, A. A.; Tauber, M. J.; Weiller, B. H.; Wasserman, E. P.; Kyle, K. R.; Moore, C. B.; Bergman, R. G. *J. Am. Chem. Soc.* **1994**, *116*, 7369-7377. (b) Bengali, A. A.; Arndtsen, B. A.; Burger, P. M.; Schultz, R. H.; Weiller, B. H.; Kyle, K. R.; Moore, C. B.; Bergman, R. G. *Pure Appl. Chem.* **1995**, *67*, 281-288. (c) Bengali, A. A.; Schultz, R. H.; Moore, C. B.; Bergman, R. G. *J. Am. Chem. Soc.* **1994**, *116*, 9585-9589.
37. Bender, B. R. unpublished results cited in reference 6b.

38. The ZPE term increases from zero to a limiting value of unity as the temperature is increased, while EXC decreases from unity to a limiting value of 1/MMI (*i.e.* 1/VP).
39. Janak, K. E.; Parkin, G. *J. Am. Chem. Soc.* **2003**, *125*, 6889-6891.
40. The additional vibrations are derived from rotational and translational degrees of freedom of the reactants.
41. (a) Hascall, T.; Rabinovich, D.; Murphy, V. J.; Beachy, M. D.; Friesner, R. A.; Parkin, G. *J. Am. Chem. Soc.* **1999**, *121*, 11402-11417. (b) Rabinovich, D.; Parkin, G. *J. Am. Chem. Soc.* **1993**, *115*, 353-354. (c) Abu-Hasanayn, F.; Krogh-Jespersen, K.; Goldman, A. S. *J. Am. Chem. Soc.* **1993**, *115*, 8019-8023. (d) Bender, B. R.; Kubas, G. J.; Jones, L. H.; Swanson, B. I.; Eckert, J.; Capps, K. B.; Hoff, C. D. *J. Am. Chem. Soc.* **1997**, *119*, 9179-9190.
42. For any given set of vibrational frequencies, the product VP•EXC approaches unity at infinite temperature. Since VP is mathematically equivalent to MMI, the product MMI•EXC likewise approaches unity (as also required for the EIE to approach SYM). However, in view of the aforementioned errors in computed frequencies (reference 30), the calculated VP term may not be *exactly* equal to the MMI term. Thus while VP•EXC approaches unity, MMI•EXC actually approaches the ratio MMI/VP. For coordination and oxidative addition of methane to $\{[H_2Si(C_5H_4)_2]W\}$, the MMI/VP ratios are 1.05 and 1.01, respectively. These discrepancies have little effect on the derived EIE.
43. In this regard, it is worth noting that oxidative addition of CH_4 to $Ir(PH_3)_2(CO)H$ is also calculated to have a normal EIE of 3.64 at 300 K. See reference 41c.

Chapter 6

Alkane C–H Bond Activation by O-Donor Ir Complexes

Gaurav Bhalla, Xiang Yang Liu, Antek Wong-Foy, C. J. Jones, and Roy A. Periana*

Department of Chemistry, Loker Hydrocarbon Institute, University of Southern California, 837 West 37th Street, LHI 122, Los Angeles, CA 90089–1661

The first examples of well-defined, O-donor ligated, Ir complexes that are competent for alkane C-H activation to generate Ir-alkyl complexes are reported. The O-donor complexes exhibit thermal and protic stability and are efficient catalysts for H/D exchange reactions with alkanes. Somewhat surprisingly, the O-donor Ir alkyl complexes with β-CH bonds are stable to generation of coordinated or uncoordinated olefinic products. Mechanistic studies suggest that while these O-donor Ir-alkyl complexes undergo β-hydride elimination reactions these reactions are reversible and unproductive.

Introduction

Catalysts based on the C-H activation reaction show potential for the development of new, selective, hydrocarbon oxidation chemistry (*1*). A central consideration in the design of such catalysts is the choice of ligands. The ligands generally acceptable for C-H activation reactions range from C-donor, (*e.g.* cyclopentadienyl ligands), to mono and multi-dentate P- or N-donor ligands, to chelating NC or PC type ligands (*2*). While O-donor ligands have been studied with early and late transition metals (*3*), to our knowledge well-defined, O-ligated, late transition metal complexes that activate alkane C-H bonds have not been reported. We have been particularly interested in O-ligated, late transition metals as such complexes could exhibit protic and oxidant stability given the lower basicity and higher electronegativity of O compared to

N, C or P. Another key reason for study is that the electronegativity and "hardness" of O-donor ligands could allow access to higher oxidation states during catalysis that could facilitate the oxidative functionalization reactions of M-R intermediates to functionalized RX products in a catalytic cycle. Given these considerations, it was important to establish whether well-defined, O-ligated late transition metal complexes could activate alkane C-H bonds. Herein, we report a well-defined, O-ligated, late transition metal Ir complex that can activate alkane C-H bonds.

Alkane C-H Activation with O-Donor Ir Complex

Recently, we demonstrated that the O-ligated complex, Ph-Ir(III)(acac-O,O)$_2$(Py), (acac-O,O = η^2-O,O-acetylacetonate, Py = pyridine), **1-Ph**, catalyzes the hydroarylation of olefins with arenes to generate alkyl benzenes (4). Herein we report that the Me-Ir(III) derivative, Me-Ir(III)(acac-O,O)$_2$(Py), **1-Me**, reacts with alkanes (RH) via C-H activation to generate the corresponding alkyl-Ir complexes, **Ir-R**, (RH = cyclohexane and n-octane). **1-Me** was synthesized from Ir(acac-C^3)(acac-O,O)$_2$(H$_2$O), **1**, by treatment with (CH$_3$)$_2$Hg or (CH$_3$)$_2$Zn followed by addition of pyridine, in good yields (70%) as shown in Eq. 1. Complex **1-Me** is air stable and was fully characterized by ^{1}H and ^{13}C-NMR spectroscopy and elemental analysis.

Heating **1-Me** in neat cyclohexane at 130°C for 3 hrs yielded the corresponding Ir-cyclohexyl complex, **1-C$_6$H$_{11}$**, as shown in Eq. 2. ^{1}H-NMR analysis of the crude reaction mixture showed that the reaction was essentially quantitative. Complex **1-C$_6$H$_{11}$**, could be isolated from the reaction mixture and has been fully characterized by ^{1}H and ^{13}C-NMR spectroscopy and elemental and X-Ray structural analyses. An ORTEP drawing of **1-R** (R = C$_6$H$_{11}$) is shown in Figure 1.

[Chemical scheme showing reaction 1-Me + RH → 1-R + CH₄ at 130°C, with RH = cyclohexane, mesitylene, n-hexane (with CH at 2-position), benzene, acetone] (2)

Consistent with the stoichiometry shown in Eq 2, when the reaction is carried out in a sealed NMR tube with cyclohexane-d_{12}, mono-deuterated methane is observed based on gas chromatography-mass spectroscopy (GC-MS) analysis. These observations unambiguously show that complexes based on the O-ligated, (acac-O,O)$_2$Ir(III) motif can activate alkane C-H bonds. To our knowledge, this is the first well-defined, late-metal, O-donor ligated complex that shows this reactivity for alkane C-H activation.

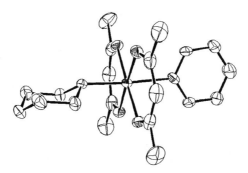

Figure 1. ORTEP drawing of **1-R**, R = Cyclohexyl. Thermal ellipsoids are at the 50% probability level. Hydrogen atoms omitted for clarity. Selected bond lengths (A°): Ir1-C4, 2.060(7); Ir1-N1, 2.225(6).

Other hydrocarbon substrates that react by C-H activation with **1-Me** are shown in Eq 2. Thus, heating a solution of **1-Me** in mesitylene at 130 °C for 3 hr results in the formation of a single new species. ^{1}H and ^{13}C NMR spectroscopy analyses of the crude mixture in CDCl$_3$ show clean formation of the Ir-mesityl species, **1-R**, (R = mesityl, Eq 2) in which only the benzylic C-H bond was activated. The reaction with benzene and acetone cleanly provided the corresponding Ir-phenyl and Ir-acetonyl derivatives. These materials have also been isolated and characterized by ^{1}H and ^{13}C NMR spectroscopy and elemental analysis. The reaction with n-alkanes, exemplified by n-octane, could not be fully characterized and ^{1}H and ^{13}C-NMR spectroscopy show that several Ir-octyl products (presumably resulting from 1° and 2° C-H bond activation) are produced that could not be separated and quantified.

The alkane C-H activation reactions (Eq 2) in the corresponding alkane solvent are retarded by added free pyridine. As shown in Figure 2, a plausible mechanism for the C-H activation from **1-R** can involve initial loss of pyridine, *trans to cis* isomerization to generate a 5 coordinate, *cis*-intermediate (***cis*-2**) that cleaves alkane C-H bonds via a 7-coordinate oxidative addition intermediate or transition state (**3**) or sigma-bond metathesis transition state (not shown) (*5*). We are currently carrying out kinetic and theoretical studies of this system to further elucidate the details of these CH activation reactions.

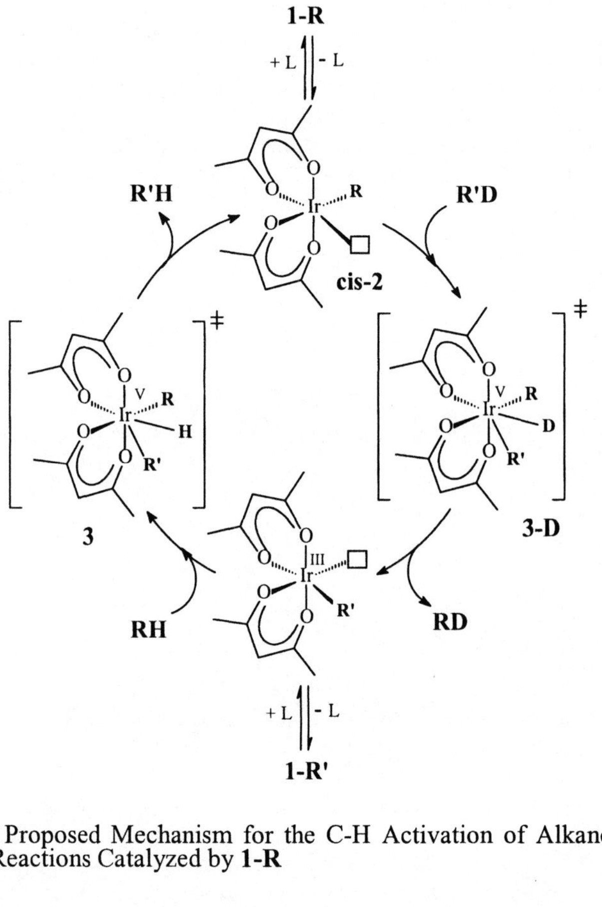

Figure 2. Proposed Mechanism for the C-H Activation of Alkanes and H/D Exchange Reactions Catalyzed by **1-R**

H-D Exchange of Alkanes Catalyzed with O-Donor Ir Complexes

Having established that O-ligated, late metal complexes can stoichiometrically activate the C-H bonds of alkanes, we have begun to examine the catalytic activity of this class of complexes with hydrocarbons.

$$RH + DY \underset{}{\overset{Cat}{\rightleftharpoons}} RD + HY \quad (3)$$

Analyses by GC/MS and NMR spectroscopy show that **1-Me** efficiently catalyzes H/D exchange between C_6D_6 and hydrocarbons, including alkanes, according to Eq 3, RH = hydrocarbon, Y = C_6D_5 (Table 1, entries 1 – 5). These reactions presumably proceed via the catalytic sequence shown in Figure 2. The reactions are clean and no catalyst decomposition is observed, showing that these systems are thermally stable and activate alkane C-H bonds reversibly. ^{1}H NMR analysis of the crude reaction mixtures after heating shows that the resting state of the catalyst in the reaction with C_6D_6 is **1-Ph-d$_5$**. Control experiments with added drops of Hg metal (to test for catalysis by reduced metals) show no change in rate. Consistent with the presumption of stoichiometric C-H activation reactions with n-octane, ^{13}C NMR analysis of the C_6D_6/n-octane reaction mixture after catalysis shows deuterium incorporation into all the positions of n-octane with higher selectivity for the 1° positions.

Consistent with the expected protic stability of O-donor ligands, preliminary results show that **1-Me** is thermally stable (to loss of the O-ligated acac ligands) in protic media such as D_2O, CH_3CO_2D, and CF_3CO_2D and remains active for C-H activation and catalysis in these media. Thus, reaction of 0.1 ml of mesitylene with 1 ml of CF_3CO_2D containing 10 mM of **1-Me** shows H/D exchange (according to Eq 3, Y = CF_3CO_2, RH = mesitylene) of only the benzylic C-H bonds with a TOF of ~10^{-3} s^{-1} at 160°C. These H/D exchange reactions in protic media are being examined in greater detail.

Table 1. H/D exchange With C_6D_6 Catalyzed by **1-Me**[a]

Entry	Substrate	TON	TOF sec^{-1}
1	Cyclohexane	240	0.010
2	Methane	123	0.0017
3	n-Octane	43	0.0029
4	Benzene	1210	0.675
5	Acetone	72	0.043

[a] All reactions were carried out at 180°C using **1-Me** as the catalyst (2 – 20mM).

Chemistry of O-Donor Ir Alkyl Complexes with β-CH Bonds

The quantitative formation of the Ir cyclohexyl complex, **1-C$_6$H$_{11}$**, on heating **1-Me** in cyclohexane and clean H-D exchange between cyclohexane and benzene catalyzed by **1-Me** suggest that the Ir-alkyl products are stable to formation of uncoordinated or Ir-coordinated olefin products that could result from possible β-hydride elimination reactions. However, while **1-C$_6$H$_{11}$** is a 6-coordinate, 18e$^-$ complex, as shown in Figure 3, it is likely that the Ir-cyclohexyl species are in equilibrium with 5 coordinate, 16e$^-$ species expected to be required for C-H activation and it could be anticipated that uncoordinated or Ir-coordinated olefinic products could be generated by β-hydride elimination reactions from these intermediates.

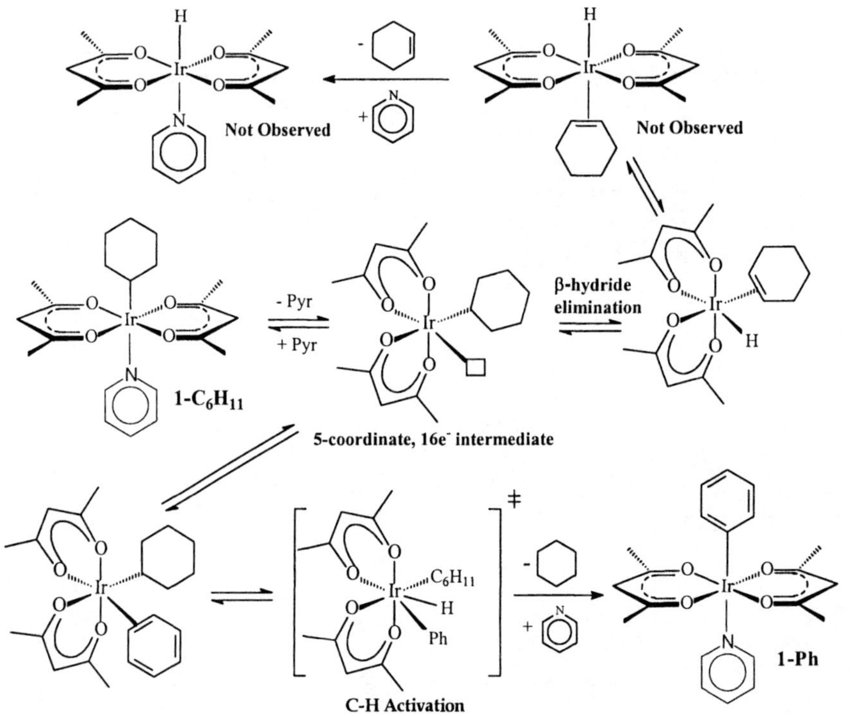

Figure 3. Plausible Mechanism for Observed Arene C-H Activation and Expected β-Hydride Elimination Products from **1-C$_6$H$_{11}$**.

To understand why no such olefinic products are observed, we examined the thermal, alkane elimination and arene CH activation chemistry of the **Ir-C$_2$H$_5$** complex in benzene-d$_6$. This complex was synthesized as shown in Eq 4.

Treatment of this complex with benzene-d_6, as in the case of **1-Me**, readily leads to quantitative formation of **1-Ph-d_5** and loss of mono-deuteroethane (C_2H_5D) by the stoichiometry shown in Eq 5. These reactions are proposed to proceed via the CH activation reaction mechanism shown in Figure 3 with ethyl replacing the cyclohexyl group. Consistent with the observations of the Ir-cyclohexyl system, GC-MS and NMR analyses of the liquid and gas phases of the reaction mixture show that no uncoordinated or Ir-coordinated ethylene products are produced during loss of ethane.

To examine this reaction of the Ir-ethyl complex more carefully, the ^{13}C-labeled complex, **Ir-$^{13}CH_2CH_3$** was prepared from the corresponding ($^{13}CH_2CH_3)_2Zn$ reagent by an analogous reaction to that shown in Eq 4. This complex was prepared to test the possibility that reversible *but unproductive* β-hydride elimination to generate cis or trans $(acac)_2Ir(H)(C_2H_4)$ intermediates could proceed from the five coordinate, 16e⁻ intermediates involved in CH activation (*i.e.* β-hydride elimination from **1-R** could occur reversibly without leading to the productive formation of olefinic products). As shown in Figure 4, if such reversible but unproductive β-hydride reactions did occur with **Ir-$^{13}CH_2CH_3$**, this could be expected to lead to isomerization of the ^{13}C-label from the α to the β-positions with formation of the **Ir-$CH_2^{13}CH_3$** regio-isotopomer. Carrying out this reaction in C_6D_6 as the reaction solvent could be expected to lead to two regio-isotopomers of ethane, $^{13}CH_3CH_2D$ and $^{13}CH_2DCH_3$ (formed by C-D activation of the C_6D_6 solvent and loss of ethane) if reversible β-hydride elimination did occur or only $^{13}CH_2DCH_3$ if no ^{13}C-migration occurred. Importantly, measurements of the rate of ^{13}C migration in the ^{13}C-labelled **Ir-**

C_2H_5 and/or the relative ratio of the two regio-isotopomers of ethane could both be expected to provide information on the relative rates of reversible β-hydride elimination versus benzene CH activation. Thus, for example, if the reversible β-hydride reaction is fast compared to benzene CH activation, then it could be anticipated that approximately equal amounts of **Ir-$CH_2^{13}CH_3$** and **Ir-$^{13}CH_2CH_3$** as well as $^{13}CH_3CH_2D$ and $^{13}CH_2DCH_3$ would be observed upon heating to facilitate the arene CH activation reaction to generate ethane and **1-Ph-d_5**.

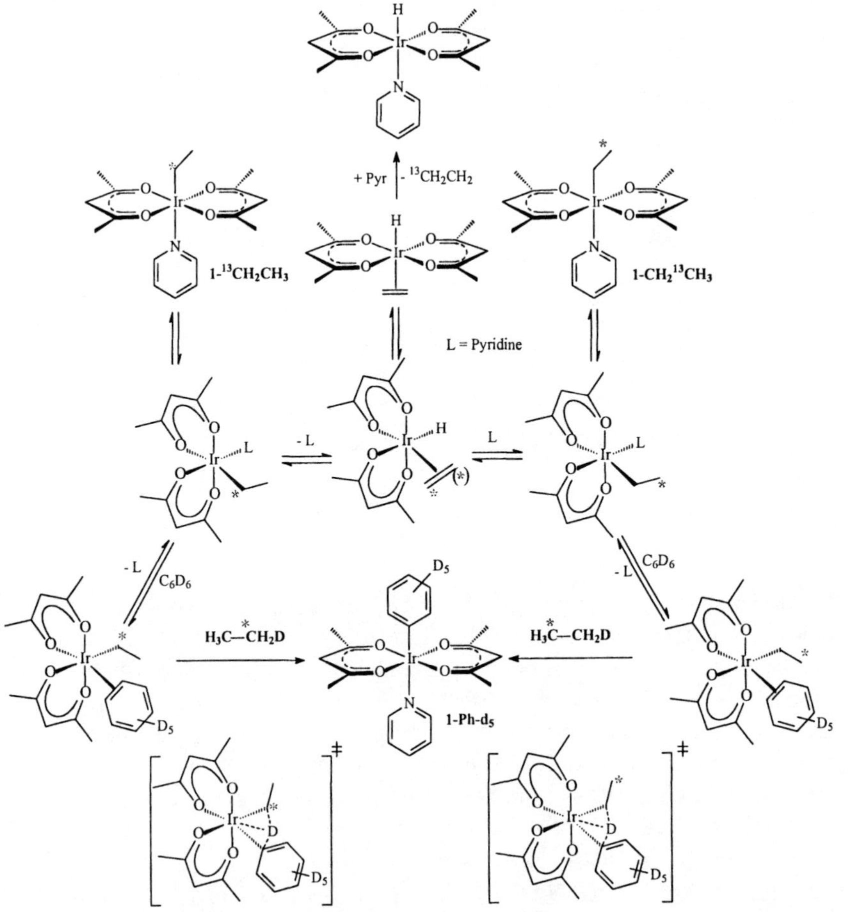

Figure 4. Possible Products Expected from Heating **1-$^{13}CH_2CH_3$** in C_6D_6 to Generate Ethane by C-H Activation.

The reaction of **1-$^{13}CH_2CH_3$** with C_6D_6 at 150°C was monitored periodically by ^1H and ^{13}C NMR spectroscopy of the liquid phase and GC-MS of the gas and liquid phases. As can be seen in Figure 5, ^{1}H decoupled ^{13}C NMR spectra (with sufficiently long relaxation delay to afford accurate integration of the ^{13}C resonances) of the reaction mixture as the reaction proceeds show that ^{13}C migration from the α-position in **1-^{13}CH$_2$CH$_3$** to the β-position in **1-CH$_2$^{13}CH$_3$** does occur. It is also clear from Figure 5, that a steady concentration of the β-isomer, **1-CH$_2$^{13}CH$_3$**, is attained that is substantially lower than the amount of **1-^{13}CH$_2$CH$_3$** present. This indicates that: A) reversible β-hydride elimination most likely occurs and accounts for the α to β-migration of the ^{13}C-label of **1-^{13}CH$_2$CH$_3$** and B) importantly, the lack of formation of equimolar amounts **1-^{13}CH$_2$CH$_3$** and **1-CH$_2$^{13}CH$_3$** as ethane is lost with concomitant CH activation of the benzene solvent, strongly indicates that the α to β-migration of the ^{13}C-label is substantially slower than the CH activation of benzene and formation of ethane and **1-Ph-d$_5$**.

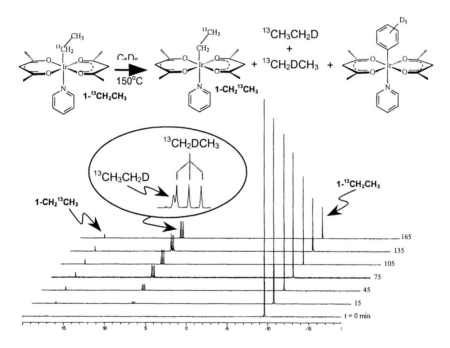

Figure 5. Time Dependent ^{13}C NMR Spectra of Reaction of **1-^{13}CH$_2$CH$_3$** with C$_6$D$_6$ at 150°C

This result is confirmed by analysis of the dissolved ethane that is produced from arene CH activation. As can be seen, the ^{13}C-resonance of ethane is not a simple singlet but is composed of a smaller singlet superimposed on a 1:1:1 triplet due to ^{2}H-^{13}C coupling. Simulation of this pattern readily shows that the predominant ethane product is $^{13}CH_2DCH_3$ with ~ 16 mol % of $^{13}CH_3CH_2D$. Analyses by ^1H NMR, while not as clear as the ^{13}C NMR analyses, confirm these results and show (on the basis of the methyl resonances due to the (acac)$_2$Ir resonances) that **1-Ph-d$_5$** is the only new (acac)$_2$Ir product formed on loss of ethane.

These results strongly indicate that these O-donor Ir-alkyls do undergo reversible β-hydride elimination reactions but that such reactions are unproductive. It is possible that the migration of the ^{13}C-label occurs via a concerted process involving synchronous β-hydrogen and carbon transfers, but initial DFT calculations indicate that such a transition state would be substantially higher in energy than that leading to β-hydride elimination. Interestingly, the results also indicate that the reversible, intramolecular β-hydride elimination reactions are slower than the intermolecular arene CH activation reactions. This is a somewhat unexpected result and it will be important to understand why β-hydride elimination reactions are not highly favorable in these O-donor systems. A likely reason is that with O-donor ligands, the metal is not sufficiently electron-rich to strongly stabilize olefinic intermediates.

Summary

In summary, we demonstrated that well-defined, late metal, O-ligated complexes are competent for alkane C-H activation, exhibit high thermal and protic stability and are efficient catalysts for H/D exchange reactions with alkanes. ^{13}C-labeling studies show that these O-donor Ir-alkyl complexes may likely undergo reversible β-hydride elimination reactions that are unproductive with respect to stable olefinic products. It will be interesting to further explore and understand the differences between these new O-donor metal complexes and the known Cp, P or N-donor systems for the alkane CH activation reaction. Given the unusual stability of these O-donor CH activation systems, we are currently investigating the oxidative functionalization of O-donor M-R complexes and new O-donor complexes that activate C-H bonds.

Acknowledgement

We thank the National Science Foundation (CHE-0328121) and Chevron Texaco Energy Research and Technology Company for financial support for this research.

References and Notes

(1) (a) Jia, C.G.; Kitamura, T.; Fujiwara, Y. *Acc. Chem. Res.* **2001**, *34*, 633 and references therein. (b) Jones, W. D. *Acc. Chem. Res.* **2003**, *36*, 140. (c) Shilov, A. E.; Shul'pin, G. B. *Activation and Catalytic Reactions of Saturated Hydrocarbons in the Presence of Metal Complexes* Kluwer; Dordrecht, **2000**. (d) Crabtree, R. H. *J. Chem. Soc., Dalton Trans.* **2001**, *19*, 2437. (e) Labinger, J. A.; Bercaw, J. E. *Nature* **2002**, *417*, 507.

(2) (a) Fulton, J. R.; Holland, A. W.; Fox, D. J.; Bergman, R. G. *Acc. Chem. Res.* **2002**, *35*, 44. (b) Jones, W.D.; Feher, F.J. *Acc. Chem. Res.* **1989**, *22*, 91. (c) Harper, T.G.P; Shinomoto, R.S; Deming, M.A; Flood, T.C. *J. Am. Chem. Soc.* **1988**, *110*, 7915. (d) Wang, C.M.; Ziller, J.W.; Flood, T.C. *J. Am. Chem. Soc.* 1995, *117*, 1647. (e) Holtcamp, M.W.; Labinger, J.A.; Bercaw, J.E. *J. Am. Chem. Soc.* **1997**, *119*, 848. (f) Periana, R. A.; Taube, D. J.; Gamble, S.; Taube, H.; Satoh, T.; Fuji, H. *Science* **1998**, *280*, 560. (g) Johansson, L.; Ryan, O.B.; Tilset, M. *J. Am. Chem. Soc.* **1999**, *121*, 1974. (h) Fekl, U.; Goldberg, K.I. *Adv. Inorg. Chem.* **2003**, *54*, 259. (i) Liu, F. C.; Pak, E. B.; Singh, B.; Jensen, C. M.; Goldman, A. S. *J. Am. Chem. Soc.* **1999**, *121*, 4086. (j) Nuckel, S.; Burger, P. *Angew. Chem.* **2003**, *42*, 1632.

(3) (a) Cinellu, M. A.; Minghetti, G. *Gold Bull.* **2002**, *35*, 11. (b) Sharp, P. R. *J. Chem. Soc., Dalton Trans.* **2000**, *16*, 2647. (c) Mayer, J. M. *Polyhedron* **1995**, *14*, 3273. (d) Bergman, R. G. *Polyhedron* **1995**, *14*, 3227. (e) Wigley, D. E. *Prog. Inorg. Chem.* **1994**, *42*, 239. (f) Fryzuk, M. D.; Montgomery, C. D. *Coord. Chem. Rev.* **1989**, *95*, 1. (g) Power, P. P. *Comments Inorg. Chem.* **1989**, *8*, 177. (h) West, B. O. *Polyhedron* **1989**, *8*, 219. (i) Bryndza, H. E.; Tam, W. *Chem. Rev.* **1988**, *88*, 1163. (j) Griffith, W. P. *Coord. Chem. Rev.* 1970, *5*, 459. (k) LaPointe, R.E.; Wolczanski, P. T.; Van Duyne, G. D. *Organometallics* **1985**, *4*, 1810. (l) Klaui, W.; Muller, A.; Eberspech, W.; Boese, R.; Goldberg, I. *J. Am. Chem. Soc.* **1987**, *109*, 164. (m) Besecker, C. J.; Day, V. W.; Klemperer, W. G. *Organometallics* **1985**, *4*, 564. (n) Grim, S. O.; Sangokoya, S. A.; Colquhoun, I. J.; McFarlane, W.; Khanna, R. K. *Inorg. Chem.* **1986**, *25*, 2699. (o) Burk, M. J.; Crabtree, R. H. *J. Am. Chem. Soc.* **1987**, *109*, 8025. (p) Tanke, R. S.; Crabtree, R. H. *J. Am. Chem. Soc.* **1990**, *112*, 7984.

(4) (a) Periana, R. A.; Liu, X. Y.; Bhalla, G. *Chem. Commun.* **2002**, *24*, 3000. (b) Matsumoto, T.; Periana, R. A.; Taube, D. J.; Yoshida, H. *J. Mol. Cat. A-Chemical* **2002**, *180*, 1. (c) Matsumoto, T.; Periana, R. A.; Taube, D. J.; Yoshida, H. *J. Catal.* **2002**, *206*, 272. (d) Matsumoto, T.; Taube, D. J.; Periana, R. A.; Taube, H.; Yoshida, H. *J. Am. Chem. Soc.* **2000**, *122*, 7414.

(5) Related Ir(V), seven coordinate intermediates have been proposed and observed in CH Activation by Ir(III) complexes. (a) Webster, C. E.; Hall, M. B. *Coord. Chem. Rev.* 2003, *238-239*, 315-331. (b) Klei, S. R.; Tilley, T. D.; Bergman, R. G. *J. Am. Chem. Soc.* **2000**, *122*, 1816.

Chapter 7

Lanthanide Complexes: Electronic Structure and H–H, C–H, and Si–H Bond Activation from a DFT Perspective

Lionel Perrin[1], Laurent Maron[2,]*, and Odile Eisenstein

[1]LSDSMS (UMR 5636), Université Montpellier 2,
34095 Montpellier Cedex 5, France
[2]Laboratoire de Physique Quantique, (UMR 5626), IRSAMC, Université Paul Sabatier, 118 Route de Narbonne, 31062 Toulouse Cedex 4, France

This paper discusses some relationships between the electronic structure and the reactivity of lanthanide complexes. The electronic structures of some representative lanthanide complexes are described. The *4f* electrons do not participate in the Ln-X bond and a comparison between lanthanide and d^0 transition metal complexes from Groups 3 and 4 highlights the dominant ionic character in the bonding to lanthanide centers. However some covalent character cannot be excluded and this covalent character rationalizes the geometry of the complexes such as the non-planar structure of LaX_3 (X = H, Me, F). The consequences of the nature of the bonding on the β Si-C agostic interaction in $La\{CH(SiMe_3)_2\}_3$ is presented. The unusual O-bonding mode of CO to $Cp*_2Yb$ is briefly summarized. The reactivity of X_2Ln-Z (X = Cp, H, Cl, effective group potential; Z = H, Me) with simple molecules like H_2, CH_4, and SiH_4 is compared. It is shown that the strong ionic character of the lanthanide bonding is key to rationalizing the selectivity of the σ-bond metathesis with alkane and to the lack of it in the case of silane. In particular, the position β to the metal center in the diamond shape *4c-4e-* transition state is not allowed for a methyl group (transition state of very high energy) whereas it is permitted to a silyl group. This is shown to be related to the relative ability of carbon and silicon to be hypervalent.

Introduction

The activation of inert bonds has long been a key concern of organometallic chemists. Oxidative addition has been recognized as an efficient route for inert bond activation and considerable effort has been spent analyzing the reactivity with a wide variety of transition metal fragments in which the metal can be oxidized (*1*). Sigma-bond metathesis, which occurs without change of oxidation state at the metal center, has also been used especially in relation to efficient catalysts for polymerization (*2*). These two types of reactions have both attracted the attention of computational chemists (*3*). It is not our intention to present an overview of this very rich field. We focus on the theoretical approach of the reactivity of lanthanocene derivatives with H_2, CH_4 and SiH_4.

Bond activation by lanthanide complexes has attracted less attention despite the key discovery by Watson of CH_4 activation by $Cp*_2Lu-CH_3$ ($Cp* = \eta^5$-C_5Me_5) (*4*). Organometallic chemistry of lanthanide complexes is the concern of a relatively small number of groups in the world despite the well known ability of lanthanide complexes to be involved in important catalytic processes including polymerization (*2,5*). In addition, only a few members of the series of lanthanide elements have been experimentally studied because of stability problems.

Computational studies of lanthanide species are still few (*6-37*) (a non-comprehensive list is given) in comparison to the increasing flow of calculations on *d* transition metals. Only recent reviews on the theoretical treatment of reactivity are cited here (*3*). A number of studies have been carried out on relatively small systems like the halide complexes (*14-17*). Some studies on *4f* element complexes have been focused on the coordination of lanthanide ions (*10,11*) in connection to extraction and separation of nuclear waste. Calculations of the organometallic molecular compounds and reactivity of lanthanide complexes are scarce (*21-25,28,29,32,33*), especially when one takes into account the fact that, in many cases, lanthanide centers have been represented by isoelectronic group 3 metal centers such as scandium and yttrium (*12,22,23*). Several factors have probably discouraged computational studies of lanthanide complexes. One major difficulty is the presence of open *4f* shells. Calculations treating explicitly the *4f* electrons by use of a small-core quasi-relativistic effective core potential (ECP) have been limited to small systems, usually of spectroscopic interest,(*13*) due to the large number of electrons (42 electrons for Yb). It has been accepted that *4f* electrons do not participate in the lanthanide-ligand bond. Therefore, most calculations have been carried out by using large-core quasi-relativistic ECP in which the *4f* electrons are included in the core leading to pseudo-closed systems. Another difficulty is associated with relativistic effects (*6*). The scalar relativistic effect is included in the small and large-core ECP's but the spin-orbit coupling is not treated. It is also currently recognized that spin-orbit coupling has no drastic influence on the structure of *4f* complexes because the *4f* electrons do not participate in bonding, unlike what has been shown for actinides (*6*). The next important difficulty in the calculations of lanthanide complexes is the size of the ligands in the coordination sphere of the metal. The lanthanide ion has a large ionic radius and thus can accommodate either a large number (up to 12) of small- to medium-size ligands or a smaller number of very bulky ligands (*5*). This introduces in the calculations, an important number of atoms and therefore a large number of degrees of freedom which makes the geometry optimization a time-consuming process. Furthermore, bulky ligands can have a large number of conformers, and there is no simple procedure to locate an absolute minimum on a potential energy surface that may have many secondary minima close in energy. Reducing the number of atoms in the ligand is a current technique among computational chemists and in many cases presents no difficulties. For instance C_5Me_5 has usually been modeled by C_5H_5 but also by Cl or H with relative success (*22-24,38*). However, the use of other models, like SiH_3 for $SiMe_3$, has lead to artifacts and structures that cannot be compared to the experimental species (*30*). Special caution is thus needed in the modeling of ligands of lanthanide complexes. The necessity to take into consideration the full nature of ligands for the study of weak interactions, such as the agostic interaction, has been noted for *d*-metal transition systems (*39*). This should be a key concern for lanthanide complexes. For this reason, the hybrid QM/MM method has been extremely useful for the study of large size complexes (*40*)

Here, we analyze first the participation of the *4f* electrons in lanthanide-ligand bonds. Computations suggest that *4f* electrons can be put in the core of the ECP. This allows the calculations of large systems with manageable computational effort. The strong ionic character of the lanthanide–ligand bond is discussed and some consequences on structures are presented. The reactivity of models of Cp_2Ln with H_2, CH_4 and SiH_4 is discussed. The trends can be rationalized using the strong ionic character of the lanthanide-ligand bonds.

The electronic structure of the Ln-X bond

Non-participation of the 4f electrons of the Ln center in the Ln-X bonds

It has been widely accepted within the "lanthanide" community that, unlike the *5f* electrons for actinides, the *4f* electrons on the lanthanide centers do not participate in bonding to the coordinated ligand. This qualitative statement needed some quantitative evaluation by theory. Earlier studies, carried out on few systems and lanthanide elements, have shown that the *4f* electrons do not participate in the lanthanide ligand bonding (*13*). We have carried out a systematic study of LnX_3 ($X = NH_2$) for the entire series of lanthanide elements, in which we have compared the results of calculations with explicit participation of the *4f* electrons (small-core ECP) with calculations in which the *4f* electrons are part of the core (large-core ECP) (*27*). The results of calculations show the non-involvement of the *4f* electrons. The Ln-N bond distance is shorter by only 0.04 Å with a small-core ECP than with a large-core ECP for any lanthanide center. This difference has been attributed to the lack of core-valence correlation with a large-core ECP and not to the participation of the *4f* electrons. The lack of participation of the *4f* electrons in bonding is also evidenced by the calculated *4f* orbital occupancy. The total number of *4f* electrons, as calculated by an NBO analysis, is equal to that in the isolated atom and determines the total spin of the molecule as indicated by its electronic configuration. In each case, the most stable spin state is that predicted by Hund's rule. The non-participation of the *4f* electrons in bonding with a ligand behaving as both σ- and π-electron-donor (NH_2) has also been established in the case of a π-accepting ligand such as CO. Calculations of $H_2Yb(CO)_{1,2}$, models for $(C_5Me_5)_2Yb(CO)_{1,2}$ show the absence of back-donation from the *4f* electrons of the Yb^{II} center to the CO ligand. The level of calculation has been validated by comparing the calculated geometry and v_{CN} stretching frequency of $Cp_2Yb(CNMe)_2$ with the corresponding experimental data on $Cp*_2Yb(CN-2,6-Me_2C_6H_3)_2$ (*31*). Calculations of $Cp_2Yb(CO)_{1,2}$ have been carried out with a large-core ECP for Yb. The observed v_{CO} stretching frequencies which are lower than that of free CO, have been interpreted as coordination of CO to Yb via the oxygen atom (*31*). Therefore, the *4f* electrons can probably be safely included in the core ECP for all types of lanthanide complexes. The lanthanide contraction (the difference between the Lu-X and La-X bond lengths) is well represented by use of the averaged relativistic electron core potential (AREP). The calculations give values varying between 0.165 and 0.185 Å depending on X (*36*).

One should highlight the limitations associated with the use of the large-core ECP in the calculations. Each large-core ECP is specific not only of a lanthanide element but to its oxidation state. Therefore, it is not possible to compare complexes with a lanthanide in different oxidation states because the total energies with different core ECP's cannot be compared. It is thus also not possible to carry out a computational study of a reaction involving a change of oxidation state at the lanthanide center in the way done currently with *d* transition metal complexes. Luckily, the σ-bond metathesis reactions, predominant in lanthanide chemistry, can be studied with this methodology.

Strong Ionic Nature of the Ln-X Bond

The nature of the Ln-X bonding has been the subject of several papers (*15-17,36*). The lanthanide elements are electropositive centers resulting in ionic Ln-

X bonding. However the covalent bonding cannot be totally excluded. It plays a key role in the non-planarity of LnF$_3$ which has interested the theoretical community for many years (*14-16,41*). We have carried out a calculation of several sets of MX$_3$ (M = Ln, Sc, Y , Ti$^+$, Zr$^+$, Hf$^+$; X = H, Me, Hal, NH$_2$). Non-planar geometries have been obtained for X = H, Me and F for earlier lanthanide and group 4 cationic complexes. Late lanthanide centers and group 3 metal complexes as well as Cl, Br, I and NH$_2$ favor a nearly planar or totally planar geometry. An NBO analysis of these complexes has shown a significant 5*d* participation in the Ln-X bonding which is known to favor pyramidalization for *d*^{0} MX$_3$ complexes (*42*). Comparison with ScX$_3$, YX$_3$, TiX$_3^+$, and HfX$_3^+$ shows that the participation of the valence *d* orbitals is considerably smaller for all lanthanide complexes than for the group 3 and 4 complexes. The 6s participation is not negligible either but also smaller than for group 3 and 4 elements while the 6p participation is almost null. The total charges remain close to the charges given by the formal oxidation number in the case of all lanthanide elements and are considerably smaller in the case of the group 3 and 4 transition metal centers. This illustrates the predominant ionic character in an Ln-X bond but shows how the small amount of covalent contribution is fundamental to rationalizing the geometry of lanthanide complexes. Similar conclusions have been reached in a study of Ln{(CH(SiMe$_3$)$_2$}$_3$ (*35,37*)

The ionic character of the Ln-X bond and the agostic interaction

The C-H agostic interaction manifests itself by a short distance between a metal center and a C-H bond, traditionally part of a ligand coordinated to the metal center. It is also manifested by an elongation of the agostic C-H bond as shown by spectroscopic or solid-state data (*43*) This interaction, which occurs when the metal is electron-deficient, has been first interpreted as the donation of electron density of the C-H bond to an empty coordination site of the metal. This interaction is weak because the C-H bond is a very poor Lewis base even if the electron deficient metal center is a reasonably powerful Lewis acid. Furthermore, this weak interaction needs some through-space overlap between the metal center and the C-H bond. This is why a C-H bond is a better candidate for an agostic bond than any C-C bond of an alkyl chain. The C-H bonds shield the C-C bond from access to the metal center which explains why there are no reported agostic C-C bonds except under some specific constraints (*44*). However, this is in contradiction with the report of agostic β Si-C bonds, suggested by the much longer Si-C distance compared to other Si-C bonds within the same molecule. Several lanthanide complexes with an elongated Si-C bond part of CH(SiMe$_3$)$_2$ or N(SiMe$_3$)$_2$ ligand have been reported (*19,37,45,46*). The absence of a γ-C-H agostic bond in these complexes is surprising because γ-agostic bonds have been reported,(*47*) and because the metal center could easily access the electron density of a C-H bond from a Me group of CH(SiMe$_3$)$_2$ or N(SiMe$_3$)$_2$.

Calculations of complexes with reported elongated Si-C bonds in the solid state structure have been carried out (*19,35,37*). The QM/MM calculations reproduce well the experimental geometrical features as shown for La{CH(SiMe$_3$)$_2$}$_3$ in Figure 1 (**1 and 2**) (*35*). Calculations show an elongation of a single β Si-C bond and the absence of any long C-H bond in the α and γ positions. The traditional interpretation of the agostic interaction cannot be satisfactory. In a study of La{CH(SiMe$_3$)$_2$}$_3$ and La{N(SiMe$_3$)$_2$}$_3$, we have suggested that the elongation of the β Si-C bond is due, in part, to a delocalization of the lone pair used for Ln-R bonding for R = CH(SiMe$_3$)$_2$ and N(SiMe$_3$)$_2$ within the ligand (*35,36*). This interpretation is not new in the literature (*48*). In key theoretical studies of Cl$_3$Ti-CH$_2$-CH$_3$, it has been suggested that the β-agostic C-H in Cl$_3$Ti-CH$_2$-CH$_3$ is due to a delocalization of the electron density of the Ti-C bond within the ethyl ligand (*48b*). Similar analysis has been reported for Li-CH$_2$-CH$_3$ (*48c*). A similar interpretation was suggested for an Yb complex prior to computational support (*48a*). What was suggested for early *d* transition metal complexes is enhanced for lanthanides complexes. The much greater ionic character of the Ln-C bond, compared to that of a Ti-C bond, makes the electron density of the Ln-C bond even more localized on C. The delocalization by negative hyperconjugation (delocalization into the σ* orbital of the adjacent bonds), well established in numerous anions, elongates all σ bonds that are vicinal to the carbanionic lone pair. This delocalization is magnified with 3rd row elements like silicon, because σ*$_{Si-C}$ is at lower energy than σ*$_{C-C}$. However the lanthanide ion plays a role. If negative hyperconjugation was the only cause for elongations of bonds, all Si-C bonds of the SiMe$_3$ groups in CH(SiMe$_3$)$_2$ would be long. This is not the case and only the β bond pointing towards the metal is elongated. The strong electrostatic field of the metal polarizes the electron density in the ligands, the σ*$_{Si-C}$ of the bond closest to the metal receives more density which in turn further elongates this bond. An immediate consequence of this interpretation is that the elongation of the Si-C bond in a N(SiMe$_3$)$_2$ ligand should not be as large as in CH(SiMe$_3$)$_2$ because a nitrogen lone pair would delocalize less than a carbanionic lone pair. This is supported by the available experimental data on the amido complexes (*46*). It should be pointed out that the traditional interpretation (donation of the electron density of the σ bond to the metal center) and the interpretation by negative hyperconjugation are interconnected and are not easy to separate. This is why the donation of the Si-C bond to the metal center has been considered as the major effect in another study of Ln{CH(SiMe$_3$)$_2$}$_3$ (Ln = La, Sm) (*37*).

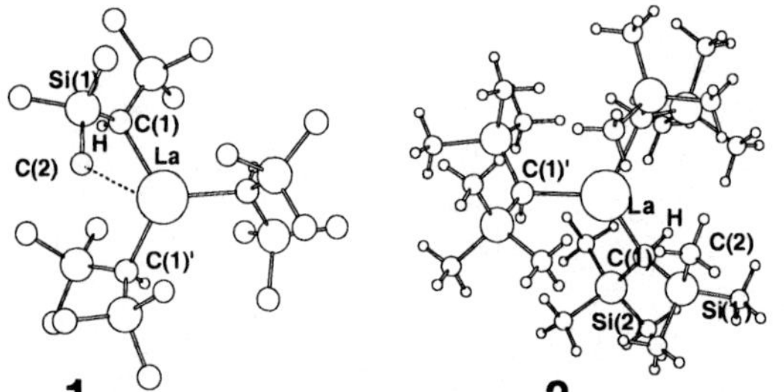

La-C(1) = 2.515(9), 2.526
La...C(2) = 3.121(9), 3.296
C(1)-S(1) = 1.841(9), 1.872
S(1)-C(2) = 1.923(11), 1.925
C(1)-La-C(1)' = 109.9(2), 111
La-C(1)-Si(1) = 102.0(4), 105
La-C(1)-Si(2) = 121.0(4), 120°
C(1)-Si-C(2)= 109.7(4), 110
La-C(1)-Si(1)-C(2) = 13°, 17

Figure 1. Experimental,(45) 1, and QM/MM,(35) 2, structures of La{CH(SiMe$_3$)$_2$}$_3$.. Distances (experimental, calculated) in Å, Angles in degrees.

Sigma Bond Activation

The discovery that Cp*$_2$Lu-CH$_3$ (Cp* = η^5-C$_5$Me$_5$) activates CH$_4$ (eq 1) has been a landmark in the search for new catalysts able to activate inert bonds especially those in alkanes (4). The Cp*$_2$Lu-CH$_3$ complex also reacts with H$_2$ to give Cp*$_2$Lu-H and CH$_4$ (eq 2), but it does not give the hydride complex and a product resulting from the formation of a C-C bond when reacting with CH$_4$ (eq 3). The hydride complex Cp*$_2$Lu-H exchanges H with H$_2$ as proven by isotope labeling (eq 4). Cp*$_2$Lu-H does not react with CH$_4$ to give back the methyl complex (eq 5) and it does not undergo H/H exchange with CH$_4$ (eq 6).

Cp*$_2$Lu-CH$_3$ + H-*CH$_3$* → Cp*$_2$Lu-*CH$_3$* + H-CH$_3$ (1)

Cp*$_2$Lu-CH$_3$ + *H-H* → Cp*$_2$Lu-*H* + *H*-CH$_3$ (2)

Cp*$_2$Lu-CH$_3$ + H-*CH$_3$* *does not give* Cp*$_2$Lu-H + H$_3$C-*CH$_3$*   (3)

Cp*$_2$Lu-H + *H-H* → Cp*$_2$Lu-*H* + H-*H*   (4)

Cp*$_2$Lu-H + H-*CH$_3$* *does not give* Cp*$_2$Lu-*CH$_3$* + H-H   (5)

Cp*$_2$Lu-H + *H*-CH$_3$ *does not give* Cp*$_2$Lu-*H* + H-CH$_3$ (6)

Several examples of reactions with a Si-Y bond (Y = H, R) and different lanthanide elements are known: synthesis of organolanthanide silyl complexes, hydrido complexes, hydrosilylation of alkenes and dehydropolymerization of silane (49,50). In these reactions, one key issue is the comparison between the silylation reaction represented with SiH$_4$ as a model silane by eq 7 and the H/H exchange reaction represented by eq 8.

$$Cp*_2Ln-H + H-SiH_3 \rightarrow Cp*_2Ln-SiH_3 + H-H \qquad (7)$$

$$Cp*_2Ln-H + H-SiH_3 \rightarrow Cp*_2Ln-H + H-SiH_3 \qquad (8)$$

These reactions need to occur by σ–bond metathesis via a *4c-4e*-transition state **3**, because there is no available *d* electron on the metal center and because the *4f* electrons are too deep in energy to be involved in redox processes. Eqs 2 and 5 suggest that formation of the hydride lanthanide complex from the alkyl complex is exothermic. The regioselectivity indicated by eqs 1 and 3 as well as eq 6 suggests that, in the *4c-4e*-transition state, the position opposite to the metal center (position β) cannot be occupied by an alkyl group but only by H. In contrast, no equivalent regioselectivity is mentioned for silane derivatives where the silicon center can occupy the α or β positions in the *4c-4e*-transition state.

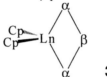

3

The reaction of eq 4 occurs at low temperature whereas the reaction of eq 1 occurs at room temperature, suggesting a higher barrier for the latter (*4*). These reactions have been run with selected lanthanide elements and there is no information on the influence of the lanthanide on these reactions. The non-implication of the *4f* electrons in Ln-X bonding suggests that all lanthanide elements could have similar reactivity. However the change in the Ln atomic radius, the related lanthanide contraction, as well as the variation in electronegativity can influence the reactivity. We thus have looked at the energy profile of eq 1 to 8 for any Ln element.

Reaction of Cp$_2$Ln-H with H$_2$

This reaction has been thoroughly studied (*21-24,28*). The Cp*$_2$Ln-H species, never isolated as a monomer, forms dimers or oligomers with bridging hydrides (*51*). It has been accepted that the monomer is the reactive species and has thus been considered in all computational studies. Modeling C$_5$Me$_5$ has been needed for saving computational time. C$_5$H$_5$ is a current model but models like H, Cl and effective group potentials (EGP) have also been used (*23,24,29,50*) DFT (B3PW91) calculations with the large core ECP from the Stuttgart-Dresden groups were carried out for all Ln (*28*). The usual oxidation state III has been considered by having neutral X$_2$Ln-H. However oxidation state IV is accessible for Ce and oxidation state II is accessible for Eu and Yb. For this reason, Cp_2Ce-H^+, Cp_2Eu-H, Cp_2Yb-H^- were considered to estimate the influence of the oxidation state of Ln. This leads to some singularities, left aside here *(28,32,33)*

The Ln-H bond in Cp_2Ln-H is not along the C_2 axis of Cp_2Ln, which indicates a preference for a pyramidal lanthanide as found for LnX_3. The H_2 molecule does not form a stable adduct with Cp_2Ln-H with the method of calculation used. The two cyclopentadienyl and the hydride ligands are probably sufficiently electron-donating to decrease significantly the Lewis acidity of the lanthanide center. The H/H exchange of eq 4 has a very low energy barrier, which varies between 0.67 to 3.1 kcal/mol depending on the Ln^{III} element. The smallest barrier occurs around Pm and Sm namely near the middle of the lanthanide series. Significantly higher barriers from 5 to 8 kcal/mol have been obtained for Ce^{IV}, Eu^{II} and Yb^{II} complexes. The very low barriers for Ln^{III} complexes represent properly a reaction observed at low temperature.

The *4c-4e*-transition state has the geometry of a diamond compressed along the Ln-Hβ direction as schematically represented in **4**. The Hα-Hβ-Hα' angle is around 156° for any Ln^{III} element. This leads to a short distance between Ln and Hβ. A charge analysis indicates that Hα and Hα' are negatively charged, Hβ positively charged and that there is no bond between Ln and Hβ. This description suggests that the transition state should be viewed as an H_3^- ligand bonded to the metal center through the wingtip Hα and Hα' centers. This interpretation is also consistent with the obtuse Hα-Hβ-Hα' angle, not too far from the 180° angle in isolated H_3^-. The short Ln-Hβ distance is the result of the short Hα–Hβ distances in the H_3^- moiety but not that of an interaction with Ln.

There is only a small influence of the lanthanide center (taken in oxidation state III). For this reason, we show the energy profile for La in Figure 2. There is no clear reason for the energy barrier to be the smallest near the middle of the lanthanide series, but this pattern has been found for reactions of eqs 1 to 8.

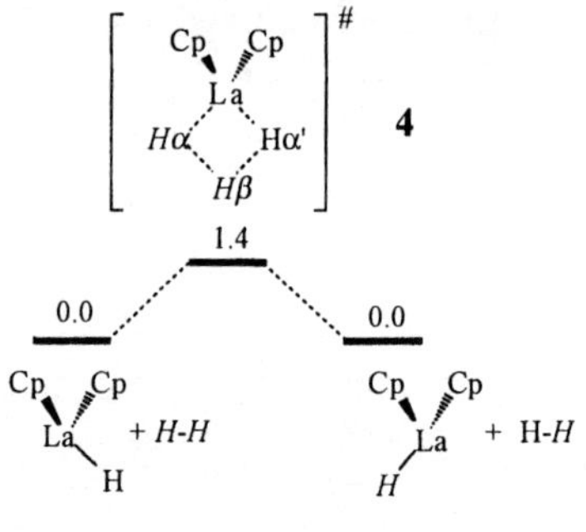

Figure 2. Potential energy profile (kcal/mol) for the reaction of eq 4 (28)

Although calculations with the explicit C_5H_5 ligand can be carried out at reasonable computational time, there is interest in comparing results with models of C_5H_5. Folga and Ziegler have considered Cl_2Lu-H with a functional different from ours (24). They have found that H_2 coordinates to Lu to form a dihydrogen adduct, only 1 kcal/mol more stable than the isolated reactants, before the H/H reaction takes place. Their calculated barrier is also considerably higher (10.6 kcal/mol above the isolated reactants). Is this due to the difference in the functional or in the modeling of the reagent? The questions have been answered via a comparison of the H/H exchange reaction with X_2Ln-H (X = H, Cl and an EGP) using the same level of calculations as for X = C_5H_5 (29).

No search for an H_2 adduct was carried out; the comparison has been focused on the height of the energy barrier and the geometry of the transition state. The energy barriers for the X_2Ln^{III}-H complexes vary between 1.8 and 4.5 kcal/mol for X = H, between 1.9 and 8.4 kcal/mol for X = Cl for the entire lanthanide family. The energy barriers are almost equal with the explicit C_5H_5 and the EGP representing this ligand. This shows that the high barrier found by Folga and Ziegler is associated, in great part, with the modeling of Cp by Cl (in the case of Lu the energy barrier is 10.6 kcal/mol for Folga and Ziegler and 8.4 kcal/mol in our work). Although the energy barrier is higher than that found with the Cp ligand, all calculations give an energy barrier characteristic of a kinetically facile reaction. Remarkably, Cl is the poorest model for Cp because it is a poor electron donor despite the presence of Cl lone pairs which can act like the degenerate e orbitals of Cp and make Cl and Cp formally isolobal. The more electron-donating hydride ligand gives numerical results closer to that with explicit C_5H_5. Furthermore the variation of the barrier within the lanthanide series is slightly different for explicit Cp and H and Cl. In particular, the lower barrier near the middle of the lanthanide series is not obtained with X = H and Cl for eq 4. The geometrical shape of the transition states is reasonably well reproduced by all models with systematic shifts in distances.

These comparisons show that modeling Cp by Cl and H is acceptable for obtaining gross features but not subtle effects. The relative energy barriers for Cp and Cl supports the interpretation of the H/H exchange as a nucleophilic substitution of H⁻ at H_2 in the coordination sphere of Ln. A more electron-donating ligand (Cp vs Cl) makes H more hydridic in Cp_2Ln-H than in Cl_2Ln-H. The energy barrier for H/H exchange is thus lower with the Cp ligand.

Reactions of X_2Ln-CH_3 with H_2 or X_2Ln-H with CH_4

The reaction of eq 2 is exothermic with a calculated energy of reaction of around 13.3 kcal/mol for all Cp_2Ln^{III} complexes. With Cl in place of Cp, the calculated energy of reaction is around 7.8 kcal/mol which is similar to the value reported by Folga and Ziegler (24). The energies of reaction of eq 2 are reasonably close for Cp and Cl ligands and for computational time-saving reasons, the full energy profiles for the entire series of lanthanide elements and for eqs 1, 2, 3, 5 and 6 were calculated with Cl in place of Cp. The calculations done for eq 4 with X = H, Cl and Cp, have given a higher energy barrier for Cl than for Cp. Similar trends are expected for the other reactions. The energy barrier does not vary much with the lanthanide element. A calculation with Cp was thus carried out only for La for eqs 2 and 5 (Figure 3).

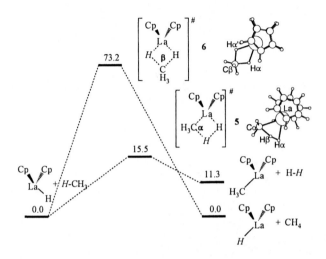

Figure 3. Potential energy profile (kcal/mol) for reactions of eqs 2 and 6 (32).

The $X_2Ln\text{-}CH_3$ complexes react with H_2 to form $X_2Ln\text{-}H$ and CH_4 (eq 2) but the $X_2Ln\text{-}H$ complexes do not undergo H/H exchange in presence of CH_4 (eq 6). The activation barrier for eq 2 varies depending on the Ln^{III} element between 5.5 kcal/mol and 9.2 kcal/mol above separated $Cl_2Ln\text{-}CH_3$ and H_2 with a minimum of 5.1 kcal/mol around Pm and Sm and the highest value for Lu. The activation barrier for eq 6, which is over 70 kcal/mol for any Ln^{III} element, shows that the H/H exchange is a kinetically forbidden reaction. The energy values for Cp_2La and Cl_2La are similar.

The two transition states, for eqs 2 and 6, differ by the position of the CH_3 in the *4c-4e*-transition states. The Me group is in the α position, _, for eq 2 and in the β position for eq 6. It has been shown that the β position in *4c-4e*-transition state is unfavorable for a CH_3 group (22,52). In the case of eq 2, Hα, Hβ and C of CH_3 are almost aligned (5) and a $CH_3^{\delta-}$ group abstracts H^+ from a strongly polarized $H^{\delta-}\text{-}H^{\delta+}$ species. This occurs in the coordination sphere of Ln. In the transition state for eq 6, a CH_5^- anion is formed (6) as shown by the square base pyramid geometry at C. It is currently accepted that CH_5^- species is an unfavorable species, especially in the absence of electronegative substituents to stabilize the hypervalent carbon. Groups favoring hypervalent anionic pentacoordination makes this transition state energetically accessible (see SiH_4).

Reactions of X_2Ln-CH_3 with CH_4

The only possible reaction between X_2Ln-CH_3 and CH_4 should be the exchange of the methyl groups (eq 1) because the Me group should not go at the β position in the *4c-4e*-transition state. This prevents the formation of X_2Ln-H and C_2H_6 (eq 3) for which an energy barrier of at least 70 kcal/mol is expected. In addition, this latter reaction is calculated to be endothermic by around 5.70 kcal/mol for X = Cl and by around 11 kcal/mol for X = Cp for any Ln^{III} elements. The energy profile for Cp_2La is shown in Figure 4 for eq 1.

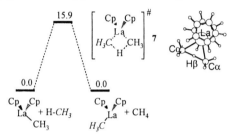

Figure 4. Potential energy profile (kcal/mol) for the reaction of eq 1 (32).

The geometry of the transition state corresponding to the methyl exchange reaction (eq 1) as observed by Watson is shown in **7**. There is essentially full alignment between the two carbons and the central hydrogen that transfers between the two methyl groups. This reaction should better be viewed as a proton transfer between two methyl anions in the coordination sphere of the lanthanide center. The activation energy for reaching this transition state varies between 15.9 and 17.3 kcal/mol with the minimum around Pm and Sm with Cl ligand. This relatively low activation energy is consistent with the experimental data. The calculated value is significantly lower than the one calculated with the same model but with another functional (26 kcal/mol) by Folga and Ziegler (*24*) The values calculated with Cl_2Ln and Cp_2Ln are close.

Reactions of Cp_2Ln-H with SiH_4

The energy patterns of the two reactions of Cp_2Ln-H with SiH_4, as represented by eq 7 and 8, are shown in Figure 5. The energy profiles of the H/H exchange reaction (eq 8) are drastically different from that with CH_4 (eq 6). In the case of the silane, the reaction starts by the formation of an SiH_4 adduct. The bond dissociation energy of SiH_4 varies between 4 and 6 kcal/mol which indicates only weak bonding of SiH_4. The most important change compared to CH_4 lies in the energy barriers. The energy barrier for H/H exchange is only between 1.7 and 1.9 kcal/mol above this adduct. This strongly contrasts with the case of CH_4 where the H/H exchange is associated with a barrier of 70 kcal/mol. The H/H exchange reaction forbidden for CH_4 is very facile for SiH_4. The geometry of the transition state (**8**) shows that the SiH_5^- moiety has the geometry of a square pyramid with the two exchanging H at the basal sites. The difference between CH_4 and SiH_4 appears to be in the relative stability of CH_5^- and SiH_5^- and not in a change in the geometry of the transition states.

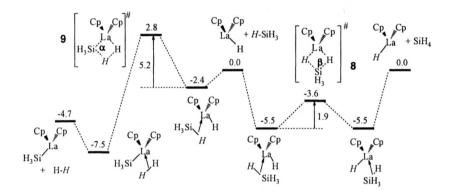

Figure 5. Potential energy profile (kcal/mol) for eqs 7 and 8 (33).

The silylation reaction (eq 7) is calculated to be energetically favorable. The reaction energy varies between 3 and 4.7 kcal/mol for all Ln^{III} centers. As shown in Figure 5, the reaction starts with the formation of a weak silane adduct (BDE around 2 kcal/mol); a transition state, 5.2 kcal/mol above the SiH_4 adduct, leads to an H_2 complex, 5 kcal/mol lower than the silane adduct. This reaction also has a low energy barrier, but it is slightly higher than the barrier for H/H exchange. The geometry of the transition state (**9**) is also a distorted diamond with a $Si\alpha\text{-}H\beta\text{-}H\alpha$ angle between 164° and 168° depending on the Ln^{III} element.

Comparison of activation of H_2, CH_4 and SiH_4

The results above show no significant influence of the lanthanide element for any of the reactions and a moderate influence of the model used for cyclopentadienyl. Consequently, we can compare the calculations done with Cp_2La which are shown in Figures 2-5.

Activation of H_2 by Cp_2La-H (eq 4) entails almost no energy barrier. Reaction of H_2 with Cp_2La-CH_3 requires a slightly higher barrier but the value of 4.2 kcal/mol is still characteristic of a kinetically facile reaction. There is a thermodynamic drive to form the hydride complex. The increase in the energy barrier when H is substituted by a Me group has been documented in the reductive elimination (*53*). A Me group has directional bonding preference whereas the spherical *1s* orbital of H allows bonding in all directions. In agreement with this fact is the higher but still accessible energy barrier (15.9 kcal/mol) for the methyl exchange process of eq 1. The higher energy barriers are in agreement with the higher temperature at which this reaction has been observed.

The reactivity of CH_4 and SiH_4 is quite different. The greater difference concerns the H/H exchange reaction. The energy barrier is inaccessibly high (73.2 kcal/mol) for CH_4 and very low (1.9 kcal/mol) for SiH_4. The reaction is best described as a nucleophilic substitution of H at either C or Si in the coordination sphere of Ln. The transition state is a pentacoordinated anionic species CH_5^- or SiH_5^- which is energetically highly unfavorable for C and much more favorable for Si. A carbon center cannot be at the β position in the *4c-4e*-transition state but the energy barrier is lowered with electronegative substituents (F), known to stabilize a hypervalent species (in progress). We have also verified vinyl and phenyl groups do not favor either the β position.

The reactions of the lanthanide-hydride complex with CH_4 and SiH_4 to form the lanthanide-methyl and lanthanide-silyl complexes also show differences. Formation of the methyl complex from the hydride complex is energetically disfavored (the reaction energy is 11.3 kcal/mol for eq 5) whereas formation of the silyl complex is energetically favored (the reaction energy for eq 7 is -4.7 kcal/mol). The endothermicity associated with eq 5 suggests that the La-H bond is stronger than the La-Me bond because the H-H and C-H bond energies are similar. No simple conclusion can be drawn from the exothermicity of eq 7. The Si-H bond in SiH_4 is weaker than the C-H bond in CH_4, which can explain the reaction exothermicity of eq 7. The calculations suggest that the Ln-silyl bond is not weak. Reactions of organosilanes with lanthanide hydride complexes can thus occur *via* reactions of eqs 7 and 8. Multiple step reactions can lead to a high variety of products observed experimentally (*49*)

Conclusions and perspectives

The theoretical studies of some complexes and reactions involving lanthanide complexes allow us to draw the following conclusions:

- The *4f* electrons do not participate in lanthanide-ligand bonding. This allows to carry out calculations with large-core ECP. Because of the dependence of the large-core ECP on the oxidation state of the lanthanide, only reactions involving no redox process at the metal center can be studied.
- The Ln-X bond has a strong ionic component, which is much larger than for isoelectronic group 3 and 4 complexes. Therefore modeling lanthanide elements by group 3 or 4 metals may not be appropriate. Despite the predominantly ionic character in Ln-X bonding, the covalent part involving a participation of the Ln *5d* orbitals is key to rationalizing the geometry of lanthanide complexes such as the non-planar geometry of LnX_3 for X = H, alkyl and F.
- The strong ionic character of the Ln-X bond suggests that the β agostic Si-C bond observed in complexes with $CH(SiMe_3)_2$ and $N(SiMe_3)_2$ ligands is in part due to a delocalization of the C or N lone pair involved in bonding with Ln in the β Si-C bond closer to the metal. The delocalization is stronger for $CH(SiMe_3)_2$ than for $N(SiMe_3)_2$, consistent with a greater Si-C elongation in the former ligand.
- The strong ionic character of the Ln-X bond brings a new perspective to the σ-bond metathesis reactions. The reaction of H_2, CH_4 and SiH_4 with Cp_2Ln-H or Cp_2Ln-CH_3 are best viewed as nucleophilic substitution by H⁻ or CH_3^- at H_2, CH_4 or SiH_4 in the coordination sphere of the lanthanide cation. This leads to a very low energy barrier for H/H exchange in the reaction of Cp_2Ln-H with H_2 and SiH_4 but to a forbidden H/H exchange in the reaction of Cp_2Ln-H with CH_4. The different behavior for CH_4 and SiH_4 is due to the formation, at the transition state, of CH_5^- and SiH_5^- respectively. Hypervalency is unfavorable for C and more favorable for Si. The reactions with alkanes are thus very selective. Only H can take the position β to the metal in the *4c-4e*-transition state. Preliminary calculations with other reactants like alkenes and arene give similar results. No such selectivity occurs with silane derivatives because the silyl group can take the positions α and β to the metal center in the *4c-4e*-transition state.
- There is no calculated significant influence of the lanthanide element for a given oxidation state (here Ln^{III}) on the energy of reaction and energy barrier for all studied reactions. One should not exclude effects that are not present here like the steric bulk of ligands, which can influence the accessibility to the metal center in the case of the late lanthanides, which have a smaller ionic radius.

To what extent are these results characteristic of lanthanide complexes? Alkali and alkali-earth elements also give ionic bonding but no similar reactivity has been reported. Is this difference in reactivity associated with the difference in ionic radius, the small covalent character in the lanthanide elements, or with other factors? Clarifying these points would help to rationalize the uniqueness of lanthanide complexes in activating inert bonds.

References

1. (a) Activation of Unreactive Bonds and Organic Synthesis, *Topics in Organometallic Chemistry*, Vol. 3 Murai, S. Ed, **1999**, Springer, Berlin. (b) Crabtree, R. H. *Chem. Rev.* **1995**, *95*, 987. (c) Hall, C.; Perutz, R. N. *Chem. Rev.* **1996**, *96*, 3125. (d) Shilov, A. E.; Shu'lpin G. B. *Chem. Rev.* **1997**, *97*, 2879. (e) Kubas, G. J. *Transition Metal σ bond and Dihydrogen Complexes* Kluwer Academic/Plenum Publishers, New-York, NY, **2001**.
2. Gladysz, J. A. *Chem. Rev.* **2000**, *100*, 1167 and other papers in this issue.
3. (a) Davidson, E. R. *Chem. Rev.* **2000**, *100*, 351 and other papers in this issue. (b) Maseras, F.; Lledós, A. Eds. *Computational Modeling of Homogeneous Catalysis* Kluwer, **2002**.
4. Watson, P. L.; Parshall, G. W. *Acc. Chem. Res.* **1985**, *18*, 51.
5. (a) Kagan, H. B. Ed. "Frontiers in Lanthanide Chemistry" *Chem. Rev.* **2002**, *102*, 1805. (b) Anwander, R. in *Lanthanides: Chemistry and Use in Organic Synthesis. Topics in Organometallic Chemistry* Vol 2 , pp 1-62 Kobayashi, S. Ed, Springer, Berlin 1999.
6. (a) Pyykkö, P. *Chem. Rev.* **1988**, *88*, 563. (b) Kaltsoyannis, N. *Chem. Soc.Rev.* **2003**, *32*, 9. (c) Pepper, M.; Bursten, B. *Chem. Rev.* **1991**, *91*, 719.
7. (a) Kaupp, M.; Schleyer, P. v. R.; Dolg, M.; Stoll, H. *J. Am. Chem. Soc.* **1992**, *114*, 8202. (b) Kaupp, M.; Schleyer, P. v. R. *Organometallics*, **1992**, *11*, 2765.
8. Strout, D. L.; Hall, M. B. *J. Phys. Chem.A* **1996**, *100*, 18007.
9. Karl, M.; Harms, K.; Seybert, G.; Massa, W.; Fau, S.; Frenking, G.; Dehnicke, K. *Z. Anorg. Allg. Chem.* **1999**, *625*, 2055.
10. Boehme, C.; Wipff, G. *J. Phys. Chem. A* **1999**, *103*, 6023.
11. (a) Cosentino, U.; Moro, G.; Pitea, D.; Calabi, L.; Maiocchi, A. *J. Mol. Struct. (Theochem)* **1997**, *392*, 75. (b) Rabbe C.; Mikhalko V.; Dognon J. P. *Theor. Chem. Acc.* **2000**, *104*, 280.
12. Wetzel, T. G.; Dehnen, S.; Roesky, P. W. *Angew. Chem. Int. Ed.* **1999**, *38*, 1086.
13. (a) Dolg, M.; Stoll, H. in *Handbooks on the Physics and Chemistry of Rare Earths*, Gscheinder, K. A. Jr.; Eyring, L. Eds Elsevier, vol 22, Amsterdam, 1996, chapter 152. (b) Hong, G.; Schautz, F.; Dolg, M. *J. Am. Chem. Soc.* **1999**, *121*, 1502. (c) Hong, G.; Dolg, M.; Li, L. *Int. J. Quantum Chem.* **2000**, *80*, 201. (d) Lu, H. G.; Li, L. M. *Theor. Chem. Acc.* **1999**, *102*, 121.
14. (a) Cundari, T. R.; Sommerer, S. O.; Strohecker, L. A.; Tipett, L. *J. Chem. Phys.* **1995**, *103*, 7058. (b) Bella, S. D.; Lanza, G.; Fragalá, I. L. *Chem. Phys. Lett.* **1996**, *214*, 598. (c) Broclawik, E.; Eilmes, A. *J. Chem. Phys.* **1998**, *108*, 3498. (d) Hartwig, R.; Koch, W. *Chem. Eur. J.* **1999**, *5*, 312. (e) Joubert, L.; Silvi, B.; Picard, G. *Theor. Chem. Acc.* **2000**, *104*, 109. (f)

Solomonik, L. V. G.; Marochko, O. Yu, *J. Struct. Chem.* **2001**, *41*, 725. (g) Jansik, B.; Sanchez de Meras, A. M. J.; Schimmelpfennig, B.; Ågren, H. *J. C. S. Dalton Trans.* **2002**, 4603.
15. (a) Adamo, C.; Maldivi, P. *Chem. Phys. Lett.* **1997**, *268*, 61. (b) Adamo, C.; Maldivi, P. *J. Phys. Chem. A.* **1998**, *102*, 6812.
16. Tsuchiya, T.; Taketsugu, T.; Nakano, H.; Hirao, K. *J. Mol. Struct. (Theochem)* **1999**, *461- 462*, 203.
17. (a) Andersen, R. A.; Boncella, J. M.; Burns, C. J.; Green, J. C.; Hohl, D.; Roesch, N. *J. Chem. Soc. Chem. Commun.* **1986**, 405. (b) DeKock, R. L.; Peterson, M. A.; Timmer, L. K.; Baerends, E. J.; Vernooijs, P. *Polyhedron* **1990**, *9*, 1919.
18. Russo, M. R.; Kaltsoyannis, N.; Sella, A. *Chem. Commun.* **2002**, 2458.
19. (a) Klooster, W. T.; Brammer, L.; Schaverien, C. J.; Budzelaar, P. H. M. *J. Am. Chem. Soc.* **1999**, *121*, 1381. (b) Niemeyer, M. *Z. Anorg. Allg. Chem.* **2002**, *628*, 647.
20. Hieringer, W.; Eppinger, J.; Anwander, R.; Hermann, W. A. *J. Am. Chem. Soc.* **2000**, *122*, 11983.
21. Rabaâ, H.; Saillard, J.-Y.; Hoffmann, R. *J. Am. Chem. Soc.* **1986**, *108*, 4327.
22. Steigerwald, M. L.; Goddard, W. A. III *J. Am. Chem. Soc.* **1984**, *106*, 308.
23. Cundari, T. R.; Stevens, W. J.; Sommerer, S. O. *Chem. Phys.* **1993**, *178*, 235. (b) Folga, E.; Ziegler, T. *Can. J. Chem.* **1992**, *70*, 33.
24. (a) Ziegler, T.; Folga, E. *J. Am. Chem. Soc.* **1993**, *115*, 636.
25. Margl, P.; Deng, L.; Ziegler, T. *Organometallics* **1998**, *17*, 933.
26. Green, J. C.; Jardine, C. N. *J. Chem. Soc. Dalton Trans* **1998**, 1057.
27. Maron, L.; Eisenstein, O. *J. Phys. Chem. A.* **2000**, *104* , 7140.
28. Maron, L.; Eisenstein, O. *J. Am. Chem. Soc.* **2001**, *123*, 1036.
29. Maron, L.; Eisenstein, O.; Alary, F.; Poteau, R. *J. Phys. Chem. A* **2002**, *106*, 1797.
30. Maron, L.; Eisenstein, O. *New J. Chem.* **2001**, *25*, 255.
31. Maron, L.; Perrin, L.; Eisenstein, O.; Andersen, R. A. *J. Am. Chem. Soc.* **2002**, *124*, 5614.
32. Maron, L.; Perrin, L.; Eisenstein, O. *J. Chem. Soc. Dalton Trans.* **2002**, 534.
33. Perrin, L.; Maron, L.; Eisenstein, O. *Inorg. Chem.* **2002**, *41*, 4355.
34. Eisenstein, O.; Maron, L. *J. Orgamet. Chem.* **2002**, *647*, 190.
35. Perrin, L.; Maron, L.; Eisenstein, O.; Lappert, M. F. *New J. Chem.* **2003**, *27*, 121.
36. Perrin, L.; Maron, L.; Eisenstein, O. *Faraday Discuss.* **2003**, 25.
37. Clark, D. L.; Gordon, J. C.; Hay, J. P.; Martin, R. L.; Poli, R. *Organometallics* **2002**, *21*, 5000.
38. Thomas, J. R.; Quelch, G. E.; Seidi, E. T.; Schaeffer, H. F. III *J. Chem. Phys.* **1992**, *96*, 6857.
39. (a) Ujaque, G.; Cooper, A. C.; Maseras, F.; Eisenstein, O.; Caulton, K. G. *J. Am. Chem. Soc.* **1998**, *120*, 361. (b) Clot, E.; Eisenstein, O.; Dube, T.; Faller, J. W.; Crabtree, R. H. *Organometallics* **2002**, *21*, 575.

40. Maseras, F. *Chem. Commun.* **2000**, 1821.
41. Hargittai, M. *Chem. Rev.* **2000**, *100*, 2233.
42. Kaupp, M. *Angew. Chem. Int. Ed.* **2001**, *40*, 3534.
43. Brookhart, M.; Green, M. L. H.; Wong, L. *Prog. Inorg. Chem.* **1988**, *36*, 1.
44. (a) Vigalok, A.; Milstein, D. *Acc. Chem. Res.* **2001**, *34*, 798. (b) Jaffart, J.; Etienne, M.; Reinhold, M.; McGrady, J. E.; Maseras, F. *Chem. Comm.* **2003**, 876.
45. Hitchcock, P. B.; Lappert, M. F.; Smith, R. G.; Bartlett, R. A.; Power, P. P. *J. Chem. Soc. Chem. Commun.* **1988**, 1007.
46. (a) Andersen, R. A.; Templeton, D. H.; Zalkin, A. *Inorg. Chem.* **1978**, *17*, 2317. (b) Tilley, T. D.; Andersen, R. A.; Zalkin, A. *Inorg. Chem.* **1984**, *23*, 2271. (c) Herrmann, W. A.; Anwander, R.; Munck, F. C.; Scherer, W.; Dufaud V.; Huber, N. W.; Artus, G. R. *Z. Naturforsch. B* **1994**, *49*, 1789. (d) Fjeldberg, T.; Andersen, R. A. *J. Mol. Struct.* **1985**, *129*, 93. (e) Rees, W. S.; Just, O.; VanDerveer, D. S. *J. Mater. Chem.* **1999**, *9*, 249. (f) Eller, P. G.; Bradley, D. C.; Hursthouse, M. B.; Meek, D. W. *Coord. Chem. Rev.* **1977**, *24*, 1. Fjeldberg, T.; Andersen, R. A. *J. Mol. Struct.* **1985**, *128*, 49. (g) Ghotra, J. S.; Hursthouse, M. B.; Welch, A. J. *J. Chem. Soc. Chem. Commun.* **1973**, 669. (h) Westerhausen, M.; Hartmann, M.; Pfitzner, A.; Schwarz, W. *Z. Anorg. Allg. Chem.* **1995**, *621*, 837. (i) Andersen, R. A.; Beach, D. B.; Jolly, W. L. *Inorg. Chem.* **1985**, *24*, 4741. (j) Brady, E.D.; Clark, D. L.; Gordon, J. C.; Hay, P. J.; Keogh, D. W.; Poli, R.; Watkin, J. G. personal communication.
47. See for instance Den Haan, K. H.; De Boer, J. L.; Teuben, J. H.; Spek, A. L.; Kojic-Prodic, B.; Hajs, G. R.; Huis, R. *Organometallics* **1986**, *5*, 1726.
48. (a) Tilley, T. D.; Andersen, R. A.; Zalkin, A. *J. Am. Chem. Soc.* **1982**, *104*, 3725. (b) Haaland, A.; Scherer, W.; Ruud, K.; McGrady, G. S.; Downs, A.; Swang, J. O. *J. Am. Chem. Soc.* **1998**, *120*, 3762. (c) Scherer, W.; Sirsch, P.; Shorokhov, D.; McGrady, G. S.; Mason, S. A.; Gardiner, M. G. *Chem. Eur. J.* **2002** *8*, 2324.
49. (a) Voskoboynikov, A. Z.; Parshina, I. N.; Shestakova, A. K.; Butin, K. P.; Belestkaya, I. P.; Kuz'mina, L. G.; Howard, J. A. K. *Organometallics*, **1997**, *16*, 4041. (b) Fu, P.-F.; Brard, L.; Li, Y.; Marks, T. J. *J. Am. Chem. Soc.* **1995**, *117*, 7157. (c) Radu, N. S.; Tilley, T. D. *J. Am. Chem. Soc.* **1995**, *117*, 5863. (d) Radu, N. S.; Tilley, T. D.; Rheingold, A. L. *J. Am. Chem. Soc.* **1992**, *114*, 8293. (e) Castillo, I.; Tilley, T. D. *Organometallics* **2000**, *19*, 4733.
50. Ziegler, T.; Folga, E. *J. Organomet. Chem.* **1994**, *478*, 57.
51. Ephritikhine, M. *Chem. Rev.* **1997**, *97*, 2193.
52. (a) Rappé, A. K. *Organometallics* **1990**, *9*, 466. (b) Thompson, M. E.; Baxter, S. M.; Bulls, A. R.; Burger, B. J.; Nolan, M. C.; Santasiero, B. D.; Schaefer, W. P.; Bercaw, J. E. *J. Am. Chem. Soc.* **1987**, *109*, 203.
53. Tatsumi, K.; Hoffmann, R.; Yamamoto, A.; Stille, J. K., *Bull. Chem. Soc. Jpn.* **1981**, *54*, 1857.

C–H Bond Activations in Organic Synthesis, Alkane Dehydrogenation, and Related Chemistry

Chapter 8

Catalytic, Thermal, Regioselective Functionalization of Alkanes and Arenes with Borane Reagents

John F. Hartwig

Department of Chemistry, Yale University, New Haven, CT 06520–8107

Work in the author's group that has led to a regioselective catalytic borylation of alkanes at the terminal position is summarized. Early findings on the photochemical, stoichiometric functionalization of arenes and alkanes and the successful extension of this work to a catalytic functionalization of alkanes under photochemical conditions is presented first. The discovery of complexes that catalyze the functionalization of alkanes to terminal alkylboronate esters is then presented, along with mechanistic studies on these system and computational work on the stoichiometric reactions of isolated metal-boryl compounds with alkanes. Parallel results on the development of catalysts and a mechanistic understanding of the borylation of arenes under mild conditions to form arylboronate esters are also presented.

1. Introduction

Although alkanes are considered among the least reactive organic molecules, alkanes do react with simple elemental reagents such as halogens and oxygen (*1,2*). Thus, the conversion of alkanes to functionalized molecules at low temperatures with control of selectivity and at low temperatures is a focus for development of catalytic processes (*3*). In particular the conversion of an alkane to a product with a functional group at the terminal position has been a longstanding goal (eq. 1). Terminal alcohols such as *n*-butanol and terminal amines, such as hexamethylene diamine, are major commodity chemicals(*4*) that are produced from reactants several steps downstream from alkane feedstocks.

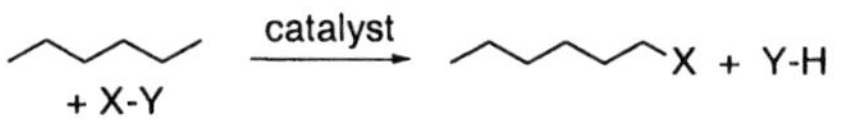

$$\text{R-H} + \text{X-Y} \xrightarrow{\text{catalyst}} \text{R-X} + \text{Y-H} \qquad (1)$$

The conversion of alkanes to functionalized products with terminal regioselectivity must confront several challenges outlined in Scheme 1. The internal C-H bond of an alkane is weaker than the terminal C-H bond,(*5*) and so reactions that are selective for one position over the other are typically selective for the internal C-H bond (*1*). Second, the products formed by functionalization are generally more reactive than the alkane starting material. Thus, alcohol products are ultimately converted to ketones, aldehydes or even CO_2, and water, while monohalo alkanes tend to undergo reaction to produce geminal dihalo alkanes (*6*). In addition to problems of selectivity are problems of conversion. Many of the desired transformations of alkanes are thermodynamically disfavored. The dehydrogenation of an alkane is, of course, uphill; hydrogenation is a well-known catalytic reaction. The carbonylation of an alkane is also uphill (*7*); decarbonylation is a known catalytic reaction (*8,9*). And a reaction related to results presented later in this manuscript – the dehydrogenative coupling of water with an alkane to form dihydrogen and a terminal alcohol – would be spectacular, but it is again thermodynamically uphill (*5*).

Scheme 1

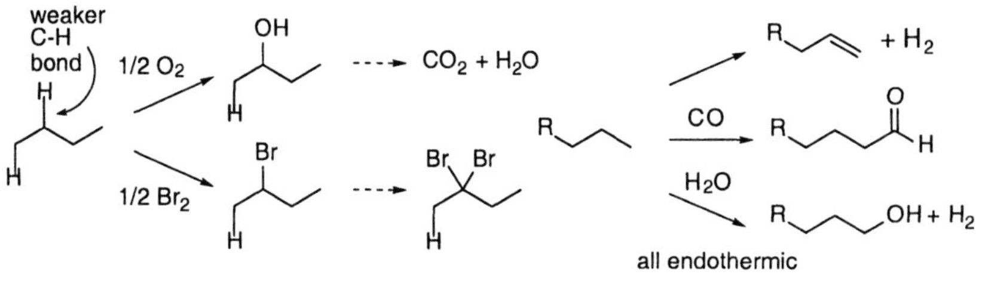

Many transition metal complexes are now known to react selectively at the terminal position of an alkane (Scheme 2). Bergman and Jones have shown that rhodium complexes will react at the terminal position of linear alkanes to generate 1-substituted rhodium alkyl products (*10,11,12,13*). In a much different type of C-H activation, Wolczanski has shown that a titanium imido complex reacts with alkanes to generate terminal titanium alkyl products (*14*). Thus, non-radical reactions of transition metal complexes with alkanes often generate the desired product from terminal activation.

Scheme 2

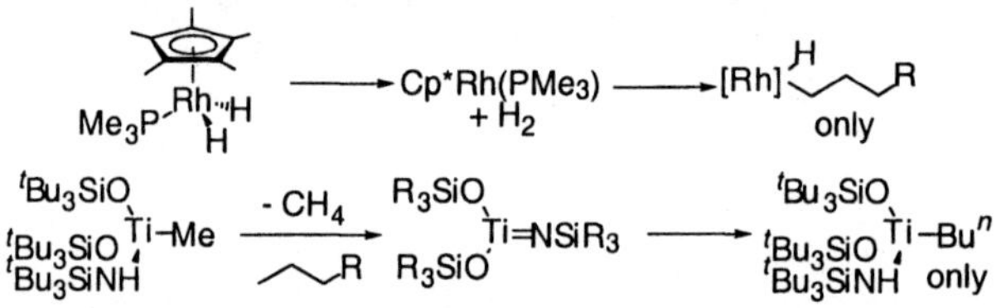

However, the reactions just described are stoichiometric. Catalytic reactions of alkanes that make use of this terminal selectivity are rare. A few examples are summarized in Scheme 3. Before any of the directly observed oxidative additions of alkanes were observed, Shilov had shown that platinum chemistry can provide more terminal product than radical chemistry does, but the selectivity for the terminal product remained lower than would be needed for synthetic purposes (*15,16*). Tanaka and Goldman studied the carbonylation of alkanes (*17,18*). Although this chemistry shows good terminal selectivity, the reaction yields are low. Products from secondary photochemistry and condensation dominate. Dehydrogenation has recently been shown by Goldman and Jensen to provide α–olefin as the kinetic product (*19*). However, the catalyst for this dehydrogenation is also a catalyst for isomerization, and a thermodynamic mixture of internal and terminal olefins is ultimately produced. This thermodynamic mixture favors the internal olefin.

Scheme 3

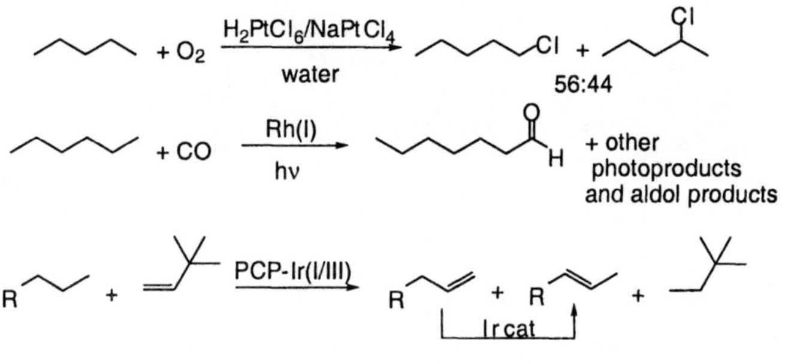

Scheme 4

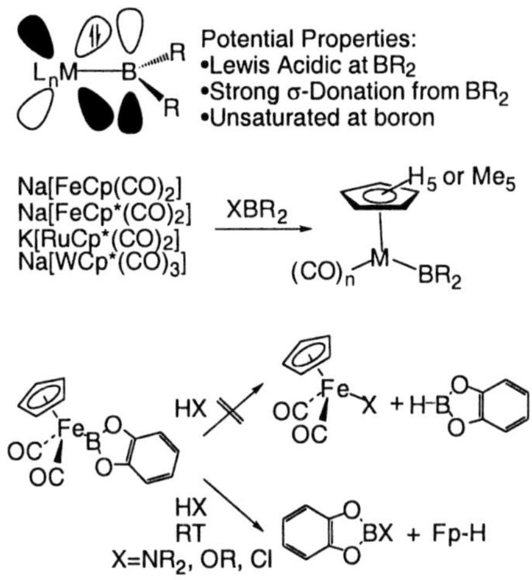

2. Development of a Catalytic Functionalization of Alkanes

2.1 Stoichiometric Studies of Metal-Boryl Complexes

Our studies on the functionalization of alkanes originated from studies of transition metal boryl complexes (20-22). We anticipated that the unoccupied p-orbital on boron would affect both structures and reactions of this class of molecule (Scheme 4, top). The boryl complexes would be strong sigma donors because of the high basicity of an anionic boryl fragment, but their unsaturation would lead to some back-donation from the metal.

We initially prepared these complexes by the reaction of classic metal carbonyl and Cp metal carbonyl anions with haloboranes (Scheme 4, middle). By this route we prepared a family of metal boryl complexes with catecholate, pinacolate, dithiocatecholate and dialkyl substituents boron (23-25).

Several years ago we found that these metal boryl complexes reacted with XH bonds in a process that placed the hydrogen on the metal and X at boron (Scheme 4, bottom) (26). Most striking, this regiochemistry was also observed from C-H activation processes. Irradiation of CpFe(CO)$_2$Bcat (Fp-Bcat, **1**) in benzene solvent generated the phenyl boronate ester PhBcat in quantitative yields, as shown at the top of Scheme 5 (24,27). Fp dimer was the transition metal product, and H$_2$ presumably formed. Because Fp complexes had shown no previous intermolecular C-H activation chemistry of hydrocarbons, this reaction demonstrated the unusual ability of a boryl ligand to trigger C-H activation.

Scheme 5

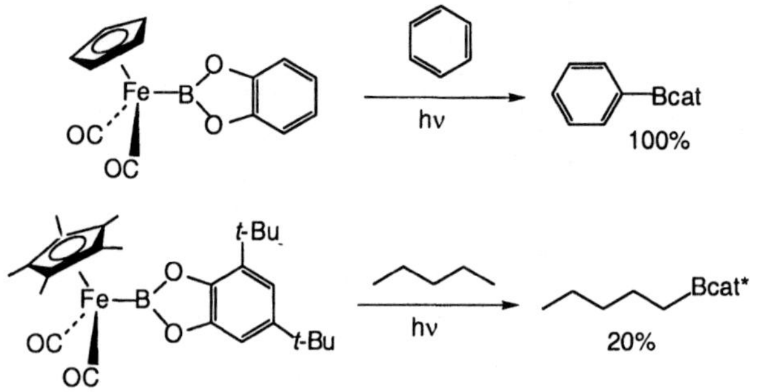

Starting with this complex that functionalized arenes, we sought to create boryl complexes that would undergo analogous reactions of alkanes to replace the hydrogen with the boryl group. If such a complex is to be generated, the reactive arene C-H bonds must be absent from the system. Thus, we made two changes to the boryl complex to observe the borylation of alkanes. First, the simple catecholboryl group was replaced by the 3,5-di-*tert*-butyl catecholboryl group to eliminate and sterically block the aromatic C-H bonds. Second, the Cp ligand was replaced with Cp* to eliminate the pseudo aromatic C-H bonds on this part of the molecule. With these two changes to the metal complex, the modified Fp boryl complex reacted with alkanes to produce functionalized products, albeit in a modest 20% yield (Scheme 5, bottom) (*28*). Yet the regiochemistry for this reaction of an alkane was perfect for placing a boryl group at the terminal position. This result indicated that even simple, classic metal fragments were capable of terminal alkane CH activation and functionalization when ligated by a boryl group.

To improve the yields of this process, we investigated second- and third-row analogs of these iron boryl complexes (*28*). The ruthenium analogs produced 40% yield of the terminal, pentylboronate ester. After a brief study with osmium analogs, we investigated the more synthetically accessible Cp*W(CO)$_3$-boryl complexes. These tungsten complexes were remarkably reactive toward borylation of terminal, alkane C-H bonds, as summarized in Scheme 6 (*28*). Irradiation of this complex in pentane for 20 min generated 85% yield of the 1-pentylboronate ester without formation of any detectable quantities of internal isomers. These metal fragments displayed low reactivity toward any hydrocarbon secondary C-H bond. For example, the analogous reaction of the tungsten complex in cyclohexane solvent generated only 22% yield of the cyclohexyl boronate ester. Reaction with ethylcyclohexane produced only the product from functionalization at the terminal position of the ethyl sidechain in 74% yield. Finally, isopentane, which contains two sterically different methyl groups, reacted in 55% yield to produce an 11:1 ratio of the product from the functionalization of the less hindered methyl group to the product from functionalization of the more hindered methyl group.

Scheme 6

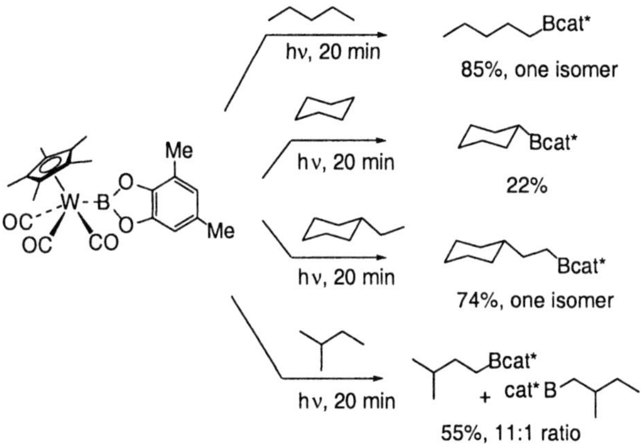

2.2. Development of a Photochemical Catalytic Functionalization of Alkanes

These results implied that boryl complexes could be used as intermediates in a catalytic functionalization of alkanes and arenes. However, a catalytic cycle involving the Cp*M(CO)$_n$-boryl complexes was difficult to imagine because these complexes were generated by substitution reactions with metal anions. Thus, we sought to use diboron compounds such as bis-pinacoldiborane(4) (B$_2$pin$_2$, pin=pinacolate) as reagents to generate the required boryl intermediates. We envisioned that oxidative addition of these diboron reagents would occur to a 16-electron Cp*M(CO)$_n$ fragment in a similar fashion to the additions to other unsaturated fragments published previously by Marder and Baker,(29-31) Miyaura,(32) Smith,(33) and us (31). Subsequent reaction of this Cp*M(CO)$_n$ bis-boryl intermediate with an alkane or arene could generate functionalized product, borane side product, and the same 16-electron fragment that added the diboron compound. If we started his process with Cp*M(CO)$_3$ (M=Mn,Re), then this reaction would occur through intermediates that are isoelectronic and nearly isostructural with the Cp*M(CO)$_n$-boryl complexes that functionalized alkanes stoichiometrically. This reaction occurred as planned (34). As shown in Scheme 7, the reaction of bis-pinacolato diboron with pentane in the presence of 2.5 mol% of the rhenium catalyst under photochemical conditions generated 95% yield of the 1-pentylboronate ester. This reaction was slow – It required 56 h for complete conversion – but the yield of functionalized product was high. This result demonstrated the potential to conduct catalytic borylation of alkanes.

Scheme 7

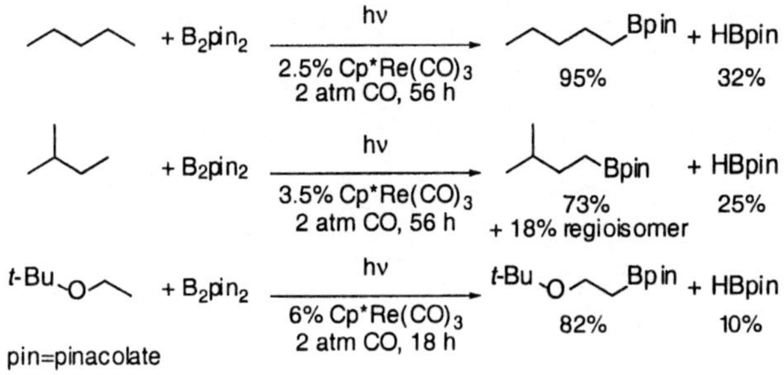

$$\text{pentyl} + B_2pin_2 \xrightarrow[\text{2 atm CO, 56 h}]{h\nu, \ 2.5\% \ Cp^*Re(CO)_3} \text{pentyl-Bpin} + HBpin$$
95%   32%

$$\text{isobutyl} + B_2pin_2 \xrightarrow[\text{2 atm CO, 56 h}]{h\nu, \ 3.5\% \ Cp^*Re(CO)_3} \text{isobutyl-Bpin} + HBpin$$
73%   25%
+ 18% regioisomer

$$t\text{-Bu-O-CH}_2\text{CH}_3 + B_2pin_2 \xrightarrow[\text{2 atm CO, 18 h}]{h\nu, \ 6\% \ Cp^*Re(CO)_3} t\text{-Bu-O-CH}_2\text{CH}_2\text{-Bpin} + HBpin$$
82%   10%

pin=pinacolate

2.3. Development of a Thermal Catalytic Functionalization of Alkanes

This success with a photochemical, catalytic functionalization suggested that we could develop a thermal process for the conversion of alkanes to alkylboronate esters. To observe catalytic chemistry under such conditions, the carbonyl ligands must be replaced with ligands that undergo thermal dissociation more readily. We initially investigated iridium complexes because of their history of success in alkane transformations (*35-37*). We envisioned that the boron reagent would abstract hydrides from Cp*IrH$_4$ and ethylene from Cp*Ir(C$_2$H$_4$)$_2$ to generate an unsaturated intermediate. Indeed, early results with these complexes, albeit modest, suggested that a catalytic process could be developed. With 10 mol % iridium over 4 d at 170 °C, the bis-ethylene complex generated 16% yield of functionalized product (eq. 2) (*38*).

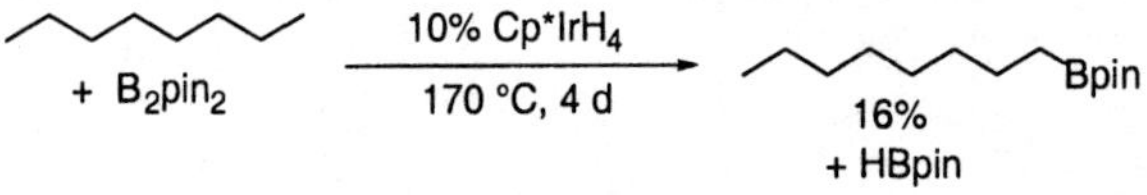

$$\text{alkane} + B_2pin_2 \xrightarrow[\text{170 °C, 4 d}]{10\% \ Cp^*IrH_4} \text{alkyl-Bpin} + HBpin \qquad (2)$$
16%

Second-row metal complexes typically react with faster rates than first-row complexes. For this reason we hoped to increase the rates of these iridium-catalyzed reactions by studying the analogous rhodium complexes. Some results from this study are shown in Scheme 8. Cp*RhH$_4$ is not known, but Cp*Rh(C$_2$H$_4$)$_2$ is well known (*39*). Thus, we tested Cp*Rh(C$_2$H$_4$)$_2$ for the borylation of alkanes with B$_2$pin$_2$ and found that the alkyl boronate ester was formed in yields over 80%. C$_2$-boryl products, however, were generated from the two ethylene ligands, and the generation of these products reduced reaction yields. To generate a catalyst without forming such organoborane byproducts, we tested the related η^4-hexamethyl benzene complex whose synthesis had been published,(*40,41*) but whose reaction chemistry had not been studied. Reaction of octane with B$_2$pin$_2$ in the presence of 0.5 mol% of Cp*Rh(η^4-C$_6$Me$_6$)$_2$ with respect to Bpin units formed 1-octylBpin in 72% yield, which corresponds to 144 turnovers (*38*). Reaction with 5 mol% Cp*Rh(η^4-C$_6$Me$_6$)$_2$ gave 88% yield of the 1-alkylboronate ester. The catalyst can also be generated from the air-stable and commercially available [Cp*RhCl$_2$]$_2$ (*42,43*). Abstraction of chloride by B$_2$pin$_2$ generates ClBpin and the active catalyst. Thus, reaction of octane with 1.25 mol% [Cp*RhCl$_2$]$_2$ gave 80% yield of the 1-octylboronate ester. Thus, regiospecific functionalization of an alkane can be conducted with a catalyst and reagent that are air-stable and commercially available.

Scheme 8

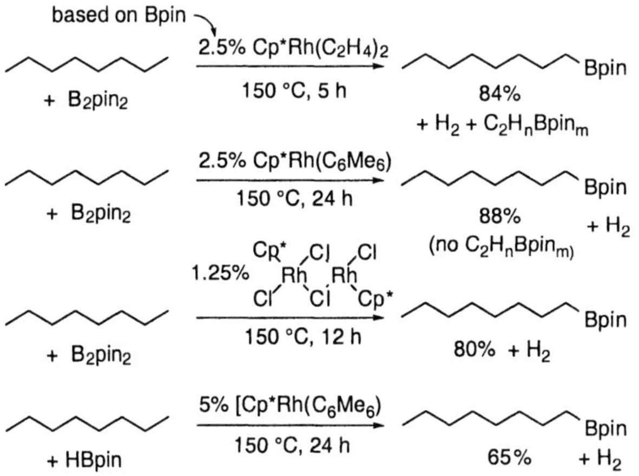

based on Bpin

octane + B$_2$pin$_2$ → (2.5% Cp*Rh(C$_2$H$_4$)$_2$, 150 °C, 5 h) → octyl-Bpin, 84% + H$_2$ + C$_2$H$_n$Bpin$_m$

octane + B$_2$pin$_2$ → (2.5% Cp*Rh(C$_6$Me$_6$), 150 °C, 24 h) → octyl-Bpin, 88% (no C$_2$H$_n$Bpin$_m$) + H$_2$

octane + B$_2$pin$_2$ → (1.25% [Cp*Rh(Cl)$_2$]$_2$, 150 °C, 12 h) → octyl-Bpin, 80% + H$_2$

octane + HBpin → (5% [Cp*Rh(C$_6$Me$_6$)], 150 °C, 24 h) → octyl-Bpin, 65% + H$_2$

In addition to reactions of diboron reagents with alkanes to form alkylboronate esters, reactions of pinacolborane with alkanes occurred in the presence of Cp*Rh(η^4-C$_6$Me$_6$). Octane reacted with pinacolborane to form 1-octylboronate ester in 80% yield in the presence of 5 mol% Cp*Rh(η^4-C$_6$Me$_6$). This process is simply the second stage of the reactions of B$_2$pin$_2$ with alkanes. The first stage of the catalytic reaction of B$_2$pin$_2$ with an alkane generates one equivalent of alkyl boronate ester and one equivalent of pinacol borane. The second stage leads to the dehydrogenative coupling of pinacolborane with the alkane to form the second equivalent of alkyl boronate ester.

Scheme 9

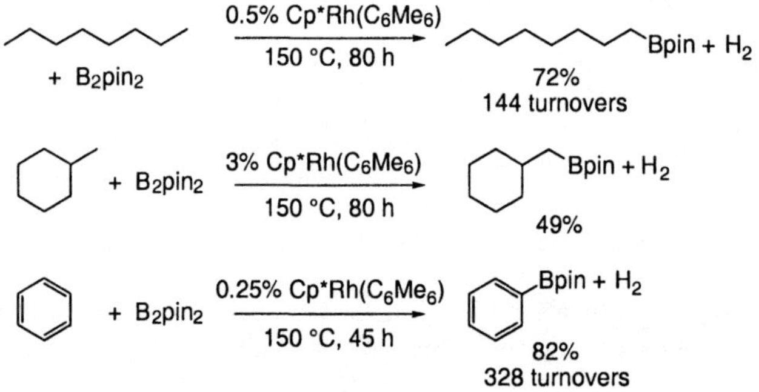

$$\text{octane} + B_2pin_2 \xrightarrow[150\,°C,\,80\,h]{0.5\%\ Cp^*Rh(C_6Me_6)} \text{octyl-Bpin} + H_2$$
72%
144 turnovers

$$\text{methylcyclohexane} + B_2pin_2 \xrightarrow[150\,°C,\,80\,h]{3\%\ Cp^*Rh(C_6Me_6)} \text{cyclohexylmethyl-Bpin} + H_2$$
49%

$$\text{benzene} + B_2pin_2 \xrightarrow[150\,°C,\,45\,h]{0.25\%\ Cp^*Rh(C_6Me_6)} \text{PhBpin} + H_2$$
82%
328 turnovers

The thermal, catalytic borylation of alkanes is sensitive to the steric properties of the alkane (Scheme 9). Reactions of linear alkanes, such as octane, occurred in the highest yields. Reactions of alkanes containing more hindered methyl groups, such as methylcyclohexane, required longer times, higher catalyst loadings, and occurred in lower yields. However aromatic hydrocarbons, such as benzene, generated the products in high yields. Reactions of benzene with B_2pin_2 in the presence of $Cp^*Rh(\eta^4\text{-}C_6Me_6)$ generated the phenylboronate ester in 82% yield with only 0.25 mol% catalyst, corresponding to 328 turnovers.

Scheme 10

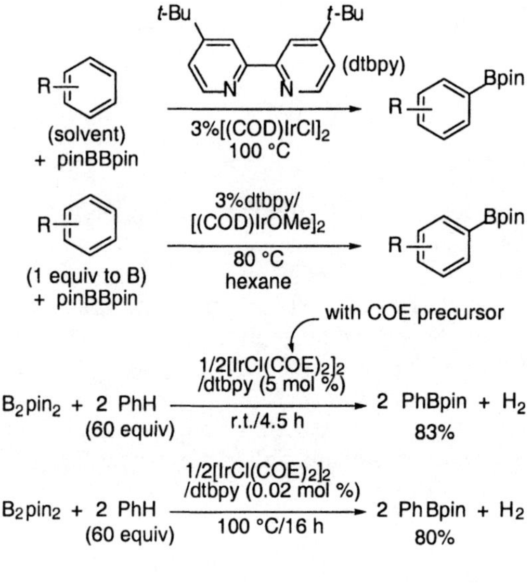

ArH (solvent) + pinBBpin $\xrightarrow[3\%[(COD)IrCl]_2,\ 100\,°C]{\text{dtbpy}}$ ArBpin

ArH (1 equiv to B) + pinBBpin $\xrightarrow[80\,°C,\ \text{hexane}]{3\%\ \text{dtbpy}/[(COD)IrOMe]_2}$ ArBpin

with COE precursor

$$B_2pin_2 + 2\ PhH\ (60\ \text{equiv}) \xrightarrow[\text{r.t.}/4.5\,h]{1/2[IrCl(COE)_2]_2/\text{dtbpy (5 mol\%)}} 2\ PhBpin + H_2$$
83%

$$B_2pin_2 + 2\ PhH\ (60\ \text{equiv}) \xrightarrow[100\,°C/16\,h]{1/2[IrCl(COE)_2]_2/\text{dtbpy (0.02 mol\%)}} 2\ PhBpin + H_2$$
80%

2.4. Development of an Efficient Functionalization of Arenes

In addition to developing catalysts for the regioselective functionalization of alkanes, we have developed efficient catalysts for the functionalization of arenes (Scheme 10). Aryl boronate esters are common reagents for Suzuki cross-coupling (*44-46*) and for oxidation to phenols. These reagents are typically prepared by halogenation of an arene, generation of the corresponding Grignard reagent, quenching of the Grignard reagent with a trialkylborate and either hydrolysis or alcoholysis with a diol. Thus, a route to aryl boronate esters directly from arenes would be much more efficient. Synthetic studies by Ishiyama and Miyaura, along with mechanistic studies by our group, led to the development of a room-temperature process for the borylation of arenes (*47-49*). The combination of 4,4'-di-*tert*-butyl bipyridine and [(COE)$_2$IrCl]$_2$ generated a species that converted arenes, including various substituted arenes, to aryl boronate esters in excellent yields. Reactions of monosubstituted arenes gave a mixture of 3-, and 4-aryl boronate esters while reactions of 1,3-disubstituted arenes gave 5-aryl boronate esters, and reactions of symmetric 1,2-disubstituted arenes generated 4-aryl boronate esters. In addition, the complex that is generated *in situ* catalyzed the borylation of arenes with remarkably high turnover numbers at elevated temperatures. Reaction of benzene with B$_2$pin$_2$ at 100 °C generated 80% yield of arylboronate ester with only 0.01 mol% catalyst, corresponding to 8,000 turnovers (*47*). This value surpasses most, if not all, turnover numbers reported for homogeneous functionalization of hydrocarbons.

$$\text{M}_n=1200 \text{ or } 37,000 \quad n/m=9 \quad \xrightarrow[\substack{5\% \text{ Cp*Rh(C}_6\text{Me}_6\text{)} \\ 150\ ^\circ\text{C}, 24\ \text{h}}]{\substack{0.025\text{-}.1 \\ \text{B}_2\text{pin}_2/\text{monomer}}} \quad \text{Bpin} + \text{H}_2 \quad \xrightarrow{\text{H}_2\text{O}_2/\text{HO}^-} \quad \text{OH} \quad (3)$$

2.5. Functionalization of Polyolefins

Polyolefins are alkanes, and selective functionalization of these high molecular weight alkanes would generate particularly valuable products. Polyolefins containing hydroxyl groups display higher glass transition temperatures, improved barrier properties, improved miscibility with polar phases, and the potential to generate graft copolymers (*50-55*). To demonstrate that polyolefins undergo selective borylation, we conducted catalytic reactions of boron reagents with hydrogenated polybutadiene (polyethyl ethylene, PEE, see eq. 1),(*56*) which contains relatively unhindered methyl groups on the polymer sidechain and which can be prepared with narrow polydispersities (*57*). The narrow polydispersities would allow the identification of chain coupling or chain scission processes that are characteristic of radical oxidation chemistry. Indeed, reaction of diboron compounds with these polyolefins in the presence of a Cp* rhodium catalyst generated PEE with boryl groups at the termini of the sidechains (eq. 1)(*58*). Subsequent standard oxidation generated the analogous hydroxylated polymer. ^{1}H NMR, ^{13}C NMR and IR spectroscopy demonstrated that hydroxylation of the termini of the ethyl sidechains had occurred. The amount of functionalization was dictated by the ratio of diboron compound to ethyl sidechains. Functionalization of up to 20% of the ethyl groups was achieved.

3. Mechanistic Studies of Alkane Borylation

3.1 Stoichiometric Alkane Borylation

Our mechanistic studies on the borylation of alkanes include stoichiometric reactions of isolated Cp*M(CO)$_n$BR$_2$ complexes and single-turnover experiments of potential intermediates in the catalytic process. To probe whether photochemical dissociation of CO occurred during reactions of the Cp* metal carbonyl complexes, we irradiated these complexes in the presence of ^{13}CO, 2 atm of CO, and PMe$_3$ (*25*). Irradiation of these complexes in the presence of ^{13}CO did not lead to incorporation of label into the starting complex. Moreover, reactions in the presence of 2 atm of CO did not occur more slowly than those conducted in the absence of added CO. Thus, dissociation of carbon monoxide either does not occur or occurs irreversibly. Reactions of the tungsten complex in the presence of PMe$_3$ demonstrated that dissociation of carbon monoxide occurred irreversibly (*25*). Irradiation of Cp*W(CO)$_3$Bcat* in the presence of 2 equiv of PMe$_3$ generated the mixture of phosphine-ligated tungsten boryl and the alkyl boronate ester in Scheme 11. Irradiation of Cp*W(CO)$_3$Bcat* with 4 equiv of PMe$_3$ generated roughly twice as much of the phosphine ligated boryl product as irradiation with 2 equiv of PMe$_3$. Thus we propose that generation of a 16-electron intermediate occurs by initial photochemical dissociation of carbon monoxide and that reaction of the 16-electron intermediate with alkane occurs faster than reassociation of the released carbon monoxide.

Scheme 11

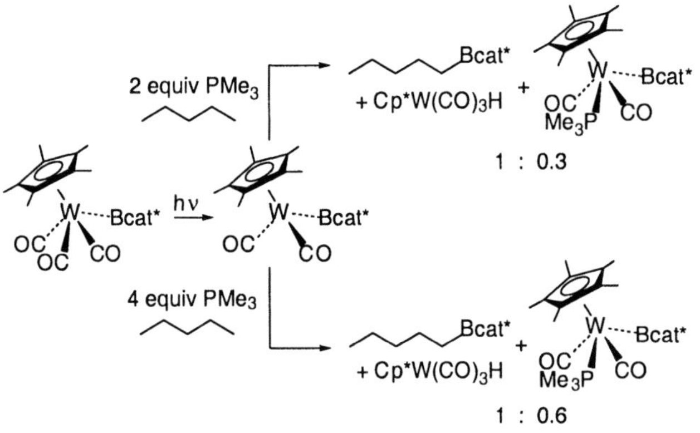
2 equiv PMe3
4 equiv PMe3
hν
Bcat*
OC
CO
+ Cp*W(CO)3H
Me3P
1 : 0.3
1 : 0.6

Scheme 12

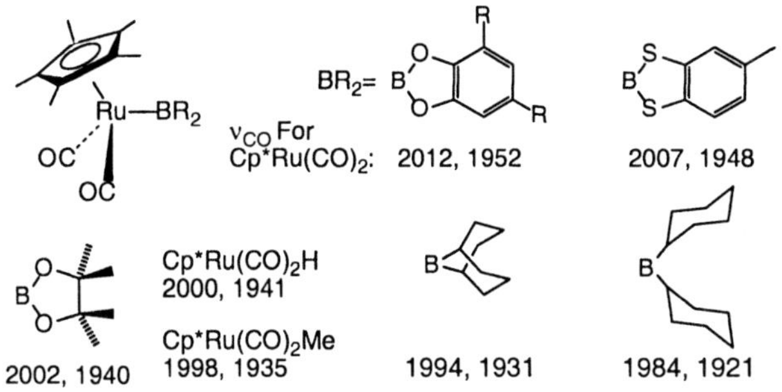
Ru—BR2
BR2=
νCO For Cp*Ru(CO)2:
2012, 1952
2007, 1948
Cp*Ru(CO)2H
2000, 1941
Cp*Ru(CO)2Me
1998, 1935
2002, 1940
1994, 1931
1984, 1921

One might propose that the boryl ligand imparts onto the transition metal unusual electronic properties and that this unusual electronic property triggers the C-H activation chemistry. To determine if the overall electron density of the metal center was unusually high because of strong sigma donation of the boryl ligand or unusually low because of back donation from the metal, we prepared several $Cp^*Ru(CO)_2$-boryl complexes and obtained infrared data (Scheme 12) (*25*). The catecholboryl complexes displayed ν_{CO} values that were 10-20 wavenumbers higher than those of the classic $Cp^*Ru(CO)_2H$(*59*). and $Cp^*Ru(CO)_2Me$(*59*). analogues. The dialkylboryl complexes displayed stretching frequencies that were 5-15 wavenumbers lower than those of the hydride and methyl analogues. Yet the ruthenium complex of the pinacolboryl group, which is the most effective boryl group for the catalytic transformations and reacts as cleanly as the catecholboryl groups in the stoichiometric chemistry, showed ν_{CO} values that were nearly identical to those of the hydride and methyl compounds. Therefore, the overall electron density of the metal center is not the driving force for the unusual C-H activation chemistry.

Instead, recent computational work in collaboration with Hall's group at Texas A&M suggests that the unoccupied p-orbital on boron triggers the C-H activation process (60). These calculations indicate that reaction occurs by the sigma-bond metathesis mechanism in Scheme 13 in which the alkane C-H bond of a boryl σ-alkane complex adds across the M-B bond to generate an alkyl σ-borane complex. After an intramolecular rearrangement to place the boryl group of the coordinated borane cis to the alkyl group, reductive elimination of alkylborane occurs. This calculated pathway is roughly 10 kcal/mol lower in energy than the oxidative addition pathway (61).

Scheme 13

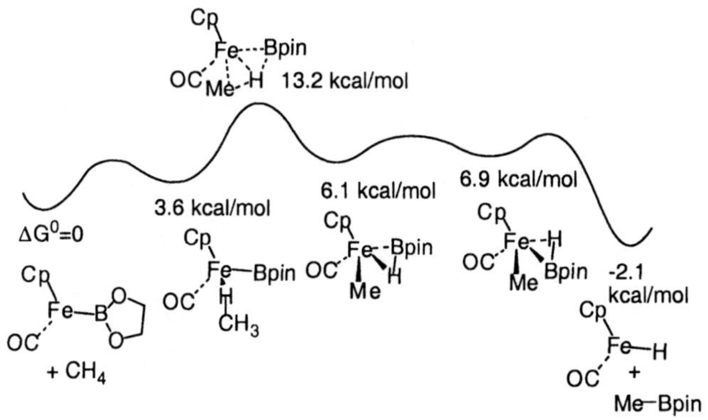

3.2 Single-Turnover Studies of the Borylation of Alkanes

Additional mechanistic studies have led to the generation of Cp*Ir(V)-boryl complexes that are potential intermediates in the iridium catalyzed borylation of alkanes and that are analogs of intermediates in the rhodium catalyzed borylation of alkanes (62). Reaction of Cp*IrH$_4$ with excess pinacolborane generated Cp*IrH$_2$(Bpin)$_2$, and lithiation of Cp*IrH$_4$,(63). followed by quenching of the anion with ClBpin, generated Cp*IrH$_3$Bpin. These iridium boryl complexes reacted with alkanes, as shown in Scheme 14, to generate alkyl boronate esters in roughly 50% yield with the same regiospecificity as observed in the catalytic process. These complexes also reacted with benzene to generate the phenyl boronate ester. Related pinacolboryl and 9-BBN complexes were also generated. The catecholboryl complexes did not react with alkanes but did react in good yields with arenes to generate aryl boronate esters. The 9-BBN complexes did not react with either alkanes or arenes.

Scheme 14

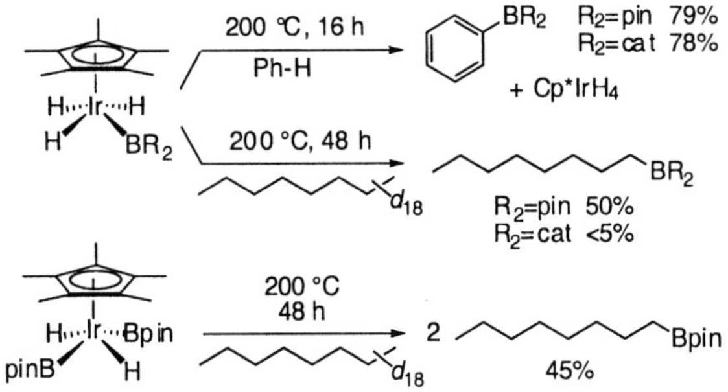

Scheme 15

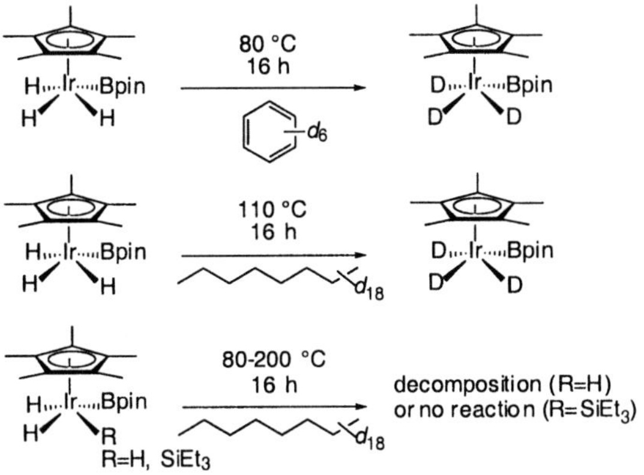

Reactions of the mono- and bis-boryl complexes of iridium required long times, even at 200 °C, to undergo complete reaction. Yet, interaction of the hydrocarbons with the iridium complexes did occur at lower temperatures, as summarized in Scheme 15. For example, the mono pinacolboryl complex underwent H/D exchange at 80 °C with benzene-d_6 to place deuterium at the hydride position. It more slowly underwent exchange at the Cp* position, presumably by an intramolecular C-H activation process of the initially formed deuteride complex (*64*). H/D exchange also occurred with alkanes at temperatures lower than those required for the functionalization process. Reaction of the mono pinacolboryl complex at 110° with octane-d_{18} also led to H/D exchange at the hydride position. This reactivity stands in contrast to the inertness of the iridium tetrahydride and iridium silyl hydride complexes toward alkanes (*65*). We found that the tetrahydride simply undergoes decomposition at about 80 °C in octane-d_{18}, while the triethylsilyl complex does not undergo decomposition or H/D exchange after 20 h at 200 °C in octane-d_{18}.

3.3. Single Turnover Experiments on the Borylation of Arenes

Iridium boryl complexes that are potential intermediates in the rapid and high-turnover borylation of arenes have also been prepared (47). Reaction of bis-cyclooctadiene iridium dichloro dimer with di-*tert*-butylbipyridine in the presence of excess of the B_2pin_2 reagent generated a tris-boryl complex (Scheme 16). Yields for the generation and isolation of this complex were only 15-30%, but a sample suitable for X-ray diffraction and samples suitable for preliminary studies on reactivity were obtained. This tris-boryl complex reacted at room temperature with neat benzene to generate in high yield 3 equiv of the phenyl boronate ester. The role of this complex in the major catalytic pathway is not yet clear, and the mechanism by which this complex reacts with benzene has not yet been determined, but the selectivity of this complex for reaction with benzene vs. benzene-d_6 and for trifluorotoluene vs. toluene was indistinguishable from that of the catalytic process.

Scheme 16

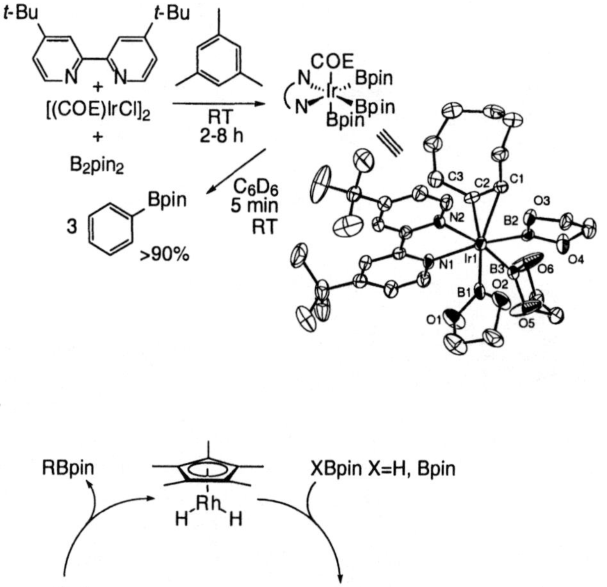

Scheme 17

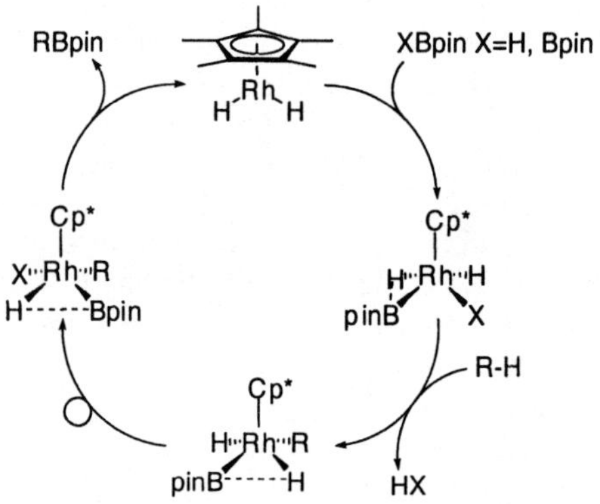

3.4 Proposed Catalytic Cycle

Based on the isolation of Cp*Ir(V)-boryl complexes and the participation of the boron p-orbital in the C-H bond cleavage process deduced by theory, we propose the mechanism in Scheme 17 as one pathway followed during the catalytic functionalization of alkanes. Clearly, further work must be conducted to test this hypothesis, but the proposal is consistent with the current data. In this mechanism, a Cp*RhH$_2$ intermediate undergoes oxidative addition of the diboron or pinacolborane reagent to generate a dihydride bis-boryl or trihydride mono-boryl intermediate. These boryl complexes would then undergo dissociation of borane or H$_2$ to generate a mono hydride, mono boryl intermediate. This 16-electron intermediate would then undergo reaction with alkane to generate an alkyl σ-borane complex that would form the alkyl boronate ester product and the starting dihydride. Several other mechanisms containing other monomeric hydride or boryl complexes as intermediates can also be drawn and may be occurring in parallel. Yet, a Rh(III)-boryl complex should be the intermediate that reacts with the alkane in any mechanism.

4. Conclusions

The first process by which alkanes are transformed regiospecifically into terminal functionalized products has been developed. Enzymes called omega-hydroxylases are also capable of terminal functionalization of alkanes. These enzymes generate alcohol with preferential reactivity at the terminal position, but the regioselectivity of the oxidation varies with alkane chain length and is much lower than the regioselectivity of the process described here (*66*). Moreover, these hydroxylases are membrane-bound and are, therefore, difficult to study and to use synthetically (*67*). Other transition metal complexes noted in the Introduction are capable of cleaving the terminal C-H bond of an alkane, but catalytic processes that involve this regioselective elementary reaction and that form functionalized product in good yield are rare.

Boranes and metal boryl complexes undergo this C-H functionalization chemistry for several reasons. First, the B-C bond is a strong 111-113 kcal/mol in MeB(OR)$_2$,(*68,69*). which makes the C-H bond of the alkane appear weak. Second, the boryl group contains no β-hydrogens, and the pinacolboryl group is remarkably stable. Finally, the unoccupied p-orbital at boron appears to assist in the C-H bond cleavage event and may also facilitate the B-C coupling. Perhaps these concepts can be used to find additional reagents and metal complexes that create similarly selective functionalization processes.

Acknowledgement

We thank the NSF and Shell chemicals for support of this work and Johnson-Matthey for a gift of precious metal salts. The author thanks the efforts and insight provided by the postdoctoral and graduate students at Yale (Karen Waltz, Clare Muhoro, Huiyuan Chen, Sabine Schlecht, Kazumori Kawamura, Natia Anstasi, Domingo Garcia, Yuichiro Kondo, Makoto Takahashi and Chulsung Bae) as well as collaborators at Hokkaido (Prof. Tatsuo Ishiyama, Prof. Norio Miyaura, and Jun Takagi), Texas A&M (Dr. Charles Edwin Webster, Dr. Yubo Fan, and Prof. Michael Hall) and the University of Minnesota (Dr. Nicole Boaen, Dr. Nicole Wagner, and Prof. Marc Hillmyer).

References

1. Fokin, A. A.; Schreiner, P. R. *Chem. Rev.* **2002**, *102*, 1551-1593.
2. Stahl, S. S.; Bercaw, J. A. L. E. *Angew. Chem. Int. Ed.* **1998**, *37*, 2181-2192.
3. Haggin, J. *Chem. Eng. News* **1993**, *71*, 23-27.
4. Chenier, P. J. *Survey of Industrial Chemistry;* 2 ed.; VCH: Weinheim, 1992.
5. McMillen, D. F.; Golden, D. M. *Ann. Rev. Phys. Chem.* **1982**, *33*, 493-532.
6. See, for example, Jones, M. J. *Organic Chemistry*; W.W. Norton and Company: New York, 1997.
7. Kunin, A. J.; Eisenberg, R. *Organometallics* **1988**, *7*, 2124-2129.
8. Beck, C. M.; Rathmill, S. E.; Park, Y. J.; Chen, J.; Crabtree, R. H. *Organometallics* **1999**, *18*, 5311-5317.
9. O'Connor, J. M.; Ma, J. *J. Org. Chem.* **1992**, *57*, 5075-5077.
10. Periana, R. A.; Bergman, R. G. *J. Am. Chem. Soc.* **1986**, *108*, 7332-7346.
11. Jones, W. D.; Feher, F. J. *Organometallics* **1983**, *2*, 562-563.
12. Flood, T. C.; Janak, K. E.; Iimura, M.; Zhen, H. *J. Am. Chem. Soc.* **2000**, *122*, 6783-6784.
13. Jones, W. D.; Hessell, E. T. *J. Am. Chem. Soc.* **1993**, *115*, 554-562.
14. Bennett, J. L.; Wolczanski, P. T. *J. Am. Chem. Soc.* **1997**, *119*, 10696-10719.
15. Shilov, A. E.; Shteinman, A. A. *Coord. Chem. Rev.* **1977**, *24*, 97-143.
16. Shilov, A. E.; Shul'pin, G. B. *Chem. Rev.* **1997**, *97*, 2879-2932.
17. Sakakura, T.; Tanaka, M. *J. Chem. Soc., Chem. Commun.* **1987**, 758-759.
18. Rosini, G. P.; Zhu, K. M.; Goldman, A. S. *J. Organomet. Chem.* **1995**, *504*, 115-121.
19. Liu, F.; Pak, E. B.; Singh, B.; Jensen, C. M.; Goldman, A. S. *J. Am. Chem. Soc.* **1999**, *121*, 4086-4087.
20. Smith, M. R., III *Prog. Organomet. Chem.* **1999**, *48*, 505-567.

21. Irvine, G. J.; Lesley, M. J. G.; Marder, T. B.; Norman, N. C.; Rice, C. R.; Robins, E. G.; Roper, W. R.; Whittell, G. R.; Wright, L. J. *Chem. Rev.* **1998**, *98*, 2685-2722.
22. Braunschweig, H. *Angew. Chem. Int. Ed. Engl.* **1998**, *37*, 1787-1801.
23. Hartwig, J. F.; Waltz, K. M.; Muhoro, C. N. In *Advances in Boron Chemistry;* W. Siebert, Ed.; The Royal Society of Chemistry: Cambridge, 1997; pp 373-380.
24. Waltz, K. M.; Muhoro, C. N.; Hartwig, J. F. *Organometallics* **1999**, *18*, 3383-3393.
25. Waltz, K. M.; Hartwig, J. F. *J. Am. Chem. Soc.* **2000**, *122*, 11358-11369.
26. Hartwig, J. F.; Huber, S. *J. Am. Chem. Soc.* **1993**, *115*, 4908-4909.
27. Waltz, K. M.; He, X.; Muhoro, C. N.; Hartwig, J. F. *J. Am. Chem. Soc.* **1995**, *117*, 11357-8.
28. Waltz, K. M.; Hartwig, J. F. *Science* **1997**, *277*, 211-213.
29. Marder, T. B.; Norman, N. C.; Rice, C. R.; Robins, E. G. *Chem. Commun.* **1997**, 53-54.
30. Dai, C.; Stringer, G.; Marder, T. B. *Inorg. Chem.* **1997**, *36*, 272-273.
31. Hartwig, J. F.; He, X. *Angew. Chem. Int. Ed. Engl.* **1996**, 315-17.
32. Ishiyama, T.; Matsuda, N.; Murata, M.; Ozawa, F.; Suzuki, A.; Miyaura, N. *Organometallics* **1996**, *15*, 713-720.
33. Iverson, C. N.; Smith, M. R., III *Organometallics* **1996**, *15*, 5155-5165.
34. Chen, H.; Hartwig, J. F. *Angew. Chem. Int. Ed. Engl.* **1999**, *38*, 3391-3.
35. Janowicz, A. H.; Bergman, R. G. *J. Am. Chem. Soc.* **1983**, *105*, 3929-3939.
36. Janowicz, A. H.; Bergman, R. G. *J. Am. Chem. Soc.* **1982**, *104*, 352-354.
37. Hoyano, J. K.; McMaster, A. D.; Graham, W. A. G. *J. Am. Chem. Soc.* **1983**, *105*, 7190-7191.
38. Chen, H.; Schlecht, S.; Semple, T. C.; Hartwig, J. F. *Science* **2000**, *287*, 1995-1997.
39. Maitlis, P. M.; Moseley, K.; Kang, J. W. *J. Chem. Soc. A* **1970**, 2875-83.
40. Bowyer, W. J.; Geiger, W. E. *J. Am. Chem. Soc.* **1985**, *107*, 5657-63.
41. Bowyer, W. J.; Merkert, J. W.; Geiger, W. E. *Organometallics* **1989**, *8*, 191-198.
42. White, C.; Yates, A.; Maitlis, P. M. *Inorg. Synth.* **1992**, *29*, 228-234.
43. Shimada, S.; Batsanov, A. S.; Howard, J. A. K.; Marder, T. B. *Angew. Chem. Int. Ed. Engl.* **2001**, *40*, 2168-2171.
44. Suzuki, A. *J. Organomet. Chem.* **1999**, *576*, 147-168.
45. Miyaura, N.; Suzuki, A. *Chem. Rev.* **1995**, *95*, 2457-2483.
46. Miyaura, N. *Top. Curr. Chem.* **2002**, *219*, 11-59.
47. Ishiyama, T.; Takagi, J.; Ishida, K.; Miyaura, N.; Anastasi, N.; Hartwig, J. F. *J. Am Chem. Soc.* **2002**, *124*, 390-391.
48. Takagi, J.; Sato, K.; Hartwig, J. F.; Ishiyama, T.; Miyaura, N. *Tetrahedron Lett.* **2002**, *43*, 5649-5651.
49. Ishiyama, T.; Takagi, J.; Hartwig, J. F.; Miyaura, N. *Angew. Chem. Int. Ed.* **2002**, *41*, 3056-3058.
50. Imuta, J.; Kashiwa, N. *J. Am. Chem. Soc.* **2002**, *124*, 1176-1177.

51. Santos, J. M.; Ribeiro, M. R.; Portela, M. F.; Bordado, J. M. *Chem. Eng. Sci.* **2001**, *56*, 4191-4196.
52. Watson, M. D.; Wagner, K. B. *Macromolecules* **2000**, *33*, 5411-5417.
53. Chung, T. C.; Lu, H. L. *J. Molecular Catalysts* **1997**, 115-127.
54. Chung, T. C.; Raate, M.; Berluche, E.; Schulz, D. N. *Macromolecules* **1988**, *21*, 1903-1907.
55. Chung, T. C. *Macromol. Symp.* **1995**, *89*, 151-162.
56. Doi, Y.; Yano, A.; Soga, K.; Burfield, D. R. *Macromolecules* **1986**, *19*, 2409-2412.
57. Hillmyer, M. A.; Bates, F. S. *Macromolecules* **1996**, *29*, 6994-7002.
58. Kondo, Y.; Garcia-Cuadrado, D.; Hartwig, J. F.; Boaen, N. K.; Wagner, N. L.; Hillmyer, M. A. *J. Am. Chem. Soc.* **2002**, *124*, 1164-1165.
59. Stasunik, A.; Wilson, D. R.; Malisch, W. *J. Organomet. Chem.* **1984**, *270*, C18-C22.
60. Webster, C. E.; Fan, Y.; Hall, M. B.; Kunz, D.; Hartwig, J. F. *J. Am. Chem. Soc.* **2002**, *124*, 858-859.
61. Lam, W. H.; Lin, Z. *Organometallics* **2003**, *22*, 473-480.
62. Kawamura, K.; Hartwig, J. F. *J. Am. Chem. Soc.* **2001**, *123*, 8422-8423.
63. Gilbert, T. M.; Bergman, R. G. *Organometallics* **1983**, *2*, 1458-1460.
64. Bulls, A. R.; Schaefer, W. P.; Serfas, M.; Bercaw, J. E. *Organometallics* **1987**, *6*, 1219-1226.
65. Ricci, J. S., Jr.; Koetzle, T. F.; Fernandez, M.-J.; Maitlis, P. M.; Green, J. C. *J. Organomet. Chem.* **1986**, *299*, 383-9.
66. Fisher, M. B.; Zheng, Y.-M.; Rettie, A. E. *Biochem. Biophys. Res. Commun.* **1998**, *248*, 352-355.
67. Shanklin, J.; Achim, C.; Schmidt, H.; Fox, B. G.; Munck, E. *Proc. Natl. Acad. Sci. USA* **1997**, *94*, 2981-2986.
68. Rablen, P. R.; Hartwig, J. F.; Nolan, S. P. *J. Am. Chem. Soc.* **1994**, *116*, 4121-4122.
69. Rablen, P.; Hartwig, J. F. *J. Am. Chem. Soc.* **1996**, *118*, 4648-4653.

Chapter 9

C–H Bond Functionalization in Complex Organic Synthesis

Dalibor Sames

Department of Chemistry, Columbia University, New York, NY 10027

Activation and functionalization of unactivated sp^3 C-H bonds have been demonstrated in complex substrates containing multiple reactive groups. These reactions were designed as key steps in syntheses of natural products (rhazinilam, teleocidins) as well as in direct hydroxylation of α-amino acids. Both stoichiometric and catalytic C-H bond functionalization methods are discussed in this chapter.

Introduction

C-H bond activation represents a chemical process of broad synthetic potential. The ability to activate ubiquitous, but often inert C-H bonds has far-reaching implications ranging from oxidation of simple hydrocarbons to the synthesis of complex organic molecules. Although C-H bond functionalization of alkanes has historically relied on radical chemistry, homogeneous transition metal complexes capable of activating C-H bonds in arenes and alkanes unlocked a new era in this field ([1]).

The inorganic and organometallic communities have generated a body of fundamental knowledge related to the feasibility and mechanistic issues of this transformation (see other chapters in this volume). However, most of these pioneering studies have been focused on simple hydrocarbons. Thus, we set out to address the considerable challenge of whether C-H functionalization could be realized in complex substrates and in a selective manner. In other words, we have been searching for new types of reagents and catalysts that are compatible with reactive functionalities. It is well-recognized that functional groups present in substrates strongly influence the reactivity of proximal C-H bonds, a fact that has been extensively studied in arene systems (e.g. *ortho*-metallation, electrophilic substitution) (2). Similarly, sp^3 C-H bonds adjacent to a heteroatom or a functional group (activated C-H bonds) are often (but not always) more reactive toward functionalization (Figure 1). At the same time, functional groups of the substrate affect the reactivity of metal catalysts (coordination chemistry). All of these issues must be considered and studied in a systematic fashion in order to develop new catalysts capable of selective functionalization of different types of C-H bonds in the presence of reactive functionalities. Such systems, applicable to a broad range of substrates, will unlock unprecedented synthetic possibilities, on both conceptual and practical level.

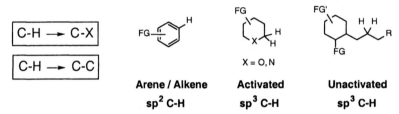

Figure 1. C-H bond functionalization in complex organic substrates. Different types of C-H bonds.

Alkane C-H Bonds: Coordination-Directed C-H Bond Activation of Alkyl Groups

As a part of our initial goals, we have focused on the possibility of functionalizing alkane segments in complex organic substrates, perhaps the most challenging aspect of this area. As a first approach we proposed to achieve this goal via *coordination-directed C-H bond activation (3, 4)*. Following this strategy, a suitable heteroatomic function is utilized to activate and direct a metal complex to a specific alkane segment of the substrate in such a way as to prevent interference by other functional groups. We set out to explore the formation and reactivity of metallacycle intermediates with the view of developing new functionalization reactions, and possibly catalytic systems (Figure 2).

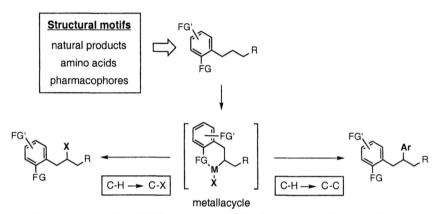

Figure 2. Selective C-H Bond functionalization via coordination-directed approach

We have demonstrated the feasibility of this approach in the context of the synthesis of the antitumor agent rhazinilam (Figure 3) (5). The pivotal step in rhazinilam assembly involved a selective functionalization of the diethyl segment in intermediate **1** via the attachment of a platinum complex. The proximity of the amino group to the ethyl groups augured well for directed C-H activation. This was achieved *via* the intermediacy of platinum complex **4** (Figure 4). Remarkably, thermolysis of complex **4** provided platinum hydride **5** as a single product in excellent yield (>90%) as determined by proton NMR spectroscopy. Indeed, activation of the desired ethyl group took place with concomitant loss of methane, followed by β-H-elimination affording alkene-hydride platinum(II) complex **5**. Decomplexation of the platinum metal *via* treatment with aqueous potassium cyanide, followed by the Schiff base hydrolysis, provided alkene **2**. Noteworthy is the fact that selective dehydrogenation of one of the ethyl groups was achieved in the presence of a variety of functional groups including an ester, pyrrole and arene rings. Furthermore, we have recently demonstrated that the platinum chemistry described above could be adapted to chiral oxazoline ligands ultimately affording alkene **2** in an optically pure form (*6*).

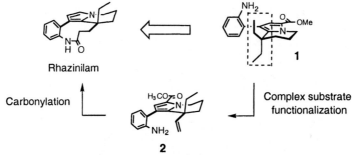

Figure 3. The total synthesis of rhazinilam was achieved via selective C-H activation (dehydrogenation)

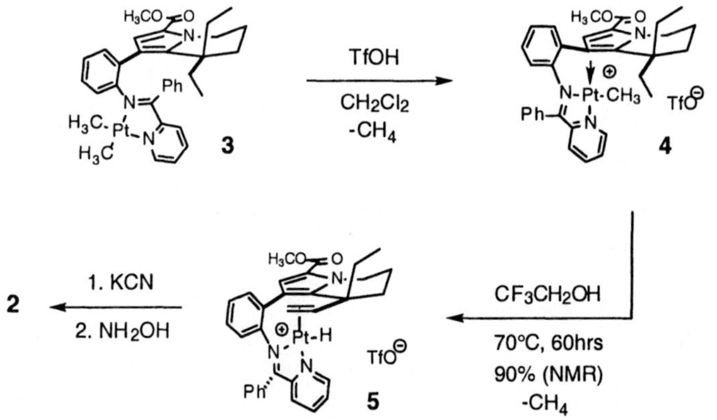

Figure 4. Selective dehydrogenation of one ethyl group was accomplished in the presence of many functional groups via coordination-directed C-H bond activation.

We have also demonstrated that the directed cyclometallation reaction may be extended to other substrates with a promising degree of generality. At the same time, we began to focus on the development of catalytic systems wherein the transition metal, following the C-H activation and functionalization step, could be recycled. In this context we became interested in the possibility of functionalizing free amino acids in water.

We found that L-valine was converted to γ-lactone **6** in water in the presence of K_2PtCl_4 (1-5 mol%) as the catalyst and $CuCl_2$ (5-7 eq) as the oxidant (Figure 5A). The highest turnover number achieved to date is twenty (1 mol% K_2PtCl_4, 5 eq $CuCl_2$) (7). A number of substrates have been investigated including norvaline, leucine, isoleucine, proline, *n*-butylamine and valeric acid. The results generated in this study uncovered regioselectivity trends for α-amino acids that were distinctly different from those for simple aliphatic amines and carboxylic acids. In the case of α-amino acids, γ-hydroxylation was the favored process over δ-oxidation. Thus, the direct functionalization of natural amino acids afforded valuable intermediates (γ-lactones) in one step without the use of organic solvents. We proposed that the reaction proceeds via coordination directed C–H bond activation yielding a putative platinacycle intermediate **8** or **9**, which collapses to afford a lactone and then regenerates platinum catalyst (Figure 5B).

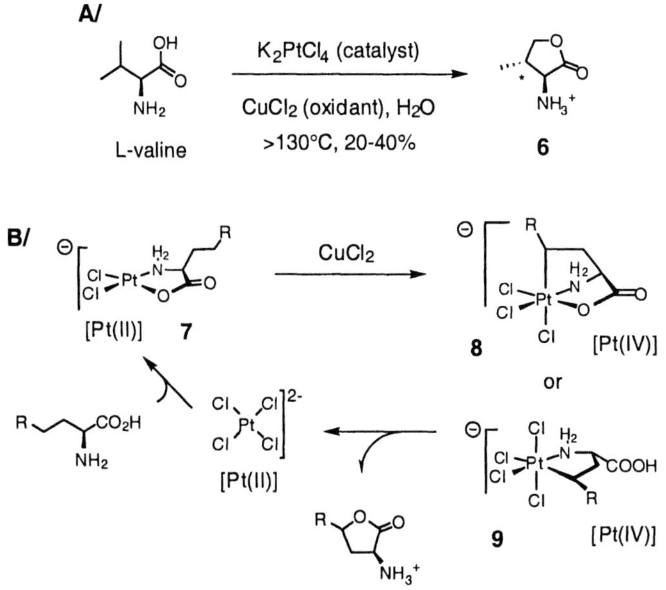

Figure 5. A/ Selective hydroxylation of α–amino acids in water via catalytic C–H bond activation. B/ Proposed catalytic cycle. Mechanistic work suggests that the reaction proceeds through a Pt(IV)-platinacycle (8 or 9).

Mechanistic studies (unpublished data) revealed important insights which may have broader applicability in the context of C-H bond functionalization of complex substrates. Namely, our data is consistent with the mechanistic outline wherein the exchange of the product for starting material at the metal center is slow. Perhaps surprisingly, both C-H activation and functionalization steps are facile transformations, whereas the catalyst regeneration step is slow. In order to achieve a practical turnover rate, the reaction must be heated to >130 °C. Thus, in the process of designing catalytic systems for C-H functionalization of complex substrates, the character and strength of the coordination bond between the functional groups and the metal catalyst must be taken into account.

Having demonstrated the feasibility of the coordination-directed activation both in stoichiometric and catalytic functionalization of alkyl groups, we subsequently turned to the development of methods for direct formation of C-C bonds from C-H bonds. In this context we were inspired by a class of natural products known as teleocidins. We envisioned that the teleocidin core, a complex fragment of a natural product containing two quaternary stereocenters and a pentasubstituted benzene ring, would be made via a series of non-traditional disconnections, serving as the platform for new reaction discovery (Figure 6) (8).

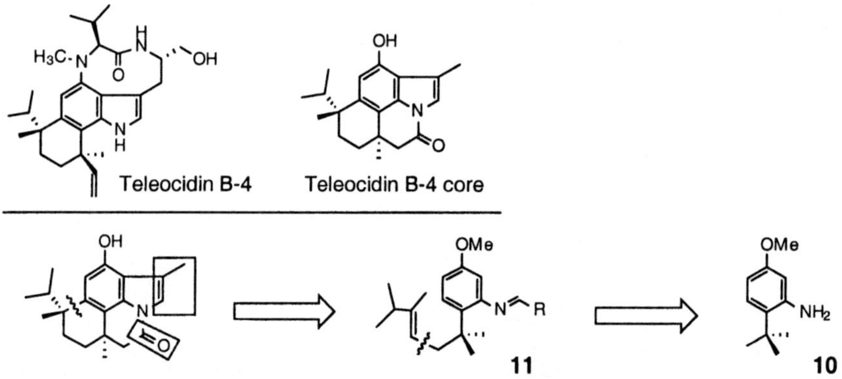

Figure 6. Retrosynthetic analysis of teleocidin natural products. Proposed non-traditional disconnections stimulated the development of new C-H activation/C-C bond forming processes.

According to the synthetic plan outlined above, we set out to prepare intermediate **11** through alkenylation of *t*-butyl aniline **10** (Figure 6). We envisioned that a new transformation of this type might be accomplished via sequential cyclometallation and transmetallation. Consequently, Schiff base **12** was prepared and submitted to a systematic screening of metal salts in the context of directed C-H bond activation (cyclometallation). We found that Pd(II) salts were the only reagents capable of furnishing desirable and stable metallacycle products (cf. **13**, Figure 7). We were delighted to find that palladacycle **13** and vinyl boronic acid **14** yielded desired alkene **15** in 86% yield in the presence of Ag$_2$O as the reagent of choice. This two-step sequence (**12**→**13**→**15**) provided not only the desired alkenylation product **15**, but moreover set the stage for the development of a new catalytic transformation (see below).

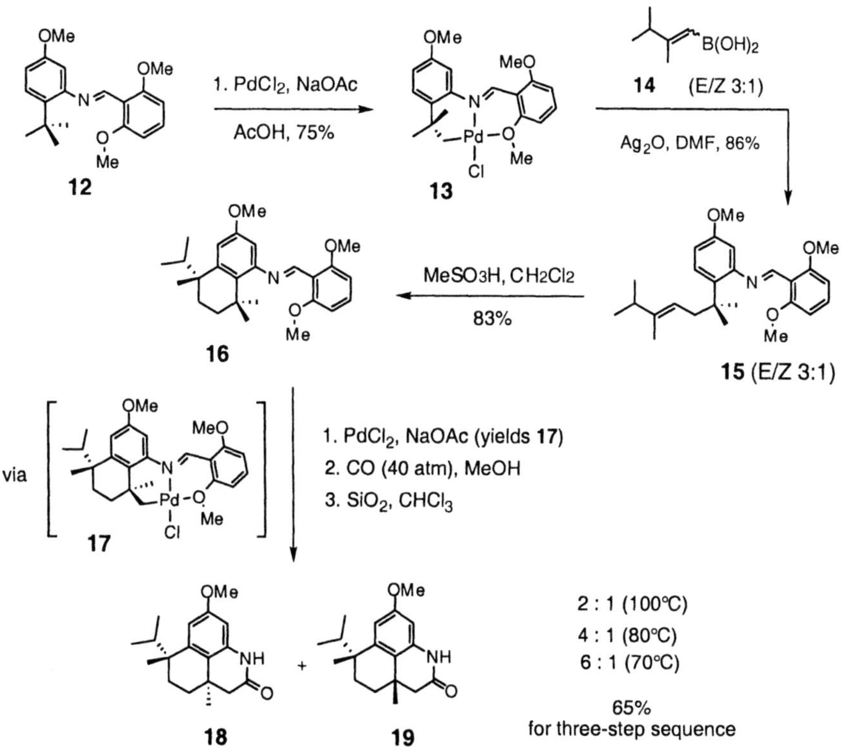

Figure 7. Two tandem cycles of C-H bond functionalization render two quaternary centers of the teleocidin core.

The subsequent step in the route centered on the closure of the cyclohexane ring through a formal hydroarylation process. In this instance, the presence of the methoxy group *meta-* to the amine facilitated the Friedel-Crafts reaction in the presence of methanesulfonic acid, providing racemic compound **16** in 83% yield. At this stage of the synthesis a diastereoselective one-carbon homologation of the methyl group that is *anti* to the isopropyl group (1,4-chiral transfer) was required (Figure 7). Thus, intermediate **16** was again treated with PdCl$_2$ and NaOAc to yield a mixture of diastereomeric palladacycles (cf. **17**), followed by addition of CO(g) and methanol. The resulting methyl esters were not isolated, but instead acidic hydrolysis of the Schiff base, accompanied by spontaneous cyclization, furnished lactams **18** and **19** (6:1 ratio at 70 °C). This three-step sequence converted compound **16** to the desired lactam without isolation of a single intermediate. Lactam **18** was subsequently converted to the final product (the teleocidin B4 core) in three steps (not shown, ref. 8)

In summary, the core of teleocidin B4 was synthesized in four C-C bond forming steps starting from *t*-butyl derivative **12** (9 steps in total). The key sequence of the synthesis consisted of two C-H bond functionalization cycles, namely alkenylation and oxidative carbonylation of a *t*-butyl group. This work demonstrated that the consideration of non-traditional disconnections in the context of synthetic strategy has significant consequences, in that not only new perspectives on organic compounds emerge, but the development of new chemical transformations is further inspired.

As the previous study led to the development of stoichiometric alkenylation and arylation methodology, the stage was set to address the next central question of whether a new catalytic system could be generated. We have put forth a mechanistic proposal of a putative catalytic cycle, which served well as a design blueprint and guide for our explorations (Figure 8).

Based on our preliminary results we focused on the possibility of generating a phenylpalladium species (*cf.* **22**), followed by the C-H activation step (Figure 9). Phenylboronates (*10*), phenylstannanes (*11*), and phenylsilanols (*11,12*) have been suggested to undergo transmetallation with Pd(OAc)$_2$ in polar solvents. Consequently, through a systematic search of reaction conditions (Pd^{2+} salts, arene donors, solvents) we uncovered an exciting lead. While PhB(OH)$_2$ and PhSnBu$_3$ failed, Schiff base **20** was transformed to compound **21** in 53% isolated yield via *direct arylation of the t-butyl group in the presence of PhSi(OH)Me$_2$ and Pd(OAc)$_2$* (Figure 9). Note that the phenyl ring is attached at the *neo*-alkyl position (*13*) and that no *bis*-arylation products were identified. Thus, the first three steps of the cycle have been synchronized and direct arylation of compound **20** in the presence of a stoichiometric amount of Pd(OAc)$_2$ was developed.

The next critical question centered on the compatibility of this system with an oxidant. Following an extensive study we were delighted to find that *the direct arylation of substrate **26** proceeded under conditions catalytic in palladium in the presence of Cu(OAc)$_2$* as the oxidant and phenyldimethylsilanol as the phenyl ring donor (Table 1).

A systematic mapping of the system included examining the Schiff base-directing element of the substrate, metal salts, oxidants and solvents. 2-Thiomethoxybenzylidene Schiff base **25** afforded consistently higher yields than the dimethoxy substrate **20**, while Pd(OAc)$_2$ proved to be a catalyst of choice and DMF the best solvent. Out of the oxidants tested, Cu(OAc)$_2$ was the most efficient. We rapidly discovered that Ph$_2$Si(OH)Me was a superior phenyl donor affording the highest yield of product **26**, with only traces of biphenyl and phenyl acetate side products (*14*).

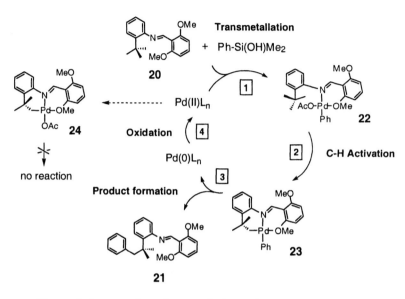

Figure 8. Proposed catalytic cycle for arylation of t-butyl group.

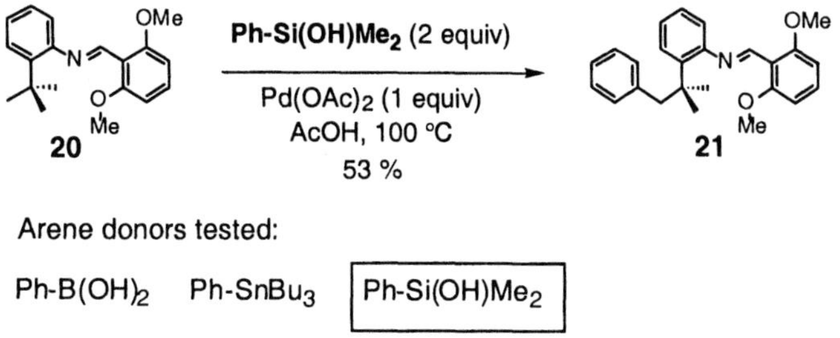

Figure 9. Development of one-pot arylation protocol

Further optimization showed that the highest isolated yield of **26** (73%) was achieved in the presence of 4 mol% of Pd(OAc)$_2$ and benzoquinone (1:1), and two equivalents of Cu(OAc)$_2$ (Table 2).

Table 1. Catalytic arylation. Optimization Studies

25 + Ph–Si → **26**
Pd(OAc)$_2$ (cat.), Cu(OAc)$_2$ (2 equiv), DMF, 100° C

silanol	Pd(OAc)$_2$ mol %	Yield (w/BQ)[a] %	TON	Ph-Ph (%)	PhOAc (%)
PhSi(OH)Me$_2$	1	17	17	8	7
	2.5	33	13	5	5
	4	37 (52)	9 (13)	5 (12)	5 (7)
	5	42 (49)	8 (10)	7 (12)	4 (8)
Ph$_2$Si(OH)Me	1	26	26	<1	<1
	2.5	51	20	<1	<1
⇨	4	58 (**73**)	15 (18)	<1	<1
	5	61 (68)	12 (15)	<1	<1
PhSi(OMe)$_3$	2.5	31	12	4	<1

(a) isolated yields. The value in parentheses represents the isolated yield in the presence of a catalytic amount of benzoquinone (1:1 ratio of Pd to BQ).

The new arylation methodology was also extended to alkenylation, as documented by the transfer of styrenyl group to substrate **25** furnishing compound **27** in 64% yield (Table 2). In addition to the *t*-butylaniline substrate, 2-pivaloylpyridine also proved to be a good substrate for both arylation and alkenylation reactions. As in the previous case, substrate **28** underwent single arylation/alkenylation at the *t*-butyl group. However, the Schiff base derived from ortho-*i*-propylaniline **31** yielded no desired material, biphenyl (22%) being the major detected product.

A concise synthesis of compound **36**, depicted in Figure 10, demonstrated the synthetic power of the new methodology, and simultaneously revealed both the remarkable selectivity as well as limitations of this system. Substrate **33** was converted to complex product **36** in three steps via catalytic alkenylation of **33**, Friedel-Crafts cyclization, and finally catalytic arylation to furnish substance **36**. Thus, tandem alkenylation-arylation of the t-butyl group provided a product of considerable complexity via a novel bond construction strategy (*14*).

Table 2. Catalytic arylation and alkenylation of selected substrates

entry	substrate	conditions	product
1	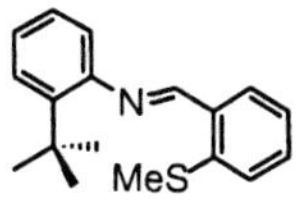 **25** (aryl imine with MeS and i-Pr groups)	Ph⁀Si(OH)Me₂ / a, 64%	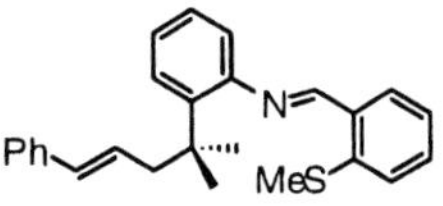 **27**
2	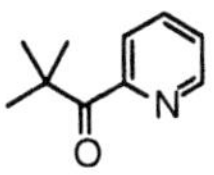 **28** (t-Bu pyridyl ketone)	Ph₂Si(OH)Me / a, 78%	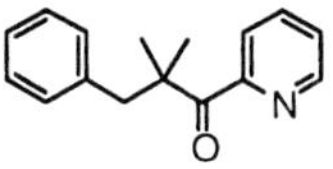 **29**
3	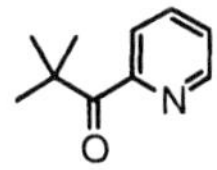 **28**	Ph⁀Si(OH)Me₂ / a, 63%	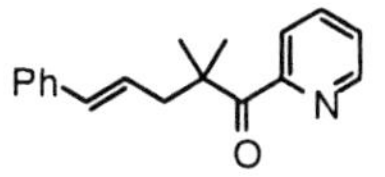 **30**
4[b]	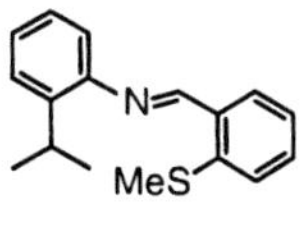 **31**	Ph₂Si(OH)Me / a, trace	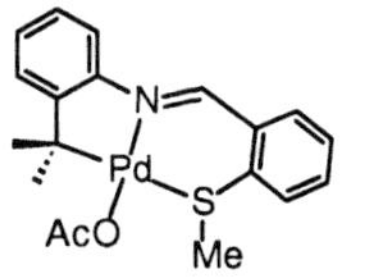 **32** (Pd complex with AcO, SMe)

(a) substrate (c = 0.02 M) in DMF, silanol (2 equiv), Pd(OAc)₂, (4 mol %), benzoquinone (4 mol %), Cu(OAc)₂ (2 equiv), 100 °C. The alkenylation reactions also yielded a side product (PhCH=CH₂)₂ in 10% yield.

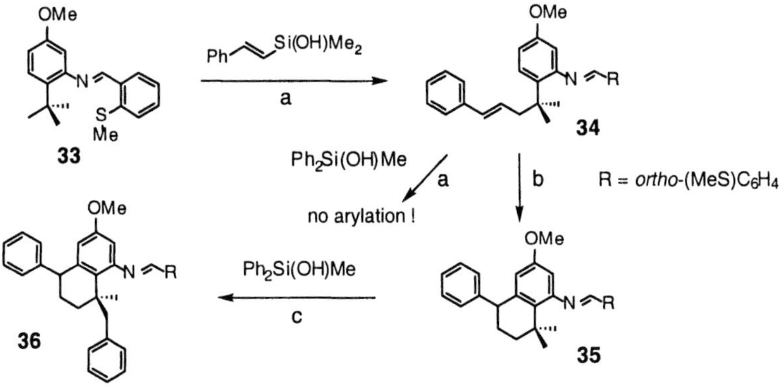

Figure 10. Tandem alkenylation-arylation of t-butyl group. Cond: (a) Pd(OAc)₂, (4 mol %), benzoquinone (4 mol %), Cu(OAc)₂ (2 equiv), 100 °C, 66 %. (b) MeSO₃H, CH₂Cl₂, 52 %. (c) Pd(OAc)₂, (8 mol %), BQ (8 mol %), Cu(OAc)₂ (2 equiv), 100 °C, 45 % (4:1 ratio of diastereomers).

In summary, a new system for the catalytic arylation and alkenylation of alkane segments has been developed. The *ortho-t*-butylaniline substrates and 2-pivaloylpyridine may be arylated and alkenylated on the *t*-butyl group while no functionalization occurred at more reactive C-H and other bonds. We hypothesize that the high selectivity of this system stems from the confluence of the directing effect of the Schiff base or pyridine moiety and the unique reactivity properties of a phenyl-palladium acetate species (Ph-Pd-OAc•L_n). Formation of palladacycle 24 represents a competitive and unproductive route as this intermediate did not undergo transmetallation (or conversion to the product) under reaction conditions (Figure 8). *ortho-i*-Propylaniline 31 appeared to favor this pathway which may in part explain the negative results with this substrate.

Summary and Future Directions

We are currently exploring this remarkable transformation (catalytic arylation) in terms of scope and mechanism. Furthermore, this system serves as an encouraging starting point for a broad program. We plan to study the reactivity of a range of aryl-metal systems (Ar-M-X_n•L_n) in the context of arylation of diverse substrates. The key issue that we intend to address is the possibility for discerning different types of C-H bonds in the presence of other reactive bonds (e.g. N-H, O-H). In addition to natural products, pharmacophore motifs will be included as substrate candidates. The selection of substrates provides the important context for a methodological program.

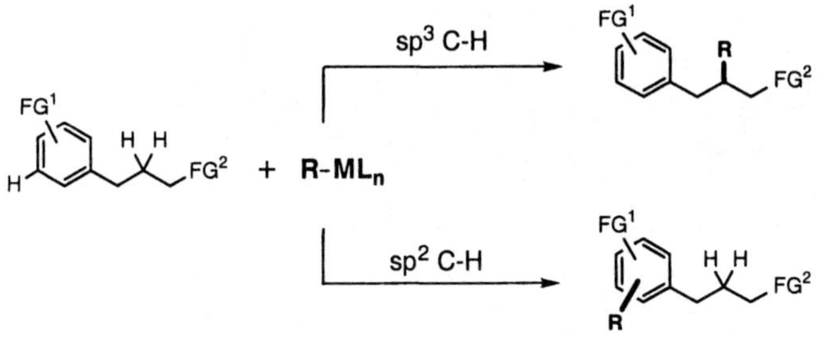

Figure 11. Future directions. Programmable targeting of different types of C-H bonds.

Acknowledgement

I am most grateful for hard work of my colleagues in the Sames groups. Outstanding contribution of those who have directly been involved in the work described in this chapter must be acknowledged (Dr. James A. Johnson, Bengü Sezen, Kamil Godula, Dr. Brain D. Dangel, and Dr. So Won Youn). I am also grateful for generous funding of this work by the National Institute of Health (NIGMS: R01 GM60326), Petroleum Research Fund, GlaxoSmithKline, BMS, Johnson & Johnson, and Merck. D. S. is a recipient of the Cottrell Scholar Award of Research Corporation, Alfred P. Sloan Fellowship, and the Camille Dreyfus Teacher-Scholar. I am deeply indebted to Dr. J. B. Schwarz (Pfizer) for continuing encouragement, advice, and editorial assistance. I would also like to thank Vitas Votier Chmelar and Gail Freeman for their endless inspiration.

References

1. (a) Arndtsen, B. A.; Bergman, R. G.; Mobley, T. A.; Peterson, T. H. *Acc. Chem. Res.* **1995**, *28*, 154-162. (b) Crabtree, R. H. *Chem. Rev.* **1995**, *95*, 987-1007. (c) Shilov, A. E.; Shul'pin, G. B. *Activation and Catalytic Reactions of Saturated Hydrocarbons in the Presence of Metal Complexes*; Kluwer Academic Publishers: Dordrecht, 2000. Catalysis by Metal Complexes, Vol. 21. Ed. James, B. R.
2. Historically, the term "C-H bond activation" carries considerable mechanistic claim while "C-H bond functionalization" simply describes a formal process. Consequently, in the case of arenes, the term "C-H activation" should be used thoughtfully since other mechanistic modes are readily available, for instance, electrophilic metallation (substitution).
3. (a) Kakiuchi, F.; Murai, S. In *Activation of Unreactive Bonds and Organic Synthesis*; Murai, S., Ed.; Springer: Berlin, 1999; pp 47-79. (b) Lenges, C. P.; Brookhart, M. *J. Am. Chem. Soc.* **1999**, *121*, 6616-6623. (c) Jun, Ch.-H.; Hong, J.-B.; Kim, Y.-H.; Chung, K.-Y. *Angew. Chem. Int. Ed.* **2000**, *39*, 3440-3442. (d) Chatani, N.; Asaumi, T.; Ikeda, T.; Yorimitsu, S.; Ishii, Y.; Kakiuchi, F.; Murai, S. *J. Am. Chem. Soc.* **2000**, *122*, 12882-12883. Thalji, R. K.; Ahrendt, K. A.; Bergman, R. G.; Ellman, J. A. *J. Am. Chem. Soc.* **2001**, *123*, 9692-9693.
4. Directed activation stoichiometric in the metal: (a) Suggs, J. W.; Jun, C. H. *J. Am. Chem. Soc.* **1984**, *106*, 3054-3056. (b) Ryabov, A. D. *Chem. Rev.* **1990**, *90*, 403-424.
5. Johnson, J. A.; Sames D. *J. Am. Chem. Soc.* **2000**, *122*, 6321-6322.
6. Johnson, J. A.; Li, N.; Sames D. *J. Am. Chem. Soc.* **2002**, *124*, 6900-6903.
7. Dangel, B. D.; Johnson, J. A.; Sames, D. *J. Am. Chem. Soc.* **2001**, *123*, 8149-8150.

8. Dangel, B. D.; Godula, K.; Youn, S. W.; Sezen, B.; Sames, D. *J. Am. Chem. Soc.* **2002**, *124*, 11856-11857.
9. Arylation of the Herrmann palladacycle with phenylstannane has been reported: Louie, J.; Hartwig, J. F. *Angew. Chem. Int. Ed.* **1996**, *35*, 2359-2361.
10. (a) Dieck, H. A.; Heck, R. F. *J. Org. Chem.* **1975**, *40*, 1083-1090. (b) Moreno-Mañas, M.; Pérez, M.; Pleixats, R. *J. Org. Chem.* **1996**, *61*, 2346-2351.
11. Hirabayashi, K.; Ando, J.; Kawashima, J.; Nishihara, Y.; Mori, A.; Hiyama, T. *Bull. Chem. Soc. Jpn.* **2000**, *73*, 1409-1417.
12. Denmark, S. E.; Sweis, R. F. *J. Am. Chem. Soc.* **2001**, *123*, 6439-6440.
13. For intramolecular functionalization of a *t*-butyl group, see: (a) Dyker, G. *Angew. Chem. Ed. Engl.* **1994**, *33*, 103-105. (b) Laaziri, H.; Bromm, L. O.; Lhermitte, F.; Gschwind, R. M.; Knochel, P. *J. Am. Chem. Soc.* **1999**, *121*, 6940-6941.
14. Sezen, B.; Franz, R.; Sames, D. *J. Am. Chem. Soc.* **2002**, *124*, 13372-13373.

Chapter 10

Pincer and Chelate *N*-Heterocyclic Carbene Complexes of Rh, Ir, and Pd: Synthetic Routes, Dynamics, Catalysis, Abnormal Binding, and Counterion Effects

Anthony R. Chianese and Robert H. Crabtree*

Department of Chemistry, Yale University, New Haven, CT 06520–8107

Mild synthetic routes to the title complexes have been developed, including activation of imidazolium C-H bonds, metallation assisted by a weak base, and transmetallation from metastable silver complexes. Abnormal binding via imidazole C5 rather than the usual C2 is found as a kinetic product in Ir(III) chelate complexes; the C2/C5 product ratio is strongly dependent on the counterion. Counterion dependence is also found for a fluxional process in a CNC pincer complex and explained on the basis of anion coordination. Electronic and steric properties of N-heterocyclic carbenes (NHCs) are documented for Rh(I) and Ir(I) complexes. C versus N binding is proposed as a possible binding mode for histidine. Computational work in collaboration with Eisenstein and Clot is a key element in understanding many of these effects. Hydrogen transfer catalysis is observed for Ir(III) complexes of bidentate NHCs, and the Heck reaction is catalyzed by CNC pincer complexes of Pd(II). The systems are surprisingly robust thermally and with respect to air.

Introduction

The first metal complexes of N-heterocyclic carbenes (NHCs) (*1, 2*), shown in Figure 1, were reported concurrently in 1968 by Öfele (**1**) (*3*), and by Wanzlick and Schönherr (**2**) (*4*). Both compounds were prepared directly from imidazolium salts. Lappert made major contributions in the 1970s (*5*), preparing a wide variety of transition metal-NHC compounds from electron-rich olefins, e.g. **3** and **4**. Nile showed that rhodium-NHC compounds, including **3**, were active catalysts for hydrosilylation (*6*). The isolation of the first stable, crystalline NHC **5** in 1991 by Arduengo et al. (*7*) spurred a recent revival of this chemistry, since metal compounds could be made by adding the neutral free carbene to appropriate metal precursors. NHCs have been found to be exceptionally strongly-binding ligands, and are increasingly used in organometallic catalysis thanks to Herrmann's early work, notably the 1995 report of Heck coupling by palladium-biscarbene compounds (*8*). Excellent results have been obtained using NHCs as spectator ligands for aryl amination

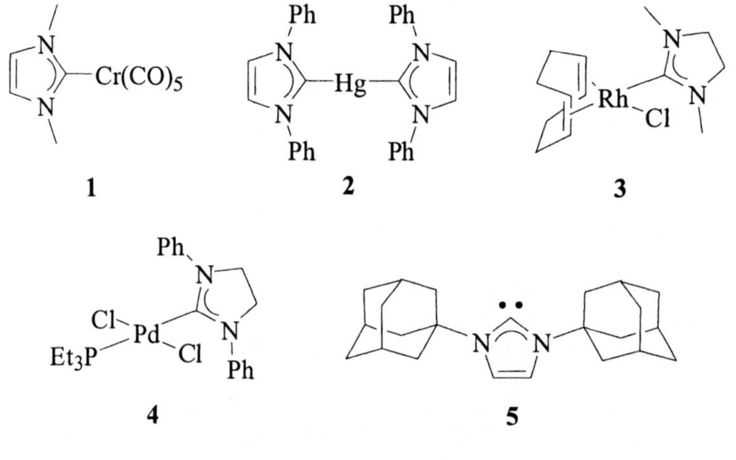

*Figure 1. Early examples of metal complexes of N-heterocyclic carbenes **1-4** and the first crystalline free carbene **5**.*

(*9*), olefin metathesis (*10, 11*), and other reactions (*12*). The ligand precursors, disubstituted azolium salts, are relatively easy to synthesize, allowing for ligands with diverse structural and electronic properties to be designed. A strong base such as potassium *tert*-butoxide or potassium hydride is usually used to prepare the free carbene, by deprotonating the azolium salt at carbon. This limits the possible range of ligand design, because acidic protons or electrophilic sites may be attacked by the base. We have recently focused on the use of multidentate NHCs as spectator ligands for late metal catalysis, and this has required the development of less harsh procedures for synthesizing metal-NHC compounds.

Activation of the Imidazolium C-H Bond

One potentially very mild way to make transition metal-NHC complexes is by oxidative addition of the azolium C-H bond to an appropriate low-valent metal center. Theoretical and experimental work by McGuinness et al. (*13*) has shown that this is a viable synthetic route. In interesting recent work, Heinekey and coworkers (*14*) found that one C-H bond of tris(1-pyrazolyl)borate was activated by iridium to produce an NHC-like metal-carbon bond, indicating that prior metal coordination by N-donor ligands may facilitate the desired C-H activation by electron-rich metal centers. We were able to synthesize the Fischer-type iridium-carbene **9** *via* the double C-H activation route shown in Figure 2 (*15*), and we observed the agostic intermediate **7** by low-temperature NMR spectroscopy. The α-elimination could be reversed easily to give **8**, a proposed intermediate in the forward reaction, by addition of acetone. This unusual result demonstrates both the heteroatom stabilization of metal-carbene species and the aid of a pendant donor group in C-H activation.

Using this strategy with a related starting material, $IrH_5(PPh_3)_2$ (**10**), we sought to prepare chelating NHC compounds by C-H activation of the imidazolium salt, as shown in Figure 3 (*16*). Much to our surprise, when the potentially bidentate ligand **11b** was reacted with **10**, the "wrong" C-H bond was apparently activated to give the C-5 metallated compound **12b**. This compound may be considered to be an abnormally bound heterocyclic carbene, in which the Fischer carbene character is probably reduced from that in the normal carbenes. The compound may also be considered as a metallated imidazolium salt. Fortunately, NHCs metallated at C-5 are easily distinguished from the normal, C-2 metallated carbenes by NMR spectroscopy, although we verified the binding mode by X-ray diffraction in several cases.

Previously, only metallation of the imidazole C-2 position had been observed, and free carbenes formed by deprotonation of imidazolium salts are always deprotonated at this position. One factor favoring the formation of the abnormal carbene in this case is steric crowding at the metal center (*17*). When

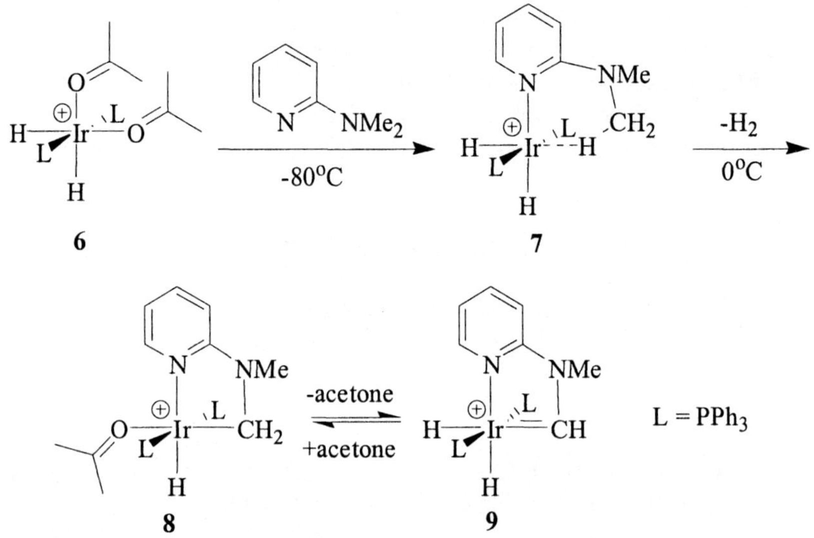

Figure 2. Fischer carbene complex prepared via double C-H activation.

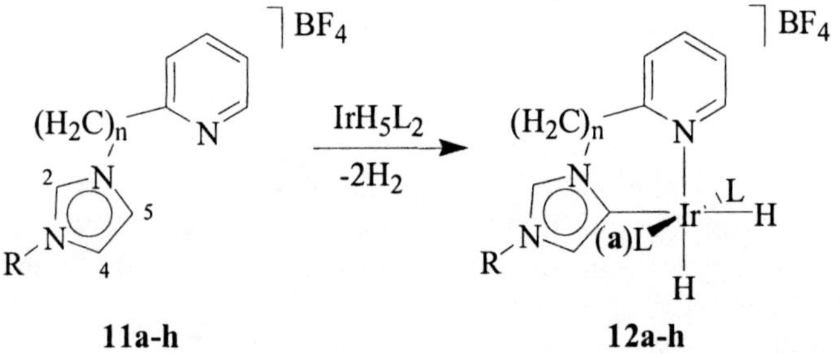

n = 1, R = mesityl (**a**), isopropyl (**b**), *n*-butyl (**c**), methyl (**d**) (55%), L = PPh$_3$
n = 0, R = mesityl (**e**), isopropyl (**f**), *n*-butyl (**g**), methyl (**h**), L = PPh$_3$

Figure 3. Abnormal carbene formed by metallation at imidazole C4.

R is bulky, the abnormal carbenes **12a-c** are formed, but for small R such as methyl, a mixture of the abnormal carbene **12d** and the standard C-2 carbene **13d** is formed (Figure 4). If the Ir ring size is decreased from 6 to 5, by omitting the methylene linker, only the abnormal carbenes (**12e-h**) are formed. The normal carbene **13i** could still be formed if the C-4 and C-5 positions were blocked by using a benzimidazolium salt. These reactions may be performed without the necessity of excluding air or water.

The modest steric size of the i-Pr group does not seem to be a sufficient

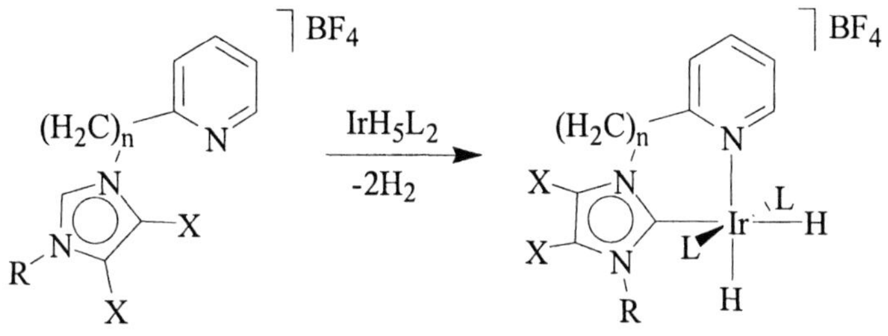

11d,i → **13d,i** (IrH$_5$L$_2$, -2H$_2$), counterion BF$_4$

d. n = 1, R = methyl (45%), X = H, L = PPh$_3$
i. n = 0, R = isopropyl, [XX] = C$_6$H$_4$, L = PPh$_3$

Figure 4. Normal carbene formed by metallation at imidazole C2.

reason to prevent normal carbene formation because many normal carbenes are formed, even in our system, with much bulkier R groups, such as mesityl. The mechanism of the apparent metallation is under study and it may turn out that the reaction in fact goes by a much more complex pathway.

Recently, we observed (18) that by modifying the counterion of the imidazolium salt, and hence also for the cationic Ir-carbene product as well, we could control the ratio of normal to abnormal carbene formed for R = isopropyl and methyl. By using increasingly strongly interacting anions in the series [SbF$_6^-$ < PF$_6^-$ < BF$_4^-$ < Br$^-$], the normal C-2 carbenes are formed to a greater extent. While the origin of this effect is not yet understood, theoretical studies (QM/MM) have indicated that the anion tends to pair specifically by hydrogen bonding with a non-coordinated imidazolium C-H bond in the products. If this is true in the transition state, which seems likely, the product distribution may be controlled by differential stabilization of the transition states by ion pairing. Work is still in progress to better determine the origin of this effect.

Very recently, in work not yet published, we have observed the acid-catalyzed isomerization from abnormal to normal carbenes. If HBF$_4$ is added to solutions of **12b-d**, complete conversion to the normally bound carbenes occurs. This indicates that the C-5 metallated products are kinetic, with the possible exception of R = mesityl (**12a**), which does not convert to the C-2 form in the presence of acid.

The formation of abnormal NHCs should serve as a caution when using imidazolium salts as NHC precursors in organometallic catalysis or when proposing mechanisms in the field. It is no longer safe to assume that the imidazolium salt can only bind to the metal at C-2, especially if very bulky N-substituents are used.

We have also used C-H activation as a route to prepare palladium-NHC complexes (*19*). The reaction of the potentially chelating imidazolium salts **14a-c** with Pd$_2$(dibenzylideneacetone)$_3$ gives the bis-carbene complexes **15a** and **16b,c**, presumably the result of two C-H oxidative additions with loss of H$_2$ (Figure 5). The mesityl N-substituent may prevent both ligands from chelating,

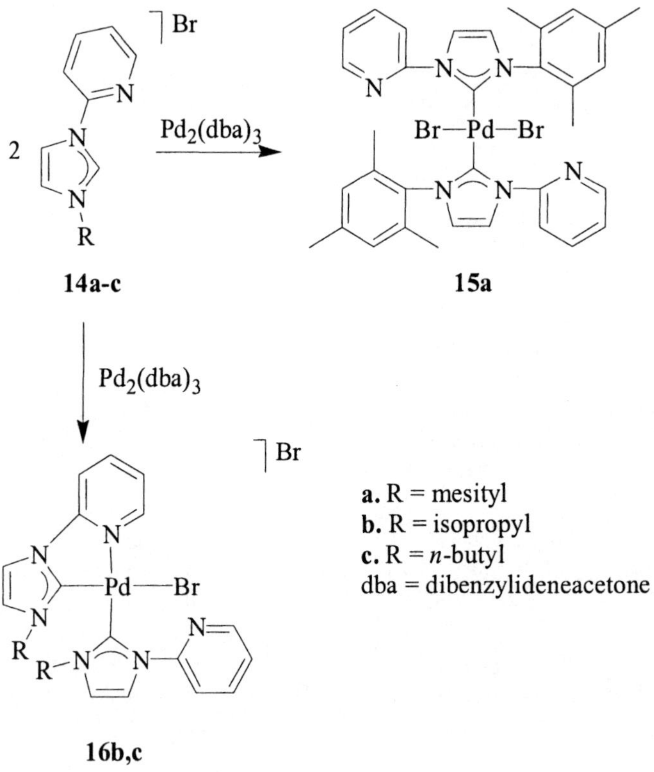

a. R = mesityl
b. R = isopropyl
c. R = *n*-butyl
dba = dibenzylideneacetone

Figure 5. Palladium bis-NHC complexes by C-H activation.

and the *trans* geometry of **15a**, though expected to be electronically disfavored for two such high *trans* effect ligands, is strongly favored on steric grounds. Regardless of the ratio of imidazolium salt added to palladium, only bis-carbene complexes have ever been observed. This implies that the Pd-monocarbene intermediate is much more reactive toward C-H oxidative addition than the starting material. We have proposed a mechanism involving initial C-H oxidative addition of **14** to Pd(0), probably with prior N-donor coordination (**17**) (Figure 6). The resulting palladium hydride species **18** could then reductively eliminate HBr and the resulting Pd(0) intermediate **19** could oxidatively add another

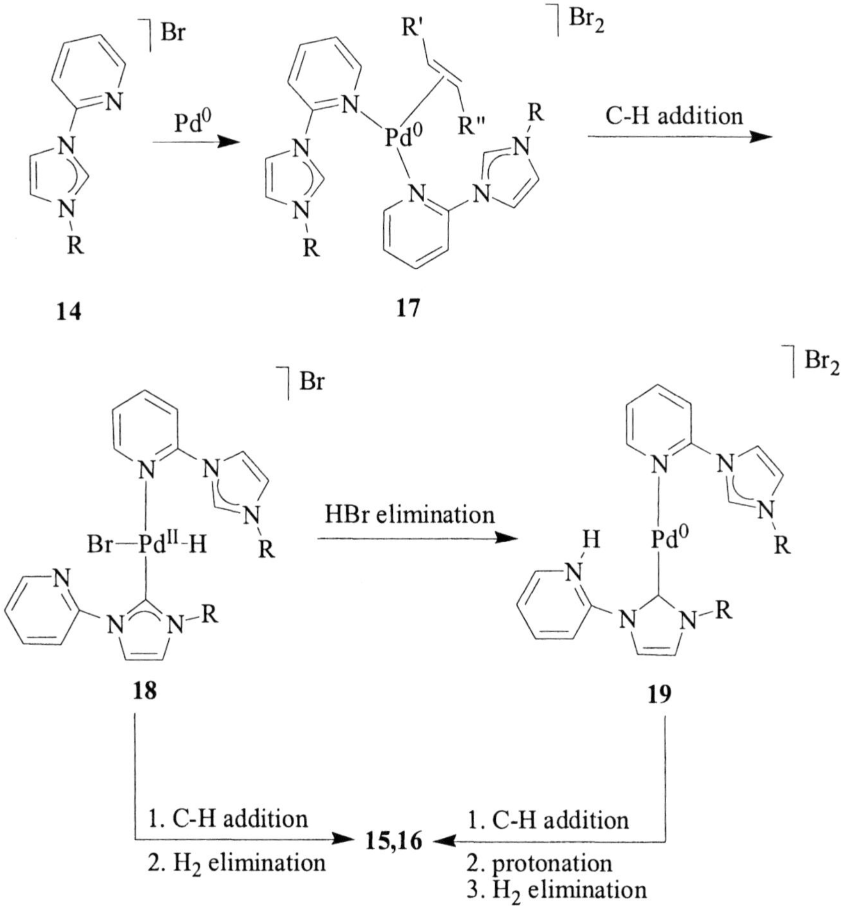

Figure 6. Proposed mechanism for formation of 15 and 16.

imidazolium C-H bond, followed by protonation and H_2 elimination to give **15,16**. Alternatively, the second C-H addition could occur directly to **18**, generating a palladium(IV) dihydride intermediate that would rapidly eliminate H_2 to give **15,16**.

For low-valent metal precursors, C-H activation is therefore a viable route to metal-NHC synthesis. In cases where the free carbene cannot be formed, this may provide a route to otherwise inaccessible complexes.

Metallation Assisted by a Weak Base

Although bases such as acetate are far too weak to deprotonate imidazolium salts directly, they may assist in metallation, as shown by Herrmann and coworkers in the reaction of $Pd(OAc)_2$ and an imidazolium salt to give a palladium bis-carbene species (*8*). This reaction may proceed through a concerted mechanism where the proton is transferred directly to metal-bound acetate. This pathway may rely on the fact that agostic C-H bonding (*20*) increases the acidity of the agostic proton. Using this method, we prepared a series (*21-23*) of palladium CNC pincer (tridentate meridional) complexes (Figure 7). These complexes, very stable to air and heat, were found to be highly active catalysts for the Heck reaction at high temperatures even in air. Compound **21b** was also shown to catalyze the related Suzuki and Sonogashira reactions.

A related compound, **23**, the first example of a CCC pincer NHC complex

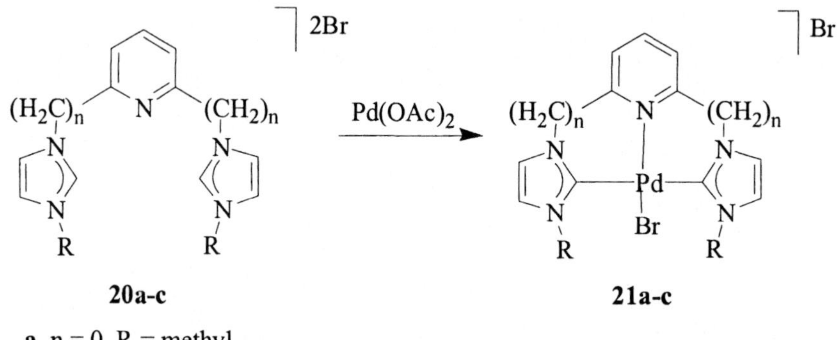

20a-c

a. n = 0, R = methyl
b. n = 0, R = *n*-butyl
c. n = 1, R = methyl

21a-c

Figure 7. Synthesis of CNC-pincer complexes of palladium.

(*22*), is formed *via* initial C-Br oxidative addition to Pd(0) (Figure 8). When used *in situ*, Na$_2$CO$_3$ is sufficiently basic to promote NHC binding via deprotonation of the imidazolium groups.

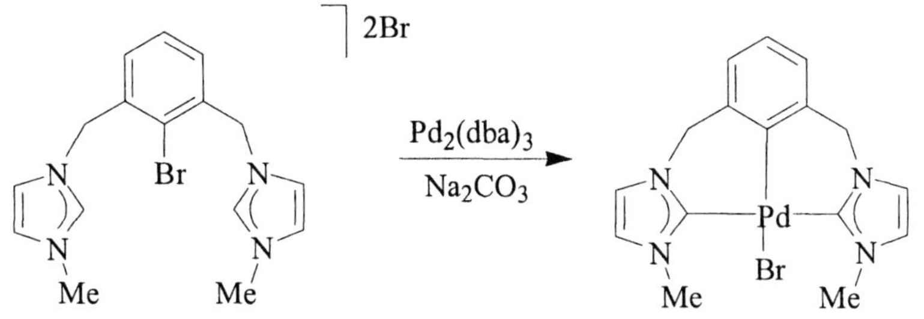

Figure 8. Synthesis of CCC-pincer complex of palladium.

While **21a,b** are flat and hence are symmetric across the Pd coordination plane, compound **21c** and **23** are puckered; the methylene protons resonate as an AB pattern in low-temperature NMR spectroscopy. At higher temperatures, the signals coalesce into a singlet. The dynamic process proposed is a double ring-flip inversion between two chiral C$_2$-symmetric atropisomers (Figure 9). We have explored the mechanism of this process in detail (*24*), by modifying the inner and outer sphere anions for the cationic **21c**. Experimental and theoretical studies indicate that two mechanisms of inversion can operate. If a coordinating anion is used, the anion displaces the pyridine functionality prior to inversion. This is an apparently unique example of organometallic fluxionality where the rate-limiting step is substitution by the outer sphere anion. If a noncoordinating anion is used, pyridine cannot be displaced. This forces an intramolecular inversion, which results in a higher energy barrier. A similarly high barrier is seen in compound **23**, where no anion is present and the aryl group is tightly bound. Compounds of this type could have applications in asymmetric catalysis if the inversion barrier were large enough to allow separation of enantiomers and to withstand the temperatures required for catalysis. Recently, Tulloch et al. (*25*) have reported complexes analogous to **21c**, with R = mesityl or 2,6-diisopropylphenyl. These complexes show no inversion up to 80°C.

Chelating carbene ligands had rarely been successfully installed on rhodium (*26, 27*), and we are unaware of any previous cases for iridium. Thanks to a collaboration with Peris and coworkers (Castellón, Spain), we were able to synthesize a series of Rh(III) (*28*) and Ir(III) (*29*) compounds with bidentate NHC ligands (Figure 10). It is unclear how M(I) is oxidized to M(III) in these reactions, which proceed similarly whether air is excluded or not. The compounds were found to be catalytically active for the transfer hydrogenation of ketones to alcohols in basic solution, *via* hydrogen donation from solvent isopropanol. Iridium was significantly more active than rhodium, with the most active N-substituent being neopentyl, perhaps because this group lacks a beta hydrogen and is inert to Hofmann degradation. Compound **25f**, loaded at 0.1 mol%, gave an impressive 50,000 turnovers per hour for the reduction of

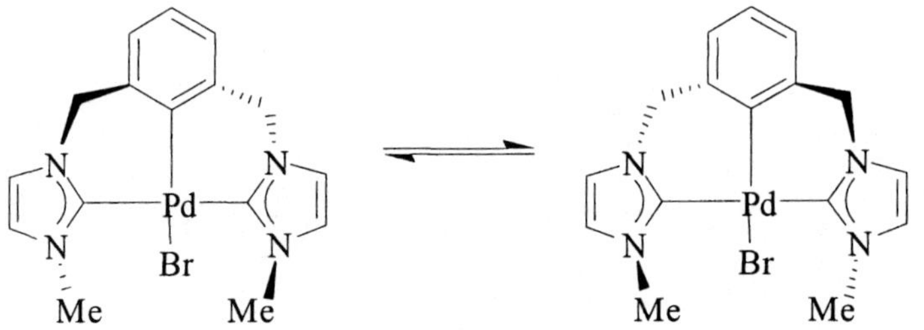

Figure 9. "Ring-flip" atropisomerism process observed by NMR spectroscopy.

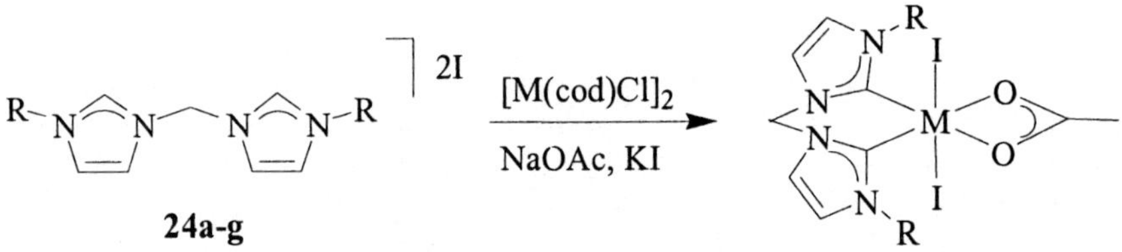

24a-g

25a-g

a. M = Rh, R = *n*-butyl
b. M = Rh, R = isopropyl
c. M = Ir, R = methyl
d. M = Ir, R = *n*-butyl
e. M = Ir, R = isopropyl
f. M = Ir, R = neopentyl
g. M = Ir, R = benzyl

Figure 10. Synthesis of Rh(III) and Ir(III) bis-NHC complexes.

benzophenone at 80°C. Labeling experiments (*29*) on **25f** support a monohydride mechanism for transfer hydrogenation.

Compounds **25** are air and water stable, and even the catalytic reactions are unaffected by the presence of air or water in the system. This stands in contrast to phosphine-based systems, where catalysis often must be performed under inert and/or dry conditions.

In an effort to compare the effects of chelating triazole-based NHCs with the more commonly used imidazole-based ligands, we found a very interesting result. The reaction (Figure 11) of ligand precursors **26a,b** with [Rh(norbornadiene)Cl]$_2$ gave some of the expected biscarbene products **27a,b**, along with a majority of the unexpected products **28a,b** (*30*). The diene ligand has rearranged to a nortricyclyl group, bound in a *fac* geometry with respect to

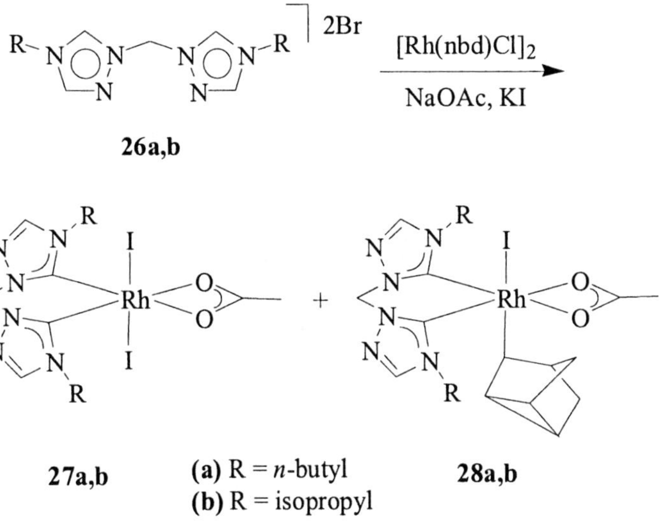

Figure 11. Metallation of bis-NHCs to give Rh-nortricyclyl complex.

the bis-NHC. Mechanistic investigations are not yet complete, but a plausible mechanism involves oxidative coupling followed by C-H reductive elimination (Figure 12). Such a mechanism could plausibly be responsible for the dissociation of cyclooctadiene in the formation of **25a-g**, and would account for the oxidation of M(I) to M(III), if initial C-H oxidative addition of an azolium salt produced the required metal hydride species.

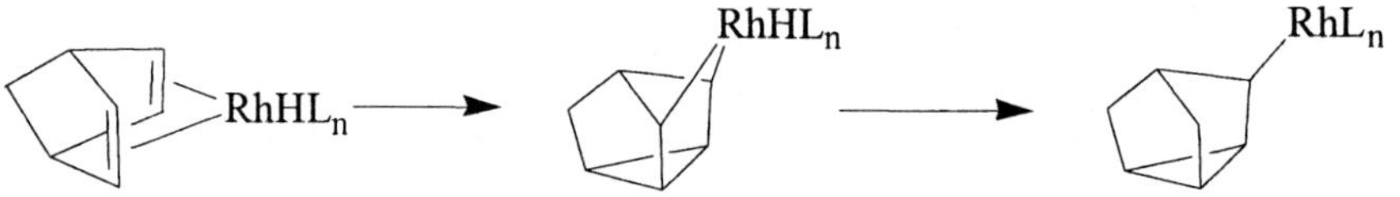

Figure 12. Proposed mechanism for formation of Rh-nortricyclyl complex.

Transmetallation from Ag-carbene Complexes

The important work of Wang and Lin (*31*) demonstrated that silver(I) complexes of NHCs could be synthesized directly from imidazolium salts and Ag_2O or Ag_2CO_3. These complexes are capable of transferring the carbene ligand to other metals, as has been demonstrated for Pd (*25, 31-33*) and Au (*31*). We have extended this methodology to rhodium and iridium (*34*) in the syntheses of the monocarbene compounds **31a,b** (Figure 13). Because the benzylic protons are diastereotopic by NMR spectroscopy, these compounds demonstrate hindered rotation about the M-carbene bond. We displaced the rather bulky cyclooctadiene ligand with 2 CO molecules to form **32a,b**. The benzylic protons now exchange on the NMR timescale, and we were able to calculate the activation free energies for M-C bond rotation. Our observations confirmed that for NHCs, the M-C bond is essentially single (*1*), although M-C rotation may be hindered sterically. IR spectroscopy of **32a,b** allowed the estimation of the donor power of the NHC ligand, which supported theoretical calculations (*35*) and previous experimental work (*36*), indicating that NHCs are more strongly donating than even alkyl phosphines.

Implications for Metalloproteins

Given the known strength of the M-C bond in imidazole-based NHCs, the question arises of whether histidine residues in proteins could bind through carbon instead of nitrogen. This has never been reported, but the hypothetical C-bound tautomer is essentially indistinguishable by X-ray diffraction from the conventional N-bound form (Figure 14). The situation here is notably different from known NHCs, because the N-substituents can both donate electron density to the ring, and more importantly, block N-M binding from occurring. In one case, Taube and coworkers (*37*) directly observed the tautomerism for a ruthenium compound, demonstrating that C-binding can be thermodynamically preferred even when N-binding is possible.

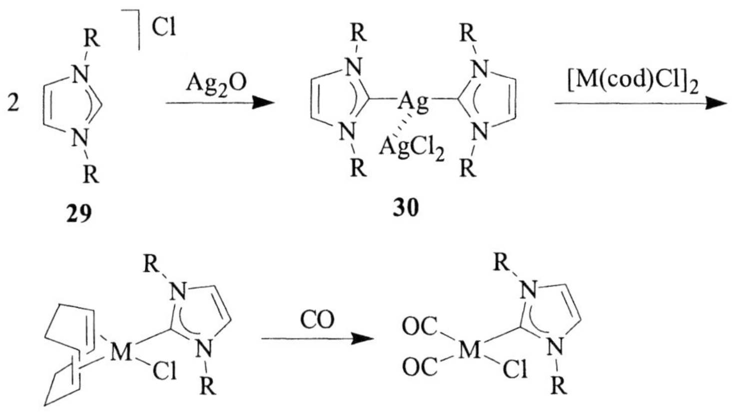

29 **30**

31a,b **32a,b**

(a) R = 4-tolylmethyl, M = Rh
(b) R = 4-tolylmethyl, M = Ir

Figure 13. Synthesis of Rh(I) and Ir(I) complexes of NHCs via transmetallation.

In collaboration with Sini and Eisenstein, the relative stability of species **33** and **34** have been calculated for a series of metal fragments (*38*). For several of these fragments, the C-bound form was found to be more stable. Moving down a group tends to increase the relative stability of the C-bound form. The presence of the potentially H-bonding ligands M-Cl *cis* to imidazole also stabilized the C-bound tautomer via N-H· · · Cl hydrogen bonds. It is plausible that an appropriately arranged metal-binding site in a metalloprotein could cause a similar effect.

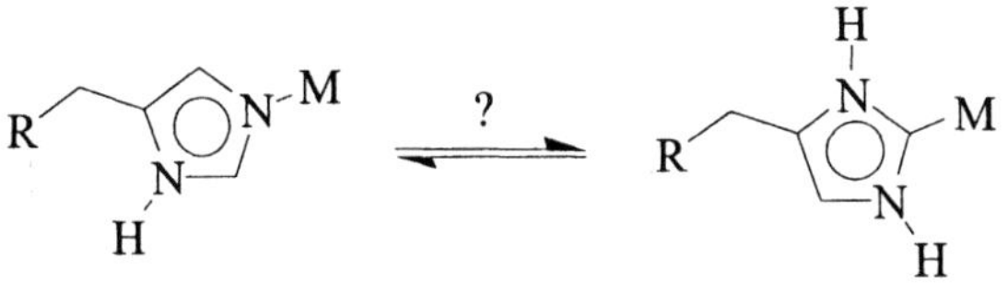

33 **34**

Figure 14. C-M versus N-M binding in histidine.

Conclusions

N-heterocyclic carbenes are proving to be very useful ligands for a wide variety of catalytic applications, owing to their strong binding, and the stability of many NHC complexes to heat and air. The new synthetic routes now available for substituted azoles permit the rapid construction of complex ligand architectures, often a relatively arduous task for phosphines. The development of these more gentle metallation protocols should allow incorporation of more elaborate functionalities, such as molecular recognition elements or pendant reactive groups.

Acknowledgments.

We gratefully thank our collaborators, Jack Faller, Odile Eisenstein, Eric Clot, and Eduardo Peris, as well as the DOE and Johnson Matthey for funding.

Literature Cited

1. Herrmann, W. A.; Köcher, C. *Angew. Chem. Int. Edit.* **1997**, *36*, 2163.
2. Bourissou, D.; Guerret, O.; Gabbaï, F. P.; Bertrand, G. *Chem. Rev.* **2000**, *100*, 39.
3. Öfele, K. *J. Organomet. Chem.* **1968**, *12*, P42.
4. Wanzlick, H. W.; Schönherr, H. J. *Angew. Chem. Int. Edit.* **1968**, *7*, 141.
5. Lappert, M. F. *J. Organomet. Chem.* **1975**, *100*, 139.
6. Hill, J. E.; Nile, T. A. *J. Organomet. Chem.* **1977**, *137*, 293.
7. Arduengo, A. J.; Harlow, R. L.; Kline, M. *J. Am. Chem. Soc.* **1991**, *113*, 361.
8. Herrmann, W. A.; Elison, M.; Fischer, J.; Köcher, C.; Artus, G. R. J. *Angew. Chem. Int. Edit.* **1995**, *34*, 2371.
9. Stauffer, S. R.; Lee, S. W.; Stambuli, J. P.; Hauck, S. I.; Hartwig, J. F. *Org. Lett.* **2000**, *2*, 1423.
10. Scholl, M.; Ding, S.; Lee, C. W.; Grubbs, R. H. *Org. Lett.* **1999**, *1*, 953.
11. Huang, J. K.; Stevens, E. D.; Nolan, S. P.; Petersen, J. L. *J. Am. Chem. Soc.* **1999**, *121*, 2674.
12. Herrmann, W. A. *Angew. Chem. Int. Edit.* **2002**, *41*, 1291.
13. McGuinness, D. S.; Cavell, K. J.; Yates, B. F.; Skelton, B. W.; White, A. H. *J. Am. Chem. Soc.* **2001**, *123*, 8317.
14. Wiley, J. S.; Oldham, W. J.; Heinekey, D. M. *Organometallics* **2000**, *19*, 1670.

15. Lee, D. H.; Chen, J. Y.; Faller, J. W.; Crabtree, R. H. *Chem. Commun.* **2001**, 213.
16. Gründemann, S.; Kovacevic, A.; Albrecht, M.; Faller, J. W.; Crabtree, R. H. *Chem. Commun.* **2001**, 2274.
17. Gründemann, S.; Kovacevic, A.; Albrecht, M.; Faller, J. W.; Crabtree, R. H. *J. Am. Chem. Soc.* **2002**, *124*, 10473.
18. Kovacevic, A.; Gründemann, S.; Miecznikowski, J. R.; Clot, E.; Eisenstein, O.; Crabtree, T. H. *Chem. Commun.* **2002**, 2580.
19. Gründemann, S.; Albrecht, M.; Kovacevic, A.; Faller, J. W.; Crabtree, R. H. *J. Chem. Soc. Dalton Trans.* **2002**, 2163.
20. Crabtree, R. H. *Angew. Chem. Int. Edit.* **1993**, *32*, 789.
21. Peris, E.; Loch, J. A.; Mata, J.; Crabtree, R. H. *Chem. Commun.* **2001**, 201.
22. Gründemann, S.; Albrecht, M.; Loch, J. A.; Faller, J. W.; Crabtree, R. H. *Organometallics* **2001**, *20*, 5485.
23. Loch, J. A.; Albrecht, M.; Peris, E.; Mata, J.; Faller, J. W.; Crabtree, R. H. *Organometallics* **2002**, *21*, 700.
24. Miecznikowski, J. R.; Gründemann, S.; Albrecht, M.; Mégret, C.; Clot, E.; Faller, J. W.; Eisenstein, O.; Crabtree, R. H. *Dalton Trans.* **2003**, 831.
25. Tulloch, A. A. D.; Danopoulos, A. A.; Tizzard, G. J.; Coles, S. J.; Hursthouse, M. B.; Hay-Motherwell, R. S.; Motherwell, W. B. *Chem. Commun.* **2001**, 1270.
26. Hitchcock, P. B.; Lappert, M. F.; Terreros, P.; Wainwright, K. P. *J. Chem. Soc. Chem. Commun.* **1980**, 1180.
27. Shi, Z. Q.; Thummel, R. P. *Tetrahedron Lett.* **1995**, *36*, 2741-2744.
28. Albrecht, M.; Crabtree, R. H.; Mata, J.; Peris, E. *Chem. Commun.* **2002**, 32.
29. Albrecht, M.; Miecznikowski, J. R.; Samuel, A.; Faller, J. W.; Crabtree, R. H. *Organometallics* **2002**, *21*, 3596.
30. Mata, J. A.; Peris, E.; Incarvito, C.; Crabtree, R. H. *Chem. Commun.* **2003**, 184.
31. Wang, H. M. J.; Lin, I. J. B. *Organometallics* **1998**, *17*, 972.
32. McGuinness, D. S.; Cavell, K. J. *Organometallics* **2000**, *19*, 741.
33. Magill, A. M.; McGuinness, D. S.; Cavell, K. J.; Britovsek, G. J. P.; Gibson, V. C.; White, A. J. P.; Williams, D. J.; White, A. H.; Skelton, B. W. *J. Organomet. Chem.* **2001**, *617*, 546.
34. Chianese, A. R.; Li, X.; Janzen, M. C.; Faller, J. W.; Crabtree, R. H. *Organometallics* **2003**, *22*, 1663.
35. Perrin, L.; Clot, E.; Eisenstein, O.; Loch, J.; Crabtree, R. H. *Inorg. Chem.* **2001**, *40*, 5806.
36. Huang, J. K.; Stevens, E. D.; Nolan, S. P. *Organometallics* **2000**, *19*, 1194..
37. Sundberg, R. J.; Bryan, R. F.; Taylor, I. F.; Taube, H. *J. Am. Chem. Soc.* **1974**, *96*, 381.
38. Sini, G.; Eisenstein, O.; Crabtree, R. H. *Inorg. Chem.* **2002**, *41*, 602.

Chapter 11

Sequential Hydrocarbon C–H Bond Activations by 16-Electron Organometallic Complexes of Molybdenum and Tungsten

Peter Legzdins and Craig B. Pamplin

Department of Chemistry, The University of British Columbia, 2036 Main Mall, Vancouver, British Columbia V6T 1Z1, Canada

Cp*M(NO)(hydrocarbyl)$_2$ complexes (Cp* = η^5-C$_5$Me$_5$) of molybdenum and tungsten exhibit hydrocarbyl-dependent thermal chemistry. Thus, gentle thermolysis of appropriate Cp*M(NO)(hydrocarbyl)$_2$ precursors (M = Mo, W) results in the loss of hydrocarbon and the transient formation of 16-electron Cp*M(NO)-containing complexes such as Cp*M(NO)(alkylidene), Cp*M(NO)(η^2-benzyne), Cp*M(NO)(η^2-acetylene), and Cp*M(NO)(η^2-allene). These intermediates effect single, double, or triple activation of hydrocarbon C-H bonds intermolecularly, the first step of these activations being the reverse of the transformations by which they were generated. This Chapter summarizes the various types of C-H activations that have been effected with these organometallic nitrosyl complexes to date and presents some of the results of kinetic, mechanistic, and theoretical investigations of these processes.

Introduction

Interest in C-H bond activations at transition-metal centers continues unabated since these processes hold the promise of permitting the more efficient utilization of the world s limited hydrocarbon resources for the production of essential chemicals (*1*). Our contribution to this area of chemistry began with our recent discovery that Cp*M(NO)(hydrocarbyl)$_2$ complexes (Cp* = η^5-C$_5$Me$_5$) of molybdenum and tungsten exhibit hydrocarbyl-dependent thermal chemistry. Thus, gentle thermolysis of appropriate Cp*M(NO)(hydrocarbyl)$_2$ precursors (M = Mo, W) results in loss of hydrocarbon and the transient formation of 16-electron organometallic complexes such as Cp*M(NO)-(alkylidene), Cp*M(NO)(η^2-benzyne), Cp*M(NO)(η^2-acetylene), and Cp*M-(NO)(η^2-allene). These intermediates first effect the single activation of hydrocarbon C-H bonds in an intermolecular manner via the reverse of the transformations by which they were generated. Some of the new product complexes formed in this manner are stable and may be isolated. Others are thermally unstable under the experimental conditions employed and react further to effect double or triple C-H bond activations of the hydrocarbon substrates. These sequential C-H bond-activating properties of the various reactive intermediates are considered in turn in the following sections, beginning with the most straightforward Cp*M(NO)(alkylidene) systems.

Cp*M(NO)(alkylidene) Intermediates

Tungsten Complexes

These intermediate complexes result from the thermal activation of bis-(alkyl) precursors, the first of which to be identified was Cp*W(NO)-(CH$_2$CMe$_3$)$_2$ (**1**). Thermal activation of **1** at 70 °C in neat hydrocarbon solutions transiently generates the highly reactive neopentylidene complex, Cp*W(NO)-(=CHCMe$_3$) (**A**), that can be trapped by PMe$_3$ as the Cp*W(NO)(=CHCMe$_3$)-(PMe$_3$) adduct in two isomeric forms (**2a-b**) which differ in the orientations of the CHCMe$_3$ ligand with respect to the Cp* group (Scheme 1). More importantly, intermediate **A** can effect the single activation of solvent C-H bonds (Scheme 1) (*2, 3*). Thus, in neat benzene and benzene-d_6 it cleanly forms the corresponding phenyl complexes, Cp*W(NO)(CH$_2$CMe$_3$)(C$_6$H$_5$) (**3**) and Cp*W-(NO)(CHDCMe$_3$)(C$_6$D$_5$) (**3**-d_6), and in Me$_4$Si it forms the unsymmetrical bis(alkyl) complex, Cp*W(NO)(CH$_2$CMe$_3$)(CH$_2$SiMe$_3$) (**4**) in high yields. Interestingly, **4** reacts further with Me$_4$Si under the thermolysis conditions, albeit very slowly, to form Cp*W(NO)(CH$_2$SiMe$_3$)$_2$ (**5**) as a second product. The relative amount of **5** gradually increases upon prolonged heating of the reaction mixture, and its formation implies the transient existence of Cp*W(NO)(=CHSiMe$_3$) (**B**) as a reactive intermediate (Scheme 1).

Scheme 1

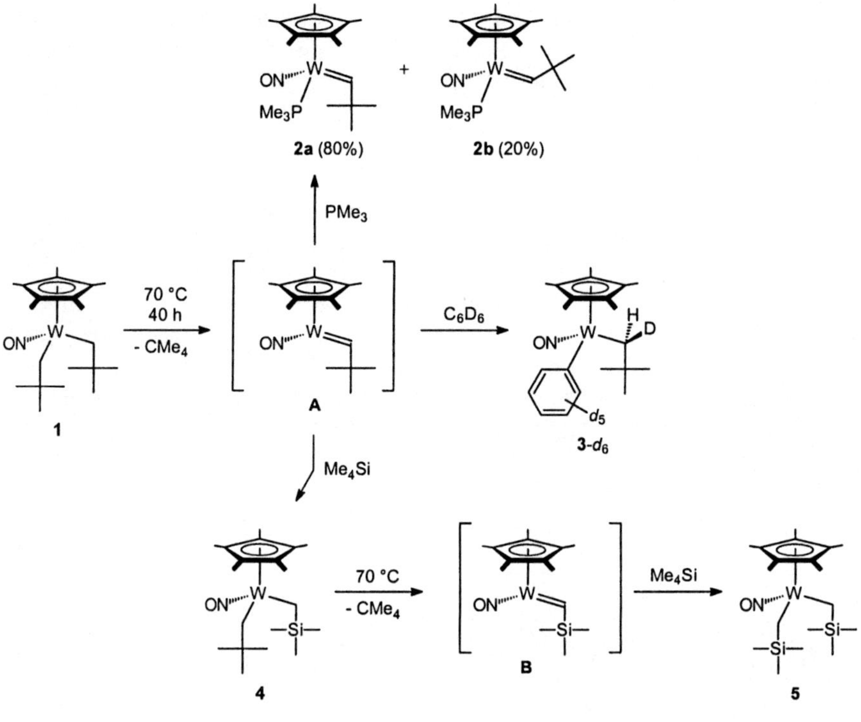

Intermediate **A** can also effect multiple C-H bond activations even in the presence of excess PMe$_3$ (Scheme 2). Thus, thermolysis of **1** in cyclohexane in the presence of PMe$_3$ yields **2a-b** as well as the olefin complex Cp*W(NO)(η^2-cyclohexene)(PMe$_3$) (**6**). It is likely that the double C-H activation of cyclohexane by **A** is in direct competition with trapping of **A** by PMe$_3$ as portrayed in Scheme 3. The competitive C-H activation by an unsaturated intermediate in the presence of a strong Lewis base is remarkable, yet it has been observed before (*4, 5*) and is aided by low concentrations of the trapping agent relative to the cyclohexane solvent. In contrast to cyclohexane, methylcyclohexane under identical experimental conditions affords trace amounts of **2a-b** and the

Scheme 2

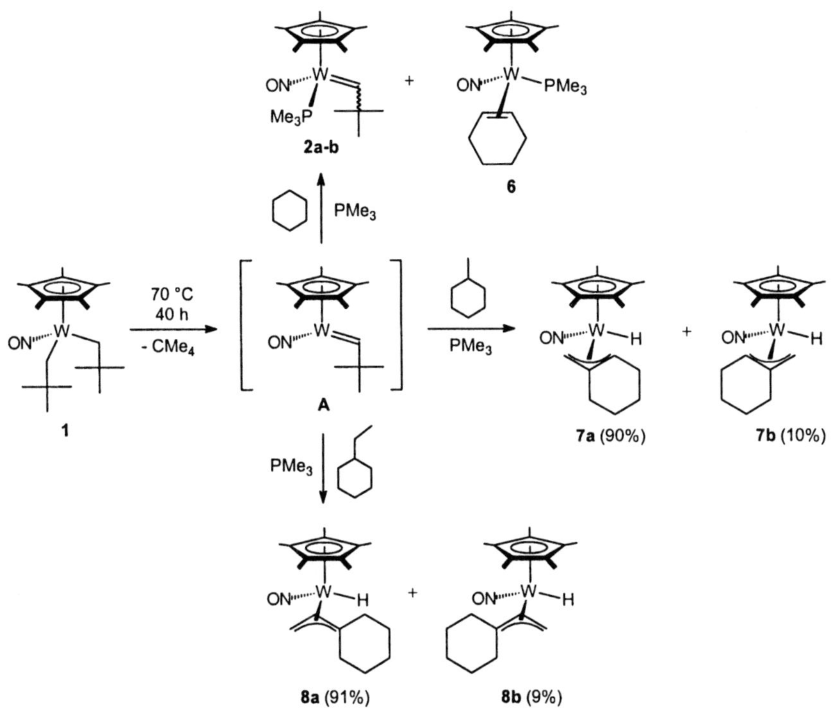

exocyclic allyl hydride complex, Cp*W(NO)(η^3-C$_7$H$_{11}$)(H), in two isomeric forms (**7a-b**) whose relative abundances are shown in Scheme 2. Ethylcyclohexane affords similar exocyclic allyl hydride products **8a-b**. The mechanisms for the transformations of **1** into **7a-b** or **8a-b** likely mirror that of cyclohexane (Scheme 3) in that activation of a solvent C-H bond by **A** is followed by β-H activation to release the second neopentyl ligand as neopentane and form a 16e alkene complex. Unlike the cyclohexene case, however, the alkene complexes in these systems evidently do not persist long enough to be trapped by PMe$_3$. Rather, there is a third γ-H activation by the metal center that results in the conversion of the alkene complexes into the isolable 18e allyl hydride complexes. Thus, these processes involve triple C-H bond activations of the original hydrocarbon substrates.

The related benzylidene complex, Cp*W(NO)(=CHPh) (**C**), results from the elimination of neopentane from Cp*W(NO)(η^2-CH$_2$Ph)(CH$_2$CMe$_3$) at 70 °C, and it exhibits C-H activating abilities similar to those summarized above in Schemes 1 and 2 for the neopentylidene intermediate **A** (*3*). The alkylidene

Scheme 3

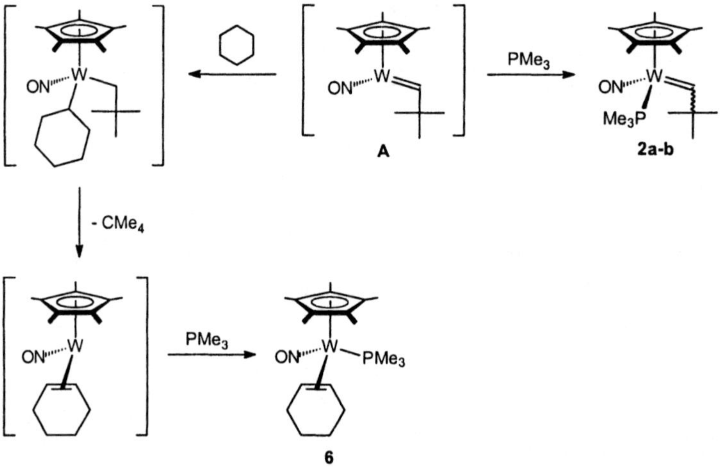

complexes **A**, **B** and **C** belong to a small class of compounds that intermolecularly activate C-H bonds via their addition across the M=C bond of an alkylidene intermediate (*6, 7, 8*). However, **A**, **B**, and **C** are unique in their ability to activate aliphatic C-H bonds to form isolable, well-defined products. On the basis of our studies to date it appears that a whole family of such C-H bond-activating tungsten alkylidene complexes with differing steric and electronic properties should be capable of existence.

Molybdenum Complexes

In general, the thermal reactivity of Cp*Mo(NO)(CH$_2$CMe$_3$)$_2$ (**9**) resembles that of its isostructural tungsten analogue (vide supra), but it occurs under the unusually mild conditions of ambient temperatures and pressures (Scheme 4). For example, the reactions of **9** with tetramethylsilane or mesitylene for 30 h at 20 °C result in the clean formation of free neopentane and Cp*Mo(NO)-(CH$_2$CMe$_3$)(CH$_2$SiMe$_3$) (**10**) or Cp*Mo(NO)(CH$_2$CMe$_3$)(η^2-CH$_2$C$_6$H$_3$-3,5-Me$_2$) (**11**), respectively (*9*). Neither organometallic product undergoes further reactivity under these mild reaction conditions. Similarly, the reaction of **9** with C$_6$D$_6$ generates predominantly Cp*Mo(NO)(CHDCMe$_3$)(C$_6$D$_5$) (**12**-*d$_6$*) with stereospecific deuterium incorporation at the synclinal methylene position of the neopentyl ligand (just as portrayed for **3**-*d$_6$* in Scheme 1).

Scheme 4

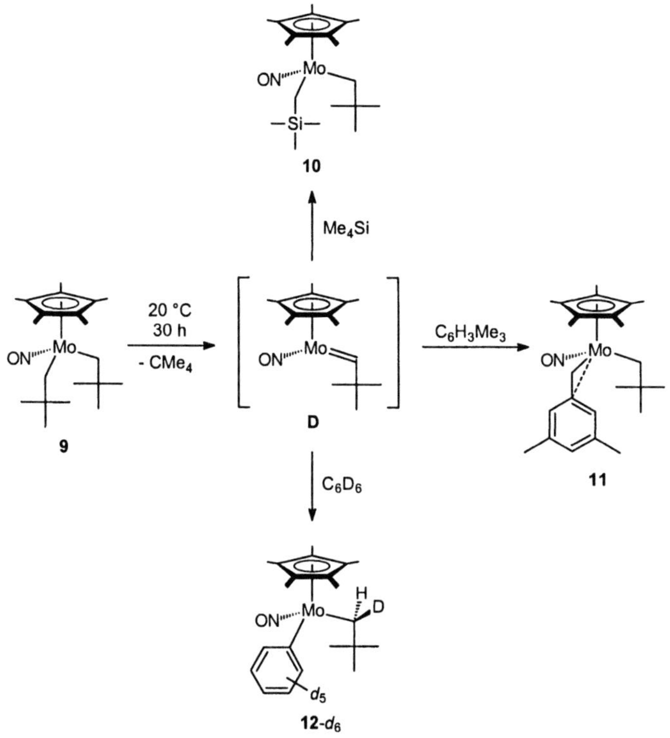

Mechanistic Aspects

Labeling, trapping, and kinetic results indicate that the C-H activation chemistry derived from **1** proceeds through two distinct steps: (1) formation of the neopentylidene complex **A** via rate-determining intramolecular α-H elimination of neopentane, and (2) a 1,2-*cis* addition of R-H across the M=CHR linkage of **A**. Similarly, a kinetic analysis of the reaction of Cp*Mo(NO)-(CH$_2$CMe$_3$)$_2$ (**9**) with C$_6$D$_6$ to form Cp*Mo(NO)(CHDCMe$_3$)(C$_6$D$_5$) (**12**-d_6) (Scheme 4) in the temperature range 26-40 °C reveals a first-order loss of organometallic reactant, and a linear Eyring plot affords $\Delta H^\ddagger = 99(1)$ kJ mol^{-1} and $\Delta S^\ddagger = -11(4)$ J mol^{-1} K^{-1} (*9*). These parameters are very similar to those determined at higher temperatures for the tungsten analogue **1** and indicate that a similar mechanism is operative for the molybdenum system.

Additional experimental and B3LYP computational mechanistic studies of the intermolecular C-H activations of hydrocarbons by the tungsten alkylidene complexes have also been carried out (*10*). In these studies it was shown that the α-deuterated derivative, Cp*W(NO)(CD$_2$CMe$_3$)$_2$ (1-d_4), undergoes intramolecular H/D exchange within the neopentyl ligands, a feature consistent with the reversible formation of σ-neopentane complexes prior to neopentane elimination. Microscopic reversibility requires that such σ-complexes also be formed during the reverse process of alkane C-H activation by the alkylidene complexes. In addition, it has been found that thermolysis of Cp*W(NO)-(CH$_2$CMe$_3$)$_2$ (1) at 70 °C for 40 h in a 1:1 molar mixture of tetramethylsilane and tetramethylsilane-d_{12} yields an intermolecular kinetic isotope effect (KIE) of 1.07(4):1. Similarly, thermolysis of 1 and Cp*W(NO)(CH$_2$CMe$_3$)(η^2-CH$_2$Ph) in 1:1 benzene/benzene-d_6 yields intermolecular KIE's of 1.03(5):1 and 1.17(19):1, respectively. The alkylidene intermediates **A** and **C** thus exhibit little preference for activating protio over deuterio substrates. Hence, the observed KIE values are inconsistent with C-H(D) bond addition to the W=C linkage being the discriminating factor in alkane and arene intermolecular competitions, but rather indicate that coordination of the substrate to the metal center is the discriminating event.

Evidence that transient π-complexes may also be formed during such processes is provided by the observation that Cp*W(NO)(CH$_2$CMe$_3$)(C$_6$D$_5$) (3-d_5) and Cp*W(NO)(η^2-CH$_2$Ph)(C$_6$D$_5$) convert to the respective H/D scrambled products under thermolytic conditions, consistent with the occurrence of reversible aromatic sp^2 C-H bond cleavage. These results suggest that the observed discrimination between the aryl and benzyl products of toluene activation by intermediate complexes **A** and **C** (*3*) originates from the coordination of toluene to the tungsten center in two distinct modes. Supporting DFT calculations (including zero-point energy and entropy corrections) on the activation of toluene by CpW(NO)(=CH$_2$) indicate that aromatic sp^2 C-H bond activation proceeds through a π-arene complex, while benzylic sp^3 C-H bond activation proceeds through a η^2(C,H)-σ-phenylmethane complex (*10*). The principal factor behind the preferential formation of the aryl products appears to be the relative ease of formation of the π-arene intermediates.

Cp*M(NO)(η^2-benzyne) Intermediates

To date only the molybdenum-containing member of these reactive intermediates has been unambiguously identified. Its existence was first inferred from the remarkable reaction of Cp*Mo(NO)(CH$_2$CMe$_3$)$_2$ (9) with C$_6$H$_6$ at room temperature that leads to the sequential activation of two molecules of benzene as depicted in Scheme 5 (*9*). These conversions resemble the sequential

Scheme 5

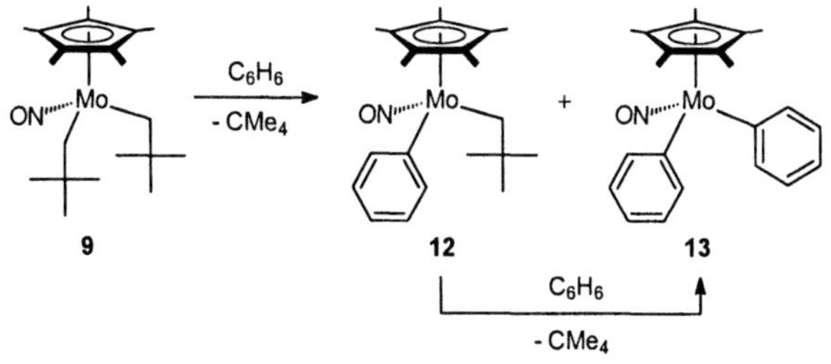

activation of two molecules of Me₄Si by the neopentylidene intermediate **A** shown at the bottom of Scheme 1. The intermediacy of a transient benzyne complex in this benzene-activation chemistry (Scheme 6) is confirmed by the fact that thermolysis of independently prepared Cp*Mo(NO)(CH$_2$CMe$_3$)(C$_6$H$_5$) (**12**) in pyridine-d_5 at room temperature for 30 h affords a mixture of the trapped η^2-benzyne complex, Cp*Mo(NO)(η^2-C$_6$H$_4$)(NC$_5$D$_5$) (**14**-d_5, 25%), and the trapped alkylidene complex, Cp*Mo(NO)(=CHCMe$_3$)(NC$_5$D$_5$) (**15**-d_5, 75%), that can be separated by fractional crystallization. As shown in Scheme 6, the 16e Cp*Mo(NO)(η^2-C$_6$H$_4$) intermediate **E** also reacts readily with the aliphatic C-H bonds of various substrates (R-H) to form the corresponding Cp*Mo(NO)-(C$_6$H$_5$)(R) products, but the exact scope of its C-H bond-activating ability remains to be delineated. However, it is not unreasonable to expect that

Scheme 6

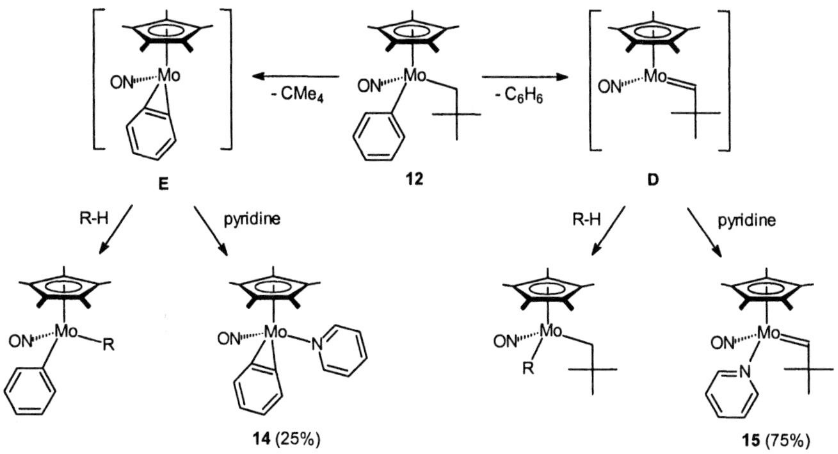

its chemistry may well resemble in some aspects that exhibited by the Cp*M(NO)(η^2-acetylene) intermediates which are considered in the next section. In this connection it may be noted that relatively few benzyne complexes have been reported to undergo intermolecular C-H bond activation processes, and then usually at temperatures in excess of 100 °C (*11, 12, 13*).

It is clear that the gentler reaction conditions required for the molybdenum system permit the isolation of the η^2-benzyne complex **14** even though it is not obtainable by the thermolysis of the diphenyl complex **13** in pyridine. We have previously proposed similar intermediates for the related tungsten systems (*14, 15*), but they have proven to be unisolable probably because of thermal decomposition under the requisite experimental conditions.

Cp*M(NO)(η^2-acetylene) Intermediates

These intermediates result from the thermolysis of alkyl vinyl precursors, the most studied of which is Cp*W(NO)(CH$_2$SiMe$_3$)(CPh=CH$_2$) (**16**). Interestingly, the NMR parameters for the CPhCH$_2$ fragment in **16** in solutions are characteristic of those of a 1-metallacyclopropene unit, a feature reflecting the metal center's attainment of the favored 18e configuration (*16*).

Thermal activation of **16** at 54 °C in neat hydrocarbon solutions transiently generates Cp*W(NO)(η^2-PhC≡CH) (**F**) which subsequently activates solvent C-H bonds (*17, 18, 19*). Examples of the single C-H activation processes that have been effected by this complex are summarized in Scheme 7. These activations of substrates R-H (R-H = benzene, methyl-substituted arenes, and (Me$_3$Si)$_2$O) yield the corresponding 18e Cp*W(NO)(R)(η^2-CPh=CH$_2$) complexes. As found with other systems (*20*) and the other nitrosyl intermediates considered in this Chapter, intra- and intermolecular selectivity studies reveal that the strongest C-H bond of R-H (yielding the stronger M-C bond) is the preferred site of reactivity of intermediate **F**. Mechanistic, labeling,

Scheme 7

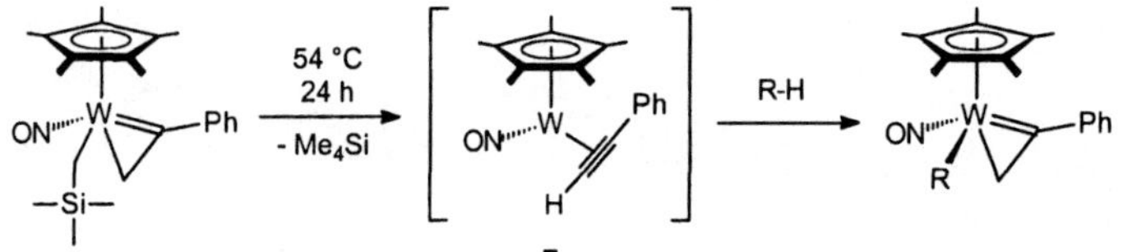

and kinetic studies of these conversions are also consistent with the rate-determining step being the reversible formation of **F** from **16** by elimination of

Me$_4$Si and with the involvement of σ- and π-hydrocarbon complexes along the various mechanistic pathways (vide supra).

The dual C-H bond activations of aliphatic hydrocarbons effected by **16** (Scheme 8) are unique to this complex, and they are particularly interesting for

Scheme 8

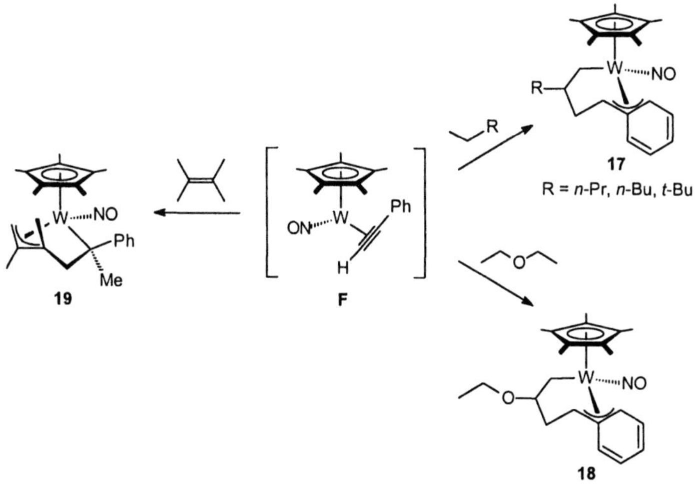

several reasons. First, these processes result in the dehydrogenation of alkanes with β-C-H bonds and their coupling with the alkenyl ligand in the tungsten's coordination sphere to form 18e η^1,η^3-metallacyclic species. Second, they selectively activate substrates that contain an ethyl substituent (e.g. the reaction with Et$_2$O to form **18**). A plausible mechanism for these dual C-H activation processes is depicted in Scheme 9 using the formation of the complexes **17** as a particular example. Interestingly, in the absence of alkyl groups with β-C-H bonds, a different mode of reactivity results, as illustrated in Scheme 8 for 2,3-dimethyl-2-butene. In this case, fusion of the olefinic fragment to the vinyl fragment occurs at the α-C of the vinyl fragment, and the two C-H bond activations now occur at the trans-γ positions in the olefinic substrate. The probable mechanism for the formation of complex **19** is depicted in Scheme 10.

Scheme 9

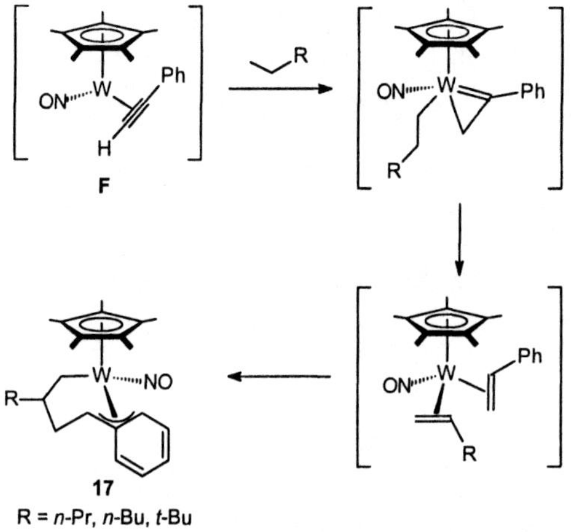

17
R = *n*-Pr, *n*-Bu, *t*-Bu

Scheme 10

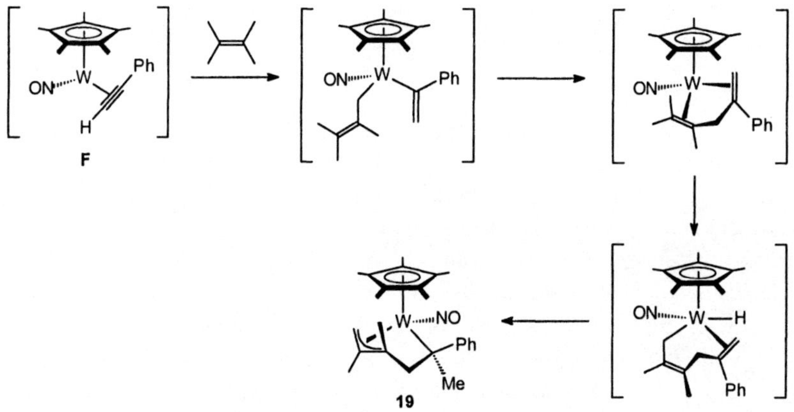

19

Cp*M(NO)(η^2-allene) Intermediates

The 16e η^2-allene intermediate species result from the thermal activation of 18e precursor complexes having η^1-alkyl and η^3-allyl ligands. The most studied member of this family of compounds is Cp*W(NO)(CH$_2$CMe$_3$)(η^3-1,1-

Me$_2$C$_3$H$_3$) (**20**), and its chemistry that we have discovered to date is summarized in Scheme 11 (*21*). Thermolysis of **20** in Me$_4$Si at 50 °C for 6 h results in the evolution of neopentane and the quantitative formation of the 18e alkyl-allyl complex, Cp*W(NO)(CH$_2$SiMe$_3$)(η^3-1,1-Me$_2$C$_3$H$_3$) (**21**). Complex **21** is thermally stable in Me$_4$Si, and the bis(alkyl) complex, Cp*W(NO)(CH$_2$SiMe$_3$)$_2$ (**4**, Scheme 1), is not formed on prolonged reaction times. In a similar manner, Cp*W(NO)-(C$_6$H$_5$)(η^3-1,1-Me$_2$C$_3$H$_3$) (**22**) is quantitatively produced via thermolysis of **20** under the same conditions in benzene solution.

Scheme 11

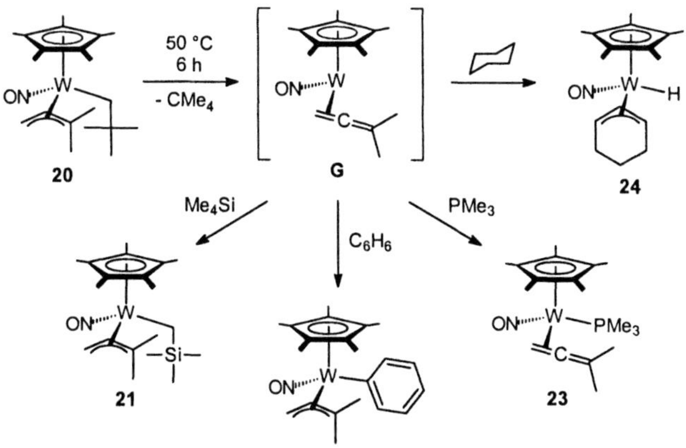

The analogous reaction in C$_6$D$_6$ leads to the formation of the corresponding deuterated complex, Cp*W(NO)(C$_6$D$_5$)(η^3-1,1-Me$_2$-allyl-d_1) (**22-d_6**) in which the deuterium atom in the allyl ligand is incorporated among all three types of carbon atoms. This conversion also exhibits first-order kinetics for the loss of **20** at 50 °C, a feature consistent with the rate-determining step being the intramolecular generation of the intermediate η^2-allene complex **G** (Scheme 11). For comparison, it may be noted that thermolyses of alkyl-allyl complexes of iridium at 120 °C are believed to form similar allene complexes. However, these intermediates have not been characterized or isolated, and they are only capable of effecting single C-H bond activations (*5*).

The thermal reaction of **20** in a solution of excess PMe$_3$ in cyclohexane cleanly affords the base-stabilized form of the allene complex, namely Cp*W(NO)(η^2-H$_2$C=C=CMe$_2$)(PMe$_3$) (**23**), which has been isolated and fully characterized. For instance, its solid-state molecular structure exhibits metrical parameters that indicate considerable back-donation of electron density from the tungsten center to the allene π* orbitals. This exceptionally strong π-donor ability of the Cp*M(NO)(PMe₃) (M = Mo, W) fragments (22) undoubtedly stabilizes all the PMe₃-trapped reactive intermediates that have been isolated during this work.

Most interestingly, the thermolysis of **20** in cyclohexane produces principally the cyclohexenyl hydrido complex, Cp*W(NO)(η^3-C₆H₉)(H) (**24**), which formally results from three C-H bond activations of the hydrocarbon solvent. Its ¹H NMR spectrum in C₆D₆ contains a distinctive hydride resonance at δ −0.57 ($^1J_{WH}$ = 131.7 Hz), and its IR spectrum as a KBr pellet exhibits ν_{WH} at 1898 cm⁻¹. GC-MS studies also reveal the presence of the coupled organic product, a dimethylpropylcyclohexane, in the final reaction mixture, but the mechanism of this transformation remains to be ascertained. Overall, the conversion of **20** into **24** constitutes a novel mode of multiple C-H activations of cyclohexane, a relatively inert solvent that has frequently been used to study the C-H activations of other hydrocarbons. For comparison, it may be noted that bis(alkyl) species such as Cp*W(NO)(CH₂CMe₃)₂ (**1**) react with cyclohexane in a different manner to afford Cp*W(NO)(η^2-cyclohexene)(PMe₃) (**6**) in the presence of PMe₃ (Scheme 2).

Summary

Our work to date with these systems has established that reactive 16-electron intermediates are generated by gentle thermolyses of appropriate Cp*M(NO)(hydrocarbyl)₂ complexes of molybdenum and tungsten. All of these intermediates effect the single C-H bond activations of substrates such as Me₄Si, benzene, and methyl-substituted arenes while exhibiting a preference for activating stronger sp² C-H bonds over sp³ C-H bonds. Furthermore, some of these intermediates also effect novel multiple C-H bond activations of hydrocarbon substrates that apparently involve the transient formation of mono- and bis(olefin) complexes. Our future investigations with these systems will focus on establishing the mechanisms of the multiple C-H bond-activation processes and on developing methods for the functionalization of the activated hydrocarbons in the metals coordination spheres.

References

1. Labinger, J. A.; Bercaw, J. E. *Nature* **2002**, *417*, 507-514.
2. Tran, E.; Legzdins, P. *J. Am. Chem. Soc.* **1997**, *119*, 5071-5072.
3. Adams, C. S.; Legzdins, P.; Tran, E. *J. Am. Chem. Soc.* **2001**, *123*, 612-624.

4. Buchanan, J. M.; Stryker, J. M.; Bergman, R. G. *J. Am. Chem. Soc.* **1986**, *108*, 1537-1550.
5. McGhee, W. D.; Bergman, R. G. *J. Am. Chem. Soc.* **1988**, *110*, 4246-4262.
6. van der Heijden, H.; Hessen, B. *J. Chem. Soc., Chem. Commun.* **1995**, 145-146.
7. Coles, M. P.; Gibson, V. C.; Clegg, W.; Elsegood, M. R. J.; Porelli, P. A. *J. Chem. Soc., Chem. Commun.* **1996**, 1963-1964.
8. Cheon, J.; Rogers, D. M.; Girolami, G. S. *J. Am. Chem. Soc.* **1997**, *119*, 6804-6813.
9. Wada, K.; Pamplin, C. B.; Legzdins, P. *J. Am. Chem. Soc.* **2002**, *124*, 9680-9681.
10. Adams, C. S.; Legzdins, P.; McNeil, W. S. *Organometallics* **2001**, *20*, 4939-4955.
11. Erker, G. *J. Organomet. Chem.* **1977**, *134*, 189-202.
12. Fagan, P. J.; Manriquez, J. M.; Maatta, E. A.; Seyam, A. M.; Marks, T. J. *J. Am. Chem. Soc.* **1981**, *103*, 6650-6667.
13. Hartwig, J. F.; Bergman, R. G.; Andersen, R. A. *J. Am. Chem. Soc.* **1991**, *113*, 3404-3418.
14. Debad, J. D. Ph.D. thesis, The University of British Columbia, Vancouver, BC, 1994.
15. Sharp, W. B. Ph.D. thesis, The University of British Columbia, Vancouver, BC, 2001.
16. Legzdins, P.; Lumb, S. A.; Rettig, S. J. *Organometallics* **1999**, *18*, 3128-3137.
17. Debad, J. D.; Legzdins, P.; Lumb, S. A.; Batchelor, R. J.; Einstein, F. W. B. *J. Am. Chem. Soc.* **1995**, *117*, 3288-3289.
18. Legzdins, P.; Lumb, S. A. *Organometallics* **1997**, *16*, 1825-1827.
19. Debad, D.; Legzdins, P.; Lumb, S. A.; Rettig, S. J.; Batchelor, R. J.; Einstein, F. W. B. *Organometallics* **1999**, *18*, 3414-3428.
20. Arndtsen, B. A.; Bergman, R. G.; Mobley, T. A.; Petersen, T. H. *Acc. Chem. Res.* **1995**, *28*, 154-162.
21. Ng, S. H. K.; Adams, C. S.; Legzdins, P. *J. Am. Chem. Soc.* **2002**, *124*, 9380-9381.
22. Burkey, D. J.; Debad, J. D.; Legzdins, P. *J. Am. Chem. Soc.* **1997**, *119*, 1139-1140.

Chapter 12

Alkane Transfer-Dehydrogenation Catalyzed by a Pincer-Ligated Iridium Complex

Alan S. Goldman, Kenton B. Renkema, Margaret Czerw, and Karsten Krogh-Jespersen

Department of Chemistry and Chemical Biology, Rutgers, The State University of New Jersey, New Brunswick, NJ 08903

The mechanism of (PCP)Ir-catalyzed transfer-dehydrogenation and a full free-energy profile have been elucidated for the substrate/acceptor couple COA/TBE (COA = cyclooctane; TBE = t-butylethylene). Stoichiometric components of the catalytic cycle have been independently observed and their kinetics have been determined. The overall catalysis has also been monitored *in situ*. Good agreement is found between rates independently obtained from stoichiometric and catalytic runs. Within the overall TBE-hydrogenation segment of the cycle, labeling experiments indicate that the rate-determining step is the C-H reductive elimination of t-butylethane from (PCP)IrH(t-butylethyl). Based on microscopic reversibility it can be then inferred that the rate-determining step for the alkane dehydrogenations is C-H addition (and not β-H elimination). In accord with this conclusion, n-alkanes are found to be more reactive than COA. A full energy profile for an n-alkane/α-olefin substrate/acceptor system can be extrapolated based on stoichiometric comparisons with the COA/TBE couple. A computed (DFT) free-energy profile generally matches well with the experimentally derived profile.

Olefins constitute the most important feedstock of the chemical industry. The dehydrogenation of alkanes is the most direct and potentially the most economical route to the production of olefins. While heterogeneous catalysts have long been used for the dehydrogenation of ethane and (less successfully) propane and isobutane (*1*), no such systems have been found capable of dehydrogenating higher alkanes to olefins with good chemoselectivity. In the early 1980's, Crabtree and Felkin developed homogeneous catalyst systems that used sacrificial hydrogen acceptors (*2*). These transfer-dehydrogenation systems showed excellent chemoselectivity but turnover numbers were limited by catalyst degradation. Systems developed subsequently have shown much higher turnover numbers and, in some cases, even high regioselectivity (*3*), but rates and conversions are still well below practical levels. Presently, the most effective such systems are those based upon pincer-ligated iridium (*4-6*), the first example of which was reported by Kaska and Jensen (*4a*). With cyclooctane/*t*-butylethylene (COA/TBE), the prototypical substrate/acceptor couple first introduced by Crabtree (*2*), it was found that (PCP)IrH$_2$ (**1**; PCP = 2,6-C$_6$H$_3$(CH$_2$P^{t}Bu$_2$)$_2$) catalyzed transfer-dehydrogenation to give cyclooctene (COE) and 2,2-dimethylbutane (*t*-butylethane; TBA) with high rates and turnover numbers (*4a*).

$$COA + TBE \xrightarrow{(PCP)IrH_2 \ (1)} COE + TBA \quad (1)$$

Herein, we report our elucidation of the mechanism of eq 1 (*7*). From this starting point we can extrapolate a full energy profile for a system in which *n*-alkane/α-olefin is the substrate/acceptor couple. The latter couple is not only of far greater potential utility (e.g. with propene as acceptor and long-chain *n*-alkane substrates), but it also presents the additional advantage of being far more amenable to computational modeling.

Stoichiometric Reactions with TBE and COA

In an effort to dissect the cycle of eq 1 into its component steps we first reacted the dihydride **1** with TBE in the absence of alkane. We thereby hoped to observe the hydrogenation segment of the cycle, which we presumed would include dehydrogenation and hydrogenation reactions.

In mesitylene (which appears to act as an inert solvent) **1** was found to react with two equiv TBE: one equiv TBE is hydrogenated and a second equiv undergoes vinylic C-H addition to give (PCP)IrH(^{tBu}Vi) (**2**); ^{tBu}Vi = *trans*-2-[*t*-butyl]vinyl; eq 2) (*8*).

$$\text{1} + 2 \, \text{tBu-CH=CH}_2 \longrightarrow \text{2} + \text{tBu-CH}_2\text{-CH}_3 \quad (2)$$

We had previously reported **2** and found that it undergoes rapid exchange with free TBE (on the NMR timescale) at ambient temperature via reversible loss of TBE (eq 3) (*9*).

$$(PCP)IrH(^{tBu}Vi) \, (\mathbf{2}) \rightleftharpoons \text{``(PCP)Ir''} + TBE \quad (3)$$

The kinetics of reaction 2 were studied by monitoring (^{1}H NMR) the disappearance of **1** and appearance of **2** at 55 °C in the presence of a slight excess of TBE. Both dependencies could be well fit with a single parameter, k_h = 0.57 M^{-1} min^{-1} where $-d[\mathbf{1}]/dt = d[\mathbf{2}]/dt = k_h[\mathbf{1}][TBE]$.

When COA is added to **2** (in *p*-xylene solvent at 55 °C) a rapid reaction ensues (eq 4).

$$\mathbf{2} + \text{COA} \longrightarrow \mathbf{1} + \text{COE} + \text{tBu-CH=CH}_2 \quad (4)$$

The sum of reactions 2 and 4 is the catalytic reaction 1. Thus, the two reactions constitute a possible mechanism for the catalytic transfer-dehydrogenation. However, before it could be concluded that the observed reactions constitute the major operative pathway, it remained to be determined if the catalytic and stoichiometric kinetics could be reconciled.

Unfortunately, whereas it was possible to isolate the suspected half of the catalytic cycle leading to TBE hydrogenation (eq 2) and then study its kinetics, the suspected second half of the cycle (eq 4) is too rapid to measure at 55 °C at low [TBE] (< ca. 0.2 M). Higher [TBE] inhibited reaction 4, but under such conditions reaction 2 then proceeds too rapidly to allow the independent measurement of the rate of reaction 4. Nevertheless, the observation of inhibition by added TBE suggests that the reaction proceeds via free "(PCP)Ir". This, in turn, suggests that the use of precursors of free (PCP)Ir other than **2** might permit study of COA dehydrogenation (as in 4) without the complications introduced by subsequent reaction of **1** with TBE.

We had previously reported a study of (PCP)IrH(Ph), an analogue of **2** (9). This complex was found to undergo dissociative hydrocarbon exchange in analogy with the behavior of **2** (eq 3). (PCP)IrH(Ph) could therefore in principle act as a "surrogate" source of (PCP)Ir, the intermediate that seemed likely to undergo the alkane C-H bond addition.

(PCP)IrH(Ph) undergoes reaction with COA (eq 5), in analogy with the reaction of **2** (eq 4).

$$\text{(PCP)IrH(Ph)} + \text{COA} \xrightarrow{k_{dh(IrPh)}} \text{(PCP)IrH}_2 + C_6H_6 \quad (5)$$

Disappearance of (PCP)IrH(Ph) and the increase of [(PCP)IrH$_2$] were monitored by ^{1}H NMR in the presence of an excess of benzene. The two resulting curves (Fig. 1) could be fit with a single parameter, $k_{dh(IrPh)}$, according to eq 6.

$$-d[(PCP)IrH(Ph)]/dt = d[\mathbf{1}]/dt = k_{dh(IrPh)}[(PCP)IrH(Ph)][COA][C_6H_6]^{-1} \quad (6)$$

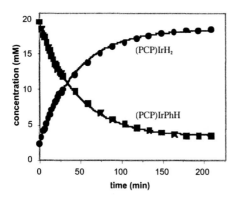

Figure 1. Concentrations of [1] and [(PCP)IrH(Ph)] vs. time (eq 5)

The value of $k_{dh(IrPh)}$ at 55 °C was determined to be 0.0023 min^{-1}. In order to extrapolate the analogous rate constant for the actual complex of interest (vinyl hydride, **2**) it is sufficient to only determine the equilibrium constant for eq 7.

$$\text{(PCP)IrH(Ph)} + {}^tBu\text{-CH=CH}_2 \xrightleftharpoons{K_{Ph/Vi}} \text{(PCP)IrH(CH=CH}^tBu) + C_6H_6 \quad (7)$$

This extrapolation is valid assuming that the dehydrogenations of reactions 4 and 5 both proceed through a pre-equilibrium involving (PCP)Ir or through *any* other common intermediate. $K_{Ph/Vi}$ (eq 7) was found to equal 0.33. Applying this value to $k_{dh(IrPh)}$, we obtain a value of 0.0070 min^{-1} for k_{dh}, the analogous rate constant for the reaction of **2** with COA (eq 4).

Monitoring the Catalytic Reactions of TBE and COA *in situ*

The catalytic COA/TBE transfer-dehydrogenation was monitored directly (*in situ*) by ^{31}P NMR and by ^{1}H NMR (7). The only ^{31}P NMR-visible species present in significant concentrations were **1** and **2**.

Based on the mechanism of eqs 2 and 4 and the corresponding rate laws, the ratio of [**1**] to [**2**] can be predicted by considering the steady state of eq 8.

$$\text{1} \xrightleftharpoons[\substack{\text{rate} = k_{dh}[\mathbf{2}][\text{COA}][\text{TBE}]^{-1} \\ + \text{COA}/- \text{TBE}/- \text{COE}}]{\substack{\text{rate} = k_h[\mathbf{1}][\text{TBE}] \\ + 2\,\text{TBE}/-\text{TBA}}} \text{2} \quad (8)$$

Thus, in the steady state, $k_h[\mathbf{1}][\text{TBE}] = k_{dh}[\mathbf{2}][\text{COA}][\text{TBE}]^{-1}$ which yields eq 9.

$$[\mathbf{2}]/[\mathbf{1}] = k_h[\text{TBE}]^2/k_{dh}[\text{COA}] \quad (9)$$

Figure 2 reveals that the measured concentrations are in excellent agreement with eq 9. The slope of the line in Figure 2 is equal to $k_h/k_{dh}[\text{COA}]$, yielding $k_h/k_{dh} = 56$ M^{-1}.

The catalytic turnover rate predicted by the mechanism of eqs 2 and 4 is simply equal to the rate of either of these two reactions under catalytic (steady-state) conditions (eq 10). Eq 9 can be re-written as eq 11; substitution into eq 10 then gives the full rate law, eq 12.

$$\text{rate} = d[\text{COE}]/dt = -d[\text{TBE}]/dt = k_h[\mathbf{1}][\text{TBE}] \quad (10)$$

$$[\mathbf{1}] = k_{dh}[\text{Ir}_{tot}][\text{COA}]/(k_{dh}[\text{COA}] + k_h[\text{TBE}]^2) \quad (11)$$

$$d[\text{COE}]/dt = k_h k_{dh}[\text{TBE}][\text{COA}][\text{Ir}_{tot}]/(k_{dh}[\text{COA}] + k_h[\text{TBE}]^2) \quad (12)$$

According to eq 12, in the limit of low [TBE] the catalytic rate is equal to $k_h[\text{TBE}][\text{Ir}_{tot}]$ (first order in [TBE]), while in the limit of high [TBE] the rate should be $k_{dh}[\text{TBE}]^{-1}[\text{COA}][\text{Ir}_{tot}]$, inverse first-order in [TBE]. Experimentally, catalytic rates were determined by *in situ* ^{1}H NMR measurement of the increase in [COE] at various initial concentrations of TBE. In accord with eq 12, the curve of rate vs. [TBE] reveals a maximum (at ca. 0.4 M TBE; Figure 3).

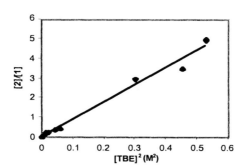

Figure 2. Relative concentrations ([2] / [1]) vs. [TBE]^2

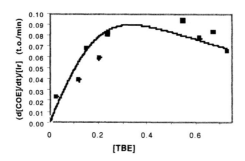

Figure 3. Rate of transfer-dehydrogenation (eq 1) vs. [TBE].

The curve shown in Figure 3 is calculated according to eq 12, generating a best fit to the data with k_h/k_{dh} held fixed at 56 M^{-1} (the value obtained from eq 9 and Figure 2). This therefore represents a one-parameter fit yielding the two rate constants, each obtained solely from *in situ* observation of the catalytic system: $k_h = 0.53$ M^{-1} min^{-1}; $k_{dh} = 0.0094$ min^{-1}.

The Full Catalytic Cycle: COA/TBE Transfer-Dehydrogenation

We now arrive at the critical question in the assessment of the proposed mechanism of eqs 2 and 4: Are the rate constants obtained from the stoichiometric experiments consistent with those obtained independently from the catalytic experiments? Considering experimental error and possible effects of different solvents (stoichiometric runs were necessarily conducted in non-alkane solvents, *p*-xylene or mesitylene) the agreement between the sets of rate constants is excellent (catalytic $k_h = 0.53$ M^{-1} min^{-1}; $k_{dh} = 0.0094$ min^{-1}; stoichiometric $k_h = 0.57$ M^{-1} min^{-1}; $k_{dh} = 0.0070$ min^{-1}). The overall set of kinetic equations is thus strongly overdetermined, and we consider these measured kinetics data to, effectively, prove the mechanisms of eqs 2 and 4 (Figure 4).

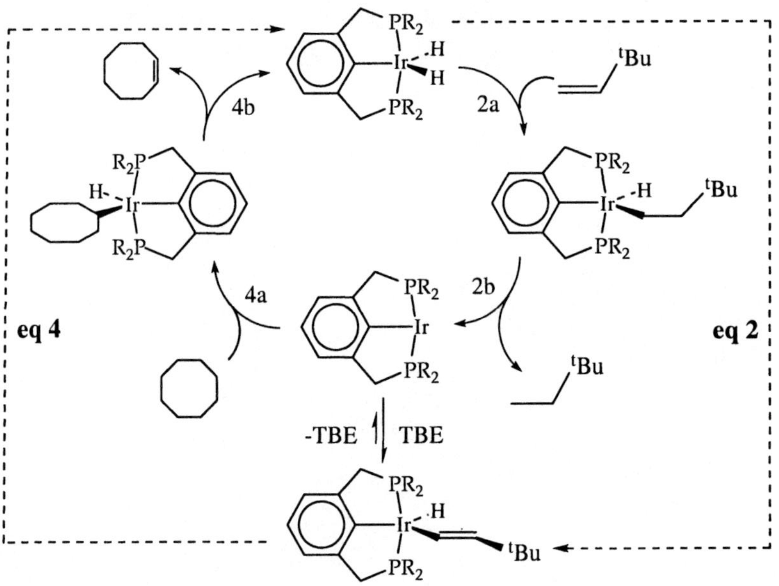

Figure 4. Mechanism of eq 1

Somewhat surprisingly, under typical conditions ([TBE] < ca. 0.3 M) the TBE-hydrogenation part of the cycle is rate-determining. We thus wished to identify the rate-determining reaction step within the overall reaction 2. Toward this end, a labeling experiment was conducted to determine if the presumed insertion of TBE into the Ir-H bond is irreversible (i.e. if $k_{2b} \gg k_{-2a}$, Figure 5).

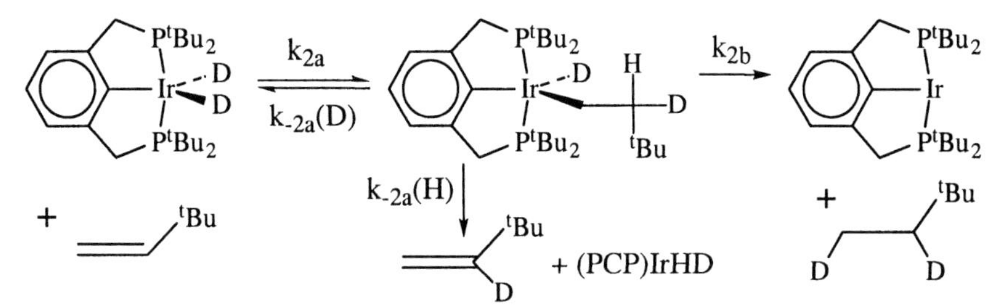

Figure 5. Reaction of 1 with TBE; competition between β-H (or β-D) elimination and C-H bond elimination

(PCP-d_{36})IrD_2 (prepared via H/D exchange of 1 with C_6D_6) was reacted with TBE. The initial rate of formation of H/D exchanged product, CH_2=CD^{t}Bu, was found to be 4.8 times greater than that of hydrogenated product, CH_2D-CHD(^{t}Bu). Neglecting isotope effects, we therefore obtain $k_{-2a}/k_{2b} = 9.6$, since statistically, formation of CH_2=CD^{t}Bu should reflect half the rate of β-H elimination from the perproteo isotopomer.

The labeling experiment of Figure 5 demonstrates that the rate-determining step of reaction 2, and thus the rate-determining step of the catalytic cycle under typical conditions, is C-H elimination of TBA (2b, Figure 4). Based on considerations of microscopic reversibility, C-H *addition* must be the rate-determining step of the reverse reaction, the (hypothetical) *dehydrogenation* of TBA. This conclusion can be further extrapolated to the terminal dehydrogenation of *n*-alkanes (note that the lack of a bulky *t*-Bu group should facilitate β-H elimination more than C-H addition) and probably to cycloalkanes, which tend to undergo C-H addition less readily and β-H elimination more readily than *n*-alkanes (*10,11*). This finding presents an important contrast to previously reported alkane dehydrogenation systems, in which β-H elimination was found to be rate-determining (*2,3*).

A combination of several parameters can be used to determine a detailed full free-energy profile of the COA/TBE transfer-dehydrogenation cycle. These parameters include:

(1) rate constants k_h, k_{dh}

(2) rates of dissociation of TBE from (PCP)Ir(*t*-butylvinyl) (eq 3)

(3) relative reductive elimination/β-H elimination rates from the labeling experiment illustrated in Fig. 5 ($\Delta\Delta G^{\ddagger} = 1.5$ kcal/mol, derived from the rate factor of 9.6)

(4) known thermodynamics of COA/TBE transfer-dehydrogenation (*12,13*)

In addition, we have determined the equilibrium constant (and thus ΔG) for eq 13 (*14*). Using $K_{Ph/Vi}$ (eq 7), K_{13} can then be used to determine the equilibrium constant and ΔG for the analogous reaction with **2** instead of (PCP)IrPhH (eq 14). This provides redundancy and an independent check on the relative thermodynamics of (PCP)IrH₂ and (PCP)IrH(*t*-butylvinyl) obtained from the parameters noted above.

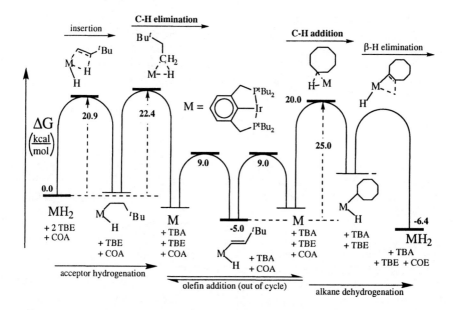

(PCP)Ir(H)(*t*-butylvinyl) + COA $\xrightleftharpoons{K}$ (PCP)IrH₂ + COE + TBE   (14)

The rate constants (converted to free energies of activation) and equilibrium constants can be pieced together to give the free-energy profile shown in Fig. 6.

Figure 6. *Free-energy profile for COA/TBE transfer-dehydrogenation cycle*

Competition Experiment: COA vs. *n*-alkane

The results discussed above indicate that C-H addition is the rate-determining step in the dehydrogenation segment of the catalytic cycle. The selectivity of addition of C-H bonds to late metals has been well studied by several groups (*10,11*). To our knowledge, addition to late metals is, in all cases, more favorable for aliphatic primary C-H bonds than for secondary. The widespread use of COA as a "model" substrate for alkane dehydrogenation is attributable to the factors that engender the anomalously favorable thermodynamics of COA dehydrogenation; the same factors come into play in a β-H elimination transition state, but not in a transition state for C–H addition. Indeed Bergman has found that COA is less reactive toward C-H addition than cyclohexane (which in turn is less reactive than the terminal groups of *n*-alkanes) (*11*). These considerations raised important questions: Is the (PCP)Ir system more reactive toward *n*-alkanes than COA? And if so, is COA more easily dehydrogenated by other catalytic systems because β-H elimination is rate-determining in those systems?

An experiment was run in which (PCP)IrH$_2$ catalyzed the competitive transfer-dehydrogenation of COA and *n*-octane, using norbornene as acceptor. In accord with our inference that C-H oxidative addition is rate-determining, we found that the *n*-alkane was significantly more reactive (by a factor of 5.7 on a per mol basis, or 15 per C-H bond if it is assumed that addition occurs exclusively at the terminal position of *n*-octane).

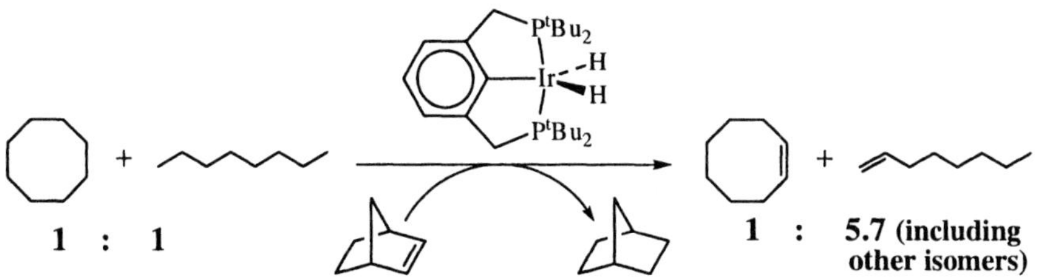

Figure 7. Competition experiment: n-octane is more reactive than COA.

Inhibition of Catalysis Due to α-Olefin Binding

Despite the higher intrinsic reactivity of *n*-alkanes illustrated in Figure 7, preliminary results reveal that catalytic rates are generally comparable or somewhat faster for runs in which COA is the only dehydrogenation substrate as compared with *n*-alkane dehydrogenation. This apparent discrepancy can be easily rationalized in terms of a mechanistic scheme that is analogous to that shown in Figure 4. As indicated in Figure 4, the only resting states are **1** and the vinyl hydride, **2**. (COE can add to (PCP)Ir but much more weakly than either TBE or norbornene (*14*).) By contrast, in the case of dehydrogenation of *n*-alkanes, preliminary experiments reveal a different resting state: (PCP)Ir(η^2-1-alkene). Thus, the relative *n*-alkane/COA rates in experiments with individual alkanes should depend strongly on conditions: in the regime where hydrogenation of acceptor is rate-determining, the nature of the substrate should have no effect. As the concentration of α-olefin increases in the *n*-alkane dehydrogenations, it will bind an increasingly large fraction of the iridium in solution and inhibit the reaction accordingly. The competition experiment of Figure 7, however, reveals the relative *intrinsic* reactivities of the two substrates.

The binding strength of α-olefin, relative to TBE C-H addition, was determined by equilibrium experiments, eq 15 (*14*).

$$\text{(PCP)Ir(H)(}^t\text{Bu)} + \text{1-hexene} \xrightleftharpoons{K} \text{(PCP)Ir(}^n\text{Bu)} + {}^t\text{Bu} \quad (15)$$

The relative rates of *n*-alkane/COA dehydrogenation and the relative α-olefin/TBE binding constants allow us to extrapolate a free-energy profile for *n*-alkane/α-olefin transfer-dehydrogenation. This extrapolation has several advantages: (i) the resulting profile describes (unlike COA/TBE) a potentially very useful system (e.g. formation of long-chain α-olefins using propene as an acceptor) (ii) the analysis is much simpler; because the reaction is degenerate, both halves are equivalent and the overall thermodynamics is obviously neutral (iii) for computational modeling there are fewer distinct intermediates and transition states due to the degeneracy, and they furthermore contain fewer atoms than either the analogous COA or TBE species (we use butane/butene as a model system). The resulting experimentally extrapolated free-energy half-profile (due to degeneracy the full profile is redundant) is shown in Figure 8.

Several points are worth noting in the context of Figure 8. *Vis-a-vis* the COA/TBE couple, *n*-alkane/α-olefin dehydrogenation is slowed by the strong binding of α-olefin (π-bonding) relative to TBE (C-H addition). The vinyl hydride species shown in Fig. 8 is never actually observed in this system; its energy is assumed to be approximately equal to the *t*-butylvinyl analog, although for steric reasons it might actually be lower. Crabtree's introduction of TBE as a hydrogen-acceptor seems particularly prescient in light of this diagram (*2*).

Figure 8 is in accord with preliminary observations that the ground state for *n*-alkane/α-olefin dehydrogenation is the π-bound α-olefin complex (*14*). However, the difference between this resting state and the dihydride is not necessarily the 7.0 kcal that might be inferred from Figure 8, since that value is based upon standard concentrations (1.0 M). Using typical concentrations the free-energy difference is substantially reduced. Further, the higher temperatures typically used for catalysis (values in Figure 8 are for 55 °C, the temperature of the mechanistic studies) favor the dihydride resting state (plus two mol free alkene) versus the vinyl hydride (plus one mol alkane) due to the entropic advantage of three free species versus two.

209

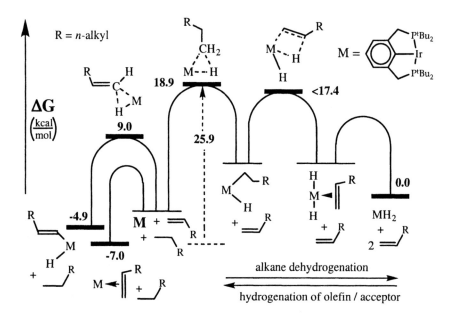

Figure 8. Free-energy profile for n-alkane/α-olefin transfer-dehydrogenation cycle, experimentally extrapolated from COA/TBE profile.

DFT Calculations and Comparison of Theory with Experiment

We have conducted extensive computational (DFT) studies in parallel with our mechanistic work. We find that the theoretical results are far more valuable when viewed in the context of the experimental results rather than in isolation.

It is important to note that all calculated energies discussed herein are (Gibbs) free energies. The greatest differences between these and the purely electronic energies (E) arise from entropy terms. Entropy considerations are particularly important when comparing states of different molecularity; at 55 °C, such entropy differences can typically yield differences between electronic and free-energies of ca. 8 kcal/mol. In extreme cases (see the following Chapter in this Volume), the differences can be nearly 20 kcal/mol. Zero-point vibrational energy terms, which are also incorporated into the calculated free energies, can contribute several kcal/mol. Other factors tend to contribute less than these two terms (*15-17*).

It should also be noted that the work discussed herein utilizes alkyl groups (rather than H atoms) on the phosphorus ligands as should, in our opinion, all computational work where a *quantitative* comparison of theory and experiment is desired. Methyl groups serve as reasonably good models for *t*-butyl groups with respect to the purely electronic effects. However, they obviously do not capture the steric effects of *t*-butyl groups, and all computed results obtained with truncated ligands must therefore be carefully interpreted with this very important caveat in mind.

Calculations with ^{Me}PCP as Model Ligand

Figure 9 illustrates the free-energy profile for *n*-alkane/α-olefin transfer dehydrogenation predicted from DFT calculations using ^{Me}PCP as a model ligand (^{Me}PCP = 2,6-$C_6H_3(CH_2PMe_2)_2$; to distinguish this ligand from the analogous complex with *t*-Bu groups used experimentally, we will in this section specify the latter as $^{t-Bu}$PCP). The computed results are in overall good agreement with the experimentally derived profile, but only in a qualitative sense. Discrepancies encountered between experimental and calculated values are very much in accord with the reduced steric demands of Me (relative to *t*-Bu) groups. Using the free energy of the dihydride as the zero point, the calculated relative free energies of all species are lower than the experimentally established values by 4-10 kcal/mol with a single exception: (PCP)Ir(vinyl)H. This is a remarkably "sensible" result: the dihydride and vinyl hydride are probably the two species which encounter negligible steric crowding due to the small size of hydride ligands and the planarity of the vinyl group which enables the latter to fit unencumbered in the cleft of the ($^{t-Bu}$PCP)Ir equatorial plane. Note that addition/elimination of the vinyl C-H bond proceeds via a distinctly non-planar TS and is accordingly more crowded than the C–H addition product.

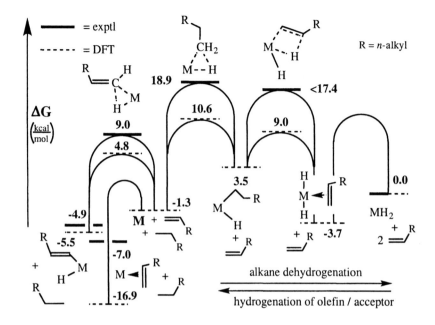

Figure 9. Free-energy profiles for n-alkane/α-olefin transfer-dehydrogenation (see Figure 8): experimental (M = (^{tBu}PCP)Ir) and DFT values (M = (^{Bu}PCP)Ir)

Of particular note is the calculated free-energy of the olefin dihydride complex, 3.7 kcal/mol below that of the dihydride. The actual olefin-dihydride species is undoubtedly of higher energy, since it never forms in observable concentration even though the actual catalytic pathway presumably proceeds through this state (in both directions). Since steric effects would be quite severe for this complex (with a $^{t-Bu}$PCP ligand), this discrepancy also makes good sense. Note that calculations ignoring entropy altogether and using the even less-bulky ^{H}PCP ligand (and C_2H_4 as model olefin)(*18*) suggest that loss of olefin from this complex would be the rate-determining step in the catalytic cycle. This suggestion is completely in discord with experimental observations, as would be expected in light of the above considerations. Using a similar approach, we obtain a relative free-energy value of –42 kcal/mol for the (PCP)Ir(olefin) complex, 49 kcal/mol less than the relative experimental free-energy!

Calculations with a Non-truncated Ligand: $^{t-Bu}$PCP

Calculations were also carried out using the full $^{t-Bu}$PCP ligand. However, due to the size of the molecular complexes involved, these calculations necessarily employed a more limited basis set on the phosphine alkyl groups than the one used in the ^{Me}PCP calculations described above.

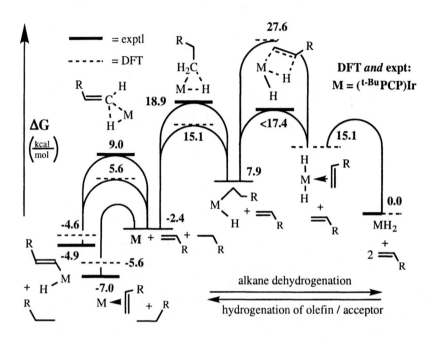

Figure 10. Free-energy profiles for n-alkane/α-olefin transfer-dehydrogenation (see Figure 8): experimental and DFT values for M = ($^{t-Bu}$PCP)Ir

The values obtained with the non-truncated $^{t-Bu}$PCP ligand were found to be in generally much better agreement with experiment (within 4 kcal/mol) than those calculated with ^{Me}PCP. The singular exception to this generalization is the TS for β-H elimination/olefin-insertion. This very crowded TS is calculated to be much higher in energy using ($^{t-Bu}$PCP)Ir than (^{Me}PCP)Ir, and the calculated free energy is significantly higher (~ 10 kcal/mol or more) than the experimental value. The high calculated value of the $^{t-Bu}$PCP TS for β-H elimination may be attributable to the pronounced tendency of DFT to overestimate steric interactions, since dispersion interactions are not included in the (B3LYP) functionals used in the calculations. Alternatively, H-atom tunneling may be contributing to an apparent reduction in the experimentally measured barrier.

Although these results illustrate some of the perils of taking at face value even the results of high-level free-energy calculations on non-truncated systems, we feel that they do also strongly suggest that DFT calculations can be of considerable value in predicting catalytic activity and catalyst design – provided the calculations are interpreted with due care. Although computational results of "chemical accuracy" are not yet obtainable, precision within ca. 4 kcal/mol is certainly of great potential value for purposes of *in silico* catalyst "screening". And while the discrepancy found for the β-H elimination TS is serious, we believe it will be mostly reproducible with other complexes of $^{t\text{-}Bu}$PCP and sterically similar ligands. Thus, to a first approximation, we can apply this derived "calibration" to future calculations on related complexes and we are, in fact, currently screening a range of pincer complexes using this approach (*19*). We are also attempting to improve the precision of our calculations via the use of mixed-mode QM/MM methodologies and different functionals.

Summary

The mechanism of (PCP)Ir-catalyzed transfer-dehydrogenation of the COA/TBE substrate/acceptor couple has been elucidated and a detailed free-energy profile has been determined. The two segments of the cycle have been observed independently, and the nature of the rate-determining step within each segment has been determined. The mechanism has several notable features:

(i) Although olefin hydrogenation is one of the most widely and readily catalyzed reactions of organometallic complexes (*11*), the rate-determining (slow) segment of the present cycle under typical conditions is hydrogenation of TBE.

(ii) Although C–H addition is often assumed to be the "difficult" step in alkane functionalizations (*10*), in this case C–H *elimination* of alkane (TBA) is apparently rate-determining.

(iii) In the regime where dehydrogenation is the rate-determining segment (very high [TBE]), C–H addition is inferred to be the rate-determining step. Nevertheless, a substantial contribution to the overall barrier derives from the thermodynamic cost of eliminating the vinyl C–H bond of TBE prior to addition of the alkane C–H bond.

Consistent with the conclusion that C–H addition is rate-determining, *n*–alkanes are found to be intrinsically more reactive than COA (generally regarded as an "easily dehydrogenated" substrate). However, this is offset by the fact that the α-olefin dehydrogenation products bind more strongly than COE and thereby inhibit catalysis more severely. The *n*-alkane /COA and α-olefin/COE competition experiments allow us to extrapolate a full energy profile for *n*-alkane /α-olefin transfer-dehydrogenation.

DFT calculations of free energies show qualitative agreement with experimental results, when ^{Me}PCP models the experimentally used $^{t\text{-}Bu}$PCP ligand. However, the energetic and structural effects of the reduced crowding exerted by methyl groups are apparent. When calculations are conducted with the non-truncated $^{t\text{-}Bu}$PCP ligand, the agreement is generally much better. The energy of the TS for β-H elimination/olefin insertion is understated for ^{Me}PCP but significantly overstated using the full $^{t\text{-}Bu}$PCP ligand. The reasons for this discrepancy remain under investigation. In the interim, we believe it is possible to use the experimental and computational values obtained in this work to "calibrate" the results of DFT calculations on related, sterically similar, systems for the purpose of computational catalyst screening.

Acknowledgement

We thank the Division of Chemical Sciences, Office of Basic Energy Sciences, Office of Energy Research, U. S. Department of Energy for support of this research.

References

1. (a) Sundarum, K. M.; Shreehan, M. M.; Olszewski, E. F. In *Kirk-Othmer Encyclopedia of Chemical Technology*; 4th ed.; Kroschwitz, J. I., Howe-Grant, M., Eds.; Wiley-Interscience: New York, 1991; Vol. 9, pp 877-915. (b) Tullo, A. H. In *Chem. Eng. News*, 2001; Vol. 79, pp 18-24. (c) Goldman, A. S. In *Encyclopedia of Catalysis*; Horvath, I., Ed.; John Wiley & Sons, 2002; Vol. 3, pp 25-33.
2. (a) Crabtree, R. H.; Mihelcic, J. M.; Quirk, J. M. *J. Am. Chem. Soc.* **1979**, *101*, 7738. (b) Crabtree, R. H.; Mellea, M. F.; Mihelcic, J. M.; Quirk, J. M. *J. Am. Chem. Soc.* **1982**, *104*, 107. (c) Baudry, D.; Ephritikine, M.; Felkin, H. *J. Chem. Soc., Chem. Commun.* **1980**, 1243–1244. (d) Baudry, D.; Ephritikine, M.; Felkin, H.; Holmes-Smith, R. *J. Chem. Soc., Chem. Commun.* **1983**, 788. (e) Burk, M. J.; Crabtree, R. H.; Parnell, C. P.; Uriarte, R. J. *Organometallics* **1984**, *3*, 816.
3. (a) Maguire, J. A.; Goldman, A. S. *J. Am. Chem. Soc.* **1991**, *113*, 6706. (b) Maguire, J. A.; Petrillo, A.; Goldman, A. S. *J. Am. Chem. Soc.* **1992**, *114*, 9492.
4. (a) Gupta, M.; Hagen, C.; Flesher, R. J.; Kaska, W. C.; Jensen, C. M. *Chem. Commun.* **1996**, 2083. (b) Gupta, M.; Hagen, C.; Kaska, W. C.; Cramer, R. E.; Jensen, C. M. *J. Am. Chem. Soc.* **1997**, *119*, 840. (c) Gupta, M.; Kaska, W. C.; Jensen, C. M. *Chem. Commun.* **1997**, 461.
5. (a) Xu, W.; Rosini, G. P.; Gupta, M.; Jensen, C. M.; Kaska, W. C.; Krogh-Jespersen, K.; Goldman, A. S. *Chem. Commun.* **1997**, 2273. (b) Liu, F.; Goldman, A. S. *Chem. Comm.* **1999**, 655.
6. Liu, F.; Pak, E. B.; Singh, B.; Jensen, C. M.; Goldman, A. S. *J. Am. Chem. Soc.* **1999**, *121*, 4086.
7. Renkema, K. B.; Kissin, Y. V.; Goldman, A. S. *J. Am. Chem. Soc.* **2003**, *125*, 7770.
8. For closely related examples of oxidative addition of vinylic C-H bonds see: Wick, D. D.; Jones, W. D. *Organometallics* **1999**, *18*, 495.

9. Kanzelberger, M.; Singh, B.; Czerw, M.; Krogh-Jespersen, K.; Goldman, A. S. *J. Am. Chem. Soc.* **2000**, *122*, 11017.
10. For some reviews of alkane C-H bond activation by organometallic complexes see: (a) Arndtsen, B. A.; Bergman, R. G.; Mobley, T. A.; Peterson, T. H. *Acc. Chem. Res.* **1995**, *28*, 154. (b) Shilov, A. E.; Shul'pin, G. B. *Chem. Rev.* **1997**, *97*, 2879. (c) Sen, A. *Acc. Chem. Res.* **1998**, *31*, 550. (d) Guari, Y.; Sabo-Etiennne, S.; Chaudret, B. *Eur. J. Inorg. Chem.* **1999**, 1047. (e) Jones, W. D. *Science* **2000**, *287*, 1942. (f) Crabtree, R. H. *Journal of the Chemical Society, Dalton Transactions* **2001**, *17*, 2437. (g) Labinger, J. A.; Bercaw, J. E. *Nature* **2002**, *417*, 507-514.
11. See reference 10; other papers that have addressed selectivity in particular, and provide good lead references, include: (a) Harper, T. G. P.; Desrosiers, P. J.; Flood, T. C. *Organometallics* **1990**, *9*, 2523. (b) Bennett, J. L.; Vaid, T. P.; Wolczanski, P. T. *Inorg. Chim. Acta.* **1998**, *270(1-2)*, 414. (c) Wick, D. D.; Jones, W. D. *Organometallics* **1999**, *18*, 495. (d) Asbury, J. B.; Hang, K.; Yeston, J. S.; Cordaro, J. G.; Bergman, R. G.; Lian, T. *J. Am. Chem. Soc.* **2000**, *122*, 12870 and references 4-11 therein. (e) Peterson, T. H.; Golden, J. T.; Bergman, R. G. *J. Am. Chem. Soc.* **2001**, *123*, 455.
12. NIST Standard Reference Database Number 69 - March, 2003 Release: http://webbook.nist.gov/chemistry/
13. Enthalpy values of -30.1 and -24.3 kcal/mol are used for hydrogenation of COE and TBE respectively (liquid phase; ref 12); thus $\Delta H°$ = -5.8 kcal/mol for the transfer-dehydrogenation. Based on respective entropy values (gas-phase value are used; ref 12) $\Delta S°$ = + 1.77 eu; giving $\Delta G°$ = -6.4 kcal/mol at 55 °C.
14. Renkema, K. B.; Goldman, A. S., to be submitted for publication.
15. (a) Abu-Hasanayn, F.; Goldman, A. S.; Krogh-Jespersen, K. *J. Phys. Chem.* **1993**, *97*, 5890. (b) Abu-Hasanayn, F.; Krogh-Jespersen, K.; Goldman, A. S. *J. Am. Chem. Soc.* **1993**, *115*, 8019.
16. (a) Schaller, C. P.; Cummins, C. C.; Wolczanski, P. T. *J. Am. Chem. Soc.* **1996**, *118*, 591. (b) Slaughter, L. M.; Wolczanski, P. T.; Klinckman, T. R.; Cundari, T. R. *J. Am. Chem. Soc.* **2000**, *122*, 7953.
17. (a) Churchill, D. G.; Janak, K. E.; Wittenberg, J. S.; Parkin, G. *J. Am. Chem. Soc.* **2003**, *125*, 1403. (b) Janak, K. E.; Parkin, G. *J. Am. Chem. Soc.* **2003**, *125*, 6889.
18. Niu, S. Q.; Hall, M. B. *J. Am. Chem. Soc.* **1999**, *121*, 3992.
19. Achord, P.; Goldman, A. S.; Krogh-Jespersen, K., to be submitted for publication.

Chapter 13

DFT Calculations on the Mechanism of (PCP)Ir-Catalyzed Acceptorless Dehydrogenation of Alkanes

Realistic Computational Models and Free Energy Considerations

Karsten Krogh-Jespersen, Margaret Czerw, and Alan S. Goldman

Department of Chemistry and Chemical Biology, Rutgers, The State University of New Jersey, New Brunswick, NJ 08903

DFT calculations have been carried out to elucidate the mechanism of acceptorless dehydrogenation of alkanes RH by "pincer"-ligated Ir-catalysts (PCP)IrH$_2$. The key alkyl-hydride intermediate, (PCP)Ir(R)(H), may be formed via dissociative (**D**) or associative (**A**) paths. The **D** path proceeds via an initial loss of H$_2$ from (PCP)IrH$_2$, followed by R-H addition to (PCP)Ir. In contrast, the **A** path involves initial R-H activation by (PCP)IrH$_2$ and subsequent formation of a "seven-coordinate" (PCP)Ir(R)(H)$_3$ complex, followed by loss of H$_2$. Free energy considerations lead squarely to the conclusion that the dissociative pathway **D** is operative under experimentally relevant conditions (high T, low pressure of H$_2$, bulky phosphine groups). Experimental results in support of this conclusion are presented briefly. To fully illustrate the distinctly different energy profiles of the two mechanisms, the calculations must employ realistic molecular models and include accurate simulations of experimental reaction conditions.

Introduction

Alkanes are the world's most abundant organic resource, but methods for their use as direct precursors to higher value chemicals are severely limited. Alkenes, on the other hand, are the most versatile and important organic feedstocks in the chemical industry.(*1*) Direct dehydrogenation of alkanes has a very large activation energy barrier, since the reaction is highly endothermic (ca. 24-32 kcal/mol) and the synchronous removal of two hydrogens from adjacent carbon atoms is symmetry forbidden. Many heterogeneous catalysts are known to effect dehydrogenation (at elevated temperatures of ca. 500 – 900 °C), but applications are generally limited to complete and unselective dehydrogenation (e.g. ethylbenzene to styrene, ethane to ethylene).(*2*) Hence the selective catalytic functionalization of alkanes to the corresponding alkenes represents a potentially highly rewarding reaction and a significant challenge to the field of catalysis. The high selectivity often displayed by homogeneous catalysts, and the remarkable selectivity of transition metal complexes with respect to stoichiometric reactions of alkanes (most notably C-H bond addition), has suggested that soluble transition-metal-based systems offer great promise in this context.(*3,4*) Indeed, considerable progress has been made toward the development of such systems for alkane dehydrogenation.(*5-8*) Most examples involve transfer of hydrogen to a sacrificial olefin (eq 1) but a smaller number of catalysts have been shown effective for "acceptorless" dehydrogenation (eq 2),(*9*) a process that is potentially simpler and economically more favorable.

$$\text{alkane + acceptor} \xrightarrow{\text{catalyst}} \text{alkene} + \text{H}_2 \cdot \text{acceptor} \quad \text{(transfer)} \quad (1)$$

$$\text{alkane} \xrightarrow{\text{catalyst}} \text{alkene} + \text{H}_2\uparrow \quad \text{(acceptorless)} \quad (2)$$

The catalysts reported to date that have proven most effective for both acceptorless and transfer-dehydrogenation are species of the type ($^{R'}$PCP)IrH$_2$, where ($^{R'}$PCP) is the tri-coordinate "pincer" ligand [η^3-2,6-(R'$_2$PCH$_2$)$_2$C$_6$H$_3$].(*7-9*)

The operative mechanisms for the acceptorless and transfer systems undoubtedly have several elementary steps in common. The key difference between the two mechanisms concerns the loss of H$_2$. In the case of the transfer systems, we have shown that the (PCP)IrH$_2$ dihydride loses hydrogen via insertion of the sacrificial acceptor into the Ir-H bond, followed by C-H elimination.(*10*) Hence, an Ir(I)/Ir(III) couple is operative in accord with independent computations by Hall and by us.(*11,12*) The acceptorless system, however, must obviously lose free dihydrogen (eq 2). This can in principle proceed either dissociatively from the resting state of the catalyst, (PCP)IrH$_2$, or associatively via a highly coordinated Ir-complex, (PCP)Ir(alkyl)H$_3$. Elucidation of the mechanism for acceptorless dehydrogenation of alkanes, eq 2, using first-principles electronic structure calculations forms the topic of this Chapter.

Overview of proposed mechanisms for acceptorless dehydrogenation of alkanes by (PCP)IrH$_2$

Three plausible mechanisms have been proposed for catalytic acceptorless dehydrogenation of alkanes by (PCP)IrH$_2$ (dissociative = ***D***; associative = ***A***; interchange = ***I***), as outlined in Scheme 1. The mechanisms all feature a (PCP)Ir-alkyl-hydride complex as a crucial, but as yet elusive, intermediate. The mechanisms differ in the initiation steps, which generate this alkyl hydride from the resting state of the catalyst, (PCP)IrH$_2$.

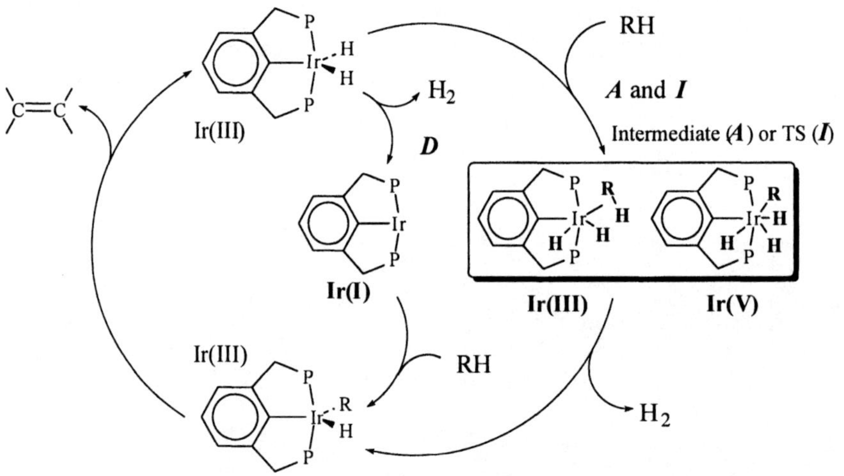

*Scheme 1. Possible mechanisms of acceptorless alkane (RH) dehydrogenation by (PCP)IrH$_2$: dissociative (**D**), associative (**A**), and interchange (**I**) pathways.*

The proposed mechanism for the dissociative pathway (***D***) invokes the following two steps to generate the alkyl hydride:

(*D1*) Reductive elimination of H$_2$ from (PCP)IrH$_2$, a process which formally changes the oxidation state of the metal from Ir(III) to Ir(I).

$$(PCP)IrH_2 + RH \longrightarrow (PCP)Ir + H_2 + RH \quad (D1)$$

(D2) Oxidative addition of the alkane RH by (PCP)Ir to form the alkyl hydride, (PCP)Ir(R)(H). The oxidation state of the metal reverts to Ir(III).

$$(PCP)Ir + H_2 + RH \longrightarrow (PCP)Ir(R)(H) + H_2 \quad (D2)$$

The associative mechanism (*A*) also uses two steps to form the intermediate alkyl-hydride complex:

(A1) Addition of alkane RH to (PCP)IrH$_2$ to form a "seven-coordinate" (PCP)Ir(R)(H)$_3$ complex.

$$(PCP)IrH_2 + RH \longrightarrow [(PCP)Ir(R)H_3] \quad (A1)$$

Formally, alkane addition oxidizes the Ir atom from Ir(III) to Ir(V) if the three hydrogen atoms coordinate as classical hydrides. If they do not, i.e., if a non-classical η^2-coordinated H$_2$ or RH molecule is present within the Ir coordination sphere, then this step involves no change in metal oxidation state.

(A2) Elimination of H$_2$ from (PCP)Ir(R)(H)$_3$ to form the alkyl hydride, (PCP)Ir(R)(H). Depending on the bonding pattern of the (PCP)Ir(R)(H)$_3$ species (three hydrides or η^2-H$_2$ or RH coordination), this may involve a reduction of the metal (Ir(V) to Ir(III)).

$$[(PCP)IrH_3R] \longrightarrow (PCP)Ir(R)(H) + H_2 \quad (A2)$$

Recently, an interchange pathway (*I*) was proposed as a possible alternative to the more conventional *D*/*A* pathways.(*13*) The *I* pathway is direct and uses only one step to obtain the alkyl hydride:

(I) Reductive removal of H$_2$ and oxidative addition of alkane occur in a concerted reaction via a single transition state.

$$(PCP)IrH_2 + RH \longrightarrow (PCP)Ir(R)(H) + H_2 \quad (I)$$

The *I* mechanism in essence requires the "seven-coordinate" (PCP)Ir(R)(H)$_3$ species to function as a TS and not as a minimum (as in the *A* mechanism).

To complete the mechanism for eq 2, the following two steps are proposed as common:

(3) β-H elimination from the alkyl group in (PCP)Ir(R)(H) to generate a coordinated alkene-dihydride species; no change in metal oxidation state occurs.

$$(PCP)Ir(R)H \longrightarrow (PCP)Ir(alkene)(H)_2 \quad (3)$$

(4) Removal of the coordinated alkene to regenerate (PCP)IrH$_2$, again with no change in metal oxidation state.

$$(PCP)Ir(alkene)(H)_2 \longrightarrow (PCP)IrH_2 + alkene \quad (4)$$

None of the intermediates proposed in Scheme 1 have been isolated for R = alkyl, but a (PCP)Ir-phenyl-hydride complex has been reported(*14*) and extensively characterized (NMR,(*14,15*) DFT calculations,(*12,14,15*) and X-ray structure determination(*16*)). Elucidation of the acceptorless alkane dehydrogenation mechanism by non-experimental means, for example by the application of modern electronic structure techniques, thus appears to be a desirable objective. Furthermore, considering that the proposed mechanisms involve only few steps, this would appear to be a relatively easy task computationally and present an excellent opportunity to test the strength of current electronic structure methods. Indeed, two computational research groups have delved extensively into this problem using very similar methods and, yet, have arrived at different conclusions with respect to the mechanism of eq 2.(*12,13,17-19*) The two groups are in accord that steps (*3*) and (*4*) represent the final steps of the mechanism for acceptorless dehydrogenation and that these steps are not rate-determining. There is discord, however, as to whether the alkyl-hydride intermediate is formed by the *D* or by the *A* (or *I*) path, and hence the nature of the rate-determining step is in dispute.

Here we focus on calculating the energetic requirements of the two initial steps composing the *D* and *A* mechanisms using electronic structure methods based on density functional theory. We use propane as a model linear alkane, and we present results obtained with three molecular models for the (^{R}PCP)Ir species: R' = H, Me, and *t*-Bu. For our model systems, we have not been able to locate a transition state appropriate to an *I* mechanism.

We shall first consider energy profiles based on the potential energy surfaces. We subsequently acquire insight into the role entropy plays on the various elementary steps of the *D* and *A* pathways by contrasting potential and free energy profiles. There are substantial differences between the two sets of profiles (*vide infra*), even under standard thermodynamic conditions (STP). We then incorporate important concentration and temperature effects to more accurately simulate the very non-standard (T, P) conditions used experimentally. We briefly present some recent experimental findings relevant to distinguishing between the proposed mechanisms. Finally, we comment on the calculation of steric effects, viz. the presence of *t*-Bu groups in ($^{t\text{-Bu}}$PCP)Ir, and on the feasibility of an *I* mechanism for eq 2.

Potential energy profiles

There are two conformations (both of C_2 symmetry) available to the (PCP)IrH$_2$ species, the resting state of the catalyst in the limit of low alkene concentrations.(*10*) Both conformers possess a d^6 metal configuration and a singlet ground state, but they differ in the occupancies of the Ir(d)-orbitals. The conformer of lowest energy is trigonal bipyramidal with a Y-shaped ligand arrangement in the equatorial plane and a narrow H-Ir-H angle near 60°. The higher energy conformer is pseudo-octahedral with a T-shaped ligand arrangement in the horizontal plane, a wide H-Ir-H angle close to 180°, and a vacant coordination site *trans* to the PCP ligand. The Y-T energy separation is about 10 kcal/mol, and the trigonal bipyramidal conformer (Y-shaped) is thus overwhelmingly dominant, even at elevated temperatures. Hall and Niu examined in detail the elementary steps of the associative and dissociative mechanisms for both conformers (using ^{H}PCPIrH$_2$ and methane as well as ethane as model alkanes) and found smaller activation barriers for the trigonal bipyramidal species.(*17*) Our own calculations agree with this assessment, and we will not consider the higher energy, T-shaped conformer further in this work.

The potential energy profile pertaining to the first two steps of the dissociative pathway for RH = propane is shown in Figure 1.

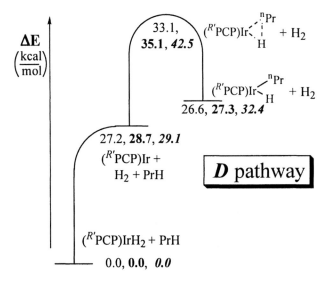

Figure 1. Potential energy profiles (kcal/mol) for the addition of propane (Pr-H) to ($^{R'}$PCP)IrH$_2$ along the dissociative pathway. The energy scale qualitatively corresponds to the data obtained for R' = t-Bu. Fonts applied: normal, R' = H; bold, R' = Me; bold italic, R' = t-Bu.

The concerted elimination of H$_2$ from (PCP)IrH$_2$ to form free H$_2$ is fully allowed under C$_2$ symmetry, and we have not been able to locate a conventional transition state (TS) for this process. There may be a solvent-created TS for removal of H$_2$ under condensed phase conditions, but in the gas phase there is none. With no TS for H$_2$ removal from (PCP)IrH$_2$ (i.e., no activation barrier for the association reaction), we find $\Delta E^{\ddagger}_{D1} = \Delta E_{D1}$ = 27.2 kcal/mol for R' = H, 28.7 kcal/mol for R' = Me, and 29.1 kcal/mol for R' = t-Bu. The activation energy for H$_2$ elimination increases with increasing electron-donating ability of the phosphines (R' = H < Me < t-Bu) and hence increasing "electron-richness" of the Ir metal. There does not appear to be any differential steric interactions associated with the removal of this small molecule from (^{R}PCP)IrH$_2$.

The activation energy for the oxidative addition of propane to the three-coordinate ($^{R'}$PCP)Ir complex is $\Delta E^{\ddagger}_{D2}$ = 5.9, 6.4, and 13.4 kcal/mol for R' = H,

Me, and t-Bu, respectively. A precursor complex may form between (PCP)Ir and the alkane, but calculations by Hall and by us indicate that such a complex should be bound only weakly, (*12,13,17-19*) and we will ignore its possible existence here. The presence of a (PCP)Ir(RH) complex will formally increase the activation energy for the C-H bond activation step slightly, but will not affect the overall energetics of the dissociative pathway. The formation of the propyl hydride intermediate from ($^{R'}$PCP)Ir and propane is a slightly exergonic reaction when R' = H (ΔE_{D2} = -0.6 kcal/mol) or R' = Me (ΔE_{D2} = -1.4 kcal/mol), but it is endergonic when R' = t-Bu, ΔE_{D2} = 3.3 kcal/mol.

Representation of the bulky phosphine substituents used experimentally, typically i-Pr or t-Bu, by hydrogen atoms is of course a very crude approximation for steric as well as electronic reasons. The *mer*-geometry attained by the (PCP)Ir fragment and the ligand enforced backbending of the P atoms (<PIrP ~ 170° in (PCP)Ir) leaves both the face (*trans* to the PCP connecting carbon atom, C_{PCP}) and sides (*cis* to C_{PCP}) of (^{H}PCP)Ir wide open for alkane attack. Methyl groups on the phosphorus atoms (i.e. (^{Me}PCP)Ir) capture most of the electron-donating nature of the substituent groups used experimentally. The methyl groups may also provide minimal steric hindrance, but they undoubtedly exert much less steric hindrance than t-Bu (or i–Pr) groups do. Substrates approaching *cis* to C_{PCP} will experience significant steric hindrance when attacking ($^{t\text{-}Bu}$PCP)Ir; an approach to the less crowded site *trans* to C_{PCP} will clearly be favored when R' = t-Bu. Increasing electron donating ability of the phosphine substituents (R' = H < Me < t-Bu) should favor both the kinetics and thermodynamics of RH addition to ($^{R'}$PCP)Ir, whereas the concomitant increase in substituent steric bulk (R' = H ~ Me << t-Bu) should distinctly disfavor both the kinetics and thermodynamics of RH addition. The computed (thermodynamic) ΔE_{D2} values (-1.4, -0.6, and 3.3 kcal/mol for R' = H, Me, and t-Bu, respectively) conform well to these expectations. The transition state energies for C-H cleavage also increase in the order (^{H}PCP)Ir ~ (^{Me}PCP)Ir < ($^{t\text{-}Bu}$PCP)Ir (5.9, 6.4, and 13.4 kcal/mol, respectively). Thus some steric interactions may be present in TS_{D2} even when R' = Me, and these unfavorable interactions most certainly intensify when R' = t-Bu. Comparison of the reaction and activation energies shows that the change from R' = Me to R' = t-Bu engenders a larger effect on $\Delta E^{\ddagger}_{D2}$ (~ 7.0 kcal/mol) than on ΔE_{D2} (~ 4.7 kcal/mol). The steric interactions are less unfavorable in the reaction product than in TS_{D2}, because the (PCP)Ir(alkyl)(hydride) adopts a square pyramidal geometry with the hydride in the apical position. This positions the propyl group *trans* to C_{PCP}, exactly in the site that is the sterically least hindered. The geometry of TS_{D2} has more trigonal bipyramidal character and the propyl unit cannot avoid steric interactions with at least one phosphino t-Bu group.

Turning to the associative pathway, there are a number of potential "seven-coordinate" intermediates of formula ($^{R'}$PCP)Ir(R)(H)$_3$. We have located five configurational isomers (**A-E**), shown schematically in Figure 2.(*18,19*) Two of these isomers (**C** and **E**) are classical structures with three hydride ligands, whereas three isomers (**A, B,** and **D**) are non-classical structures featuring an

η^2-coordinated H_2 molecule. The energies of isomers **A-E** are all in the range 13-16 kcal/mol above the reactants, (PCP)IrH$_2$ plus propane, when R' = H or Me. Significant steric "Pr/'Bu interactions destabilize those structures that have the "Pr group *cis* to C_{PCP} (structures **A**, **B**, and **C**) by approximately 20 kcal/mol relative to ($^{t\text{-}Bu}$PCP)IrH$_2$ plus propane. The relative increases are only about half as large (~ 12 kcal/mol) for the two isomers in which the "Pr group is approximately *trans* to C_{PCP} (**D** and **E**). Intermediates **D** and **E** are essentially isoenergetic, independently of R', but they lie 5-10 kcal/mol below **A**, **B** and **C** when R' = *t*-Bu. Classical and non-classical structures (e.g. **D** and **E**) may interconvert with only small activation barriers (~ 2 kcal/mol), since the potential energy surface is very flat with respect to ligand movements in the horizontal plane. Also, the barrier for internal H$_2$ rotation around the Ir-H$_2$ axis in the non-classical structures is very low (~ 1 kcal/mol).*(12)*

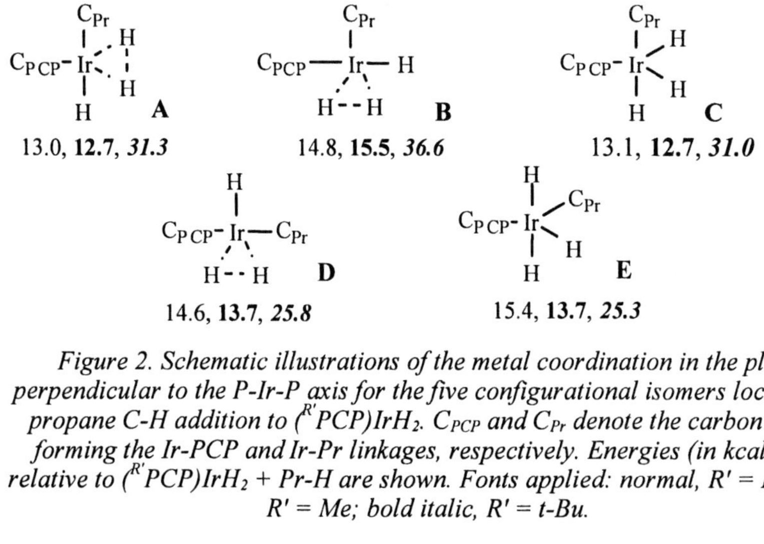

A 13.0, **12.7**, *31.3*

B 14.8, **15.5**, *36.6*

C 13.1, **12.7**, *31.0*

D 14.6, **13.7**, *25.8*

E 15.4, **13.7**, *25.3*

Figure 2. Schematic illustrations of the metal coordination in the plane perpendicular to the P-Ir-P axis for the five configurational isomers located for propane C-H addition to ($^{R'}$PCP)IrH$_2$. C_{PCP} and C_{Pr} denote the carbon atoms forming the Ir-PCP and Ir-Pr linkages, respectively. Energies (in kcal/mol) relative to ($^{R'}$PCP)IrH$_2$ + Pr-H are shown. Fonts applied: normal, R' = H; bold, R' = Me; bold italic, R' = t-Bu.

We have located (Figure 3) three transition states for formation of isomers **A-E**.*(18,19)* Intrinsic reaction coordinate calculations show that all three transition states connect to classical product isomers. The TS in which the propane molecule enters *trans* to the PCP unit between the hydrides connects to isomer **E** and hence indirectly also to **D** (**TS**$_{D/E\text{-endo}}$). Approach from the side, *cis* to C_{PCP} may lead to isomer **C** and hence to the **A**, **B**, **C** structural manifold (**TS**$_{A/B/C}$). A different initial propyl orientation, but still involving an attack from the *cis* side, also produces isomer **E** (**TS**$_{D/E\text{-exo}}$). The transition state energies fall in the narrow range of 18-22 kcal/mol (relative to ($^{R'}$PCP)IrH$_2$ plus propane) when R' = H or Me. When R' = *t*-Bu, large steric interactions occur in all three TS's. The lowest energy TS emerges clearly as the one in which the propane molecule executes a "frontal" attack, i.e. **TS**$_{D/E\text{-endo}}$. **TS**$_{D/E\text{-endo}}$ lies about 5 kcal/mol below the other two transition states, but approximately 32 kcal/mol above the reactants (($^{t-Bu}$PCP)IrH_2 plus propane).

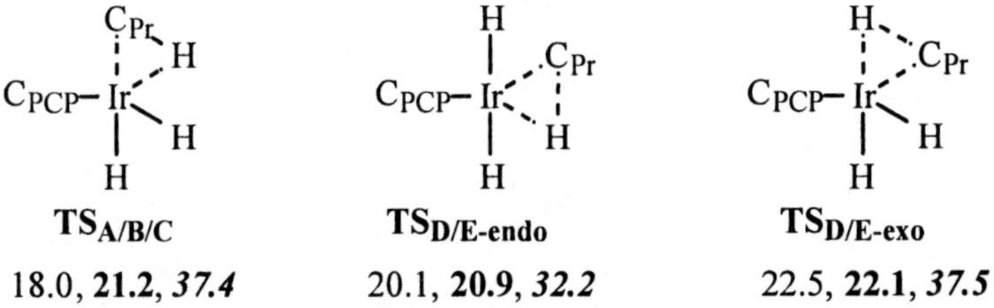

TSA/B/C
18.0, **21.2**, ***37.4***
TSD/E-endo
20.1, **20.9**, ***32.2***
TSD/E-exo
22.5, **22.1**, ***37.5***

Figure 3. Schematic illustrations of themetal coordination in the plane perpendicular to the P-Ir-P axis for the three transition states for propane C-H addition to (^{R}PCP)IrH_2. C_{PCP} and C_{Pr} are the carbon atoms forming the Ir-PCP and Ir-Pr linkages, respectively. Fonts applied: normal, R' = H; bold, R' = Me; bold italic, R' = t-*Bu.*

ΔE (kcal/mol)
TSD/E-endo
20.1, **20.9**, ***32.2***
26.6, **27.3**, ***32.4***
+ H2
A pathway
15.4, **13.7**, ***25.3***
(R'PCP)IrH2 + PrH
0.0, **0.0**, ***0.0***

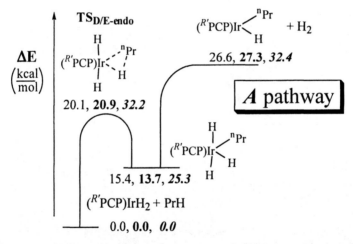

Figure 4. Potential energy profiles (kcal/mol) for the addition of propane (Pr-H) to ($^{R'}$PCP)IrH_2 along the associative pathway. The energy scale qualitatively corresponds to the data obtained for R' = t-Bu. Fonts applied: normal, R' = H; bold, R' = Me; bold italic, R' = t-Bu.

The second step of the associative mechanism is loss of an H_2 molecule from the "seven-coordinate" intermediates (**A-E**). As in the case of (PCP)IrH_2, removal of H_2 from **E** (or from **D**) appears to proceed smoothly to the products ((PCP)Ir(Pr)(H) + H_2) without the appearance of a conventional transition state. Measuring from intermediate **E**, we calculate $\Delta E^{\ddagger}_{A2} = \Delta E_{A2}$ = 11.2, 13.7, and 7.2 kcal/mol for R' = H, Me, and *t*-Bu, respectively. The low activation energy when

R' = t-Bu reflects steric interactions between R' and the *cis* hydrogens in **E** (or the η^2-coordinated H$_2$ species in **D**).

We illustrate in Figure 4 the potential energy profile for the associative pathway, using the transition state (**TS**$_{D/E\text{-endo}}$) and the intermediate (**E**) of lowest energy (R' = t-Bu) as reference points. Joint consideration of Figures 1 and 4 might seem to indicate that the favored mechanism for acceptorless dehydrogenation of linear alkanes by (PCP)IrH$_2$ should be associative, since the *A* pathway (Figure 4) clearly offers the lower activation energy barrier on the potential energy surface ($\Delta E^\ddagger_A(\Delta E^\ddagger_D)$) = 26.6(33.1), 27.3(35.1), and 32.4(42.5) kcal/mol, for R' = H, Me, and t-Bu, respectively). This conclusion, tentative (and furthermore erroneous; see later) because it is based on consideration of only potential energies, and our numerical data for R' = H agree well with results obtained by Hall and Niu on (''PCP)Ir and smaller substrates (methane and ethane).(*17*)

Considerations based on enthalpy rather than energy would not substantially alter the appearances of Figures 1 and 4. The potential energy-to-enthalpy conversion consists primarily of vibrational zero-point energy corrections. These corrections reach their largest differential effect ($\sim$ 3 kcal/mol) on the elementary steps, where an H$_2$ molecule is formed (*D1*, *A2*), but they remain small compared to the relative energy differences shown in Figs. 1 and 4.

Free energy considerations at STP

But reaction rates and equilibria are determined by differences in Gibbs free energy (G = H-TS), not by changes in potential energy (E) or enthalpy (H = E + PV). Thus, we must combine the energies (enthalpies) obtained from the electronic structure calculations with accurate estimates for the entropies of the various species involved in the catalytic process. The elementary reaction steps under consideration here include changes in molecularity and will thus show Gibbs free energy profiles that appear considerably different from the potential energy profiles shown in Figs 1 and 4. Estimates of molecular entropies may be made using standard expressions derived from statistical mechanics. The electronic structure calculations provide the fundamental quantities (vibrational and rotational frequencies) required to evaluate the partition functions, when the nature of a particular stationary point on the potential energy surface (minimum or transition state) routinely is ascertained by normal-mode analysis. It is conventional to evaluate the partition functions assuming ideal gas behavior at standard thermodynamic conditions (STP): partial pressures of P = 1 atm for all species and room temperature, T = 25 °C (298 K).

Computed free energy profiles for the associative and dissociative mechanisms at STP are shown in Figure 5. In constructing Figure 5 for the associative process, we have used the molecular species from Figure 4 (**E** and **TS**$_{D/E\text{-endo}}$, in particular) after verifying that none of the other "seven-coordinate" intermediates or transition states become competitive.

226

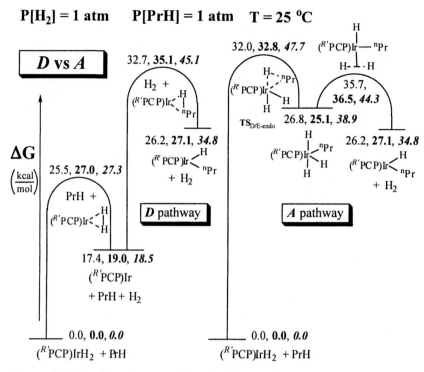

Figure 5. Gibbs free energy profiles (kcal/mol) for the addition of propane to $(^{R'}PCP)IrH_2$ along the **D** and **A** pathways at STP. The energy scale qualitatively corresponds to the data obtained for R' = t-Bu. Fonts applied: normal, R' = H; bold, R' = Me; bold italic, R' = t-Bu.

It is most important to note that Figure 5 shows barriers in free energy for the elimination of H_2 from both (PCP)IrH$_2$ and (PCP)Ir(^{n}Pr)H$_3$, even though no conventional TS for H_2 removal could be located on the potential energy surfaces (Figs. 1 and 4). The reason behind this is as follows: As the H_2 molecule is being formed and removed from the Ir center, the free energy of the system must increase because energy (enthalpy) is required for H_2 elimination. This increase in enthalpy continues until the H_2 molecule is fully formed and is released (into solution). The significant increase in entropy (and commensurate decrease in free energy) is realized only after H_2 has been produced as a free species. The free energy barrier height for H_2 elimination is thus approximately equal to the enthalpy difference between the reacting species ((PCP)IrH$_2$ or (PCP)Ir(^{n}Pr)H$_3$) and the products ((PCP)Ir + H_2 or (PCP)Ir(^{n}Pr)H + H_2). For the loss of H_2 from $(^{R'}PCP)IrH_2$, we compute ΔH = 25.5, 27.0, and 27.3 kcal/mol for R' = H, Me, and t-Bu, respectively. The corresponding values for loss of H_2 from $(^{R'}PCP)Ir(^nPr)H_3$ are considerably smaller, and we obtain ΔH = 8.9, 11.4, and 5.3

kcal/mol for R' = H, Me, and *t*-Bu, respectively. Experimentally, activation entropies near zero have been measured for loss of H_2 from several complexes closely related to the ones under consideration here,(*20*) providing support for the enthalpy/entropy view just presented.

The data presented in Figure 5 provide no particular insight into which mechanism is favored at STP. The free energy barriers for both steps on the ***A*** pathway are almost identical and they are, furthermore, essentially identical to the barriers computed for the C-H activation step (step *D2*) on the ***D*** pathway. These (coincidental) similarities are independent of the phosphine substituent (R'). The free energy barriers for either pathway are computed to be in the range 32-36 kcal/mol for R' = H and Me, but increase to 45-48 kcal/mol when R' = *t*-Bu. The free energy of activation for the loss of H_2 from (PCP)IrH$_2$ (step *D1*) is well below these values at STP.

Free energy considerations at catalytic conditions

Fortunately, considering that our goal is the elucidation of the mechanism for acceptorless dehydrogenation of alkanes by (PCP)IrH$_2$, the actual experimental conditions are far removed from the standard thermodynamic conditions applied above. The catalytic acceptorless dehydrogenation process is conducted in solution at elevated temperatures *with continuous removal of the H_2 gas produced.* One of the most attractive features of the pincer-Ir catalysts is their high thermal stability; experimental operating temperatures are commonly in the range 150-200 °C. Temperature changes are trivially incorporated into the statistical mechanical computations and influence both the enthalpy and entropy contributions to the free energy. Increases in temperature (above 25 °C) will of course preferentially favor the steps with positive entropy changes, i.e. those liberating or containing free H_2.

The corrections necessary to simulate the effective alkane and dihydrogen concentrations are intriguing. The alkane substrate (*n*-octane, cyclooctane, etc.) is also typically the solvent for the catalytic reaction and hence its concentration is on the order of 10 M rather than the 0.04 M implied by the STP partial pressure of 1 atm. We must also include a statistical factor correcting for the number of hydrogens present in the alkane and capable of undergoing reaction. For propane, this statistical factor effectively increases the concentration of potentially active C-H bonds to six times the molarity of propane. The effective concentration of C-H bonds (converted to an effective pressure using the ideal gas law) contributes a correction to the chemical potential (Gibbs free energy) of the alkane. Note that this correction will significantly lower the free energies of the TS's for steps *D2*, *A1*, and *A2* (relative to (PCP)IrH$_2$ + propane) – in all three cases by rigorously equal amounts, since all three TS's involve the loss of 1 mol free propane.

Finally, *and most importantly*, it would be impossible to dehydrogenate alkanes without the use of a sacrificial acceptor, if the partial pressure of H_2

were equal to 1 atm. The standard free energy change for alkane dehydrogenation is of the order of 20 kcal/mol, and hence the concentration of alkene product would be miniscule at STP. In order to obtain an appreciable concentration of alkene, it is necessary to operate at high T and to remove H_2 from the solution as soon as it is produced. Hence, under catalytic conditions, the pressure of H_2 must be of the order 10^{-5} atm or less.(*19*) This large correction in H_2 concentration (relative to STP) affects *only the free energy of the TS for step D2*, the only TS where free H_2 is present.

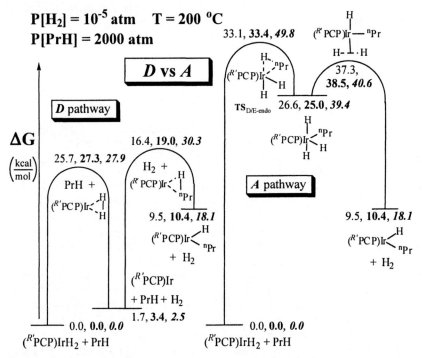

Figure 6. Gibbs free energy profiles (kcal/mol) for the addition of propane to ($^{R'}$PCP)IrH$_2$ along the **D** and **A** pathways simulated at experimental catalytic conditions. The energy scale qualitatively corresponds to the data obtained for R' = t-Bu. Fonts applied: normal, R' = H; bold, R' = Me; bold italic, R' = t-Bu.

In Figure 6, we show the free energy profiles for **A** and **D** pathways for acceptorless dehydrogenation of propane by (PCP)IrH$_2$ at the simulated experimental conditions T = 200 °C, P[H_2] = 10^{-5} atm, and P[propane] = 2,000 atm. For the **D** path we obtain free energy barriers of 25.7, 27.3, and 30.3 kcal/mol for R' = H, Me, and *t*-Bu, respectively. In contrast, the free energy barriers along the **A** path are much higher (by ca. 20 kcal/mol for R' = *t*-Bu) at 37.3, 38.5, and 49.8 kcal/mol for R' = H, Me, and *t*–Bu, respectively. The rate determining step on the D pathway is predicted to be elimination of H_2 from $(^RPCP)IrH_2$, when R' = H or Me. When R' = t-Bu, the C-H activation step has the slightly larger computed barrier ($\Delta G^{\ddagger}_{D2}$ = 30.3 kcal/mol vs. $\Delta G^{\ddagger}_{D1}$ = 27.9 kcal/mol) at 200 °C and 10^{-5} atm H_2. However, the magnitude of $\Delta G^{\ddagger}_{D2}$ depends on the concentration of H_2 (whereas $\Delta G^{\ddagger}_{D1}$ does not) and a concentration of H_2 lower than 10^{-5} atm will further reduce the magnitude of $\Delta G^{\ddagger}_{D2}$. Thus, rapid and effective expulsion of the generated H_2 is an essential experimental condition, which the calculations must take into account. Furthermore, the degree of efficiency of H_2 expulsion from solution may influence the identity of the rate-determining step within the D pathway.

Experimental support for the dissociative pathway

We have shown elsewhere that the use of a cyclic alkane, in particular cyclohexane, in DFT calculations analogous to those presented above leads to potential and free energy profiles that are almost superimposable on the ones for the linear alkane propane.(*18,19*) Thus, free energy calculations simulating catalytic conditions squarely predict that the acceptorless dehydrogenation of (linear and cyclic) alkanes by $(PCP)IrH_2$ proceeds along the D pathway. It was noted above that the associative pathway involves non-classical intermediates in which an η^2-coordinated H_2 molecule has a very low barrier to internal rotation (~ 2 kcal/mol).(*12*) If step *A1* is at all feasible, the use of deuterated hydrocarbons should lead to incorporation of deuterium into the (PCP)Ir species. Indeed, perdeuterated benzene, mesitylene, and *n*-decane do appear to add to $(^{t-Bu}PCP)IrH_2$ with a free energy of activation near 30 kcal/mol (as evidenced by H/D exchange).(*18*) However, even after one week at 180 °C, no deuterium exchange between $(^{t-Bu}PCP)IrH_2$ and cyclohexane-d_{12} could be detected. The derived activation energy for the (hypothetical) C-D activation process is greater than 36 kcal/mol and thus *larger* than the *measured* activation energy for the catalytic acceptorless dehydrogenation of cyclohexane by $(^{t-Bu}PCP)IrH_2$ (ca. 31 kcal/mol). Step *A1* can therefore not play a role in the dehydrogenation of cycloalkanes. Thus, in the case of the cycloalkanes there is strong experimental support for the D mechanism. Based on the extensive similarities between propane and cyclohexane observed in the calculations, we do strongly believe that the D mechanism is operative for linear alkanes as well.

Alternative mechanism I

The I mechanism was invoked in a discussion of the acceptorless dehydrogenation of cyclododecane by the (PCP)Ir pincer catalyst analog "anthraphos".(*13*) Calculations on model species (experimental phosphino *t*-Bu groups replaced by hydrogens, ethane used as model alkane) were presented in support of the I mechanism being favored at moderate temperatures. At high T

(250 °C), all three mechanisms (**D**, **A**, and **I**) were found to have similar free energies of activation (partial pressures = 1 atm). We have analyzed in some detail the TS initially proposed as representative of the **I** process and have found that the purported TS$_I$ is actually a TS for intramolecular rearrangement between two "seven-coordinate" species (analogues of our isomers **B** and **C**, Figure 2).(*18*) Furthermore, the steric effects of the *t*-Bu groups presented here cast additional doubt on the feasibility of an **I** process. "TS$_I$" will necessarily involve a crowded "seven-coordinate" species experiencing substantial steric interactions between the alkane and the bulky phosphino alkyl group, probably to an extent very similar to what we have computed above for TS$_{A1}$ (Figure 6). Importantly, the entropy loss arising from addition of the alkane in "TS$_I$" cannot be mitigated by the entropy gain arising from the H$_2$ molecule being formed, because this latter entropy is not realized before the H$_2$ molecule is released into solution. Thus, an **I** pathway does not appear to be energetically competitive with the **D** pathway. Indeed, even if it is supposed that an **I** pathway existed, its activation enthalpy must of course be at least as great as the overall reaction enthalpy to give ($^{R'}$PCP)Ir(R)(H) (26.6, 27.3 and 32.4 kcal/mol for R' = H, Me, and *t*–Bu, respectively). Thus, due to the same entropic factors that preclude the A path, even if we assume the lowest possible activation enthalpy (i.e. $\Delta H^{\ddagger} = 0$ in the reverse direction) an **I** pathway would still engender a free energy of activation no lower than that of step *A2* (35.7, 36.5 and 44.3 kcal/mol for R' = H, Me, and *t*–Bu, respectively).

Concluding Remarks

We have presented results from extensive sets of electronic structure calculations on the mechanism of acceptorless dehydrogenation of alkanes by pincer ($^{R'}$PCP)IrH$_2$ catalysts. We have analyzed two plausible mechanisms, dissociative (**D**) and associative (**A**), using propane as a model linear alkane and catalyst species exerting different steric demands (R' = H, Me, and *t*-Bu). We find that, independent of R', the **D** pathway is favored with a free energy barrier at least 10 kcal/mol lower than that for the **A** pathway. C-H activation by late metals, and numerous important reactions catalyzed by Group 9 metals in particular, are generally assumed to occur via low metal oxidation states.(*3a,e*) In accord with this view, in the **D** mechanism the metal shuttles between Ir(III) and Ir(I) oxidation states, while the **A** mechanism would involve Ir(III) and Ir(V) oxidation states. Bergman, however, has demonstrated that cationic Cp*Ir(III)Me can activate alkane C-H bonds, (*21*) and calculations show the mechanism to proceed via an Ir(V) intermediate.(*22*) But unlike the (PCP)Ir catalysts, an Ir(I) pathway is not available in the Cp*Ir(III)Me case. The deuterium exchange experiments alluded to above presumably also involve Ir(V) intermediates. Thus, Ir(V) C-H activation intermediates may certainly exist, but they are too high in free energy to play a role in the case of acceptorless dehydrogenation of alkanes by (PCP)Ir.

The data reported in Figures 1, 4, and 5 demonstrate that the magnitudes of the computed barriers depend strongly on the bulkiness of the phosphino alkyl groups. The experimentally used R' = t-Bu generates C-H activation barriers that are larger by 10 kcal/mol or more than those calculated when R' = H or Me. All data reported here are based on first principles quantum mechanical calculations on the entire molecule. It is certainly possible (indeed very likely) that DFT calculations, in combination with the small basis sets necessarily applied to cover the t-Bu groups, overestimate the steric effects of the bulky groups; note, however, that the separation in the ***D/A*** free energies under catalytic conditions is distinct, even when R' = H or Me (Figure 6). Steric effects from phosphine substituents larger than Me can only further favor ***D*** preferentially over ***A***.

Incorrect conclusions would be drawn from considerations based only on potential energy profiles. Gibbs free energies must be evaluated carefully, because the elementary steps in the mechanisms involve changes in molecularity and take place under reaction conditions that are far removed from STP. Efforts must be made to accurately simulate the actual conditions employed in the experiments, in particular the high T and low [H_2]. Experimental evidence shows that the ***D*** mechanism is necessarily operative in the dehydrogenation of cycloalkanes. The computed free energy profiles for cycloalkane and propane are nearly identical, as are the overall experimental catalytic barriers, providing strong support that the same conclusion applies for n-alkanes.

Computational Details

All calculations employed DFT with the B3LYP combination of exchange-correlation functionals.(*23*) For species containing (^{H}PCP)Ir and (^{Me}PCP)Ir, geometries were fully optimized with the Ir LANL2DZ ECP and corresponding basis set (*24a*); D95(d) basis sets on the second and third row elements (*24b*); the 311G(p) basis set for dihydrogen and hydrogen atoms in the alkane, which formally become hydrides in the product complexes;(*24c*) and a 21G basis set for regular hydrogen atoms (BasisA).(*24d*) Additional single-point calculations used a more extended basis set for Ir (BasisB).(*24e,f*).

Standard statistical mechanical expressions were used to convert from purely electronic reaction or activation energies (ΔE, $\Delta E^\ddagger$; T = 0 K, no ΔZPE) to enthalpies and free energies ($\Delta H°$, $\Delta H°^\ddagger$; $\Delta G°$, $\Delta G°^\ddagger$; ΔZPE included, T = 298 K, P = 1 atm). Standard formulas were also applied to obtain ΔG and $\Delta G^\ddagger$ at other temperature/pressure combinations. Energies for species containing (^{H}PCP)Ir and (^{Me}PCP)Ir are based on ΔE ($\Delta E^\ddagger$) values from B3LYP/BasisB calculations with the electronic energy-enthalpy-free energy corrections made as appropriate in an additive fashion with data derived at the B3LYP/BasisA level. To arrive at energy data for the ($^{t-Bu}$PCP)Ir containing species, we used the following approach: the energies of the relevant ($^{t-Bu}$PCP)Ir and corresponding (^{Me}PCP)Ir species were evaluated at geometries (re)optimized using B3LYP/BasisA, except that the C and H-atoms in the phosphine alkyl groups had only a minimal STO-3G basis set (BasisC). A "best" set of energies for species containing t-Bu groups were then obtained in an additive fashion by matching the (^{Me}PCP)Ir data computed at BasisB and BasisC levels.

Acknowledgement.

We gratefully acknowledge support from the Division of Chemical Sciences, Office of Basic Energy Sciences, Office of Energy Research, U. S. Department of Energy for financial support of this work.

References

1. Lappin, G. R.; Nemec, L. H.; Sauer, J. D.; Wagner, J. D. In *Kirk-Othmer Encyclopedia of Chemical Technology*; 4th ed.; Kroschwitz, J. I., Howe-Grant, M., Eds.; Wiley-Interscience: New York, 1996; Vol. 17, pp 839-858.
2. Goldman, A. S. In *Encyclopedia of Catalysis*; Horvath, I., Ed.; John Wiley & Sons, 2002; Vol. 3, pp 25-33.
3. Some recent reviews of alkane C-H bond activation by organometallic complexes: (a) Sen, A. *Acc. Chem. Res.* **1998**, *31*, 550. (b) Guari, Y.; Sabo-Etiennne, S.; Chaudret, B. *Eur. J. Inorg. Chem.* **1999**, 1047. (c) Jones, W. D. *Science* **2000**, *287*, 1942. (d) Crabtree, R. H. *J. Chem. Soc., Dalton Trans.* **2001**, *17*, 2437. (e) Labinger, J. A.; Bercaw, J. E. *Nature* **2002**, *417*, 507-514.
4. See reference 3; other papers that have addressed selectivity in particular, and provide good lead references, include: (a) Bennett, J. L.; Vaid, T. P.; Wolczanski, P. T. *Inorg. Chim. Acta.* **1998**, *270(1-2)*, 414 (b) Wick, D. D.; Jones, W. D. *Organometallics* **1999**, *18*, 495. (c) Asbury, J. B.; Hang, K.; Yeston, J. S.; Cordaro, J. G.; Bergman, R. G.; Lian, T. *J. Am. Chem. Soc.* **2000**, *122*, 12870 and references 4-11 therein. (d) Peterson, T. H.; Golden, J. T.; Bergman, R. G. *J. Am. Chem. Soc.* **2001**, *123*, 455.
5. (a) Crabtree, R. H.; Mihelcic, J. M.; Quirk, J. M. *J. Am. Chem. Soc.* **1979**, *101*, 7738. (b) Crabtree, R. H.; Mellea, M. F.; Mihelcic, J. M.; Quirk, J. M. *J. Am. Chem. Soc.* **1982**, *104*, 107. (c) Baudry, D.; Ephritikine, M.; Felkin, H. *Chem. Comm.* **1980**, 1243-1244. (d) Burk, M. J.; Crabtree, R. H.; Parnell, C. P.; Uriarte, R. J. *Organometallics* **1984**, *3*, 816.
6. (a) Maguire, J. A.; Goldman, A. S. *J. Am. Chem. Soc.* **1991**, *113*, 6706. (b) Maguire, J. A.; Petrillo, A.; Goldman, A. S. *J. Am. Chem. Soc.* **1992**, *114*, 9492.
7. (a) Jensen, C. M. *Chem. Comm.* **1999**, 2443. (b) Gupta, M.; Hagen, C.; Kaska, W. C.; Cramer, R. E.; Jensen, C. M. *J. Am. Chem. Soc.* **1997**, *119*, 840. (c) Gupta, M.; Kaska, W. C.; Jensen, C. M. *Chem. Comm.* **1997**, 461.

8. (a) Wang, K.; Goldman, M. E.; Emge, T. J.; Goldman, A. S. *J. Organomet. Chem.* **1996**, *518*, 55. (b) Liu, F.; Pak, E. B.; Singh, B.; Jensen, C. M.; Goldman, A. S. *J. Am. Chem. Soc.* **1999**, *121*, 4086.
9. (a) Xu, W.; Rosini, G. P.; Gupta, M.; Jensen, C. M.; Kaska, W. C.; Krogh-Jespersen, K.; Goldman, A. S. *Chem. Comm.* **1997**, 2273. (b) Liu, F.; Goldman, A. S. *Chem. Comm.* **1999**, 655.
10. Renkema, K. B.; Kissin, Y. V.; Goldman, A. S. *J. Am. Chem. Soc.* **2003**, *125*, 7770 -7771.
11. Li, S.; Hall, M. B. *Organometallics* **2001**, *20*, 2153.
12. Krogh-Jespersen, K.; Czerw, M.; Kanzelberger, M.; Goldman, A. S. *J. Chem. Inf. Comput. Sci.* **2001**, *41*, 56.
13. Haenel, M. W.; Oevers, S.; Angermund, K.; Kaska, W. C.; Fan, H.-J.; Hall, M. B. *Angew. Chem., Int. Ed.* **2001**, *40*, 3596.
14. Kanzelberger, M.; Singh, B.; Czerw, M.; Krogh-Jespersen, K.; Goldman, A. S. *J. Am. Chem. Soc.* **2000**, *122*, 11017.
15. Krogh-Jespersen, K.; Czerw, M.; Zhu, K.; Singh, B.; Kanzelberger, M.; Darji, N.; Achord, P.; Renkema, K. B.; Goldman, A. S. *J. Am. Chem. Soc.* **2002**, *124*, 10797.
16. Kanzelberger, M.; Renkema, K. B.; Emge, T. J.; Krogh-Jespersen, K.; Goldman, A. S., manuscript in preparation.
17. (a) Niu, S. Q.; Hall, M. B. *J. Am. Chem. Soc.* **1999**, *121*, 3992. (b) Niu, S. Q.; Hall, M. B. *Chem. Rev.* **2000**, *100*, 353.
18. Krogh-Jespersen, K.; Czerw, M.; Summa, N.; Renkema, K. B.; Achord, P.; Goldman, A. S. *J. Am. Chem. Soc.* **2002**, *124*, 11404.
19. Krogh-Jespersen, K.; Czerw, M.; Goldman, A. S. *J. Mol. Cat., A* **2002**, *189*, 95.
20. Activation entropies for H_2 loss ranging from 0.7±1.0 eu to 3.7±1.1 eu have been determined for a series of "seven-coordinate" iridium complexes, $(PR_3)_2IrX(H)_2(H_2)$, which are closely related to the non-classical intermediates calculated in this work. (a) Hauger, B. E.; Gusev, D.; Caulton, K. G. *J. Am. Chem. Soc.* **1994**, 208. (b) Bakhmutov, V. I.; Vorontsov, E. V.; Vymenits, A. B. *Inorg. Chem.* **1995**, *34*, 214. See also (c) Halpern, J.; Cai, L.; Desrosiers, P. J.; Lin, Z. *J. Chem. Soc. Dalton Trans.* **1991**, 717.
21. Tellers, D. M.; Bergman, R. G. *J. Am. Chem. Soc.* **2000**, *122*, 954.
22. Niu, S.; Hall, M. B. *J. Am. Chem. Soc.* **1998**, *120*, 6169.
23. Becke, A. D. *J. Chem. Phys.* **1993**, *98*, 5648. (b) Lee, C.; Yang, W.; Parr, R. G. *Phys. Rev. B* **1988**, *37*, 785.
24. (a) Hay, P. J.; Wadt, W. R. *J. Chem. Phys.* **1985**, *82*, 299. (b) Dunning, T. H.; Hay, P. J. In *Modern Theoretical Chemistry*; Schaefer, H. F., Ed.; Plenum: New York, 1976, pp 1-28. (c) Krishnan, R.; Binkley, J. S.; Seeger, R.; Pople, J. A. *J. Chem. Phys.* **1980**, *72*, 650. (d) Binkley, J. S.; Pople, J. A.; Hehre, W. J. *J. Am. Chem. Soc.* **1980**, *102*, 939. (e) Couty, M.; Hall, M. B. *J. Comp. Chem.* **1996**, *17*, 1359. (f) Ehlers, A. W.; Böhme, M.; Dapprich, S.; Gobbi, A.; Höllwarth, A.; Jonas, V.; Köhler, K. F.; Stegmann, R.; Veldkamp, A.; Frenking, G. *Chem. Phys. Lett.* **1993**, *208*, 111.

Chapter 14

Double Cyclometalation: Implications for C–H Oxidative Addition with PCP Pincer Compounds of Iridium

Hani A. Y. Mohammad[1], Jost C. Grimm[1], Klaus Eichele[1],
Hans-Georg Mack[2], Bernd Speiser[1], Filip Novak[1],
William C. Kaska[3], and Hermann A. Mayer[1],*

[1]Institut für Anorganische Chemie, and Institut für Organische Chemie,
Auf der Morgenstelle 18, 72076 Tübingen, Germany
[2]Institut für Physikalische und Theoretische Chemie, Auf der Morgentselle 8, 72076 Tübingen, Germany
[3]Department of Chemistry, University of Santa Barbara, Santa Barbara, CA 93106

Treatment of 4-X-C_6H_3-2,6-$(CH_2P^tBu_2)_2$ (X = H (**1**), NO_2 (**2**), MeO (**3**)) with $IrCl_3 \cdot nH_2O$ in 2-propanol/water generates the corresponding cyclometalated pincer chlorohydrido iridium complexes (**1a** – **3a**). In the case of **3**, a second intramolecular oxidative addition reaction of one of the *t*-butyl C-H bonds to the Ir(III) center, followed by the reductive elimination of H_2, gives the novel doubly metalated compound **3b**, which is stable to air and water. Comparative electrochemical studies of **1a** and **3a** establish an equilibrium between the pincer chlorohydrido compound and the doubly metalated complexes **1b**, **3b** and H_2. A square scheme can be used to describe the relationship between the redox couples and solution equilibria. Interestingly the oxidation product **1b**$^+$ undergoes a faster follow-up reaction as **3b**$^+$. The unusual stability of the doubly metalated complex was supported by DFT calculations at different levels of theory on model compounds.

Introduction

Meridional, anionic or neutral tri-dentate ligands (*1,2*) are often called pincer compounds (*3*). Although these ligands were first reported in the 1970's (*4,5*), their design has been recently expanded to include a variety of donor atoms and spacer backbones (*6*). Specifically, the ligands contain donor atoms that are able to coordinate a central atom in a *trans* fashion. Moreover, these donor atoms are anchored to spacer groups as a backbone. If the backbone is rigid, and/or planar, metal complexation can generate a thermally very stable system (*7*). The popularity of the ligands has increased because they have favorable properties when treated with transition metal compounds (*8*). The ligands can have variable bite angles which can control the steric environment and the shape of the frontier orbitals, they can synthetically self-assemble different metals (*8,9*), and they have catalytic potential (*10,11*).

Over the years many research groups have exploited the synthesis and chemistry of various (PCP)[M] pincer metal complexes ([M] = Rh, Ir, Ru, Pt fragments) (*8*), e.g. the complexation of C_6H_4-1,3-$(CH_2P^tBu_2)_2$ to a metal fragment generates two internal five-membered metallacycles and in the case of a d^6 Ir(III) metal center a square-pyramidal coordination geometry (*8*). One of the catalytic pathways for which such complexes show considerable promise, is alkane dehydrogenation, in particular, with anthraphos (PCP)Ir(III)H_2 (*7*) and (PCP)Ir(III)H_2 compounds (*11*). Theoretical calculations seem to be controversial, whether an associative or a dissociative mechanism is operative (*7, 12-14*). Consequently, an investigation of a model methoxy-functionalized PCP system is reported in this account in order to ascertain the importance of electronic and steric factors in the oxidative addition reactions of these complexes (*15,16*).

Results

Intramolecular C-H oxidative addition to Ir(III)

The pincer ligands **1** (*4*), **2** (*17*), and **3** (*15*) (Chart 1) have been synthesized according to procedures which have been described in the literature. The pincer complex **2a** (Scheme 1) was prepared in an analogous way as **1a** (*4*) by heating a 2-propanol/water solution of **2** in the presence of $IrCl_3 \cdot nH_2O$ to 80 °C. As revealed by NMR spectroscopy and X-ray structure analysis the red complex **2a** forms a square-pyramidal arrangement of the iridium bound atoms with the hydride in the apical position (*17*). Thus the structure is fully compatible with that of **1a** (Scheme 1).

Chart 1

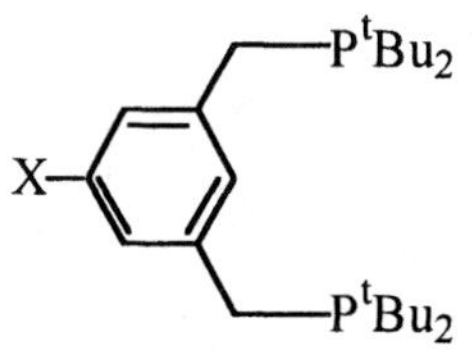

X—⟨benzene ring⟩—CH₂—P^{t}Bu$_2$ / CH₂—P^{t}Bu$_2$

X = H (**1**); NO$_2$ (**2**); OMe (**3**)

In contrast to this, treatment of the methoxy functionalized ligand **3** with IrCl$_3$·nH$_2$O in 2-propanol/water gave a reaction mixture from which deep red crystals were isolated and identified as the doubly cyclometalated Ir(III) complex **3b** (Scheme 1). In addition, **3a** was also present in the reaction mixture. Changing to a non-polar reaction medium by treating **3** with stoichiometric amounts of [Ir(COE)$_2$Cl]$_2$ in toluene also leads to reaction mixtures which contain both **3a** and **3b**. The ratio of **3a** : **3b** varies with reaction time and temperature and the amount of hydrogen present in the system. Thus, if hydrogen is allowed to escape from the reaction mixture, only **3b** is observed. On the other hand, if **3b** is treated with hydrogen, **3a** is formed exclusively. This behavior indicates that the chlorohydrido complex **3a** is in equilibrium with the doubly cyclometalated complex **3b** + H$_2$. The deep red complex **3b** can be handled and stored under air for weeks and melts at 200 °C without decomposition.

The oxidative addition of one *t*-butyl C-H bond to the iridium in **3a** causes the formation of an additional four-membered ring and the loss of any symmetry in the molecule. Thus the ^{31}P{^{1}H} NMR spectrum of **3b** shows two doublets at δ 50.8 and 9.8 which are assigned to P^{t}Bu$_2$ and P^{t}Bu, respectively. The large phosphorus-phosphorus coupling of 351.1 Hz agrees with the mutual *trans* position of the two phosphine groups. Three resonances, which are split by the interaction with the corresponding adjacent phosphorus nucleus, are observed for the intact *t*-butyl groups in the ^{1}H and ^{13}C{^{1}H} NMR spectra. The resonance of the metal bound CH$_2$ group is shifted to δ −6.7 in the ^{13}C{^{1}H} NMR spectrum and is split into a doublet of doublets by a large and a small phosphorus carbon interaction. There is no indication of either hydride or hydrogen weakly bound to iridium in the ^{1}H NMR spectra at different temperatures.

The stereochemistry of **3b** was established by an X-ray structure analysis in the solid state and is compatible with the structure shown in Scheme 1 (*15*).

Scheme 1

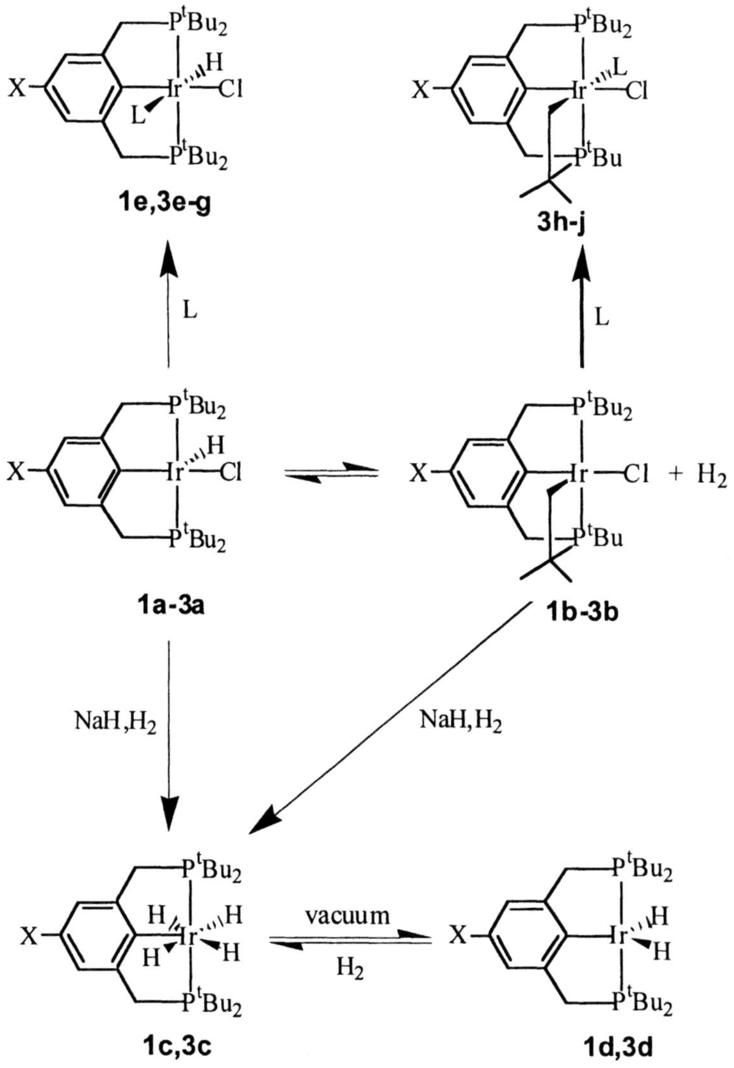

L = CO (**e,h**); NH$_3$ (**f,i**); CH$_3$CN (**g,j**)
X = H (**1**); NO$_2$ (**2**); OMe (**3**)

Complex **3b** crystallizes in the monoclinic space group $P2_1/\underline{n}$. The coordination geometry around iridium is best described as distorted square-pyramidal. The iridium atom deviates only by 0.016 Å from the basal plane. Interestingly, the bond lengths and angles involving the atoms in this plane are comparable to those found for **2a** (*15,17*). Due to the constraints of the four-membered ring the two Ir-P distances differ considerably. The four-membered ring is located nearly perpendicular to the basal plane.

Note when a sample of **2a** is stored over a longer period of time (> 9 months) it is completely converted into **2b** (*18*).

Reactivity of complexes 3a,b

Reactions of complexes **3a,b** are presented in Scheme 1. Treatment of solutions of either **3a** or **3b** in THF with NaH under a hydrogen atmosphere gives iridium tetrahydride **3c** in nearly quantitative yield. If the solvent is removed under vacuum or by a stream of argon the iridium dihydride **3d** is obtained exclusively. The hydrides **3c,d** (*15,16*) were characterized by their NMR spectra which are fully compatible with those of the analogous PCP pincer complexes **1c,d** (*19,20*).

Different products are observed when the deep red solutions of **3a** and **3b** in benzene are treated with carbon monoxide, ammonia and acetonitrile, respectively. In all cases the color changes immediately to pale red (**3f,g**; **3i,j**) or even off white solutions (**3e**, **3h**). The spectroscopic data of the complexes **3e-j** agree with the structures displayed in Scheme 1 and in the case of **3e** are compatible with those reported for $\{C_6H_3\text{-}2,6\text{-}(CH_2P^tBu_2)_2\}Ir(CO)(H)Cl$ (**1e**) (*4*). The structure of **3h** was established by an X-ray analysis (Figure 1). The AB pattern in the $^{31}P\{^1H\}$ NMR spectra, the number of resonances and their multiplicities in the ^1H and $^{13}C\{^1H\}$ NMR spectra of the complexes **3h-j** are comparable with those of the educt **3b** and demonstrate that the four-membered ring is maintained. Characteristic absorptions in the infrared are observed at 1987 cm^{-1} (**3e**) and 1990 cm^{-1} (**3h**) for the CO stretching vibration.

In summary these investigations demonstrate that bonding of any donor ligand at **3b** is preferred at the empty coordination side *trans* to the CH_2 group. Stable complexes (**3h**, **3i**) are formed depending on the donor ligand. The CH_3CN molecule is only weakly bound which prevents the isolation of **3j**. It can be concluded from this that the weak donor H_2 coordinates, in the back reaction of **3b** + H_2, to the empty coordination site before the final conversion to **3a** takes place.

Electrochemical studies

The redox properties of **1a**, **3a** and **3b** were investigated by cyclic voltammetric experiments in a dichloromethane/tetra-*n*-butylammonium

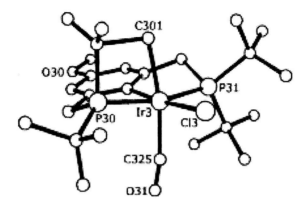

Figure 1. Ball and stick presentation of one of the four independent molecules in the crystal structure of 3h (21).

hexafluorophosphate 0.1 M electrolyte at a Pt-electrode at various scan rates. All compounds are oxidized at potentials a few hundred mV positive of the ferrocene standard potential (22).

The electrochemical behavior of **3a** and **3b** was already described qualitatively (15). While **3b** exhibits a reversible one-electron wave, which indicates oxidation to a Ir(IV) complex **3b⁺** and re-reduction of the latter species (Figures 2a,b), chlorohydride **3a** shows a more complicated voltammetric signal (Figures 2c,d).

The overlapping oxidation peaks at slow scan rates v and the variation of the reverse peaks with v indicate that oxidation of **3a** follows a square-scheme in which two electron transfers and two chemical equilibrium reactions are coupled (23,24). Comparison of the cyclic voltammograms of **3a** and **3b** suggests that the current/potential curves are easily explained if **3a** is in equilibrium with **3b** + H_2. Then **3b** gives rise to the peak couple at ≈ 0.18 V vs. the ferrocene standard, while **3a** undergoes one-electron oxidation at ≈ 0.27 V. In the oxidized state, we find an equilibrium between **3a⁺** and **3b⁺** + H_2.

The experimental curves were modelled with the DigiSim simulation program (25) and the data could be reproduced under the assumption that the equilibrium in the Ir(III) state is in favor of **3a** (in accordance with the observations discussed in earlier paragraphs of this paper), while it is shifted to the side of **3b⁺** in the Ir(IV) state. Thus, electron transfer allows switching of the relative stabilities of the chlorohydrido and the doubly cyclometalated complex. Moreover, the computer simulations indicate that **3b⁺** undergoes a slow follow-up reaction to form an electroinactive product **X** (Scheme 2a). Details of the simulations and a quantative discussion will be published separately (26).

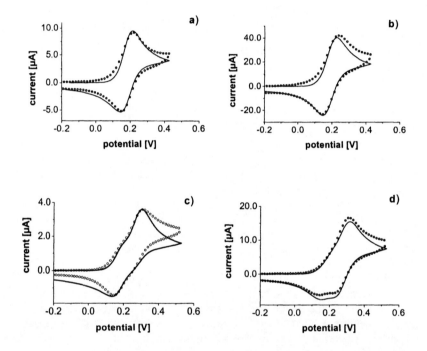

*Figure 2. Experimental voltammograms (circles) and computer simulations (lines) of **3a** (c = 0.2 mM) and **3b** (c = 0.6 mM) in CH_2Cl_2. a) **3b**, v = 100 mV, b) **3b**, v = 2V/s, c) **3a**, v = 100 mV/s, d) **3a**, v = 2 V/s.*

Scheme 2

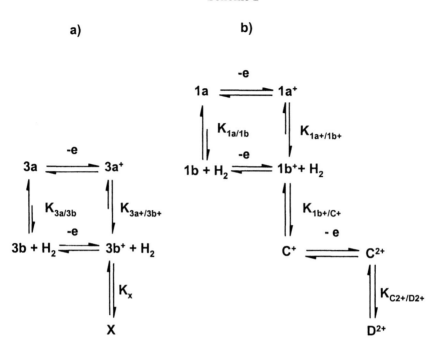

Qualitatively, the cyclic voltammograms of **1a** (Figure 3) are similar to those of **3a** in particular as regards overlapping of the oxidation signals and the behavior of the reduction peaks. However, in accordance with the electron-donating properties of the methoxy substitutent in **3a**, **1a** is oxidized at considerably higher potentials than **3a**. Furthermore, the intensity of the oxidation signal is enhanced in **1a** as compared to the case of **3a**.

The simulations of the voltammograms indicate the presence of a further oxidation step starting from the species C^+ formed as a product of the follow-up reaction of $1b^+$ (Scheme 2b). A reaction to species D^{2+} had to be included into the calculations. The identity of species C^+, C^{2+}, and D^{2+} is not yet defined. However, possibly the additional oxidation step leads to an Ir(V) complex. Thus the oxidation current increases over that expected from the square scheme (Scheme 2a).

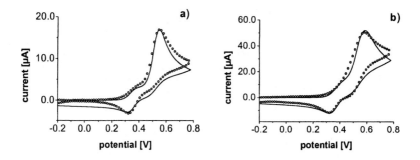

*Figure 3. Experimental voltammograms (circles) and computer simulations (lines) of **1a** (c = 0.44 mM) in CH$_2$Cl$_2$. a) v = 100 mV/s, b) v = 2 V/s.*

Quantum Chemical Calculations

The unusual finding of the stable complex **3b** prompted us to perform density functional calculations at various levels of theory for the model compounds shown in Figure 4 as well as for complex **3b** presented in Scheme 1. Because X-ray structures of the square-pyramidal complex **2a** and the doubly metalated complex **3b** have been reported (*17*), experimental and theoretical data for the two types of structures represented by **3a** and **3b** can be compared. In the case of the model complexes only one *t*-butyl group was retained during the calculations, the other three groups were replaced by hydrogen atoms in order to reduce the computational effort. The geometric structures of **2a',b',k'** and **3a',b',k'** were optimized in the local density approximation (SVWN (*27*)) as well as with the gradient-corrected B3LYP (*27*) hybrid functional. For this purpose, the effective core potential SDD (*27*) and LACVP* (*28*) basis sets were applied using the programs Gaussian 98 (*27*), Jaguar (*28*) and Titan (*29*). In the case of **3b**, structure optimizations were performed at the SVWN/LACVP* and B3LYP/LACVP* levels of approximation.

The relative thermodynamical stabilities of the model complexes are presented in Figure 4. Almost parallel trends in the theoretical energy differences are obtained for the methoxy and the nitro compounds. All computational procedures predict the square-pyramidal chlorohydrido complexes **2a'**, **3a'** to be most stable. The structures **2k'**, **3k'** are higher in energy by ca. 16 to 30 kcal/mol, depending on whether the SVWN or the B3LYP approach is applied. The B3LYP method leads to pronounced larger energy differences (Figure 4).

The doubly cyclometalated **2b'** and **3b'** (+ H$_2$) are predicted to be highest in energy: they are energetically less stable than **2a'**, **3a'** by about 34 (**2a'**,

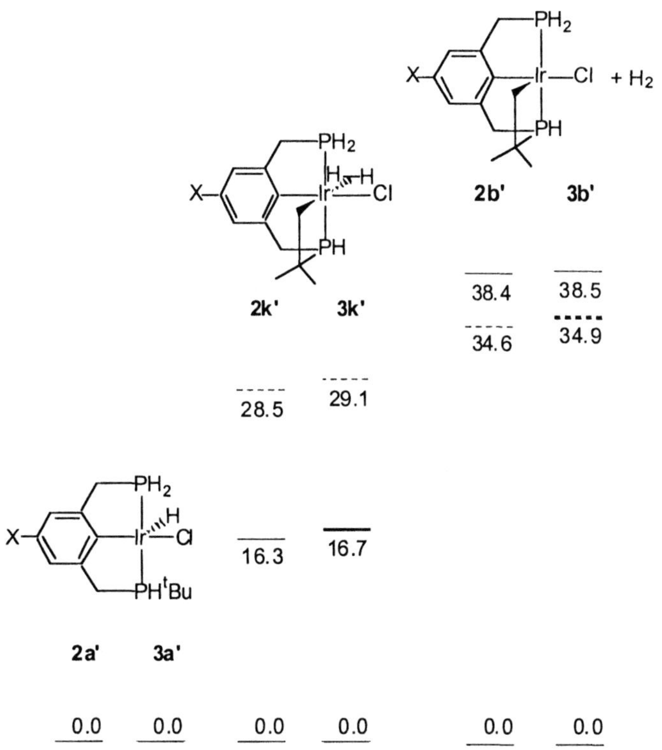

Figure 4. Computed thermodynamic stabilities ([kcal/mol]) for the model complexes (— SVWN, ---- B3LYP).

B3LYP/LACVP*) to 39 kcal/mol (**3a'**, SVWN/SDD(f)). Here, higher relative energies are obtained in the SVWN approximation.

In the case of **2k'**, **3k'** the bonding of the two hydrogen atoms to iridium may be best described as η^2-coordinated H$_2$. The calculated H-H and Ir...H$_{av.}$ distances in these complexes drastically depend on the computational approach. At the SVWN level of theory longer H-H bonds (0.883 to 0.902 Å) and shorter Ir...H$_{av.}$ distances (1.769 to 1.807 Å) are obtained as compared to the B3LYP calculations (H-H: 0.789 to 0.799 Å, Ir...H$_{av.}$: 1.905 to 1.930 Å). In any case, the H-H distances in **2k'**, **3k'** are longer than the theoretical bond length predicted for the H$_2$ molecule (0.743 (B3LYP/LACVP*) and 0.763 Å (SVWN/LACVP*), respectively). Likewise, the Ir⋯H$_{av.}$ distances in these complexes are much longer than the Ir-H bonds in **2a', 3a'** (1.555 (B3LYP/SDD(f), **2a'**) to 1.587 Å (SVWN/LACVP*, **3a'**)). These remarkable differences in structural features for **2k', 3k'** as obtained by the SVWN and B3LYP methods, respectively, correspond to the trends in the relative energies (Figure 4). The SVWN approximations predict **2k', 3k'** to be closer in energy to **2a', 3a'** as compared to the B3LYP results. On the other hand, the B3LYP energies are closer to those obtained for complexes **2b', 3b'** (+ H$_2$) than the energies calculated at the SVWN level of theory (Figure 4).

Relevant structural parameters of **3b** as obtained by X-ray diffraction and density functional calculations (B3LYP/ LACVP*, SVWN/LACVP*) have been compared (*15*). The same distorted square-pyramidal coordination geometry around iridium as observed in the experiment is predicted by the computational procedures. The overall agreement between the experimental crystal structure and the gas-phase structures as calculated for **3b** by both theoretical procedures is very good. The energy profile displayed in Figure 4 agrees with that one calculated for the intermolecular C-H activation by {C$_6$H$_3$-2,6-(CH$_2$P^{t}Bu$_2$)$_2$}IrH$_2$ (*12,13*).

Discussion

Complex **3b** is the result of two consecutive intramolecular C-H bond activations (Scheme 1) (*30,31*). We had previously shown that in the case of the PCP pincer complex {^{t}Bu$_2$PCH$_2$CH$_2$CHCH$_2$CH$_2$P^{t}Bu$_2$}IrH$_2$ there is an equal ease with which *t*-butyl and the ligand backbone C-H bonds can be activated (*32*). In contrast to this all experimental and theoretical data agree with the phenyl iridium bond formation being preferred over the *t*-butyl C-H activation in the case of **3b**. This way a bis-chelate with two thermodynamically stable internal five-membered rings is generated. The analogy to the Thorpe-Ingold effect, where geminal dialkyl groups enhance ring closure, was emphasized by Shaw (*5,33*). The formation of the phenyl iridium bond occurs first, despite the greater number of methyl C-H bonds, because this bond is stronger than the cyclometalated alkyl iridium bond from the *t*-butyl group (*34*). Thus, calculations sustain the observation that **2a', 3a'** are thermodynamically the most stable complexes.

Interestingly enough, the oxidative addition of a C-H bond of one of the *t*-butyl groups to the metal center in complexes of the type **1a** to generate an isolable complex has not been reported before. For this reason, it is difficult to explain why **3b**, with two cyclometalated bonds and easily melting at 200 °C, forms at all, since the calculations show that it is thermodynamically less stable than **3a** and **3k'** (Figure 4). Furthermore, why is this reaction not observed for **1a**?

The most likely attack of a *t*-butyl methyl C-H bond is at the free coordination site opposite to the hydride ligand in **3a**. This will generate a seven coordinate Ir(V) dihydride chloride complex, which rearranges to give the Ir(III) dihydrogen complexes **3k'**. Although **3k'** was not observed spectroscopically, its theoretical existence was demonstrated by quantum chemical calculations (Figure 4). Comparable η^2-H_2 complexes are well established (*35*). As demonstrated by the electrochemical studies, complex **3b** + H_2 is in equilibrium with **3a**, the equilibrium being on the side of the chlorohydrido complex **3a**. This explains why there is no indication of an η^2-H_2 complex in the ^{1}H NMR spectra. The loss of H_2 from **3k'** leads to the thermodynamically least stable product **3b**. This is only achieved if H_2 is allowed to escape from the reaction mixture, which renders the overall process irreversible. Moreover, this is consistent with the observation that the presence or absence of hydrogen in the reaction mixture regulates the amounts of **3a** and **3b** regardless of different solvents and temperatures during the reaction. Thus, if H_2 is added to pure **3b**, the hydrochloride complex **3a** is formed exclusively. In this case the attack of hydrogen will occur at the empty coordination site, which is *trans* to the iridium bound CH_2 group, and the η^2-H_2 complex **3k'** will be formed. In order to reductively eliminate the *t*-butyl C-H bond, the dihydrogen complex has to rearrange to give a Ir(V) dihydride intermediate with at least one hydride ligand *cis* to the CH_2 group bound to iridium. This is the microscopic reverse of the *t*-butyl C-H activation (*vide infra*). That the empty coordination site *trans* to the CH_2 group is easily accessible to donor ligands was demonstrated by the reaction of **3b** with CO, ammonia and acetonitrile which gave the complexes **3h-j**.

The electrochemical data indicate that one-electron transfer oxidation of both **1a** and **3a** reverses the stability of the two central species related to C-H activation by the Ir-PCP-pincer complexes. While the open chlorohydride form is favored in the Ir(III) oxidation state, after oxidation the doubly-cyclometalated species prevailed. In contrast to **3a**, however, where the one-electron oxidation product **3b$^+$** of the square scheme reacts only slowly, **1b$^+$** quickly undergoes further transformation. This leads to further oxidation and might be a reason why the intramolecular C-H insertion product was never observed when working with unsubstituted **1a**.

Conclusion

The isolation and X-ray crystal structure of **3b** shows that pincer complexes with pendant P^{t}Bu$_2$ groups can undergo thermally stable multiple C-H insertion reactions by consecutive cyclometalations. The experimental data reveal that in solution there is an equilibrium between **3a** and **3b** + H_2 which implies C-H oxidative addition via an Ir(V) oxidation state. Loss of hydrogen is the thermodynamic driving force for the isolation of **3b**. Electrochemical oxidation accelerates the C-H activation via an Ir(IV) state. Although our calculations show hardly any differences in the case of **2a,b** and **3a,b** a more detailed analysis of substituent effects in pincer complexes display a reduced reactivity of [{X-C$_6$H$_2$-2,6-(CH$_2$PMe$_2$)$_2$}Ir] fragment towards C-H bonds when going from X = MeO > H > NO$_2$ (*16*) which is in agreement with the results discussed here.

Acknowledgments

The support of this research by the Deutsche Forschungsgemeinschaft (Forschergruppe, Grant FOR 184/3-1 and Graduiertenkolleg, Grant GK – GRK 441/1-00), Bonn/Bad Godesberg, and by the Fonds der Chemischen Industrie, Frankfurt/Main, is gratefully acknowledged. WCK acknowledges the University of California Energy Institute, the University of California Santa Barbara and the NSF. HGM thanks the HLRS, University of Stuttgart, for access to substantial computer time. We thank M. Gloria Quintanilla, Dpto. Quimica Organica, Edificio de Farmacia, Universidad de Alcalá, for her support of the electrochemical studies.

References

1. Eller, P. G.; Bradley, D. C.; Hursthouse, M. B.; Meek, D. W. *Coord. Chem. Rev.* **1977**, *24*, 1.
2. Alvarez, S. *Coord. Chem. Rev.* **1999**, *195*, 13.
3. van Koten, G. *Pure Appl. Chem.* **1989**, *61*, 1681.
4. Moulton, C. J.; Shaw, B. L. *J. Chem. Soc., Dalton Trans.* **1976**, 1020.
5. Shaw, B. L. *J. Organomet. Chem.* **1980**, *200*, 307.
6. Bergbreiter, D. E.; Osburn, P. L.; Liu, Y. S. *J. Am. Chem. Soc.* **1999**, *121*, 9531.
7. Haenel, M. W.; Oevers, S.; Angermund, K.; Kaska, W. C.; Fan, H.-J.; Hall, M. B. *Angew. Chem. Int. Ed.* **2001**, *40*, 3569.
8. Albrecht, M.; van Koten, G. *Angew. Chem. Int. Ed.* **2001**, *40*, 3750.
9. Albrecht, M.; Lutz, M.; Spek, A. L.; van Koten, G. *Nature* **2000**, *406*, 970.
10. Ohff, M.; Ohff, A.; van der Boom, M. E.; Milstein, D. *J. Am. Chem. Soc.* **1997**, *119*, 11687.
11. Jensen, C. M. *Chem. Commun.* **1999**, 2443.
12. Niu, S. Q.; Hall, M. B. *J. Am. Chem. Soc.* **1999**, *121*, 3992.
13. Krogh-Jespersen, K.; Czerw, M.; Kanzelberger, M.; Goldman, A. S. *J. Chem. Inf. Comput. Sci.* **2001**, *41*, 56.
14. Krogh-Jespersen, K.; Czerw, M.; Summa, N.; Renkema, K. B.; Achord, P. D.; Goldman, A. S. *J. Am. Chem. Soc.* **2002**, *124*, 11404.
15. Mohammad, H. A. Y.; Grimm, J. C.; Eichele, K.; Mack, H.-G.; Speiser, B.; Novak, F.; Quintanilla, M. G.; Kaska, W. C. *Organometallics* **2002**, *21*, 5775.

16. Krogh-Jespersen, K.; Czerw, M.; Zhu, K. M.; Singh, B.; Kanzelberger, M.; Darji, N.; Achord, P. D.; Renkema, K. B.; Goldman, A. S. *J. Am. Chem. Soc.* **2002**, *124*, 10797.
17. Grimm, J. C.; Nachtigal, C.; Mack, H.-G.; Kaska, W. C.; Mayer, H. A. *Inorg. Chem. Commun.* **2000**, *3*, 511.
18. Grimm, J., PhD Thesis, Universität Tübingen 2001.
19. Nemeh, S.; Jensen, C.; Binamira-Soriaga, E.; Kaska, W. C. *Organometallics* **1983**, *2*, 1442.
20. Gupta, M.; Hagen, C.; Kaska, W. C.; Cramer, R. E.; Jensen, C. M. *J. Am. Chem. Soc.* **1997**, *119*, 840.
21. Mohammad, H. A. Y.; Eichele, K.; Mayer, H. A., Personal communication.
22. Gritzner, G.; Kuta, J. *Pure Appl. Chem.* **1984**, *56*, 461.
23. Evans, D. H. *Chem. Rev.* **1990**, *90*, 739.
24. Valat, A.; Person, M.; Roullier, L.; Laviron, E. *Inorg. Chem.* **1987**, *26*, 332.
25. Rudolph, M.; Reddy, D. P.; Feldberg, S. W. *Anal. Chem.* **1994**, *66*, 589.
26. Novak, F.; Mayer, H. A.; Speiser, B., Manuscript in preparation.
27. Frisch, M. J., et al., Gaussian 98, Revision A.7, 1998, Gaussian, Inc., Pittsburgh, PA (USA).
28. Jaguar 3.5, 1998, Schrödinger Inc., Portland, OR (USA).
29. Titan 1.0.5, 2000, Wavefunction, Inc., Irvine, CA, and Schrödinger Inc., Portland, OR, (USA).
30. van der Boom, M. E.; Liou, S. Y.; Shimon, L. J. W.; Ben-David, Y.; Milstein, D. *Organometallics* **1996**, *15*, 2562.
31. Ryabov, A. D. *Chem. Rev.* **1990**, *90*, 403.
32. McLoughlin, M. A.; Flesher, R. J.; Kaska, W. C.; Mayer, H. A. *Organometallics* **1994**, *13*, 3816.
33. Shaw, B. L. *J. Am. Chem. Soc.* **1975**, *97*, 3856.
34. Jones, W. D.; Feher, F. J. *Acc. Chem. Res.* **1989**, *22*, 91.
35. Jessop, P. G.; Morris, R. H. *Coord. Chem. Rev.* **1992**, *121*, 155.

Activation of C–H Bonds by Pt(II) (Shilov) Chemistry

Chapter 15

C–H Bond Activation at Pt(II): A Route to Selective Alkane Oxidation?

Alan F. Heyduk, H. Annita Zhong, Jay A. Labinger*, and John E. Bercaw*

Arnold and Mabel Beckman Laboratories of Chemical Synthesis, California Institute of Technology, Pasadena, CA 91125

A hypothetical catalytic cycle for the selective oxidation of alkanes to alcohols by dioxygen may be based on the activation of C-H bonds at Pt(II) centers, coupled with additional known organoplatinum chemistry. Mechanistic studies on the C-H activation process offer guides for designing potential catalytic species.

Introduction

The selective functionalization of alkanes has been identified, in an oft-cited review article (*1*), as one of the Holy Grails (*2*) of C—H bond activation research. The activation of alkane C—H bonds under mild conditions has long been considered a challenging goal. Halpern's recognition that the problem has more to do with thermodynamics — designing a system such that the product of C—H activation is sufficiently stable to observe — than with the kinetic inertness of alkanes (*3*) played a major role in inspiring new approaches, beginning in the early 1980s. The ensuing two decades have generated a wealth of examples, well represented by the contents of the present volume.

The practical value of direct functionalization of alkanes lies in their abundance and low cost as feedstocks. Just to cite one example, 1,4-butanediol

(BDO), used in polymers and other applications (over 300,000 metric tons per year in the US alone), is currently produced by the condensation of acetylene with formaldehyde followed by hydrogenation or by the hydroformylation of allyl alcohol — both routes involving expensive precursors. The cost of the acetylene plus formaldehyde in the first route (not counting the hydrogen) is about 32¢ per pound of BDO. If it could be made instead by direct oxidation of butane, the cost of the latter is only about 6¢/lb BDO — a substantial difference. Another example would be a low-cost method for conversion of methane to a liquid product, leading to potentially huge improvements in utilization of remote natural gas deposits (*4*).

To be sure, organometallic C—H activation is not the only approach to this problem; but the main alternatives do not appear all that attractive, at least for finding general solutions. It is difficult to imagine achieving selective oxidation of butane to BDO by traditional high-temperature heterogeneous catalysis, since the product will inevitably be much more reactive than the starting material. That is not to say that *no* selective heterogeneously catalyzed alkane oxidations are possible — indeed, the oxidation of butane to maleic anhydride is an important currently practiced example. That reaction can be carried out with excellent selectivity, in large part because reactivity is significantly determined by C—H bond strength, and the C—H bonds of maleic anhydride (in contrast to those of BDO, for example) are stronger than those of butane. But in that case the chemistry is dictating the choice of product, not us. Obviously we would prefer to be able to have it the other way around.

Biological and biologically-inspired routes are also the subjects of intense interests, especially the methane monooxygenase enzymes found in methanotrophic bacteria that oxidize methane to methanol with high selectivity (*5*). But there are serious obstacles to the economic application of biological catalysts to large-scale industrial processes such as these (*6*); and while much progress has been made towards designing artificial mimics, we are still far from a practical catalyst for selective oxidation of methane or other alkanes. Hence our attention returns to organometallic C—H activation as a more fruitful realm for exploration, at least in the short term.

However, successful elaboration of alkane *activation* into practical schemes for alkane *functionalization* has remained, tantalizingly, just out of reach, primarily because of an inherent dilemma. Most of the organometallic systems demonstrated to effect C—H bond activation are rapidly destroyed under oxidizing conditions that would be needed for the sorts of conversions described above; while most of the transformations that *would* be compatible with such species, such as dehydrogenation or carbonylation, are thermodynamically uphill at the temperatures at which the species are stable. There are some intriguing examples of non-oxidative reactions that evade thermodynamic restrictions, such as dehydrogenation (*7*) and borylation (*8*); it remains to be demonstrated whether either could ultimately be the basis of a practical process. We have chosen to take the other tack, and investigate organometallic C—H activation by oxidation-tolerant systems.

The Shilov System

The reaction shown in Eq. 1 was first reported about 30 years ago by Shilov and coworkers, following earlier reports of platinum salt-catalyzed H/D exchange in alkanes (9). This system presents a number of features that appear quite promising for our goals. In particular, the complexes tolerate oxidizing conditions, including dioxygen itself, and directly produce oxygenated products. Furthermore, selectivity patterns are very different from those exhibited by heterogeneous catalysts and other radical-based pathways: there is some preference for terminal oxidation, and reaction products are frequently *less* reactive than the hydrocarbons they are derived from. Some illustrations are depicted in Eqs. 2-4 (*9b,10,11*).

$$RH + PtCl_4^{2-} + PtCl_6^{2-} \xrightarrow[120°]{H_2O} ROH + RCl \qquad (1)$$

$$n\text{-pentane} \longrightarrow n\text{-pentyl chloride (56\%)} \qquad (2)$$

$$CH_3CH_2CH_2OH \longrightarrow HOCH_2CH_2CH_2OH \qquad (3)$$

$$CH_3CH_3 \longrightarrow HOCH_2CH_2OH \qquad (4)$$

On the other hand, there are severe practical limitations to this system. First and foremost, the reaction is not catalytic in platinum; or more precisely, it is *both* stoichiometric and catalytic in platinum. A variety of studies (*12*) have established the three-step mechanism shown in Scheme 1, consisting of i) electrophilic displacement of a proton from R—H by Pt(II), generating RPt(II); ii) oxidation of RPt(II) to RPt(IV) by [PtCl$_6$]$^{2-}$, and iii) nucleophilic cleavage of the C—Pt bond by H$_2$O or Cl$^-$ (only the former is shown), affording ROH or RCl respectively, and regenerating Pt(II) as the nucleofuge. That sums up to a net overall reaction in which one equivalent of Pt(IV) is consumed per equivalent of alkane oxidized, while Pt(II) acts as catalyst (and co-product). Scheme 1 would seem to suggest that an alternate oxidant might be substituted for Pt(IV), and indeed there has been some limited success along these lines, using electrochemistry (*13*), heteropolyacids (*14*), or O$_2$/Cu(II) (*15*) to recycle the catalyst; but in all such cases only a limited number of turnovers could be achieved before the system inactivates.

Furthermore, the reaction is fairly slow, and inefficient. And lastly, the catalytic species eventually precipitate as metallic platinum, because the Pt(II)-Pt(0) redox potential is very close to that of the Pt(IV)-Pt(II) couple. This turns out to be a double whammy: not only does activity for alkane oxidation die out with loss of soluble catalyst, but the metallic platinum is known to be an excellent catalyst for overoxidizing the desired alcohol products (*16*), so selectivity suffers as well.

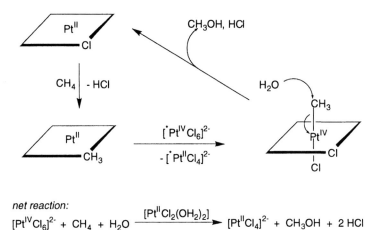

Scheme 1. Mechanism of Shilov system.

Nonetheless, this is still remarkable chemistry. How do we get clean organometallic chemistry in water, a medium in which metal-carbon bonds are often quite unstable? Why is Pt(II), which is not normally considered a particularly "hot" reaction center — neither highly electron rich, as are many of the species employed in the earliest well-defined C—H activations, such as [Cp*(PMe$_3$)Ir(I)], nor electron deficient, like the d^0 early transition metal and lanthanide complexes that effect sigma-bond metathesis of alkanes — reactive towards alkanes? And finally, if our mechanistic investigations provide some answers to these questions, can they also point out the directions we must follow to overcome the disadvantages enumerated above and achieve a practical alkane oxidation catalyst?

Taking the first question first: an organo-Pt(IV) complex (the *second* putative intermediate in Scheme 1) can be readily isolated and fully characterized as [(CH$_3$)PtCl$_5$]$^{2-}$, which is resistant to protonolytic cleavage of the Pt—C bond not only in water, but even in 1*M* acid (*12,17*). However, its presumed precursor, [(CH$_3$)PtCl$_3$]$^{2-}$, has never been obtained cleanly: attempts to generate it by reduction of the methyl-Pt(IV) complex with Sn(II) or other reductants in water lead to liberation of methane instead. That observation implies that the proposed electrophilic reaction of Pt(II) with a C—H bond, liberating a proton, is (not surprisingly) highly reversible. But when Shilov oxidation is carried out in D$_2$O, only a very small amount of deuterium incorporation in products (and unreacted substrates) is observed. Hence the second step of Scheme 1, oxidation of $(CH_3)Pt(II)$ to $(CH_3)Pt(IV)$ by $[PtCl_6]^{2-}$, must be even faster than protonolysis.

The last claim has indeed been demonstrated, even in the absence of an authentic sample of $[(CH_3)PtCl_3]^{2-}$, by two different methods. Reduction of $[(CH_3)PtCl_5]^{2-}$, with Cp_2Co in THF affords a solid, characterized as a mixture of $[(CH_3)_xPtCl_{(4-x)}]^{2-}$ (x = 0-2), which as expected liberates methane (CH_3D) when thrown into D_2O; but if the water contains $[PtCl_6]^{2-}$, some $[(CH_3)PtCl_5]^{2-}$, as well as methane is produced (Eq. 5). The variation of product composition with pD leads to the estimation that at room temperature the ratio of rate constants for oxidation and deuterolysis is about 3 (18). Alternatively, $[(CH_3)PtCl_3]^{2-}$, can be generated at 95 °C as the leaving group in nucleophilic displacement by Cl^- on $[(CH_3)_2PtCl_4]^{2-}$ (Eq. 6), and the same partition experiment carried out (19), leading to a ratio of $k_{oxidation}/k_{protonolysis}$ of about 18. Clearly the activation enthalpy for oxidation is considerably larger than that for protonolysis, so that at the Shilov operating temperature (120 °C), oxidation will dominate completely.

$$[(CH_3)PtCl_3]^{2-} \xrightarrow[{[PtCl_6]^{2-}}]{D_2O} [(CH_3)PtCl_5]^{2-} + CH_3D \quad (5)$$

$$[(CH_3)_2PtCl_4]^{2-} \xrightarrow{Cl^-} [(CH_3)PtCl_3]^{2-} + CH_3Cl \quad (6)$$

The oxidation step (ii) of the overall mechanism is therefore *remarkably fast*; since protonolysis of $[(CH_3)PtCl_3]^{2-}$ is too fast to measure, we can place a lower limit on the rate constant for Eq. 7, which turns out to be at least three orders of magnitude faster than that of the self-exchange reaction of Eq. 8 (20). Why so fast? Probably two factors are at work here: replacement of a chloride by a methyl makes oxidation of the Pt(II) center thermodynamically more favorable; and the stronger *trans* effect of the methyl ligand will accelerate formation of a chloride-bridged intermediate, required for the inner-sphere redox mechanism commonly observed in Pt(II)-Pt(IV) chemistry.

$$[(CH_3)PtCl_3]^{2-} + [PtCl_6]^{2-} \longrightarrow [(CH_3)PtCl_5]^{2-} + [PtCl_4]^{2-} \quad (7)$$

$$[PtCl_4]^{2-} + [*PtCl_6]^{2-} \longrightarrow [*PtCl_4]^{2-} + [PtCl_6]^{2-} \quad (8)$$

This finding suggests *part* of the answer to the second question: what is so special about Pt(II)? In analogy to Halpern's suggestion about alkane activation in general, it may be not so much the ability to effect C—H activation *per se*, but rather the mechanism whereby the product of that activation is (kinetically) protected against reversal by protonolysis. But a more complete answer, as well as leads for improving activity and moving towards a viable catalytic system, require more specific understanding of the actual C—H activation process —step i of the mechanism — and direct study of the Shilov process does not readily yield any such information. To get the details we need, we must turn to model systems.

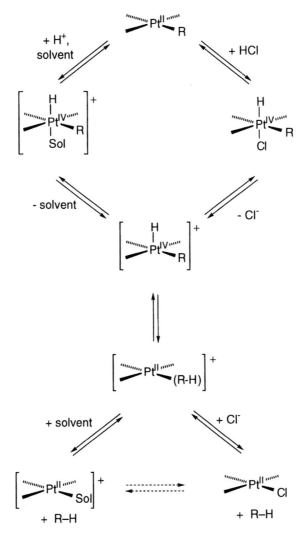

Scheme 2. Mechanism for R-Pt protonolysis (reading down)/C-H activation (reading up)

Studies on ligated Pt(II) complexes

The mechanism of Scheme 2 has been proposed to account for the results of an intensive program of experiments on the protonolysis of (tmeda)Pt(CH$_3$)Cl and related complexes (tmeda = N,N,N',N'-tetramethylethylenediamine); and, invoking the principle of microscopic reversibility, the same mechanism (reading bottom to top) should apply for C—H activation (21). It should be noted in particular that i) ligand substitution and dissociation plays a key role at several points, and ii) the actual C—H bond cleavage is an oxidative addition step. Subsequently it was postulated that similar complexes could carry out (stoichiometric) C—H bond activation as shown in Scheme 3, if a solvent (sol) could be found to satisfy several requirements: it must support dissolution of ionic species, coordinate relatively weakly, and not be subject itself to facile C—H activation. Two different systems were successfully designed, with sol = pentafluoropyridine (22) and 2,2,2-trifluoroethanol (TFE) (23) respectively.

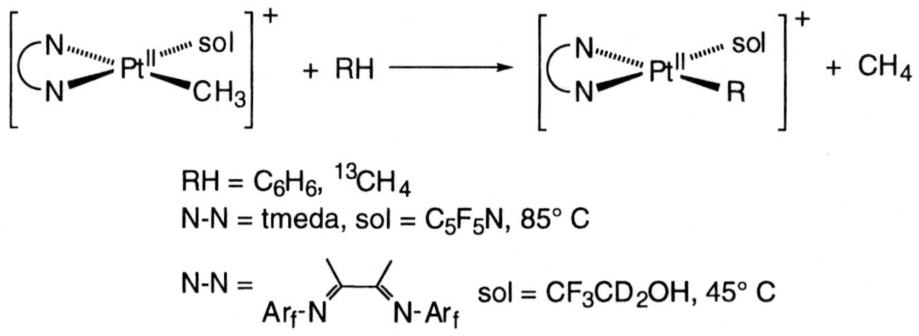

$$\left[\begin{array}{c} N_{\cdots\cdots}Pt^{II\cdots\cdots}sol \\ N^{\diagup}\diagdown CH_3 \end{array}\right]^+ + RH \longrightarrow \left[\begin{array}{c} N_{\cdots\cdots}Pt^{II\cdots\cdots}sol \\ N^{\diagup}\diagdown R \end{array}\right]^+ + CH_4$$

RH = C$_6$H$_6$, ^{13}CH$_4$
N-N = tmeda, sol = C$_5$F$_5$N, 85° C

N-N = Ar$_f$-N⟩—⟨N-Ar$_f$ sol = CF$_3$CD$_2$OH, 45° C

Scheme 3. Stoichiometric C-H activation by ligated Pt(II)

Can these ligated complexes help overcome the problems inherent in the original Shilov system? In each case the answer appears to be a somewhat qualified yes. First, with regard to replacing Pt(IV) as stoichiometric oxidant, we have found that certain Pt(II) complexes, of the form (N-N)PtMe$_2$, can be oxidized to Pt(IV) by dioxygen, according to Eq. 9 (24). Second, with regard to rate, the reactions of Scheme 3 can be operated under substantially milder conditions — some even at room temperature (23) — than the 120 °C required for the ligand-free aqueous system. And finally, many of the ligated complexes are also more resistant to decomposition to platinum metal. The most dramatic case is Periana's bipyrimidine complex, which efficiently catalyzes oxidation of methane to methyl bisulfate by sulfuric acid (thereby providing an alternative to Pt(IV) as stoichiometric oxidant as well); here the complex is not only kinetically but thermodynamically stable relative to Pt(0), as it spontaneously reforms by heating the ligand with the metal in sulfuric acid (25,26).

$$\underset{N}{\overset{N}{\diagdown}}Pt^{II}\underset{CH_3}{\overset{CH_3}{\diagup}} + 1/2\ O_2 \xrightarrow{CH_3OH} \underset{N}{\overset{N}{\diagdown}}Pt^{IV}\underset{CH_3}{\overset{OCH_3\ \ CH_3}{\diagup}} \quad (9)$$

Given these findings, we can construct (on paper) a catalytic cycle for the oxidation of alkane to alcohol by dioxygen, shown in Scheme 4. Close analogs of each of the first three steps have been demonstrated for (NN)-ligated platinum complexes. The first, C—H activation, operates as in Scheme 3; the second, oxidation, can be carried out in water, using sulfonated ligands (27), as well as in methanol; and the third has been observed, for example, with [PtMeCl$_2$(H$_2$O)(tmeda)]Cl (12). But no single complex has yet been found that effects *all* the steps, and indeed there are problems that need to be overcome with each to close a catalytic cycle. The first step would need to be a *net* alkane activation, liberating a proton as shown, rather than an alkyl exchange reaction (28); oxidation would need to be fast compared to the reverse of step 1, as is the case for the original Shilov system (see above); and the third step, nucleophilic cleavage of the Pt—C bond, may well require geometric rearrangement, as studies on the Shilov mechanism (12) and related systems (29) strongly indicate that dissociation of a ligand *trans* to the alkyl group precedes nucleophilic attack. The fourth step amounts to simple ligand substitution and should not pose any serious problems (30).

Much of our recent effort has been focused on understanding the factors controlling the rate of C—H activation, so that designing a complex capable of performing the first step of the cycle productively and rapidly can be pursued rationally. We recently reported (31) the results of extensive studies into the reaction shown in Scheme 5, where by varying aryl groups attached to N and the backbone R groups of the diimine we were able to achieve a wide range in ligand electronic and steric character. The findings can be summarized as follows:

1. Complexes obtained in this manner by protonolysis with aqueous acid exist as an equilibrium mixture of aquo and solvento complexes; for all complexes, the rate is proportional to the ratio $[C_6H_6]/[H_2O]$, and more electron-rich complexes (as measured by v_{CO} in $[(N-N)Pt(Me)(CO)]^+$) react faster.
2. For ortho substituted aryls: the measured KIE (C_6H_6 vs. C_6D_6) is very close to 1, and complete statistical scrambling of H/D in the evolved methane and Pt-Ph is observed with C_6D_6 or $C_6H_3D_3$. In one case protonolysis of (N-N)Pt(Me)(Ph) at low temperature led to observation of a π-benzene adduct by NMR. A related adduct has since been isolated and characterized crystallographically (32).

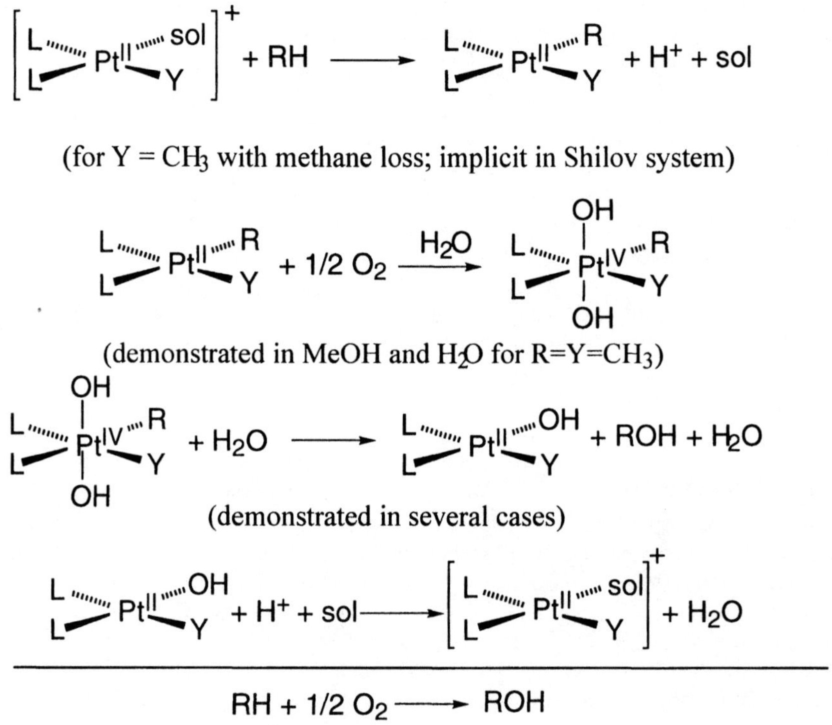

$$\left[\overset{L_{\prime\prime\prime\prime\prime}}{L}Pt^{II}\overset{sol}{\underset{Y}{}}\right]^{+} + RH \longrightarrow \overset{L_{\prime\prime\prime\prime\prime}}{L}Pt^{II}\overset{R}{\underset{Y}{}} + H^{+} + sol$$

(for Y = CH$_3$ with methane loss; implicit in Shilov system)

$$\overset{L_{\prime\prime\prime\prime\prime}}{L}Pt^{II}\overset{R}{\underset{Y}{}} + 1/2\ O_2 \xrightarrow{H_2O} \overset{L_{\prime\prime\prime\prime\prime}}{L}\underset{OH}{\overset{OH}{Pt^{IV}}}\overset{R}{\underset{Y}{}}$$

(demonstrated in MeOH and H$_2$O for R=Y=CH$_3$)

$$\overset{L_{\prime\prime\prime\prime\prime}}{L}\underset{OH}{\overset{OH}{Pt^{IV}}}\overset{R}{\underset{Y}{}} + H_2O \longrightarrow \overset{L_{\prime\prime\prime\prime\prime}}{L}Pt^{II}\overset{OH}{\underset{Y}{}} + ROH + H_2O$$

(demonstrated in several cases)

$$\overset{L_{\prime\prime\prime\prime\prime}}{L}Pt^{II}\overset{OH}{\underset{Y}{}} + H^{+} + sol \longrightarrow \left[\overset{L_{\prime\prime\prime\prime\prime}}{L}Pt^{II}\overset{sol}{\underset{Y}{}}\right]^{+} + H_2O$$

$$RH + 1/2\ O_2 \longrightarrow ROH$$

Scheme 4. Hypothetical catalytic cycle for alkane oxidation

3. For non-ortho substituted aryls: reactions are 10-40 times faster than 2,6-disubstituted Ar analogs; KIEs are on the order of 2; and only partial scrambling of H/D is found.

These results imply that there is a switch in rate-determining step (rds), depending upon steric crowding. For the more crowded complexes with 2,6-disubstituted aryls, the initial π-benzene coordination is rate-limiting, leading to a KIE ~ 1 and to extensive isotopic scrambling, as the processes responsible for the latter — reversible C—H bond breaking and making — are fast compared to the rates of hydrocarbons entering and departing from the coordination sphere. In contrast, for the less hindered non-ortho-substituted aryl complexes, C-H cleavage becomes the rds, signaled by a higher KIE and only partial H/D exchange.

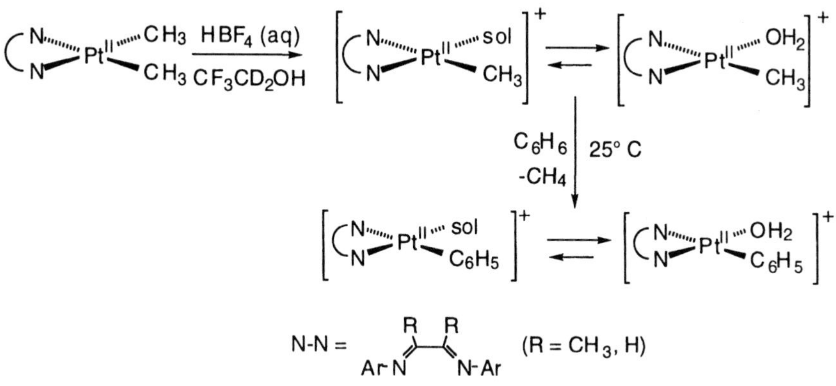

Scheme 5. Diimine-ligated Pt complexes for C-H activation study

Surprisingly, though, the electronic effects appear to be quite similar for either rds. We interpret this in terms of a ground state effect: the electronic nature of the ligand operates primarily to affect the equilibrium between aquo and solvento complexes, with more electron-withdrawing ligands favoring the (unreactive) former. Indeed, within a group of sterically similar complexes, the equilibrium and rate constants correlate well with one another (*31*). The electronic effect on the actual transition state for the reaction must be relatively small, then, but we cannot even tell if there is one and in which direction it lies, as it is masked by the trend in stabilization/destabilization of the aquo complexes.

An obvious way around this problem, which should at the same time lead to more reactive systems, would be to generate the methyl-Pt(II) cations by a rigorously anhydrous route. This has now been achieved by treating the corresponding dimethyl complexes with a neutral tris(pentafluorophenyl)boron reagent in TFE (Eq. 10) (*33*). The reaction could in principle proceed either by methide abstraction or by protonolysis by the *in situ*-generated acid $H^+[B(OCD_2CF_3)(C_6F_5)_3]^-$; the latter must be the case, since only methane and no $[BMe(C_6F_5)_3]^-$ is observed as co-product.

When water-free methyl-Pt(II) cations, thus prepared, are reacted with benzene in TFE, we do observe the expected accelerations relative to the corresponding reactions in the presence of water. However, there is no statistically significant dependence of rate on the electronic properties of the diimine ligand (Table 1). Clearly, then, in the present system the latter exert an effect *solely* through manipulation of ligand substitution — specifically on the relative thermodynamic stability of the aquo complex — and not on the actual process of C—H bond cleavage.

[Scheme showing Pt(II) diimine complex reacting with $B(C_6F_5)_3$ in TFE] (10)

Table 1. Apparent second-order rate constants for the reaction of methyl-Pt(II) cations with benzene, in the presence and absence of water, as a function of diimine ligand.

Ar group	k_2, 20°C, 0.05 M H_2O (31)	k_2, 25°C, water-free (33)
3,5-$C_6H_3(CF_3)_2$	8.4(8) x 10^{-4} $M^{-1}s^{-1}$	1.3(3) x 10^{-2} $M^{-1}s^{-1}$
3,4,5-$C_6H_2(OMe)_3$	4.0(4) x 10^{-3} $M^{-1}s^{-1}$	1.6(2) x 10^{-2} $M^{-1}s^{-1}$
3,5-$C_6H_3R_2$	6.2(6) x 10^{-3} $M^{-1}s^{-1}$ [a]	1.6(2) x 10^{-2} $M^{-1}s^{-1}$ [b]

[a] R = t-Bu. [b] R = Me.

Conclusion

Can a practical route to selective alkane oxidation be based on this Pt(II) chemistry? *Collectively*, the complexes we have studied effect all the individual steps that would be needed for catalysis, with dioxygen as the terminal oxidant. We now need to construct a *single* complex that will perform them all, closing a catalytic cycle, and at the same time get reaction rates up to practical levels, while still maintaining (or, better, improving upon) the selectivity patterns that have been demonstrated for the original Shilov system. The recent findings reported above highlight the central role played by ligand substitution processes — at least comparably important to that of actual C—H bond breaking — in regulating C—H activation reactions. Fortunately we appear to have considerable control over those processes by manipulating ligand steric and electronic properties, solvent, and other reaction parameters. Whether that will be sufficient to overcome the remaining obstacles, and put all the pieces together, remains to be seen.

Acknowledgments

We thank all those — students, postdocs, colleagues, collaborators — who have previously and/or are currently contributing to this program: Lily Ackerman, Tom Baker, Christoph Balzarek, Jeffrey Byers, Charles Carter, Andy Herring, Mike Freund, Matt Holtcamp, Lars Johansson, David Lyon, Gerrit Luinstra, Rebekah Main, Jonathan Owen, Seva Rostovtsev, Joseph Sadighi, John Scollard, Shannon Stahl, Mats Tilset, Lin Wang, David Weinberg, and Antek Wong-Foy. We thank BP for ongoing support.

References

1. Arndtsen, B. A.; Bergman, R. G.; Mobley, T. A.; Peterson, T. H. *Acc. Chem. Res.* **1995**, *28*, 154.
2. Actually we had been laboring under the impression that there is only *one* Holy Grail; but a scholarly source (the same one quoted in reference 1) provides textual evidence that the plural may in fact be appropriate: (Graham Chapman, as King Arthur) "Go and tell your master that we have been charged by God with a sacred quest. If he will give us food and shelter for the night, he can join us in our quest for the Holy Grail." (Michael Palin, in outr-r-r-rageous French accent) "Well, I'll ask him, but I don't think he'll be very keen--he's already got one, you see?"....(Chapman) "Are you sure he's got one? (Palin) "Oh yes, it's very nice."
3. Halpern, J. *Inorg. Chim. Acta* **1985**, *100*, 41-48.
4. Gradassi, M. J.; Green, N. W. *Fuel Proc. Technol.* **1995**, *42*, 65-83.
5. Brazeau, B. J.; Lipscomb, J. D. In *Enzyme-Catalyzed Electron and Radical Transfer;* Holzenburg, A.; Scrutton, N. S., Eds.; Kluwer: New York, 2000; pp. 233-277.
6. Duetz, W. A.; van Beilen, J. B.; Witholt, B. *Curr. Opin. Biotech.*, **2001**, *12*, 419-425.
7. See Chapter in this volume: Goldman, A. S.; Renkema, K. B.; Czerw, M.; Krogh-Jespersen, K. In *Activation and Functionalization of C-H Bonds*; Goldman, A. S., Goldberg, K. I., Eds.; American Chemical Society: Washington, DC.
8. See Chapter in this volume: Hartwig, J. F. In *Activation and Functionalization of C-H Bonds*; Goldman, A. S., Goldberg, K. I., Eds.; American Chemical Society: Washington, DC.
9. a) Shilov, A. E.; Shulpin, G. B. *Russ. Chem. Rev.* **1987**, *56*, 442-464; b) Shilov, A. E. *Activation of Saturated Hydrocarbons by Transition Metal Complexe;* Reidel: Boston, 1984, and references cited therein.

10. Labinger, J. A.; Herring, A. M.; Lyon, D. K.; Luinstra, G. A.; Bercaw, J. E.; Horváth, I.; Eller, K. *Organometallics* **1993**, *12*, 895-905.
11. Sen, A.; Benvenuto, M. A.; Lin, M.; Hutson, A. C.; Basickes, N. *J. Am. Chem. Soc.* **1994**, *116*, 998-1003.
12. Luinstra, G. A.; Wang, L.; Stahl, S. S.; Labinger, J. A.; Bercaw. J. E. *J. Organometal. Chem.* **1995**, *504*, 75-91, and references cited therein.
13. Freund, M. S.; Labinger, J. A.; Lewis, N. S.; Bercaw, J. E. *J. Molec. Catal.* **1994**, *87*, L11-L16.
14. Geletti, Y. V.; Shilov, A. E. *Kinetics and Catalysis, Engl. Trans.* **1983**, *24*, 413-416.
15. Lin, M.; Shen, C.; Garcia-Zayas, E. A.; Sen, A. *J. Am. Chem. Soc.* **2001**, *123*, 1000-1001.
16. Sen, A.; Lin, M.; Kao, L.-C.; Hutson, A. C. *J. Am. Chem. Soc.* **1992**, *114*, 6385-6392.
17. (a) Zamashchikov, V. V.; Kitaigorodskii, A. N.; Litvinenko, S. L.; Rudakov, E. S.; Uzhik, O. N.; Shilov, A. E. *Izv. Akad. Nauk. SSSR, Ser. Khim.* **1995**, *8*, 1730. (b) Zamashchikov, V. V.; Litvinenko, S. L.; Shologon, V. I. *Kinet. Katal.* **1987**, *28*, 1059, and references cited therein.
18. Wang, L.; Stahl, S. S.; Labinger, J. A.; Bercaw, J. E. *J. Mol. Catal.*, **1997**, *116*, 269-75.
19. Zamashchikov, V. V.; Popov, V. G.; Litvinenko, S. L. *Russ. Chem. Bull.* **1993**, *42*, 352.
20. Rich, R. L.; Taube; H. *J. Am. Chem. Soc.* **1954**, *76*, 2608-2611.
21. (a) Stahl, S. S.; Labinger, J. A.; Bercaw, J. E. *J. Am. Chem. Soc.* **1996**, *118*, 5961–5976; (b) Stahl, S. S.; Labinger, J. A.; Bercaw, J. E. *J. Am. Chem. Soc.* **1995**, *117*, 9371–9372.
22. (a) Holtcamp, M. W.; Henling, L. M.; Day, M. W.; Labinger, J. A.; Bercaw, J. E. *Inorg. Chim. Acta* **1998**, *270*, 467–478; (b) Holtcamp, M. W.; Labinger, J. A.; Bercaw, J. E. *J. Am. Chem. Soc.* **1997**, *119*, 848–849; (c) Holtcamp, M. W.; Labinger, J. A.; Bercaw, J. E. *Inorg. Chim. Acta* **1997**, *265*, 117–125.
23. Johansson, L.; Ryan, O. B.; Tilset, M. *J. Am. Chem. Soc.* **1999**, *121*, 1974–1975.
24. a) Rostovtsev, V. V.; Labinger, J. A.; Bercaw, J. E.; Lasseter, T. L;. Goldberg, K. I. *Organometallics* **1998**, *17*, 4530-1. b) Rostovtsev, V. V.; Henling, L. M.; Labinger, J. A.; Bercaw, J. E. *Inorg. Chem.* **2002**, *41*, 3608-19.
25. Periana, R. A.; Taube, D. J.; Gamble, S.; Taube, H.; Satoh, T.; Fujii, H. *Science* **1998**, *280*, 560–564.
26. Since the sulfur dioxide produced can be readily reoxidized with dioxygen and methyl bisulfate can be hydrolyzed to methanol, an overall catalytic process for the reaction of methane with dioxygen to methanol is thus

available; unfortunately it does not appear to be economically viable, because of the large quantities of sulfuric acid that must be carried around the cycle.
27. Balzarek, C.; Labinger, J. A.; Bercaw, J. E., to be submitted for publication.
28. An example of such a (stoichiometric) activation of benzene has recently been reported: Harkins, S. B.; Peters, J. C. *Organometallics* **2002**, *21*, 1753-1755.
29. Goldberg, K. I.; Yan, J. Y.; Breitung, E. M. *J. Am. Chem. Soc.* **1995**, *117*, 6889-96.
30. Though it probably will, should we manage to solve all the others, that being the way of Nature....
31. Zhong, H. A.; Labinger, J. A.; Bercaw, J. E. *J. Am. Chem. Soc.* **2002**, *124*, 1378-99.
32 Reinartz, S.; White, P. S.; Brookhart, M.; Templeton, J. L. *J. Am. Chem. Soc.* **2001**, *123*, 12724-5.
33. Heyduk, A. F.; Labinger, J. A.; Bercaw, J. E. *J. Am. Chem. Soc.*, submitted for publication.

Chapter 16

Competitive Trapping of Equilibrating Pt(IV) Hydridoalkyl and Pt(II) Alkane Complexes: A Valuable Tool for the Mechanistic Investigation of C–H Activation Reactions

Mats Tilset, Lars Johansson, Martin Lersch, and Bror J. Wik

Department of Chemistry, University of Oslo, P.O. Box 1033, Blindern, N–0315 Oslo, Norway

Rapid equilibria between M(σ-hydrocarbon) and M(H)(hydrocarbyl) species is a distinct feature of C–H activation reactions at transition-metal complexes of great structural diversity. In this contribution, we summarize recent work that has been done in order to better understand intimate details about important steps in C–H activation reactions at cationic Pt diimine complexes. An experimental design involving competitive trapping of the rapidly interconverting intermediates is utilized. The results provide new information about the processes by which hydrocarbons enter the coordination sphere of Pt, and about the kinetically preferred site of protonation of (diimine)Pt(CH$_3$)$_2$ complexes.

The seminal discoveries by Shilov during the late 1960's and 70's demonstrated that solutions of Pt(II) salts were capable of activating and functionalizing, even catalytically, normally unreactive hydrocarbon C–H bonds.[1] However, the catalytic performance – in terms of rates as well as overall turnover numbers – rendered the system impractical. Nevertheless, the discoveries have had a great impact because they demonstrated that transition metal complexes in solution may be capable of mediating economically and technologically very important hydrocarbon transformation reactions.[2] The "Shilov system" has more recently been investigated in closer detail. A substantial contribution to the current understanding of the system has been made with the use of organometallic complexes that are more amenable to mechanistic studies than the original Shilov system.

The accumulated evidence suggests that the catalytic Shilov system operates according to the mechanism that is summarized in Scheme 1.[3] The overall catalytic cycle consists of three major steps, (a) C–H activation and deprotonation, (b) oxidation, and (c) functionalization. Each of these processes may furthermore be described in terms of several steps, providing an intriguing complexity to the chemistry of the deceptively simple hydrocarbons.

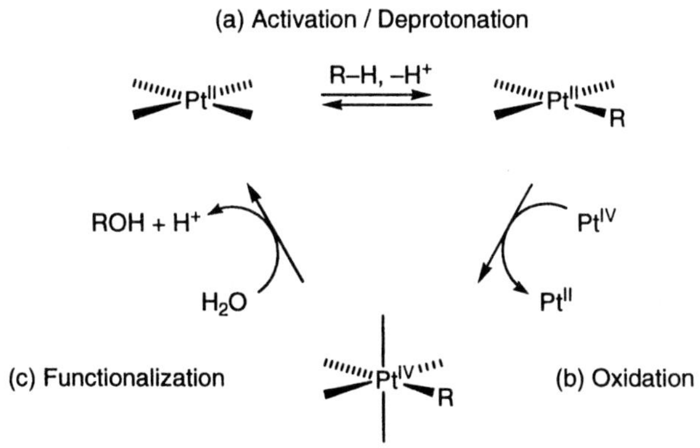

Scheme 1

In this contribution, we will summarize the results of recent efforts that have provided novel insight into important aspects of the first step (a) of the proposed mechanism. This important step is considered (Scheme 2) to involve (d) coordination of the hydrocarbon R–H at the metal center to give a Pt(II) alkane σ complex, (e) oxidative addition of the R–H bond within the σ complex to yield a Pt(IV) hydridoalkyl (or aryl) species, and (f) deprotonation. Two pertinent questions that will be addressed are: (1) How does the hydrocarbon enter the Pt(II) coordination sphere? (2) Does the deprotonation take place from the

Pt(IV) hydridoalkyl species or from the Pt(II) alkane σ complex, which is often in rapid equilibrium with the Pt(IV) hydride?

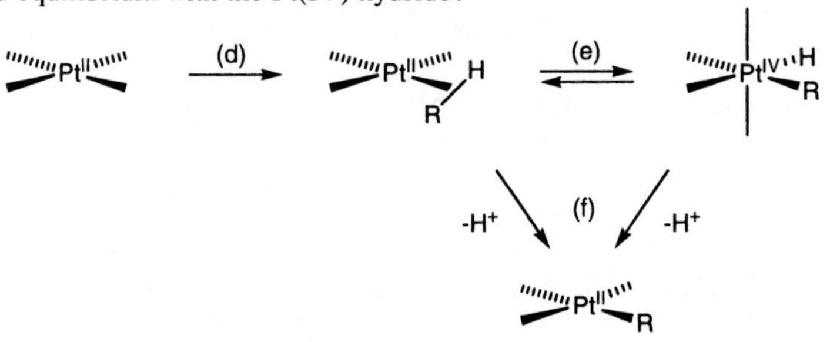

Scheme 2

C–H Activation at Cationic Pt(II) Diimine Complexes

Shortly after the report by Bercaw[4] on benzene and methane C–H activation at cationic (tmeda)Pt(II) complexes and by Periana[5] on the catalytic functionalization of methane by (bipyrimidine)Pt(II) species in fuming H_2SO_4, we communicated[6a] that protonolysis of $(N^f-N^f)Pt(CH_3)_2$ (see Table 1 for overview of diimine ligand types and abbreviations used in this contribution) in wet CH_2Cl_2 yielded the isolable cationic Pt(II) aqua complex $(N^f-N^f)Pt(CH_3)(H_2O)^+$. The aqua complex was capable of activating C–H bonds of benzene within hours at ambient temperature, and of methane within a couple of days at 45 °C (Scheme 3). Reactions with CD_4 and C_6D_6 led to multiple H/D exchange into the methane that was generated, usually taken[7] to indicate the involvement of Pt-(σ-methane) and Pt-(σ/π-benzene) intermediates.

Table 1. Overview of diimine ligands and their abbreviations

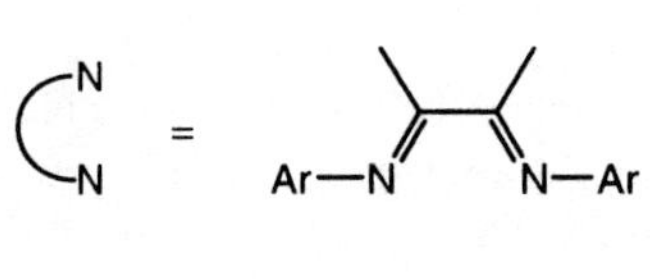

	Abbreviation	N-Ar group
	N–N	Unspecified
	N^f-N^f	$3,5\text{-}(CF_3)_2C_6H_3-$
	$N'-N'$	$2,6\text{-}(CH_3)_2C_6H_3-$
	$N^{Tol}-N^{Tol}$	$4\text{-}(CH_3)C_6H_4-$
	$N^{iPr}-N^{iPr}$	$2,6\text{-}^iPr_2C_6H_3-$

Interestingly, the reactions proceeded in the hydroxylic solvent 2,2,2-trifluoroethanol (TFE). TFE is a strongly ionizing, but weakly nucleophilic or coordinating solvent that has been extensively used in organic chemistry.[8] The aqua complexes $(N-N)Pt(CH_3)(H_2O)^+$ are in equilibrium with the corresponding TFE species $(N-N)Pt(CH_3)(TFE)^+$ in solution. The equilibrium constants have been recently evaluated for a range of substituted diimine complexes,[9] and are found to always favor the aqua species, as would be expected on steric and electronic grounds. The methane and benzene C–H activation reactions were inhibited when water was deliberately added to the medium, suggesting that a preequilibrium loss of water must occur before the rate-determining step.

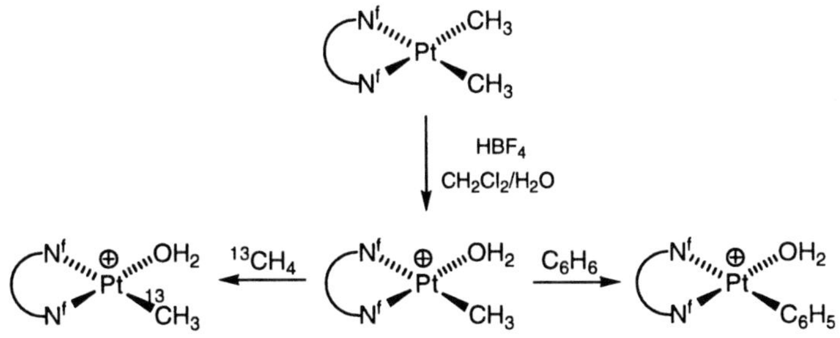

Scheme 3

The methane activation reaction[6] has not been amenable to kinetic studies, but a discussion of some mechanistic details is warranted. Methane can enter the coordination sphere of Pt to yield the crucial σ-methane complex by at least three different mechanisms that should give inhibition by water. The first mechanism is a two-step *dissociative* preequilibrium displacement of water from the aqua complex, yielding a 3-coordinate $(N^f-N^f)Pt(CH_3)^+$ intermediate at which methane coordinates in a second step (eq 1 with R = CH_3). Analogous 3-coordinate species have frequently been implied as possible intermediates in Pt(II) mediated hydrocarbon activation, although direct evidence for the involvement of such species remains elusive.

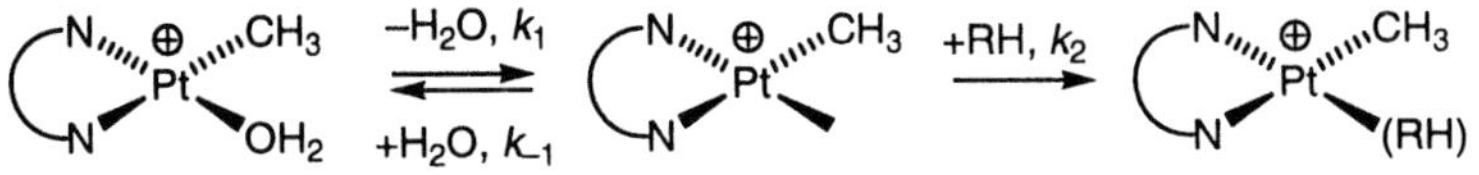

(1)

A two-step, solvent-assisted *associative* mechanism (eq 2) provides an alternative to eq 1. TFE first undergoes a preequilibrium associative displacement of water, and TFE is in turn associatively displaced by methane. The formation of the three-coordinate $(N-N)Pt(CH_3)^+$ intermediate is avoided.

$$\text{(N-N)Pt}(CH_3)(OH_2)^+ \underset{+H_2O\ -TFE,\ k_{-3}}{\overset{-H_2O\ +TFE,\ k_3}{\rightleftharpoons}} \text{(N-N)Pt}(CH_3)(TFE)^+ \xrightarrow{+RH,\ k_4} \text{(N-N)Pt}(CH_3)(RH)^+ \quad (2)$$

A third possible mechanism that can account for the inhibition by water is a one-step direct, associative preequilibrium displacement of water by methane. This scenario can be ruled out, since the methane binding step must be rate limiting in order to account for the fact that multiple H/D exchange occurs even at early stages of the reaction. This means that once CD_4 coordinates at Pt, multiple H/D exchange occurs before methane is released from the metal again. Interestingly, similar observations were made for the original Shilov system![10]

Associative Coordination of Benzene at Pt

The observed inhibition by water on the C–H activation of benzene by (N–N)Pt$(CH_3)(H_2O)^+$ can also be rationalized by the mechanisms in eqs 1 and 2, with R = C_6H_5. The product of the coordination of benzene at Pt presumably is a benzene complex (N–N)Pt$(CH_3)(\eta^2$-$C_6H_6)^+$. These fluxional π-benzene complexes have been observed by NMR when (N–N)Pt$(CH_3)(C_6H_5)$ precursors are protonated in CD_2Cl_2 at –70 °C,[11a] and when (N–N)Pt$(CH_3)_2$ species are protonated at low temperatures immediately followed by addition of benzene.[11b] Arene-Pt complexes bearing the hydridotris(pyrazolyl)borate ligand systems have also been recently reported.[12]

The activation of benzene by (N–N)Pt$(CH_3)(H_2O)^+$ systems in TFE has been the subject of kinetic and mechanistic investigations.[9,11a] The reactions were first-order in [Pt] and [C_6H_6], and inversely first-order in [H_2O], over a wide range of [H_2O] and [C_6H_6]. This is consistent with both mechanisms (eqs 1 and 2). A substantially negative entropy of activation (–67 J K^{-1} mol^{-1})[11a] for the reaction of (N'–N')Pt$(CH_3)(H_2O)^+$ argues against the dissociative pathway in eq 1, although solvent effects may complicate the issue. Recently, a variable-pressure investigation of the reaction kinetics at pressures up to 180 MPa revealed a strongly negative pressure of activation.[13] This finding further supports the notion that benzene enters the coordination sphere of Pt in an associative manner (eq 2). The accumulated data imply that the putative 3-coordinate cations (N–N)Pt$(CH_3)^+$ are *not involved* in the benzene activation. Additional support for this is provided by DFT calculations: The calculated binding energies of TFE and H_2O at the model Pt fragment (HN=C–C=N)-Pt$(CH_3)^+$ are 148 and 158 kJ/mol, respectively; the energy required for producing the 3-coordinate species from solvento precursors will be prohibitively high compared to the overall kinetics of the C–H activation reactions.[6b]

Competitive Trapping of Equilibrating Intermediates

The σ-methane and π-benzene complexes as well as the related Pt(IV) hydridoalkyl and -aryl species that result from hydrocarbon activation are too unstable to be directly observed under actual C–H activation reaction conditions. Investigating such species obviously necessitates alternative approaches. It has become commonplace to study the reverse reactions – the elimination of methane or other hydrocarbons generated by protonation of suitable Pt(II) precursors. If the liberation of the hydrocarbon proceeds by an associative mechanism, then – by the principle of microscopic reversibility – the coordination of the hydrocarbon should also proceed associatively. In the following, three case studies in which we have employed competitive trapping of equilibrating intermediates will be discussed. The basic principles of the method will be briefly described at first.

Consider a reaction of a substrate **S** that eventually yields two different products **P(1)** and **P(2)** (Scheme 4). Intermediates **I(1)** and **I(2)** are the immediate precursors for **P(1)** and **P(2)**, respectively. The reaction is initiated by a reaction of **S** to give the intermediates **I(1)** and/or **I(2)**. An *intramolecular* equilibrium (k_7, k_{-7}) exists between the two intermediates. The two intermediates are *intermolecularly* trapped (k_5 and k_6, respectively) by the trap **T** to give the two products. It must be verified that the initial reaction of **S** to give **I(1)** and **I(2)**, as well as the trapping reactions to give **P(1)** and **P(2)**, are effectively irreversible under the reaction conditions, i.e. their reverse reactions are very slow on the experimental timescale. The rate of the intramolecular equilibration will be independent of the concentration of the trap **T**, whereas both trapping reactions will exihibit a first-order dependence on [**T**].

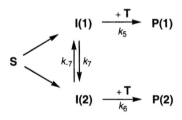

Scheme 4

The relative yields of the two products will be kinetically determined and will depend on the relative rates of the equilibration of the intermediates and the two trapping reactions. Three special cases can be considered:

(i) if k_7 and k_{-7} are $\gg k_5[\mathbf{T}]$ and $k_6[\mathbf{T}]$, the **P(1):P(2)** ratio will be independent of [**T**] and of whether **I(1)** or **I(2)** is formed from **S**. This is a situation where the Curtin-Hammett principle applies,[14,15] and the product ratio will be determined solely by the relative transition-state energies associated with k_5 and k_6.

(ii) if k_7 and k_{-7} << k_5[T] and k_6[T], the P(1):P(2) ratio will reflect the relative occurrence of generation of the two intermediates from S, but the ratio will still be independent of [T].

(iii) if the interconversion between I(1) and I(2) and trapping by T occur at comparable rates, the P(1):P(2) ratio will depend on [T]: If only intermediate I(1) is formed from S, an increase of [T] will trap I(1) more efficiently, arresting the interconversion of I(1) to I(2). Increasing [T] therefore increases the P(1):P(2) ratio. Conversely, initial generation of I(2) will cause the P(1):P(2) ratio to decrease with increasing [T].

Associative Coordination of Methane at Pt

Relatively persistent Pt(IV) methyl hydrido complexes can be prepared by protonation of Pt(II) precursors when a suitable ligand is available for occupation of the coordination site trans to the hydride.[16] Such species have been extensively used to study the mechanism for reductive elimination of methane (Scheme 5). The need to dissociate the ligand trans to the hydride (a) before reductive elimination occurs has been experimentally demonstrated and is supported by calculations. The resulting 5-coordinate intermediate undergoes C–H coupling (b) to form a methane σ complex before loss of methane and ligand reattachment (c). We have performed experiments that help establish whether the last of these steps proceeds associatively or dissociatively.[17]

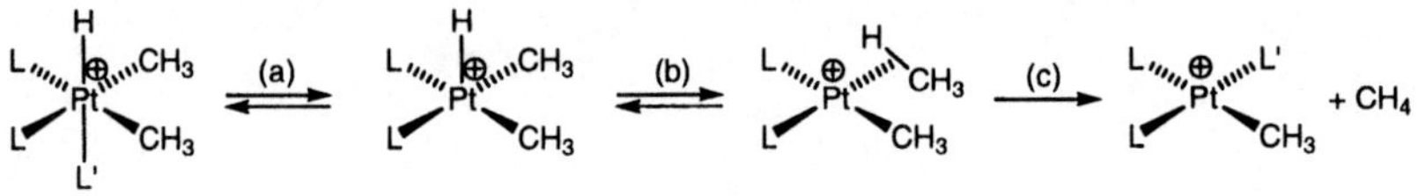

Scheme 5

Treatment of (N–N)Pt(CH$_3$)$_2$ with DOTf (deuterated triflic acid, CF$_3$SO$_3$D) generates a Pt(IV) deuteride (N–N)Pt(CH$_3$)$_2$(D)$^+$ (Scheme 6)[18] which is in equilibrium with the σ-CH$_3$D complex (N–N)Pt(CH$_3$)(CH$_3$D)$^+$. Note that an alternative protonation at a methyl group – vide infra – rather than at Pt will generate the same σ-methane species and thence, the analysis and conclusion remain invariant as to the exact identity of the protonation site. "Rotation" of the σ-bonded CH$_3$D ligand at Pt exposes a C–H, rather than the C–D, bond to the metal, and oxidative insertion into the C–H bond generates the Pt(IV) hydride (N–N)Pt(CH$_3$)(CH$_2$D)(H)$^+$ which may undergo reductive C–H coupling to generate either the σ-CH₃D complex or the isotopically scrambled isomer (N–N)Pt(CH₂D)(CH₄)⁺.

The σ-methane complex eventually liberates methane. The ratio of CH₃D to CH₄ (and of Pt-CH₃ to Pt-CH₂D) products will depend on the rate at which the σ-methane complex undergoes isotope scrambling relative to the rate of methane loss. In the event that H/D scrambling is much slower than methane loss, the CH₃D:CH₄ ratio will approach 100:0. On the other hand, if the H/D scrambling is fast (Curtin-Hammett conditions), a statistical 4:3, or 57:43, ratio of CH₃D and CH₄ will be obtained. (The D label can be distributed over 7 sites, 4 in σ-methane and 3 in a Pt-methyl group). The quantitative consideration strictly applies only when H/D kinetic isotope effects are absent, but the qualitative arguments will be equally valid when isotope effects are taken into account.

The above considerations for very slow or very rapid H/D scrambling apply irregardless of whether methane loss from the σ-methane intermediate occurs associatively or dissociatively. On the other hand, *the CH₃D:CH₄ ratio will depend on [L] if an associative displacement occurs at a rate that is comparable to the rate of isotope scrambling.* In this case, high [L] will serve to more efficiently displace CH₃D before scrambling occurs, causing the CH₃D:CH₄ ratio to increase with increasing [L].[19]

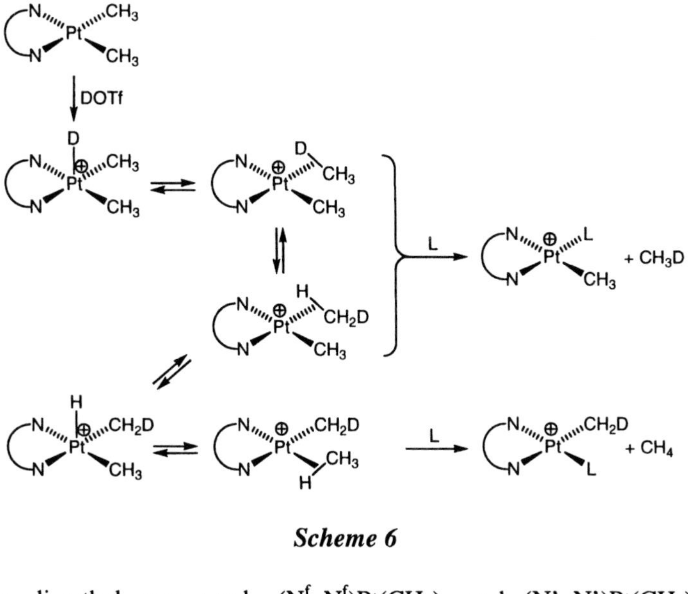

Scheme 6

The dimethyl compounds (N^f–N^f)Pt(CH₃)₂ and (N'–N')Pt(CH₃)₂ were treated with DOTf in TFE-d_3 in the presence of variable amounts of acetonitrile, selected as the trapping ligand L. Control experiments established that under the reaction conditions, the protonation is effectively irreversible (only singly deuterated products were observed), and trapping is irreversible (no exchange between coordinated CH_3CN and CD_3CN in the $(N-N)Pt(CH_3)(NCMe)^+$ products). Figure 1 shows ^1H NMR spectra for the protonation of $(N'-N')Pt(CH_3)_2$. The spectra show well-resolved signals for the isotopomers CH_3D and CH_4. It is evident that the $CH_3D:CH_4$ ratio increases with increasing [MeCN]. Consistent with this, the $(Pt-CH_3):(Pt-CH_2D)$ ratio increases with increasing [MeCN]. The quantitative results are summarized in Table 2. In the absence of CD_3CN, essentially complete scrambling occurred for $(N^f-N^f)-Pt(CH_3)_2$, and nearly complete scrambling was seen for $(N'-N')Pt(CH_3)_2$. The observed increase in the $CH_3D:CH_4$ ratio with increasing [MeCN] is most readily explained by an associative mechanism for the displacement of methane by acetonitrile at $(N-N)Pt(CH_3)(\sigma-CH_4)^+$. Then, by the principle of microscopic reversibility, the *substitution of methane for MeCN at $(N-N)Pt(CH_3)(NCMe)^+$ must occur associatively, and the 3-coordinate intermediate that would result from MeCN predissociation is not required for the C–H activation to proceed.*

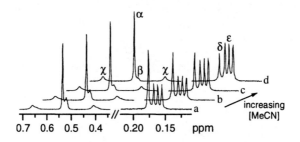

Figure 1. Pt-methyl and methane parts of 300 MHz ^1H NMR spectra of products from treatment of $(N'-N')Pt(CH_3)_2$ with 1 equiv of DOTf in: (a) TFE-d_3. CD_3CN added after reaction. (b) 0.5 M CD_3CN/TFE-d_3. (c) 2.0 M CD_3CN/TFE-d_3. (d) 6.0 M CD_3CN/TFE-d_3. Legend: (α) $PtCH_3$; (β) $PtCH_2D$; (χ) overlapping $^{195}PtCH_3$ and $^{195}PtCH_2D$ satellites; (δ) CH_4; (ε) CH_3D. Reproduced with permission from ref 17.

The actual C–H activations were performed in TFE-d_3 containing traces of water, and a comment regarding the relevance of the above reactions that were performed in the presence of additional acetonitrile is warranted. The strength of Pt–L bonding at $(N-N)Pt(CH_3)(L)^+$ increases in the order L = CH_4 < TFE < H_2O < MeCN,[6b,11a] i.e. MeCN is a considerably poorer leaving group than H_2O and TFE. In a reaction of $(N^f-N^f)Pt(CH_3)_2$ with DOTf in TFE-d_3 containing 0.5 M D_2O, the $CH_4:CH_3D$ ratio was essentially identical to that in neat TFE-d_3, i.e. complete scrambling was observed.[20] It appears that not only TFE, but also water, are insufficiently nucleophilic towards Pt to arrest the H/D scrambling reactions. The findings are consistent with both mechanisms, eqs 1 and 2, that may exhibit inhibition by water for the C–H activation. Because of the many similarities between the methane activation and the more thoroughly studied benzene activation reactions, we do however favor the solvent-assisted associative mechanism for the methane activation in TFE.

Table 2. Methane isotopomer mixtures (%) formed in the reactions between (N–N)Pt(CH$_3$)$_2$ and DOTf.

Solvent	(N–N) = (N^f–N^f)		(N–N) = (N'–N')	
	CH$_4$	CH$_3$D	CH$_4$	CH$_3$D
TFE-d_3	41	59	37	63
0.5 M CD$_3$CN/TFE-d_3	35	65	33	67
2.0 M CD$_3$CN/TFE-d_3	31	69	28	72
6.0 M CD$_3$CN/TFE-d_3	28	72	15	85
0.5 M D$_2$O/TFE-d_3	40	60	–	–

Associative Decoordination of Toluene from Pt

We have applied the experimental design that was described above in an investigation of reactions related to C–H activation of toluene and *p*-xylene.[21] These substrates raise interesting question regarding selectivity issues (benzylic vs. aromatic; ortho vs meta vs para activation). Parts of the results will be discussed in the following.

The protonation of each of the three (N^f–N^f)Pt(tolyl)$_2$ complexes (tolyl = *o*-, *m*- or *p*-tolyl) with HBF$_4$·Et$_2$O was investigated at 25 °C. The reactions were performed in neat TFE (followed by MeCN trapping after complete reaction), in TFE containing 1.7 M MeCN, and in neat MeCN. The product distributions are summarized in Table 3. Certain interesting features are apparent. First, extensive *o*/*m*/*p* isomerization of the Pt-tolyl group is observed in the product mixtures.[22] The extent of isomerization increases with decreasing acetonitrile concentrations. Second, when the protonations are conducted in neat TFE, the product distributions are identical in all cases. Third, in neat TFE, a rather unusual, significantly greater than statistical *m*/*p* ratio is seen.

The C–H activation of C$_6$D$_6$ at (N^f–N^f)Pt(CH$_3$)(H$_2$O)$^+$ in TFE led to scrambling of the D label between the Pt-phenyl and methane products. This suggests that a fairly rapid equilibrium exists between appropriately D-containing (N^f–N^f)Pt(CH$_3$)(C$_6$H$_6$)$^+$, (N^f–N^f)Pt(C$_6$H$_5$)(CH$_4$)$^+$, and (N^f–N^f)Pt(CH$_3$)-(C$_6$H$_5$)(H)$^+$ species in solution. An analogous equilibrium between (N^f–N^f)Pt(*o*-tolyl)(toluene)$^+$, (N^f–N^f)Pt(*m*-tolyl)(toluene)$^+$, and (N^f–N^f)Pt(*p*-tolyl)(toluene)$^+$ complexes (Scheme 7) provides a rationale for the isomerizations in TFE.

Table 3. Product distributions (%) in the protonations of (N^f–N^f)Pt(tolyl)$_2$ complexes with HBF$_4$·Et$_2$O at 25 °C.

Reactant	Solvent [a]	Products (%) o-tolyl	m-tolyl	p-tolyl
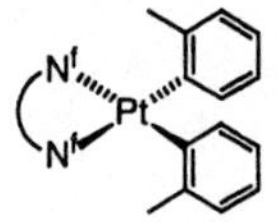 o-tolyl complex	A	–	77	23
	B	–	65	35
	C	7	57	36
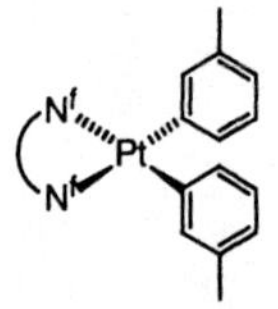 m-tolyl complex	A	–	77	23
	B	–	93	7
	C	–	100	–
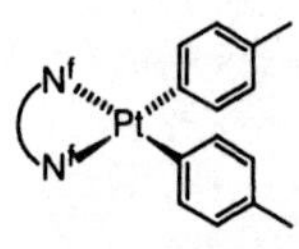 p-tolyl complex	A	–	77	23
	B	–	22	78
	C	–	4	96

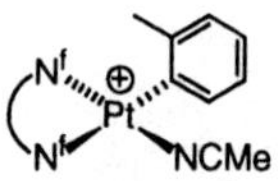

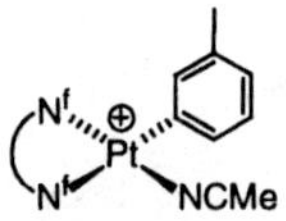

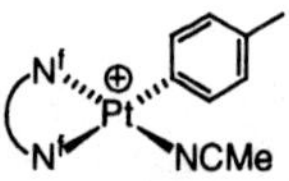

[a] (A) TFE only. Addition of MeCN after completed reaction. (B) 1.7 M MeCN in TFE. (C) MeCN.

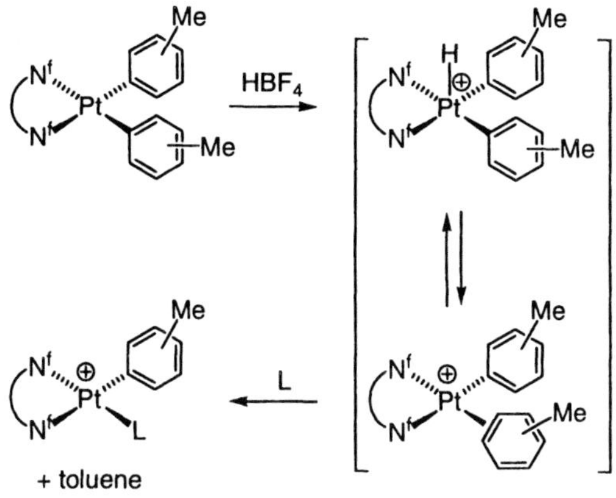

Scheme 7

The product distribution that is observed is governed by the relative rates of the product-forming and isomerization steps. If isomerization is fast, the Curtin-Hammett principle applys and the product ratio is determined by the relative transition-state energies of the product-forming steps. This must be the case for the protonations in neat TFE. The isomeric intermediates enclosed in brackets are at equilibrium, and the product distribution reflects the true kinetic selectivity for toluene elimination from these intermediates. Thus, the transition-state energies for toluene elimination must increase in the order $(N^f-N^f)Pt(m\text{-tolyl})(\text{toluene})^+ < (N^f-N^f)Pt(p\text{-tolyl})(\text{toluene})^+ << (N^f-N^f)Pt(o\text{-tolyl})(\text{toluene})^+$. When acetonitrile is added, the isomerization process is arrested before equilibrium is attained. This is a strong indication that again, hydrocarbon release from the Pt(II) intermediates proceeds by an *associative* mechanism. A dissociative mechanism should not exhibit this dramatic response to the presence of acetonitrile. The product-forming steps may pass through 5-coordinate, perhaps trigonal bipyramidal, transition states in which the incoming ligand (acetonitrile or TFE) is bonded, as indicated in Figure 2.

Steric effects presumably cause the transition state for formation of the *o*-tolyl product to be disfavored, but cannot explain why *m* is preferred over *p* to an extent that is even greater than statistical. Simple electronic arguments seem insufficient to explain the observed meta selectivity for a dissociative mechanism, since the *p*-tolyl group should better stabilize the positive charge in the putative 3-coordinate intermediate. Unfortunately, a straightforward explanation of this effect for the associative pathway is also difficult to find. We propose that the *m*-tolyl ligand in **B** is less apt at delocalizing the positive charge at Pt than is the *p*-tolyl ligand in **C**. This leaves a greater positive charge at the metal in **B**, rendering the metal more susceptible to nucleophilic attack by allowing tighter binding of L in the transition state, and paves the way for the observed meta preference.

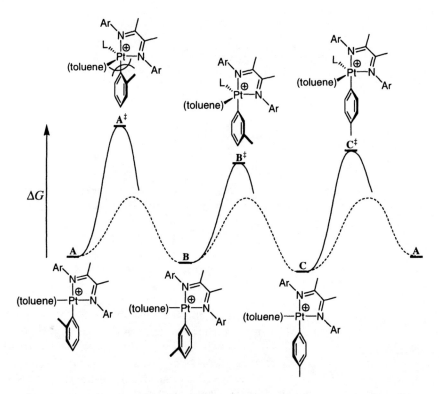

*Figure 2. Qualitative reaction coordinate diagram for the reaction path from the isomeric aryl-arene $(N^f-N^f)Pt(tolyl)(toluene)^+$ intermediates **A–C** to the corresponding 5-coordinate transition states for a ligand-assisted associative mechanism for the elimination of toluene. Dotted line: o/m/p isomerization. Full line: Toluene displacement by TFE.*

What is the Kinetic Site of Protonation in Pt Dimethyl Complexes?

This is a relevant question considering the general C–H activation mechanisms depicted in Schemes 1 and 2.[23] In order to resolve this question, we have again utilized competitive trapping experiments.[24] The strategy is delineated in Scheme 8. The protonation of (N–N)Pt(CH$_3$)$_2$ may occur at the metal to furnish a transient pentacoordinate Pt(IV) hydride that is irreversibly trapped by MeCN to yield the hexacoordinate Pt(IV) product that is relatively stable at low temperatures. Alternatively, protonation may occur at a methyl group, providing a Pt(II) σ-methane complex. From the previous discussion, it is known that acetonitrile irreversibly displaces the σ-methane ligand by an associative mechanism to give the Pt(II) product. The scheme is completed by adding the reversible isomerization that occurs between the pentacoordinate Pt(IV) hydride and the Pt(II) σ-methane complex.

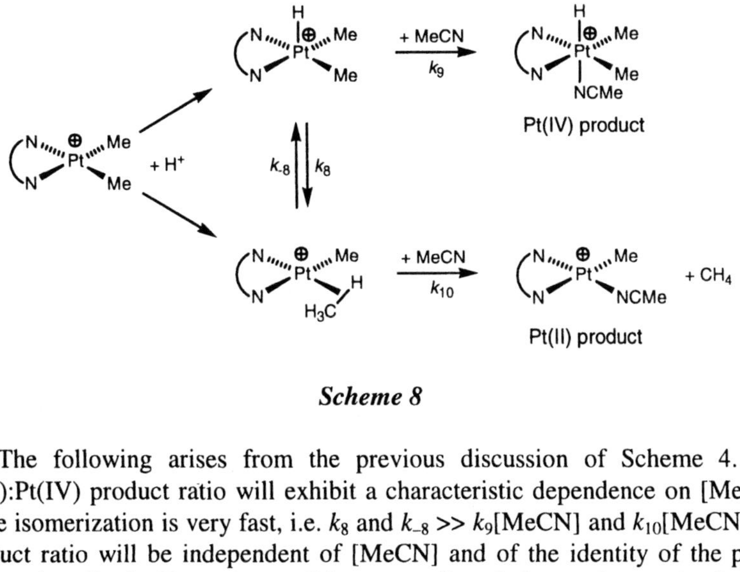

Scheme 8

The following arises from the previous discussion of Scheme 4. The Pt(II):Pt(IV) product ratio will exhibit a characteristic dependence on [MeCN]. If the isomerization is very fast, i.e. k_8 and $k_{-8} \gg k_9$[MeCN] and k_{10}[MeCN], the product ratio will be independent of [MeCN] and of the identity of the protonation site. Curtin-Hammett conditions prevail and the product ratio depends only on the relative transition-state energies of the product-forming steps, in other words on the relative magnitudes of k_9 and k_{10}. On the other hand, if k_8 and $k_{-8} \ll k_9$[MeCN] and k_{10}[MeCN], the product distribution will be independent of [MeCN] and reflects the relative importance of Pt vs. methyl protonation. Finally, if the σ-methane/hydridomethyl interconversion and the trapping reactions by MeCN occur at comparable rates, the product ratio will be [MeCN] dependent. If initial protonation occurs at Pt to give the 5-coordinate hydride species, an increase of [MeCN] will trap the hydride more efficiently, arresting the hydridomethyl to σ-methane interconversion before equilibrium is reached. An increase in [MeCN] consequently will lead to a lower Pt(II):Pt(IV) product ratio. By analogous arguments, initial protonation at a methyl group to give the σ-methane intermediate will show the opposite response to changes in [MeCN], i.e. the Pt(II):Pt(IV) product ratio will increase with increasing [MeCN].

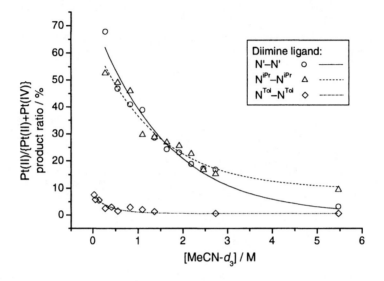

Figure 3. Pt(II)/{Pt(II)+Pt(IV)} product ratio resulting from protonation of (N–N)Pt(CH$_3$)$_2$ with HBF$_4$•Et$_2$O in CD$_2$Cl$_2$ at –78 °C.

The complexes (N'–N')Pt(CH$_3$)$_2$, (N^{Tol}–N^{Tol})Pt(CH$_3$)$_2$, and (N^{iPr}–N^{iPr})Pt-(CH$_3$)$_2$ were sufficiently soluble in CD$_2$Cl$_2$ at –78 °C so that protonation with HBF$_4$•Et$_2$O could be done on a precooled, homogeneous solution containing variable amounts of MeCN. At this temperature, the two trapping reactions were irreversible. The results (each data point represents an individually prepared NMR sample) are summarized in Figure 3, which shows the Pt(II)/{Pt(II)+Pt(IV)} product ratio as a function of [MeCN]. For all the substrates, there is a clear trend that increasing amounts of acetonitrile causes the Pt(IV) producing reaction (k_9) to be favored relative to the Pt(II) producing reaction (k_{10}). This result is only consistent with kinetically controlled protonation at Pt, rather than at a methyl ligand. The much greater Pt(II) product yields for the N'–N' and N^{iPr}–N^{iPr} subtrates, compared to N^{Tol}–N^{Tol}, may be rationalized in terms of steric effects. The 2,6-substituents on the diimine aryl groups are oriented essentially perpendicularly with respect to the coordination plane. The substituents effectively shield the apical coordination sites of the metal center, blocking access to this site by the incoming MeCN ligand.

By extrapolation of the finding that protonation occurs at Pt and by the principle of microscopic reversibility, the deprotonation step that is needed in the methane activation and functionalization at Pt in the Shilov system takes place from a Pt(IV) hydridomethyl species and not from a Pt(II) σ-methane complex. These results strictly apply to the particular reaction conditions that were used. The conditions for C–H activation by such complexes are quite different, and caution must be exerted when results are extrapolated from one set of conditions to another – and especially to the original Shilov system.

Concluding Remarks

In this contribution, we have focused on work that has been done in order to better understand the intimate details about important steps in C–H activation reactions at Pt complexes. The experimental design involving competitive trapping of interconverting intermediates has been very useful in providing novel mechanistic insight.

The results of competitive trapping experiments suggest that hydrocarbon coordination at the Pt-diimine systems proceeds in an associative fashion. This conclusion is consistent with other kinetic and mechanistic evidence. Ligand substitution reactions at d^8 square planar complexes in general follow associative mechanisms, but since hydrocarbons are expected to be very weak nucleophiles, this general rule might not apply here. Romeo and coworkers have shown that solvent exchange reactions at $(N-N)Pt(CH_3)(DMSO)^+$ and related complexes generally follow associative pathways.[25a] A mechanistic changeover to dissociative substitutions has been recognized for electron-rich, neutral systems that tend to thwart the possibility for nucleophilic attack at Pt, and enhance the stability of 3-coordinate, 14-electron intermediates.[25b] Interestingly, Krogh-Jespersen and Goldman recently concluded, on the basis of combined theoretical and experimental work, that hydrocarbon coordination/H_2 expulsion at neutral Ir "pincer" complexes $(PCP)IrH_2$ that are very promising acceptorless alkane dehydrogenation catalysts appears to proceed in a dissociative fashion.[26a] On the other hand, evidence for associative hydrocarbon ligation has been presented for a related acceptor-mediated dehydrogenation system.[26b] It appears that the mechanism of hydrocarbon coordination at metal complexes is quite sensitive to the identity of the metal complex and to the reaction conditions, and therefore any attempt at making generalizations should be done with great care.

The protonation of $(N-N)Pt(CH_3)_2$ complexes with $HBF_4 \cdot Et_2O$ in CD_2Cl_2 appears to occur at the metal. In contrast, protonation of $CpW(CO)_2(PMe_3)H$ occurs at the hydride ligand to produce a transient η^2-H_2 complex that rapidly transforms into the thermodynamically more stable dihydride. It is not yet clear whether these differences in behavior of methyl vs. hydride ligands is a general phenomenon. Clearly, more experimental data are needed to clarify the issue.

Acknowledgement

We are grateful for the support of this research by the Norwegian Research Council (NFR) with generous stipends to BJW, LJ, and ML.

References

1. (a) Goldshlegger, N. F.; Eskova, V. V.; Shilov, A. E.; Shteinman, A. A. *Zhur. Fiz. Khim.* **1972**, *46*, 1353. (b) Shilov, A. E.; Shulpin, G. B. *Chem. Rev.* **1997**, *97*, 2879. (c) Shilov, A. E.; Shul'pin, G. B. *Activation and Catalytic Reactions of Saturated Hydrocarbons in the Presence of Metal Complexes*; Kluwer: Boston, 2000.
2. (a) Crabtree, R. H. *Chem. Rev.* **1995**, *95*, 987. (b) Arndtsen, B. A.; Bergman, R. G.; Mobley, T. A.; Peterson, T. H. *Acc. Chem. Res.* **1995**, *28*, 154. (c) Shilov, A. E.; Shulpin, G. B. *Chem. Rev.* **1997**, *97*, 2879. (d) Labinger, J. A.; Bercaw, J. E. *Nature* **2002**, *417*, 507.
3. Stahl, S. S.; Labinger, J. A.; Bercaw, J. E. *Angew. Chem., Int. Ed. Engl.* **1998**, *37*, 2180.
4. Holtcamp, M. W.; Labinger, J. A.; Bercaw, J. E. *J. Am. Chem. Soc.* **1997**, *119*, 848.
5. Periana, R. A.; Taube, D. J.; Gamble, S.; Taube, H.; Satoh, T.; Fujii, H. *Science* **1998**, *280*, 560.
6. (a) Johansson, L.; Ryan, O. B.; Tilset, M. *J. Am. Chem. Soc.* **1999**, *121*, 1974. (b) Heiberg, H.; Johansson, L.; Gropen, O.; Ryan, O. B.; Swang, O.; Tilset, M. *J. Am. Chem. Soc.* **2000**, *121*, 10831.
7. Northcutt, T. O.; Wick, D. D.; Vetter, A. J.; Jones, W. D. *J. Am. Chem. Soc.* **2001**, *123*, 7257, and references cited.
8. (a) Evans, D. F.; McElroy, M. I. *J. Soln. Chem.* **1975**, *5*, 405. (b) Schadt, F. L.; Bentley, T. W.; Schleyer, P. v. R. *J. Am. Chem. Soc.* **1976**, *98*, 7667. (c) Lowry, T. H.; Richardson, K. S. *Mechanism and Theory in Organic Chemistry*, 3rd Ed.; Harper and Row: New York, 1987; pp. 335-340.
9. Zhong, H. A.; Labinger, J. A.; Bercaw, J. E. *J. Am. Chem. Soc.* **2002**, *124*, 1378.
10. Goldschlegger, N. F.; Tyabin, M. B.; Shilov, A. E.; Shteinman, A. A. *Zhur. Fiz. Khim.* **1969**, *43*, 2174.

11. (a) Johansson, L.; Tilset, M.; Labinger, J. A.; Bercaw, J. E. *J. Am. Chem. Soc.* **2000**, *122*, 10846. (b) Lersch, M.; Wik, M.; Tilset, M. Unpublished results.
12. (a) Reinartz, S.; White, P. S.; Brookhart, M.; Templeton, J. L. *J. Am. Chem. Soc.* **2001**, *123*, 12724. (b) Norris, C. M.; Reinartz, S.; White, P. S.; Templeton, J. L. *Organometallics* **2002**, *21*, 5649.
13. Procelewska, J.; Zahl, A.; van Eldik, R.; Zhong, H. A.; Labinger, J. A.; Bercaw, J. E. *Inorg. Chem.* **2002**, *41*, 2808.
14. Seeman, J. I. *Chem. Rev.* **1983**, *83*, 83.
15. For an insightful discussion about the Curtin-Hammet principle in the context of C–H activation at cationic Ir(III) complex, see Peterson, T. H.; Golden, J. T.; Bergman, R. G. *J. Am. Chem. Soc.* **2001**, *123*, 455.
16. For some key references, see ref 4 and (a) Fekl, U.; Zahl, A.; van Eldik, R. *Organometallics* **1999**, *18*, 4156. (b) Stahl, S. S.; Labinger, J. A.; Bercaw, J. E. *J. Am. Chem. Soc.* **1995**, *117*, 9371. (c) Hinman, J. G.; Baar, C. R.; Jennings, M. C.; Puddephatt, R. J. *Organometallics* **2000**, *19*, 563.
17. Johansson, L.; Tilset, M. *J. Am. Chem. Soc.* **2001**, *123*, 739.
18. The hexacoordinate acetonitrile adduct of the analogous hydride can be observed by ^{1}H NMR at low temperatures.
19. Competitive capture of a *dissociatively* generated 3-coordinate intermediate (N–N)Pt(CH$_3^+$) by L and CH$_4$ could in principle also give an [L] dependent product ratio. However, it can be ruled out that this occurs under the reaction conditions. Methane coordination is rate limiting in the C–H activation process in TFE-d_3 and occurs during a couple of days at 45 °C at a methane pressure of 20-25 bar, resulting in ca. 20 equiv of methane in solution. In contrast, methane loss is irreversible on the experimental time scale (minutes at ambient temperature in TFE-d_3 with no methane pressure applied). Therefore, under the actual conditions methane is irreversibly displaced and there can be no competitive trapping of methane and L.
20. At [D$_2$O] = 6.0 M, the full range of methane isotopomers CH$_n$D$_{4-n}$ (n = 0-4) was observed. Multiple D incorporation must result from reversible protonations, a process that is most likely facilitated by the higher basicity of water relative to TFE and MeCN.
21. Johansson, L.; Ryan, O. B.; Rømming, C.; Tilset, M. *J. Am. Chem. Soc.* **2001**, *123*, 6579.
22. With the N'–N' ligand, isomerization was seen to occur even to the benzylic position. See ref 21.
23. (a) Zamashchikov, V. V.; Popov, V. G.; Rudakov, E. S.; Mitchenko, S. A. *Dokl. Akad. Nauk SSSR* **1993**, *333*, 34. (b) Zamashchikov, V. V.; Popov, V. G.; Rudakov, E. S. *Kinet. Katal.* **1994**, *35*, 436.
24. Wik, B. J.; Lersch, M.; Tilset, M. *J. Am. Chem. Soc.* **2002**, *124*, 12116.

25. (a) Romeo, R.; Scolaro, L. M.; Nastasi, N.; Arena, G. *Inorg. Chem.* **1996**, *35*, 5087. (b) Romeo, R.; Grassi, A.; Scolaro, L. M. *Inorg. Chem.* **1992**, *31*, 4383.
26. (a) Krogh-Jespersen, K.; Czerw, M.; Summa, N.; Renkema, K. B.; Achord, P. D.; Goldman, A. S. *J. Am. Chem. Soc.* **2002**, *124*, 11404. (b) Belli, J.; Jensen, C. M. *Organometallics* **1996**, *15*, 1532.

Chapter 17

Mechanisms of Reactions Related to Selective Alkane Oxidation by Pt Complexes

Jennifer L. Look, Ulrich Fekl, and Karen I. Goldberg*

Department of Chemistry, University of Washington, Seattle, WA 98195–1700

Some of the most promising homogeneous systems for the selective activation and functionalization of alkanes are based on platinum. Proposed key reaction steps at Pt(II) and Pt(IV) metal centers have been directly observed and studied using model compounds. Investigations of oxidative addition and reductive elimination reactions of carbon-hydrogen, carbon-carbon and carbon-heteroatom bonds have provided detailed information about the intermediates and kinetic barriers involved in these bond cleavage and formation reactions. In addition, reactions of platinum alkyl complexes with dioxygen, an ideal oxidant for alkane functionalization, have been studied.

Introduction

Over thirty years ago, Shilov and co-workers reported the remarkable discovery that alkanes could be selectively oxidized in aqueous solution in the presence of Pt salts ($PtCl_4^{2-}$ and $PtCl_6^{2-}$); methane was converted to methanol and methyl chloride (*1,2*). This reaction was catalytic in Pt(II) but required Pt(IV) as a stoichiometric oxidant. Clearly, the high cost of a platinum compound as an oxidant makes this particular system impractical. However, the mere demonstration of this unprecedented transformation inspired many researchers, including us, to pursue research programs in platinum-catalyzed alkane oxidation. Perhaps the oxidant in the Shilov system can be replaced with a more suitable oxidant and a commercially viable platinum catalyzed process can be developed. Toward this end, significant effort has been directed toward understanding the mechanistic details of Shilov's alkane oxidation system. Strong support has been presented for the pathway shown in Scheme 1 (2). In step I, a Pt(II) species activates the alkane C-H bond to generate a Pt(II) alkyl complex. Then in step II, the Pt(II) alkyl is oxidized by Pt(IV) to form a Pt(IV) alkyl complex. In step III, nucleophilic attack of either water or chloride at the Pt-bonded alkyl forms the functionalized organic product and regenerates the active catalyst. While the basic cycle pictured in Scheme 1 is now generally accepted, the intimate mechanistic details of each of these reaction steps are research topics of significant current interest. Understanding the mechanism and energetics of each step – C-H bond activation, oxidation, and C-heteroatom bond formation – is expected to aid in the development of more practical alkane functionalization processes.

Scheme 1

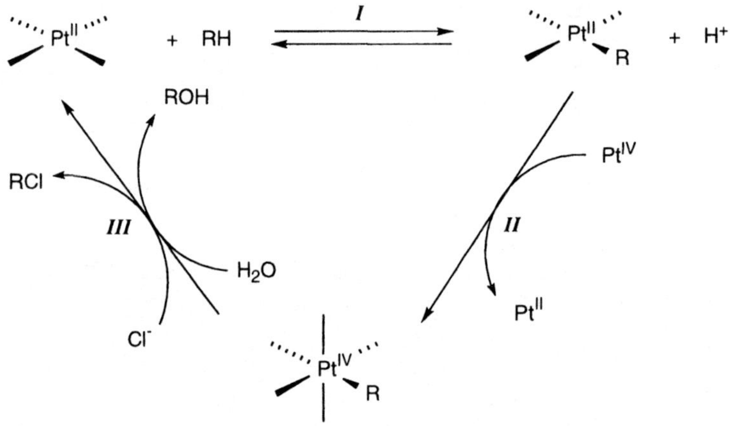

For the past several years, our research group has been carrying out mechanistic investigations of reactions related to those depicted in Scheme 1. Using model platinum complexes, we have directly observed and studied C-H bond oxidative addition reactions to Pt(II) and reductive elimination reactions from Pt(IV) (3-5). Thorough mechanistic studies of C-C and C-O reductive elimination reactions from Pt(IV) have also been carried out (5,6). In addition, inspired by the potential of utilizing dioxygen as a replacement for the Pt(IV) oxidant required in the Shilov system, we have explored the reactivity of Pt(II) and Pt(IV) alkyl complexes with dioxygen (7,8). In this contribution, we share the results of our investigations of these reactions and comment on some of the key elements to be considered in developing a Pt catalyst to convert alkanes to alcohols.

C-O Reductive Elimination

We will begin our discussion with the last reaction shown in Scheme 1 (step III). This is the product-forming step of the reaction sequence; C-X bond formation yields the alcohol or alkyl chloride product. Significant evidence has been presented for this step occurring via nucleophilic attack of a heteroatom group (Cl^- or H_2O) on an electrophilic Pt(IV) alkyl (*2*). In the related Catalytica process for oxidation of methane to methyl bisulfate using a platinum bipyrimidine catalyst, carbon-oxygen reductive elimination from a Pt(IV) methyl bisulfate intermediate was proposed as the product-forming step (*9*). Until recently (*6*), there were no directly observed examples of high-yield carbon-oxygen reductive elimination from Pt(IV). We were thus interested in finding examples of such reactivity and studying the intimate mechanism of these reactions. In particular, our goal was to understand some of the factors (e.g., solvent, substituents on oxygen, etc.) that favor carbon-oxygen reductive elimination from Pt(IV).

The Pt(IV) complexes *fac*-L_2PtMe_3OR (Scheme 2, L_2 = dppe (bis(diphenylphosphino)ethane), dppbz (bis(*o*-diphenylphosphino)benzene); OR = carboxylate, aryloxide) were prepared and characterized (*6*). It was anticipated that these complexes might undergo C-O reductive elimination since they are analogous to the Pt(IV) iodide species, *fac*-$dppePtMe_3I$, which had been previously observed to undergo competitive C-I and C-C coupling upon thermolysis (*10*). Indeed, thermolysis of the *fac*-L_2PtMe_3OR complexes proceeded to form both C-O and C-C reductive elimination products (Scheme 2), and provided us with an excellent system from which to study the intimate mechanism of C-O reductive elimination from Pt(IV) (*6*).

Scheme 2

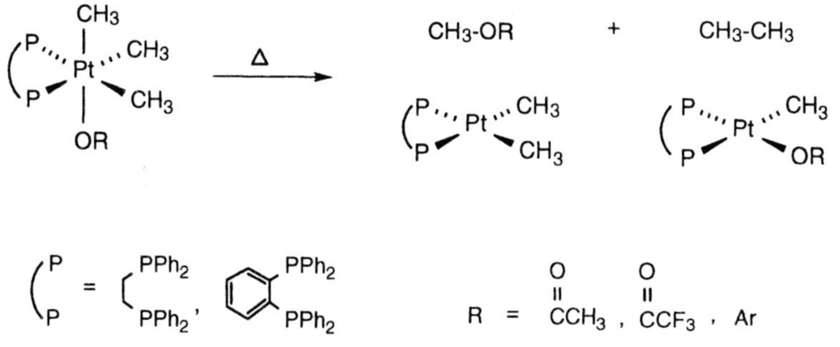

The mechanisms proposed for the competitive C-O and C-C reductive elimination reactions from fac-L_2PtMe_3OR are shown in Scheme 3 (6). Dissociation of OR⁻ forms the five-coordinate Pt(IV) cation, $L_2PtMe_3^+$ (Intermediate **A**). C-C coupling to form ethane proceeds directly from **A** in a concerted fashion. In contrast, nucleophilic attack by OR⁻ on a methyl group of the Pt(IV) intermediate **A** generates the methyl carboxylate or methyl aryl ether product and L_2PtMe_2. It is interesting to note that these mechanisms for C-O and C-C bond formation are analogous to those previously proposed for competing C-I and C-C reductive elimination reactions from fac-$dppePtMe_3I$ (10).

Scheme 3

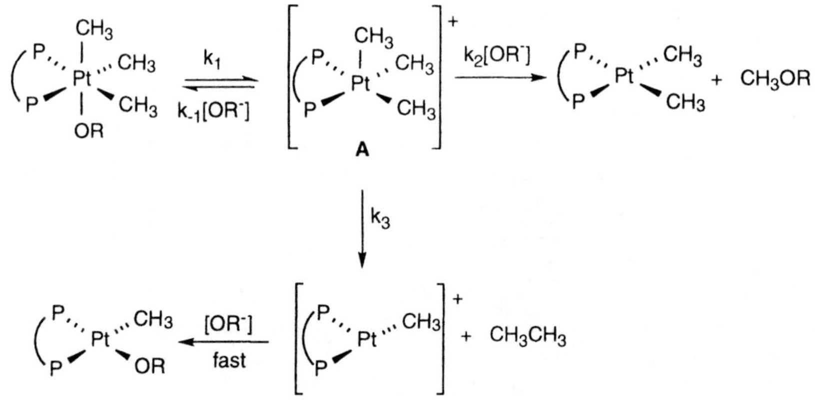

As expected for reactions that involve ionic intermediates, the thermolyses of these complexes show a significant solvent effect. For example, the rate of the disappearance of $(dppe)PtMe_3OAc$ at 99 °C increased by two orders of magnitude on going from benzene-d_6 (k_{obs} = 1.3 x 10^{-5} s⁻¹) to nitrobenzene-d_5 (k_{obs} = 1.4 x 10^{-3} s⁻¹). The ratios of the products (C-O versus C-C coupling) also varied drastically with the polarity of the solvent; 90+ % C-O coupling was observed in benzene-d_6 and less than 1% was observed in nitrobenzene-d_5. When the rates of the individual reactions to form C-O or C-C bonds are examined, it can be seen that the rate of C-C reductive elimination increases substantially more with the polarity of the solvent than the C-O coupling rate. For example, the rate constant for C-O coupling from $(dppe)PtMe_3OAc$ increased only slightly from k_{C-O} = 1.2 x 10^{-5} s⁻¹ in benzene-d_6 to k_{C-O} = 4.1 x 10^{-5} s⁻¹ in acetone-d_6, whereas the rate constant for C-C coupling increased from 1 x 10^{-6} s⁻¹ to 6-10 x 10^{-4} s⁻¹ in the same solvents. These solvent effects are consistent with the mechanistic scheme shown (Scheme 3). The first step of both coupling reactions (k_1, formation of ionic intermediates **A** and OR⁻) should be accelerated by a more polar solvent. However, the second and rate-determining step of the C-O reductive elimination (k_2, formation of neutral products from charged intermediates) should be inhibited by polar solvents. In contrast, the second step of the C-C coupling reaction (k_3) occurs intramolecularly from the cationic intermediate **A**. Thus, changes in solvent polarity should have a much larger effect on C-C reductive elimination. A similar situation is observed with the addition of small amounts of the conjugate acid, HOR, to the thermolyses. Both C-O and C-C reductive elimination reactions are accelerated in the presence of HOR but the increase in the C-C coupling rate is greater. The added HOR assists the OR⁻ group in leaving the metal (k_1) but by binding to the OR⁻ group also inhibits to some extent the C-O bond-forming step (k_2).

When OR⁻ anion is added to the thermolysis reaction, the rate of the C-O reductive elimination is unaffected. In contrast, the C-C elimination is completely inhibited. This behavior is to be expected based on the mechanisms shown in Scheme 3. The first step (k_1) of the C-O and C-C reductive eliminations is inhibited by OR⁻. For the C-O reductive elimination, this inhibition in the presence of OR⁻ is balanced by an equal acceleration of the second step. In the case of C-C reductive elimination, the second step is unaffected by the added OR⁻ and so an overall inhibition by added OR⁻ is observed. Additional support for the two-step mechanisms involving initial OR⁻ dissociation is that the reductive elimination reactions are significantly affected by modifying the electron-withdrawing ability of the OR⁻ group. More electron-withdrawing groups accelerate both C-O and C-C reductive elimination reactions, albeit to different extents. For example, replacing acetate with trifluoroacetate (dppePtMe$_3$(O$_2$CCF$_3$)), leads to a 12-fold increase in the rate of C-O reductive elimination and a 200-fold increase in the rate of C-C reductive elimination. The C-O reductive elimination experiences a lesser acceleration because unlike the C-C reductive elimination, the second step involves nucleophilic attack (k_2) and as such would be inhibited by the presence of electron-withdrawing groups on OR⁻.

In summary, our results support the formation of a five-coordinate cationic intermediate **A** in both the C-O and C-C coupling reactions from Pt(IV). Thus, circumstances that accelerate formation of this five-coordinate intermediate – polar solvents, acids, and electron-withdrawing groups on OR⁻ – act to increase the rates of both C-O and C-C reductive elimination. Of note is that the successful Pt alkane oxidation systems reported to date do in fact operate in polar, acidic media (*2,9,11*). The increase in rate observed with electron-withdrawing groups on R may at first seem counterintuitive for a mechanism involving a nucleophilic attack by OR⁻(Scheme 1). However, this effect is a direct consequence of the involvement of the five-coordinate intermediate (Scheme 3), and leads to the suggestion that most useful coupling reactions in platinum-catalyzed alkane functionalization processes may be those involving good leaving groups such as trifluoroacetate, sulfonates and bisulfates. Particularly interesting in this regard is that the use of such groups is also attracting attention as a methodology for preventing overoxidation of the alcohol derivative product (*2b,9,12*).

C-H Bond Activation and Related Reactions

The first step of the reaction sequence for platinum-catalyzed alkane functionalization, as shown in Scheme 1, is the activation of the C-H bond by a Pt(II) species (*2*). Studying this reaction directly has been challenging as model Pt(II) complexes which are capable of activating alkane C-H bonds have only recently been prepared (*13*). However, a significant amount of insight into the mechanism of C-H bond activation reactions by Pt(II) centers has been garnered from the study of alkane C-H reductive elimination reactions from Pt(IV) and the application of the principle of microscopic reversibility (*2,4,5,14-16*). Over the past 30 years, considerable mechanistic evidence has been collected on C-C and C-H reductive elimination reactions from Pt(IV) centers. In almost every case (vide infra for the first exception (*5*)), the data have supported dissociation of a ligand from the six-coordinate starting complex to form a five-coordinate Pt(IV) intermediate prior to C-C or C-H bond formation. This general mechanism is shown in Scheme 4. C-C or C-H coupling occurs from the five-coordinate Pt(IV) intermediate to form a σ-C-H alkane species. The observation of isotopic scrambling and inverse kinetic isotope effects in C-H reductive elimination from Pt(IV) has been offered as support for such σ-alkane intermediates in Pt(IV) C-H reductive elimination reactions (*4,14a,c,17,18*). The alkane product is released from the σ-alkane species by either a dissociative or associative ligand substitution to form the final Pt(II) product (*19*). Thus, although it may seem that a very simple coupling of R-R' could take place from

Scheme 4

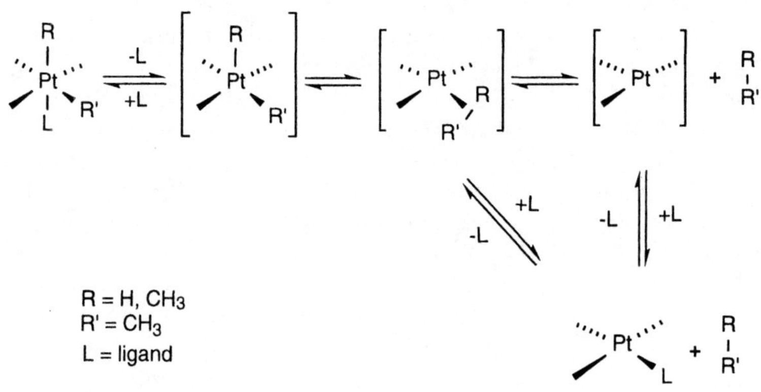

R = H, CH$_3$
R' = CH$_3$
L = ligand

the starting six-coordinate Pt(IV) complex to form the Pt(II) product and alkane, the most common mechanism for this reaction instead involves initial loss of ligand followed by recoordination of ligand at the end of the reaction.

For alkane functionalization, the critical reaction step is clearly not C-H reductive elimination from Pt(IV) but rather C-H activation by Pt(II). Yet, the principle of microscopic reversibility tells us that a C-H oxidative addition pathway will involve the same intermediates and transition states as the reverse reductive elimination reaction (Scheme 4). Thus, a Pt(II) complex expected to undergo facile C-H bond oxidative addition should have at least one weakly bound ligand. Indeed, this appears to be the case, as the examples of Pt(II) complexes exhibiting C-H bond activation reactivity do in fact contain a weakly coordinating ligand (e.g. OTf$^-$, NC$_5$D$_5$, H$_2$O) (*13,20*). Yet most of these examples of model Pt(II) complexes which activate hydrocarbons simply exchange one hydrocarbyl group for another, and only Pt(II) (and not Pt(IV)) species are detected. The five-coordinate Pt(IV) intermediate shown in Scheme 4 would be expected to be highly reactive toward reductive elimination. A good ligand, however, should be able to trap this species to form a six-coordinate Pt(IV) alkyl hydride. The key would be to use a trapping ligand that would not be an effective ligand to Pt(II) so as not to inhibit the initial hydrocarbon coordination and activation steps. Such a strategy is described in the following section.

Investigations With Tp^{Me2} Ligands

As shown in Scheme 5, when the potentially facially coordinating ligand Tp^{Me2} (Tp^{Me2} = hydridotris((3,5-dimethyl)pyrazolyl)borate) is employed, alkanes

Scheme 5

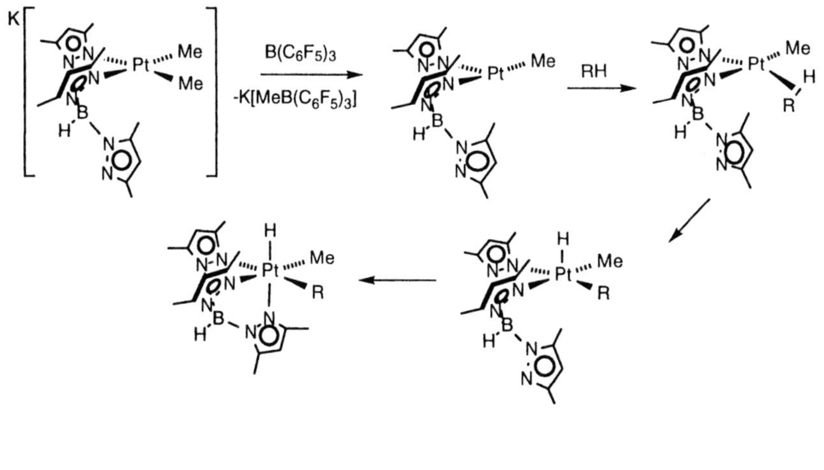

can add to a Pt(II) center to produce stable octahedral Pt(IV) alkyl hydride complexes (*3*). The initially pendent pyrazolyl "arm" coordinates in the Pt(IV) product. As detailed in Scheme 5, it was proposed that methide abstraction from KTp^{Me_2}PtMe$_2$ using B(C$_6$F$_5$)$_3$ generates the open site needed for alkane coordination. Then oxidative addition of RH from the σ-alkane complex intermediate yields a five-coordinate Pt(IV) alkyl hydride which is trapped by the free pyrazolyl arm. However, while the observation of Pt(IV) alkyl hydride products directly demonstrated that alkanes can oxidatively add to Pt(II) centers, support for the proposed mechanism was lacking. Mechanistic studies were hampered by the low solubility of the anionic Pt(II) precursor in hydrocarbon solvents, the requisite bimolecular activation step with B(C$_6$F$_5$)$_3$, and the occurrence of competing side reactions. Instead, detailed information about C-H bond cleavage in this Tp^{Me_2} system was obtained from studies of the microscopic reverse reaction, reductive elimination from Tp^{Me_2}PtMe$_2$H (*4*).

Thermolysis of Tp^{Me_2}PtMe$_2$H in benzene at 110 °C results in reductive elimination of methane and oxidative addition of benzene to form Tp^{Me_2}Pt(Me)(Ph)H. Reductive elimination of a second equivalent of methane then leads to activation of a second equivalent of benzene to yield Tp^{Me_2}PtPh$_2$H. An inverse kinetic isotope effect ($k_H/k_D = 0.8$) for the C-H reductive elimination reaction was observed. It was also demonstrated that thermolysis of Tp^{Me_2}PtMe$_2$D at 60 °C leads to scrambling of deuterium into the methyl positions, without loss of methane. The observations of an inverse kinetic isotope effect and H/D scrambling can be most directly accommodated by the involvement of a σ-CH$_4$ complex intermediate prior to methane elimination (*18*) (Scheme 6).

Scheme 6

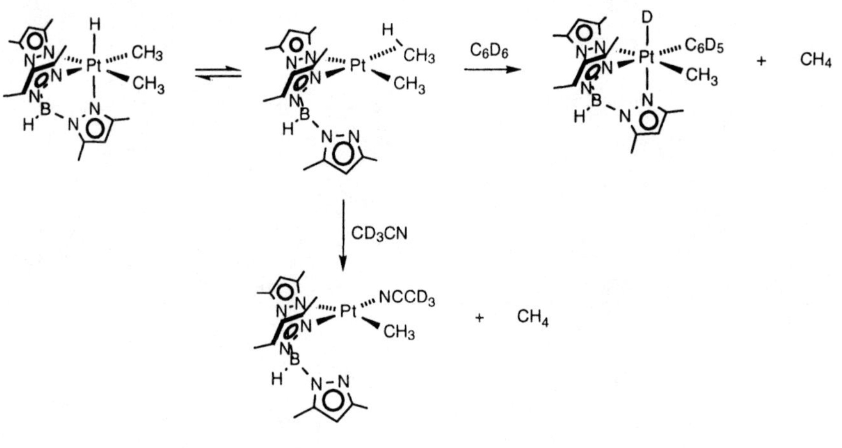

When $Tp^{Me_2}PtMe_2H$ is heated at 110 °C in the presence of acetonitrile, the Pt(II) species resulting from methane loss is trapped (Scheme 6). The rate of formation of $\kappa^2\text{-}Tp^{Me_2}PtMe(CD_3CN)$ is independent of the concentration of acetonitrile and is much slower than the rate of H/D scrambling (see above), indicating that methane loss from the σ-CH_4 complex is dissociative (19,21). Combining this mechanistic information with the activation parameters obtained for both the scrambling and the reductive elimination reactions allows construction of an energy profile of the reaction (Figure 1) and the calculation that the enthalpy of alkane dissociation from the σ-complex must be at least 9 kcal/mol (4).

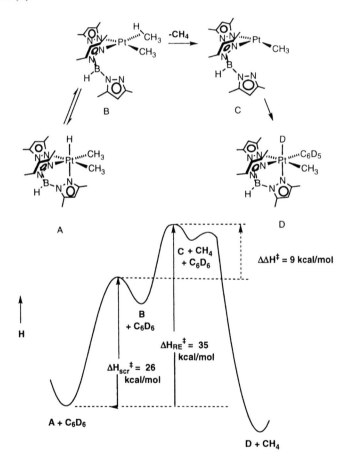

Figure 1. Enthalpy diagram for the reductive elimination of methane from $Tp^{Me_2}PtMe_2H$

This study of C-H bond formation and cleavage from $Tp^{Me2}PtMe_2H$ emphasizes the importance of unsaturated species in oxidative addition/reductive elimination at Pt(II)/Pt(IV). Earlier studies of both C-C and C-H reductive elimination from Pt(IV) have found support for the involvement of five-coordinate Pt(IV) intermediates. Here we find evidence, in a particular Pt system, for the involvement of an unsaturated three-coordinate Pt(II) intermediate (κ^2-$Tp^{Me2}PtMe$) on the reaction pathway (22).

Investigations With 'nacnac' Ligands

The virtually ubiquitous nature of five-coordinate Pt(IV) intermediates in platinum(II)/(IV) bond cleavage and formation reactions led us to search for stable five-coordinate Pt(IV) complexes to model these proposed intermediate species. Bis(o-diisopropylaryl)β-diketiminate (nacnac) was employed to synthesize the first stable five-coordinate Pt(IV) alkyl complex (Scheme 7) (23). Our strategy was to use a chelating anionic ligand to produce a charge neutral complex. In this way, we would avoid coordination of a counterion in the sixth position, as had been previously observed (24). In addition, bulky groups on the aryl rings were incorporated to shield the open site to some extent and also to allow solubility in alkanes. The new compound was fully characterized and was shown to have a square pyramidal structure in the solid state by X-ray crystallography (23).

Scheme 7

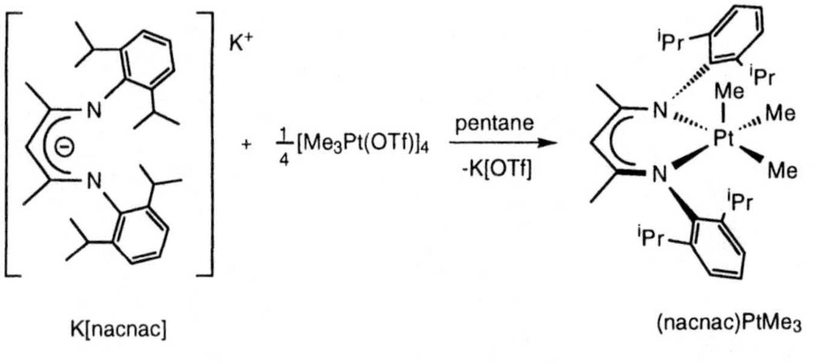

K[nacnac] + ¼[Me₃Pt(OTf)]₄ →(pentane, -K[OTf]) (nacnac)PtMe₃

The next question was whether this compound would serve as a model for the intermediates in reductive elimination from Pt(IV) and actually undergo C-C coupling. Indeed, reductive elimination of ethane from (nacnac)PtMe₃ was observed at elevated temperature in hydrocarbon solvent (25). Methane and a Pt(II) product, a cyclometallated olefin hydride (**X** in Scheme 8), were also produced in this reaction. The formation of these products is readily explained in the proposed mechanism shown in Scheme 8. Reductive elimination of ethane from (nacnac)PtMe$_3$ leads to a three-coordinate Pt(II) species (*26*). As discussed above, this unsaturated intermediate should readily activate C-H bonds. The proximity of the isopropyl groups results in intramolecular activation to produce a five-coordinate cyclometallated Pt(IV) methyl hydride complex. The five-coordinate Pt(IV) methyl hydride intermediate undergoes facile reductive elimination of methane. β-Hydrogen elimination from the cyclometalated group then leads to the Pt(II) olefin hydride complex **X**. Intermolecular C-H bond activation of benzene and alkane solvents by the Pt(II) olefin hydride complex **X** was also demonstrated. Deuterium incorporation into all of the isopropyl positions, the isopropenyl group, and the hydride position formed **X-*d*$_{27}$** when **X** was heated in benzene-*d*$_6$ or pentane-*d*$_{12}$. Deuterium labeling studies showed an 85% selectivity for the labeling of the terminal positions of pentane. The mechanism suggested for this process is that reversible β-hydride elimination allows olefin insertion from **X** to generate a three-coordinate Pt(II) species which undergoes intermolecular C-H activation (*25*).

Scheme 8

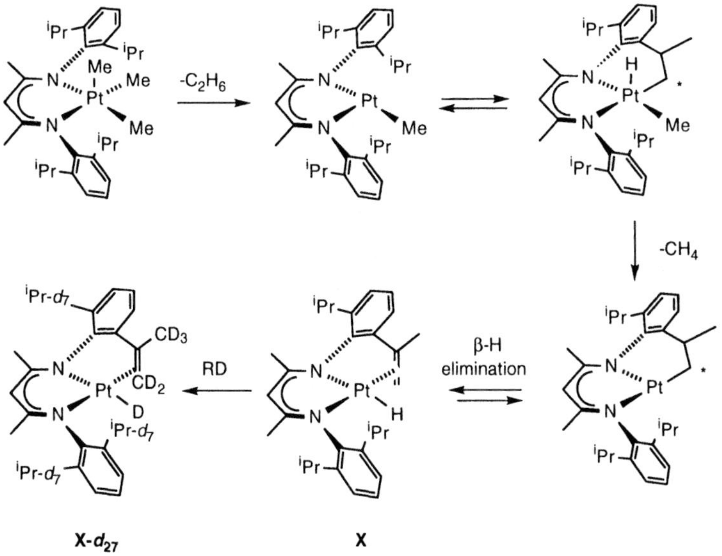

* activation occurs at both methyl and methine positions of ^{i}Pr residue

This work involving the preparation, characterization and reactivity studies of the first stable five-coordinate Pt(IV) alkyl complex has provided further insight into C-H activation chemistry at Pt(II). Support has again been put forth for the involvement of unsaturated five-coordinate Pt(IV) and three-coordinate Pt(II) species in C-H activation reactions. A unique and successful approach to generating the requisite site for alkane coordination (and activation) at Pt(II) via olefin insertion was introduced.

Investigations With Phosphine Ligands

Although we have detailed the importance of five-coordinate Pt(IV) intermediates in reductive elimination reactions from Pt(IV), we have also found that one must exercise caution in presuming that all such reactions involve five-coordinate Pt(IV) intermediates. We have recently presented the first evidence for a C-H reductive elimination from octahedral Pt(IV) that proceeds directly without ligand dissociation (5). The novel phosphine-ligated Pt(IV) alkyl hydride complexes, L_2PtMe_3H (L_2 = dppe (bis(diphenylphosphino)ethane), dppbz (bis(o-diphenylphosphino)benzene)) were prepared and characterized. These complexes undergo C-H reductive elimination to form methane at room temperature (Scheme 9). The complexes, which bear chelating phosphine ligands of very different rigidity, react at virtually the same rate (at 50 °C, k_{obs} = 1.3(1) x 10^{-4} s^{-1} and 1.2(1) x 10^{-4} s^{-1} for the dppe and dppbz derivatives, respectively). The activation parameters are also very similar. No H/D scrambling into the methyl groups or the methane is observed when the complexes L_2PtMe_3D are subjected to thermolysis, and normal kinetic isotope effects are observed for the reactions of both complexes. The lack of any difference between the reaction rates for the dppe derivative versus the more rigid dppbz complex is not consistent with a mechanistic pathway involving phosphine predissociation. Instead, reductive elimination directly from a six-coordinate Pt(IV) complex is the favored mechanism.

In contrast, the mechanism for C-C reductive elimination from the analogous dppe complex, (dppe)$PtMe_4$ is consistent with a phosphine dissociation pathway to form a five-coordinate intermediate prior to C-C coupling. The dppbz analog, dppbz$PtMe_4$ eliminated methane at 165 °C with a rate at least two orders of magnitude slower than the dppe complex. It was also shown that phosphine chelate opening is kinetically viable at high temperatures. When the related complex dppePtMe$_3$Et is heated at 165 °C, β-hydride elimination occurs to produce ethylene. This indicates that an open site, generated by phosphine arm dissociation, must be accessible.

These studies of C-C and C-H reductive elimination from Pt(IV) mark the first time that Me-Me and Me-H coupling reactions from the same metal–ligand environment have been directly compared (5). To achieve comparable rates of reaction, a temperature difference of almost 150 °C was required. Perhaps even

Scheme 9

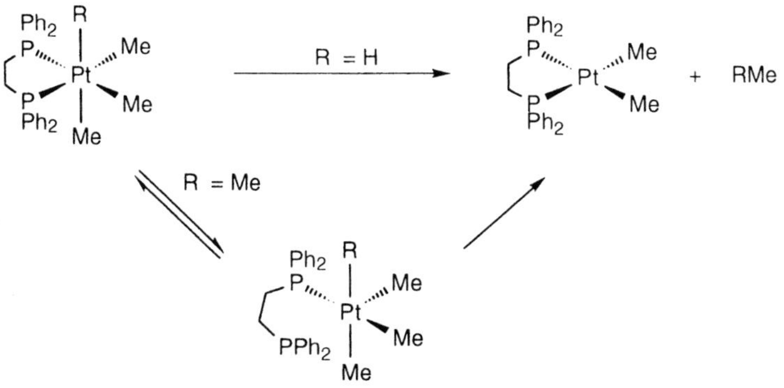

[Scheme 9: Pt complex with Ph2P chelating ligand, R, and two Me groups; R = H pathway gives cis-(Ph2P)2Pt(Me)2 + RMe; R = Me pathway goes through five-coordinate intermediate with dissociated PPh2 arm bearing R, two Me, and Me groups.]

more significant, however, is that the C-C and C-H reductive eliminations proceed via different mechanisms. C-H reductive elimination proceeds via a direct pathway and C-C reductive elimination proceeds via ligand dissociation and a five-coordinate intermediate (Scheme 9). The observation of these distinct mechanisms in such closely related systems emphasizes the need for caution when making analogies between C-C and C-H bond-forming reactions.

Oxidation by O_2

Activation of C-H bonds is necessary, but not sufficient for alkane functionalization. Oxidation of the platinum alkyl complex must precede elimination of the functionalized product. In the original Shilov system, Pt(IV) is consumed as a stoichiometric oxidant (Scheme 1, step II). More recently, systems which use Cl_2, SO_3, and H_2O_2 as oxidants have been explored (9,11a,b). The use of O_2 and co-catalysts like $CuCl_2$ or heteropolyacids has also been reported (11c,d). Despite some progress with these alternative oxidants, a commercially viable system for selective alkane functionalization has yet to emerge. When one considers economic, environmental, and availability issues, the best oxidant for this reaction would certainly be molecular oxygen from air. We have been investigating the concept of using oxygen directly (without a co-catalyst) for this alkane functionalization and have been studying the reaction chemistry between platinum alkyl complexes and O_2.

Alkane activation by Pt(II) species can produce Pt(II) alkyl or Pt(IV) alkyl hydride products. Thus, the reaction of each type of Pt alkyl product with O_2 merits investigation. Several years ago, together with Bercaw, Labinger and co-workers, our group reported that L_2PtMe_2 complexes (L_2 = tmeda, bpy, phen) react with oxygen in methanol to form Pt(IV) methyl hydroxide methoxide complexes (Scheme 10) (7). This reaction is similar to the oxidation step in the Shilov system in that a Pt(II) alkyl is directly oxidized by the oxidant, in this case dioxygen, to a Pt(IV) species. Bercaw, Labinger and co-workers have recently investigated the mechanism of this reaction with O_2 and have identified a Pt(IV) hydroperoxide intermediate that reacts with starting material to ultimately produce two equivalents of Pt(IV) hydroxide product (27).

Scheme 10

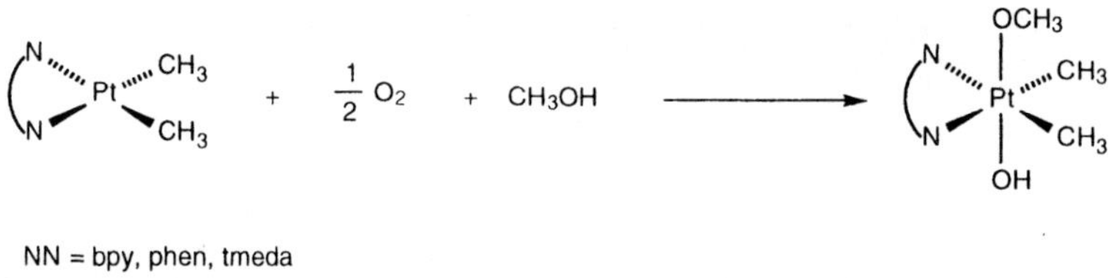

NN = bpy, phen, tmeda

Another potentially useful reaction with dioxygen that we recently discovered is shown in Scheme 11. Formal insertion of dioxygen into the Pt-H bond of $Tp^{Me_2}PtMe_2H$ cleanly generates the hydroperoxide complex, $Tp^{Me_2}PtMe_2OOH$ (8). This is exciting because oxidative addition of C-H bonds has been observed with this same $Tp^{Me_2}Pt$ system (see above) (3,4). Now two of the three desired steps – C-H bond activation and utilization of O_2 – have been observed within the same type of ligand environment.

Scheme 11

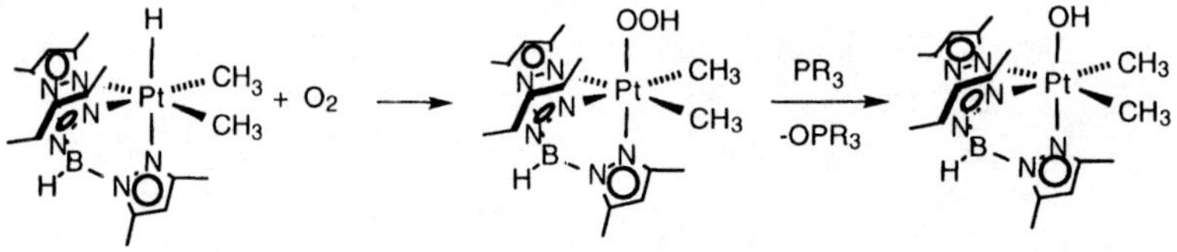

The rate of the reaction of $Tp^{Me_2}PtMe_2H$ with O_2 increases in the presence of light or AIBN (azobisisobutyronitrile, a radical initiator) and decreases in the presence of radical inhibitors (e.g. 1,4-cyclohexadiene). These observations suggest a mechanism involving radical species. A mechanism analogous to the well-studied radical chain pathway for autoxidation of hydrocarbons (28) was inititally proposed (Scheme 12). The initiation step generates a radical, which abstracts the hydrogen atom attached to platinum. The Pt-based radical reacts with oxygen to form a platinum peroxyl radical. This species abstracts a hydrogen atom from another molecule of starting material to form the hydroperoxide complex and the chain-carrying Pt-based radical. Eventually, two radicals combine and terminate the chain.

Scheme 12

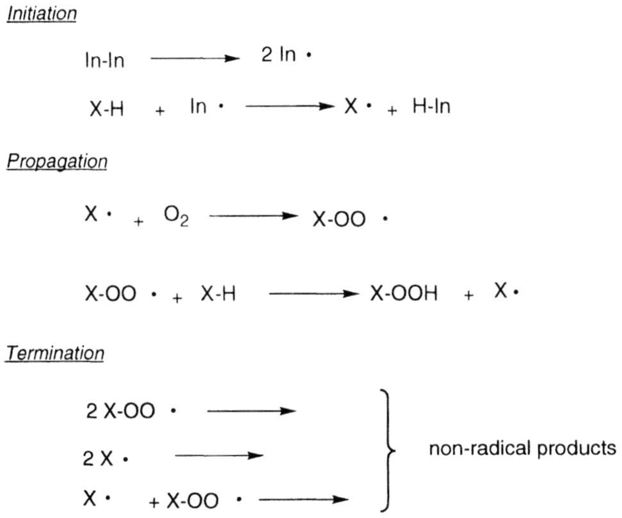

Initiation

$$In\text{-}In \longrightarrow 2\ In\cdot$$

$$X\text{-}H + In\cdot \longrightarrow X\cdot + H\text{-}In$$

Propagation

$$X\cdot + O_2 \longrightarrow X\text{-}OO\cdot$$

$$X\text{-}OO\cdot + X\text{-}H \longrightarrow X\text{-}OOH + X\cdot$$

Termination

$$\left.\begin{array}{l} 2\ X\text{-}OO\cdot \longrightarrow \\ 2\ X\cdot \longrightarrow \\ X\cdot + X\text{-}OO\cdot \longrightarrow \end{array}\right\} \text{non-radical products}$$

X-H = oxidizable hydrocarbon or $Tp^{Me2}PtMe_2H$

Consistent with the proposal of a Pt-based radical intermediate, radical-based substitution at Pt(IV) can be observed under photolytic conditions. When $Tp^{Me2}PtMe_2H$ is dissolved in CCl_4 and exposed to light, $Tp^{Me2}PtMe_2Cl$ is formed cleanly and quantitatively (29). Furthermore, benzyl radicals generated in the presence of $Tp^{Me2}PtMe_2H$ abstract a hydrogen atom to form toluene (30). At sufficient O_2 pressure, the rate of oxidation is independent of O_2 concentration. The kinetic behavior of the reaction is also consistent with autocatalysis by the product $Tp^{Me2}PtMe_2OOH$ (29). Similar observations have also been noted in the autoxidation of hydrocarbons but a detailed kinetic analysis of the reaction of $Tp^{Me2}PtMe_2H$ and O_2 indicates that a more complex mechanism than that presented in Scheme 12 may be operating (29).

When the Pt(IV) hydroperoxide species $Tp^{Me2}PtMe_2OOH$ is allowed to react with a good oxygen atom acceptor such as trimethylphosphine, the Pt(IV) hydroxide complex $Tp^{Me2}PtMe_2OH$ is formed (Scheme 11). $Tp^{Me2}PtMe_2OH$ is also obtained by thermolysis of the hydroperoxide complex. Note that if carbon-oxygen reductive elimination were to take place from the Pt(IV) hydroxide complex $Tp^{Me_2}PtMe_2OH$, the product would be methanol. Since C-H activation and oxygen incorporation, the first two steps of the desired alkane functionalization cycle, were observed with this same ligand set, this is an exciting possibility as a methane to methanol transformation can be envisaged. However, such C-O bond formation is not observed to take place from $Tp^{Me_2}PtMe_2OH$. Yet, our earlier work on C-O reductive elimination from Pt(IV) (6) may shed some light on why this is the case. Note that if the oxygen group were to dissociate, it would not be possible to attack a methyl group that is trans to an open site. Thus, although the facially coordinating Tp^{Me_2} ligand supports the first two steps of C-H activation and O_2 consumption, the C-O reductive elimination step is problematic in this environment.

Summary

Key reaction steps for platinum systems capable of oxidatively functionalizing alkanes have been examined. We have found Pt(II) complexes which can oxidatively add alkane C-H bonds and studied the mechanism of this reaction via investigation of C-H reductive elimination from Pt(IV). Reactions of Pt(II) alkyl and Pt(IV) alkyl hydride complexes with dioxygen have also been discovered. Finally, the mechanism of C-O reductive elimination reaction from Pt(IV) has been investigated. These studies have emphasized the importance of unsaturated intermediates in both bond-forming and bond-cleavage reactions at platinum.

Coordination of alkane to form a Pt(II) σ−complex appears to be an important prerequisite for hydrocarbon activation. This σ−complex can either be formed by the reaction of alkane with a three-coordinate Pt(II) species or by an associative substitution reaction at a four-coordinate Pt(II) complex. Oxidative addition of the C-H bond then occurs from the σ-alkane species to form a five-coordinate Pt(IV) alkyl hydride complex. The five-coordinate intermediate can be trapped by ligand to generate a stable Pt(IV) alkyl hydride. Alternatively, deprotonation of the five-coordinate Pt(IV) alkyl hydride would form a Pt(II) alkyl.

Examples of both Pt(II) alkyls and Pt(IV) alkyl hydrides have been shown to react directly with dioxygen to yield Pt(IV) complexes. In the case of Pt(II) dimethyl complexes, Pt(IV) dimethyl methoxide hydroxide complexes are formed in methanol solution. With a Pt(IV) dimethyl hydride complex, we found that the reaction with O_2 leads to a Pt(IV) dimethyl hydroperoxide complex which can subsequently be converted to a Pt(IV) dimethyl hydroxide species. Finally the reductive elimination of C-O bonds from Pt(IV) occurs via a two-step process. Dissociation of the oxygen group is followed by nucleophilic attack at a five-coordinate Pt(IV) alkyl to form the C-O coupled product.

It is anticipated that a detailed understanding of these important reactions at Pt(II)/Pt(IV) centers will allow the design of a ligand capable of stabilizing the required intermediates such that a complete cycle for efficient alkane functionalization can be realized.

Acknowledgements

We gratefully acknowledge the graduate students, postdoctoral research associates, visiting professors and undergraduate researchers in our laboratory who have contributed to this research project: Scott Williams, Dawn Crumpton-Bregel, April Getty, Nicole Smythe, Andy Pawlikowski, Douglas Wick, Mike Jensen, William Dasher, Tami Lasseter, Andy Holland and Darcie Porter. We also thank Professor Joe Templeton and Stefan Reinartz (University of North Carolina) for an enjoyable and productive collaboration (reference 4). Finally, we thank the National Science Foundation and the donors of the Petroleum Research Fund, administered by the American Chemical Society, for funding of this work.

References

1. (a) Gol'dshleger, N. F.; Es'kova, V. V.; Shilov, A. E.; Shteinman, A. A. *Russ. J. Phys. Chem.* **1972**, *46*, 785. (b) Shilov, A. E.; Shteinman, A. A. *Coord. Chem. Rev.* **1977**, *24*, 97. (c) Kushch, L. A.; Lavrushko, V. V.; Misharin, Y. S.; Moravsky, A. P.; Shilov, A. E. *Nouv. J. Chim.* **1983**, *7*, 729.
2. Recent reviews: (a) Shilov, A. E.; Shul'pin, G. B. *Chem. Rev.* **1997**, *97*, 2879. (b) Stahl, S. S.; Labinger, J. A.; Bercaw, J. E. *Angew. Chem. Int. Ed.* **1998**, *37*, 2180. (c) Shilov, A. E.; Shul'pin, G. B. *Activation and Catalytic Reactions of Saturated Hydrocarbons in the Presence of Metal Complexes*; Kluwer: Boston, 2000. (d) Fekl, U.; Goldberg, K. I. *Adv. Inorg. Chem.* **2003**, *54*, 259.
3. Wick, D. D.; Goldberg, K. I. *J. Am. Chem. Soc.* **1997**, *119*, 10235.
4. Jensen, M. P.; Wick, D. D.; Reinartz, S.; White, P. S.; Templeton, J. L.; Goldberg, K. I.; *J. Am. Chem. Soc.* **2003**, *125*, 8614.
5. Crumpton-Bregel, D. M.; Goldberg, K. I. *J. Am. Chem. Soc.* **2003**, *125*, 9442.

6. (a) Williams, B. S.; Holland, A. W.; Goldberg, K. I. *J. Am. Chem. Soc.* **1999**, *121*, 252. (b) Williams, B. S.; Goldberg, K. I. *J. Am. Chem. Soc.* **2001**, *123*, 2576.
7. Rostovtsev, V. V.; Labinger, J. A.; Bercaw, J. E.; Lasseter, T. L.; Goldberg, K. I. *Organometallics* **1998**, *17*, 4530.
8. Wick, D. D.; Goldberg, K. I. *J. Am. Chem. Soc.* **1999**, *121*, 11900.
9. Periana, R. A.; Taube, D. J.; Gamble, S.; Taube, H.; Satoh, T.; Fuji, H. *Science* **1998**, *280*, 560.
10. (a) Goldberg, K. I.; Yan, J. Y.; Breitung, E. M. *J. Am. Chem. Soc.*, **1995**, *117*, 6889. (b) Goldberg, K. I.; Yan, J. Y.; Winter, E. L. *J. Am. Chem. Soc.* **1994**, *116*, 1573.
11. (a) Horváth, I. T.; Cook, R. A.; Millar, J. M.; Kiss, G. *Organometallics* **1993**, *12*, 8. (b) DeVries, N.; Roe, D. C.; Thorn, D. L. *J. Mol. Catal. A* **2002**, *189*, 17. (c) Lin, M.; Shen, C.; Garcia-Zayas, E. A.; Sen, A. *J. Am. Chem. Soc.* **2001**, *123*, 1000. (d) Geletii, Y. V.; Shilov, A. E. *Kinet. Catal.* **1983**, *24*, 413.
12. Sen, A. *Acc. Chem. Res.* **1998**, 31, 550.
13. (a) Holtcamp, M. W.; Labinger, J. A.; Bercaw, J. E. *J. Am. Chem. Soc.* **1997**, *119*, 848. (b) Wick, D. D.; Goldberg, K. I. *J. Am. Chem. Soc.* **1997**, *119*, 10235. (c) Holtcamp, M. W.; Henling, L. M.; Day, M. W.; Labinger, J. A.; Bercaw, J. E. *Inorg. Chim. Acta* **1998**, *270*, 467. (d) Johansson, L.; Ryan, O. B.; Tilset, M. J. *J. Am. Chem. Soc.* **1999**, *121*, 1974. (e) Heiberg, H.; Johansson, L.; Gropen, O.; Ryan, O. B.; Swang, O.; Tilset, M. *J. Am. Chem. Soc.* **2000**, *122*, 10831. (f) Johansson, L.; Tilset, M.; Labinger, J. A.; Bercaw, J. E. *J. Am. Chem. Soc.* **2000**, *122*, 10846. (g) Johansson, L.; Ryan, O. B.; Rømming, C.; Tilset, M. *J. Am. Chem. Soc.* **2001**, *123*, 6579. (h) Zhong, H. A.; Labinger, J. A.; Bercaw, J. E. *J. Am. Chem. Soc.* **2002**, *124*, 1378.
14. (a) Stahl, S. S.; Labinger, J. A.; Bercaw, J. E. *J. Am. Chem. Soc.* **1996**, *118*, 5961. (b) Jenkins, H. A.; Yap, G. P. A.; Puddephatt, R. J. *Organometallics* **1997**, *16*, 1946. (c) Fekl, U.; Zahl, A.; van Eldik, R. *Organometallics* **1999**, *18*, 4156. (d) Reinartz, S.; White, P. S.; Brookhart, M.; Templeton, J. L. *Organometallics* **2000**, *19*, 3854. (e) Reinartz, S.; Baik, M.-H.; White, P. S.; Brookhart, M.; Templeton, J. L. *Inorg. Chem.* **2001**, *40*, 4726. (f) Norris, C.M.; Reinartz, S.; White, P. S.; Templeton, J. L. *Organometallics* **2002**, *21*, 5649. (g) Prokopchuk, E. M.; Puddephatt, R. J. *Organometallics* **2003**, *22*, 787. (h) Reinartz, S.; White, P. S.; Brookhart, M.; Templeton, J. L. *J. Am. Chem. Soc.* **2001**, *123*, 12724. (i) Reinartz, S.; White, P. S.; Brookhart, M.; Templeton, J. L. *Organometallics* **2001**, *20*, 1709. (j) Prokopchuk, E. M.; Puddephatt, R. J. *Organometallics* **2003**, *22*, 563.

15. (a) Hill, G. S.; Rendina, L. M.; Puddephatt, R. J. *Organometallics* **1995**, *14*, 4966. (b) Hinman, J. G.; Baar, C. R.; Jennings, M. C.; Puddephatt, R. J. *Organometallics* **2000**, *19*, 563. (c) Johansson, L.; Tilset, M. *J. Am. Chem. Soc.* **2001**, *123*, 739. (d) Wik, B. J.; Lersch, M.; Tilset, M. *J. Am. Chem. Soc.* **2002**, *124*, 12116.
16. Puddephatt, R. J. *Coord. Chem. Rev.* **2001**, *219-221*, 157.
17. (a) Lo, H. C.; Haskel, A.; Kapon, M.; Keinan, E. *J. Am. Chem. Soc.* **2002**, *124*, 3226. (b) Iron, M. A.; Lo, H. C.; Martin, J. M. L.; Keinan, E. *J. Am. Chem. Soc.* **2002**, *124*, 7041.
18. The interpretation of kinetic isotope effects reported in ref 17 has been reexamined. (a) Churchill, D.G.; Janak, K. E.; Wittenberg, J.S.; Parkin, G. *J. Am. Chem. Soc.* **2003**, *125*, 1403. (b) Jones, W. D. *Acc. Chem. Res.* **2003**, *36*, 140.
19. In contrast to the results of our study (*4*), substitutions at Pt(II) are generally found to proceed via associative mechanisms. For evidence of associative substitution mechanisms involving hydrocarbons as ligands at Pt(II) centers, see: (a) Johansson, L.; Tilset, M. *J. Am. Chem. Soc.* **2001**, *123*, 739. (b) Procelewska, J.; Zahl. A.; van Eldik, R.; Zhong, H. A.; Labinger, J. A.; Bercaw, J. E. *Inorg. Chem.* **2002**, *41*, 2808.
20. (a) Brainard, R. L.; Nutt, W. R.; Lee, T. R.; Whitesides, G. M. *Organometallics* **1988**, *7*, 2379. (b) Thomas, J. C.; Peters, J. C. *J. Am. Chem. Soc.* **2001**, *123*, 5100. (c) Harkins, S. B.; Peters, J. C. *Organometallics* **2002**, 21, 1753. (d) Konze, W. V.; Scott, B. L.; Kubas, G. J. *J. Am. Chem. Soc.* **2002**, *124*, 12550.
21. It is possible that the uncoordinated pyrazolyl ring provides anchimeric assistance in the loss of methane from κ_2-Tp^{Me2}PtMe(CH$_4$). Such an interaction would not be detected in our kinetic experiments.
22. The "uncoordinated" pyrazolyl ring may interact with the Pt center to provide some additional stabilization to the "three-coordinate" Pt(II) species.
23. (a) Fekl, U.; Kaminsky, W.; Goldberg, K. I. *J. Am. Chem. Soc.* **2001**, *123*, 6423. (b) Simultaneous discovery of a five-coordinate Pt(IV) dihydrido silyl complex: Reinartz, S.; White, P. S.; Brookhart, M.; Templeton, J. L. *J. Am. Chem. Soc.* **2001**, *123*, 6425.
24. Counterion coordination to cationic complexes has been prohibitive for the synthesis of five-coordinate cationic Pt(IV) alkyls. Hill, G. S.; Yap, G. P. A.; Puddephatt, R. J. *Organometallics* **1999**, *18*, 1408.
25. Fekl, U.; Goldberg, K. I. *J. Am. Chem. Soc.* **2002**, *124*, 6804.
26. Three-coordinate intermediates pictured in Scheme 8 may be stabilized by agostic donations from ligand C-H bonds to the open site.

27. Rostovtsev, V. V.; Henling, L. M.; Labinger, J. A.; Bercaw, J. E. *Inorg. Chem.* **2002**, *41*, 3608.
28. (a) Foote, C. S.; Valentine, J. S.; Greenberg, A.; Liebman, J. F.; Editors *Active Oxygen in Chemistry. [In: Struct. Energ. React. Chem. Ser., 1995; 2]*, **1995**. (b) Kochi, J. K.; Editor *Free Radicals, Vol. 2*, **1973**. (c) Howard, J. A. *Advan. Free-Radical Chem.* **1972**, *4*, 49-173.
29. Look, J. L.; Goldberg, K. I., manuscript in preparation.
30. (a) Kolwaite, D.S.; Franz, J.A. unpublished results. (b) methodolgy similar to Franz, J. A. L., J.C.; Birnbaum, J.C.; Hicks, K. W.; Alnajjar, M.S. *J. Am. Chem. Soc.* **1999**, *121*, 9824-9830.

Chapter 18

Hydrocarbon C–H Bond Activation with Tp'Pt Complexes

Cynthia M. Norris and Joseph L. Templeton*

Department of Chemistry, University of North Carolina, Chapel Hill, NC 27599-3290

Stabilization of platinum complexes resembling key intermediates in C-H bond activation has been achieved through the use of the strongly electron-donating hydridotris(3,5-dimethylpyrazolyl)borate (Tp') ligand. Isolation of stable alkyl(hydrido)platinum(IV) complexes, Tp'PtMe$_2$H and Tp'PtMeH$_2$, allowed us to investigate the mechanism of reductive alkane elimination from these complexes via (1) thermolysis, (2) Lewis acid addition, and (3) low temperature protonation. By replacing the alkyl group with SiR$_3$ and C$_6$R$_5$ in the low temperature protonation of Tp'PtRH$_2$, isolation of five-coordinate Pt(IV) complexes and Pt(II) η^2-arene adducts, respectively, was achieved. These Pt(II)/Pt(IV) interconversions provide insight into the mechanisms by which these reagents activate strong C-H bonds and also provide a foundation for future plans to functionalize hydrocarbons.

Organometallic reagents offer promising routes toward selective functionalization of alkanes (*1-15*). A key discovery by Shilov and co-workers in 1972 was platinum(II) catalyzed oxidation of methane to methanol (*11,16*). This discovery sparked extensive exploration of C-H bond activation chemistry at platinum centers. Although the Shilov oxidation remains one of the most well known examples of homogeneous C-H activation, this system consumes stoichiometric platinum(IV) as the oxidant, an economically and environmentally unacceptable restriction. Advancements have been made toward replacement of platinum(IV) by less expensive oxidants, such as SO_3 (*8*), O_2 with copper salts (*17*), or heteropoly acids (*18*), however, an economical process that catalyzes methane oxidation to methanol has yet to be reported. The mechanism of the Shilov oxidation has been extensively studied by Bercaw, Labinger, and co-workers in hopes of gaining insight to guide the design of catalysts capable of C-H activation and functionalization (*7,15,19-23*).

Most of our current mechanistic understanding comes from investigations of the microscopic reverse of alkane activation, namely reductive alkane elimination from platinum(IV) alkyl hydride complexes. The proposed mechanism for C-H activation involves three steps: (1) electrophilic activation of an alkane C-H bond by Pt(II) to form a platinum(II)-alkyl species; (2) oxidation to a platinum(IV)-alkyl complex; (3) reductive elimination of R-X (X= Cl or OH) from a five-coordinate intermediate to release the functionalized alkane and to regenerate the Pt(II) catalyst (*7,11,19*). Dissection of the Shilov reaction into individual reaction steps has provided detailed mechanistic information. We summarize here our advancements with Tp'Pt reagents that are relevant to the ultimate goal of selective hydrocarbon functionalization.

Stable Platinum(IV) Akyl Hydride Complexes

Platinum(IV) alkyl hydride complexes, often proposed as intermediates in alkane C-H activation by Pt(II) complexes (*23,24*), remained elusive until the mid-1990's (*25*). Diimine (*26,27*), diamine (*28*), and trialkylphosphine (*21*) alkyl(hydrido)platinum(IV) complexes readily reductively eliminate alkane at or below room temperature. Alkane elimination in such systems has been proposed to occur through a five-coordinate intermediate generated by loss of a labile ligand. Stabilization of the octahedral geometry favored for the platinum(IV) d^6 configuration through the use of strong electron donating, *fac-*

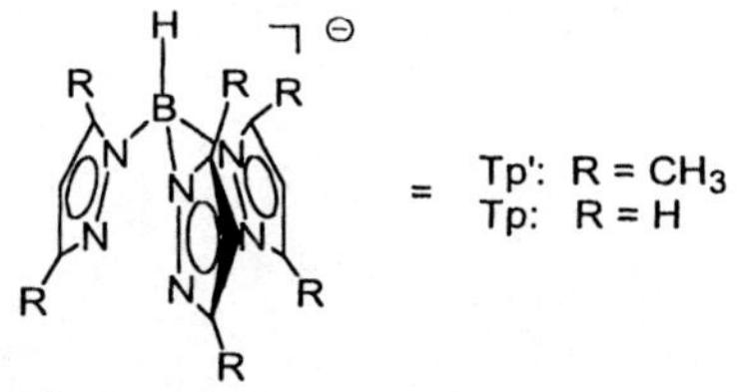

Tp': R = CH_3
Tp: R = H

Figure 1: *Structure of hydridotris(3,5-dimethylpyrazolyl)borate (Tp') and tris(pyrazolyl)borate (Tp)*

tridentate pyrazolylborate (29) ligands (Figure 1) allowed the isolation of stable alkyl(hydrido)platinum(IV) complexes (30,31).

These platinum(IV) complexes can be synthesized by protonation of the anionic intermediates, [K][Tp'PtMe$_2$] and [K][TpPtMe$_2$] [Tp' = hydridotris(3,5-dimethylpyrazolyl)borate and Tp = tris(pyrazolyl)borate] (29), with weak acids, including ammonium chloride and acetic acid, to give Tp'PtMe$_2$H (**1**) (30) and TpPtMe$_2$H (**2**) (31), respectively (eq 1). Protonation of the anionic Pt(II) precursor [K][Tp'PtMe$_2$] was shown to occur exclusively at metal to form the dimethylhydrido complex **1** suggesting a greater basicity at platinum than at either a pyrazole nitrogen or a platinum-hydrocarbon bond (30). Further support for the thermodynamic basicity of the Pt(II) center comes from the reaction of cis-PtMe$_2$(SMe$_2$)$_2$ with the N-protonated Tp' ligand (HTp') to give complex **1** instead of the platinum(II) product [(HTp')PtMe$_2$]. The first example of alkane activation at a neutral Pt(II) center to form a stable platinum(IV) alkyl hydride complexes was demonstrated by Goldberg and Wick (32). They showed that addition of the Lewis acid, B(C$_6$F$_5$)$_3$ (33), to the anionic Pt(II) precursor, [K][Tp'PtMe$_2$], resulted in methide abstraction and C-H activation of the hydrocarbon solvent to give alkyl hydride complex Tp'PtMe(R)H (32).

The analogous Tp'PtMeH$_2$ (**3**) (34) complex was isolated from the reaction of Tp'Pt(Me)(CO), which exists in solution in both square-planar and trigonal-bipyramidal geometries (35), with water in a basic acetone/water mixture (eq 2). This procedure was a slightly modified form of the synthesis of TpPtMeH$_2$ from TpPtMe(CO) via reaction with water that was reported by Keinan (36). Assuming that the reaction involves nucleophilic attack by water on the CO ligand added base should accelerate the reaction. Indeed in the absence of base the reaction was substantially slower. The final complex in the Tp'Pt(IV) alkyl hydride series Tp'PtMe$_{3-n}$H$_n$ with n = 0,1,2,3, the trihydride complex Tp'PtH$_3$ (**4**), was prepared by refluxing Tp'Pt(SiEt$_3$)(H)$_2$ in a mixture of methanol and methylene chloride (10:1) (37). The importance of alkyl(hydrido)platinum(IV) complexes as potential intermediates in alkane C-H activation by Pt(II) complexes prompted the study of alkane reductive elimination, the microscopic reverse, from these complexes.

Reductive Methane Elimination from Tp'PtMe$_2$H via Thermolysis

Thermolysis of the dimethyl hydride complex **1** in C$_6$D$_6$ induces reductive elimination of methane and activation of solvent to give Tp'Pt(CH$_3$)(C$_6$D$_5$)D, which then reductively eliminates CH$_3$D and activates a second solvent molecule to give the final Tp'Pt(C$_6$D$_5$)$_2$D (**5-d**) product (*38*). The presence of methane isotopomers (CH$_3$D, CH$_2$D$_2$, and CHD$_3$) was evident in the ^{1}H NMR spectrum, and these were attributed to isotopic scrambling through a putative σ-bound methane intermediate, Pt^{II}(CH$_3$-H). Similar σ-alkane intermediates have been proposed to explain isotopic scrambling in late metal alkyl hydride complexes (*20,28,39-51*). To further support the presence of a σ-methane intermediate, thermolysis of Tp'PtMe$_2$D (**1-d**) in toluene-d_8 at temperatures below those required for methane loss was monitored for several hours (*38*). As expected, deuterium incorporation to form the stronger C-D bond and appearance of a Pt-H ^{1}H NMR signal was observed. This isotopic scrambling most likely proceeds by thermally induced dissociation of the pyrazole arm trans to the hydride ligand to give a five-coordinate intermediate, followed by reductive elimination of methane to give a σ-bound alkane intermediate; subsequent C-H bond activation leaves deuterium in the platinum bound methyl group (eq 3). Similar isotopic scrambling in the related TpPtMe(H)$_2$ complex was observed by Keinan and co-workers without elimination of methane (*52-54*).

The study of reductive methane elimination from complex **1** has provided insight into the mechanism of the microscopic reverse process, namely alkane C-H oxidative addition. The temperature dependence of this reaction was monitored by ^{1}H NMR to give activation parameters for the first reductive methane elimination of $\Delta H^{\ddagger} = 35.0 \pm 1.1$ kcal/mol and $\Delta S^{\ddagger} = 13 \pm 3$ cal/(mol K). Rates of reductive methane elimination from **1** are independent of solvent and added ligand, suggesting a dissociative mechanism for the loss of methane to give a reactive three-coordinate intermediate which reacts rapidly with solvent (eq 4) (*38*).

Reductive Methane Elimination from Tp'PtMe$_2$H via Protonation

Protonation of Tp'PtMe$_2$H (**1**) induces reductive methane elimination at low temperatures, and thus is complementary to methane loss promoted by thermolysis above 100 °C (vide supra) (*35*). The use of [H(OEt$_2$)$_2$][BAr'$_4$] [BAr'$_4$ = tetrakis(3,5-trifluoromethylphenyl)borate] (*55*), a strong acid with a weakly coordinating anion, proved to be beneficial in studying Pt(II)/Pt(IV) interconversions. Complex **1** contains three distinct potential protonation sites: the platinum metal center, a nitrogen atom of a pyrazole ring, or a platinum-alkyl bond. To address the question of protonation site in the case of Tp'Pt(IV) complexes, complex **1** was treated with the deuterated acid, [D(OEt$_2$)$_2$][BAr'$_4$]. The result was formation of the cationic platinum(II) solvent complex, [(DTp')Pt(Me)(solv)][BAr'$_4$] [solv = CD$_2$Cl$_2$ or Et$_2$O] and elimination of methane (*35*). No deuterium incorporation into the remaining platinum bound methyl group or into the eliminated methane was observed: protonation occurs exclusively at the pyrazole nitrogen atom.

Low temperature protonation of Tp'PtMe$_2$D (**1-*d***) with [H(OEt$_2$)$_2$][BAr'$_4$] results in elimination of CH$_3$D and CH$_4$ and subsequent formation of Pt(II) solvent cations [(HTp')Pt(CH$_3$)(solv)][BAr'$_4$] (**6**) and [(HTp')Pt(CH$_2$D)(solv)][BAr'$_4$] (**6-*d***) [solv = CD$_2$Cl$_2$ or Et$_2$O], respectively (*35*). Addition of excess acetonitrile results in the formation of the stable cationic platinum(II) acetonitrile adducts [(HTp')Pt(CH$_3$)(NCCH$_3$)][BAr'$_4$] (**7**) and [(HTp')Pt(CH$_2$D)(NCCH$_3$)][BAr'$_4$] (**7-*d***). Isotopic scrambling from the hydride position into the platinum bound methyl position indicates reversible formation of a σ-methane adduct prior to methane elimination (Scheme 1) consistent with results from thermolysis of **1** and **1-*d***. Although the formation of a σ-methane adduct is consistent with both methane loss induced thermally and by protonation, the mechanism for methane loss induced by protonation was not determined and may differ from the dissociative process observed in the thermally induced methane elimination. A variety of stable cationic platinum(II) adducts of the form [(HTp')Pt(CH$_3$)(L)][BAr'$_4$] were synthesized via protonation of complex **1** and subsequent addition of a trapping ligand, L = SMe$_2$, CH$_2$=CH$_2$, NC$_5$H$_5$, CO, PMe$_2$Ph, and CN^{t}Bu (35).

Scheme 1

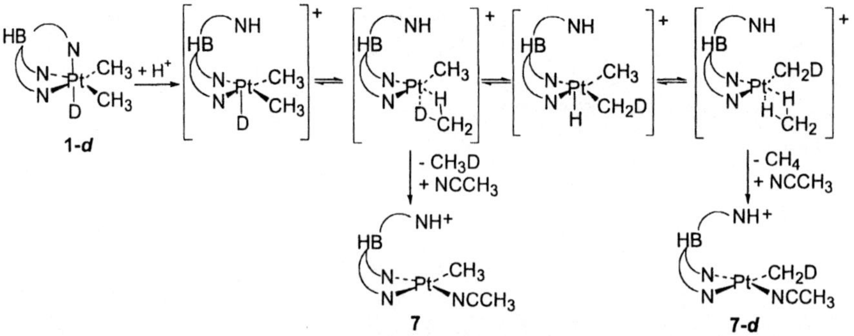

Direct observation of C-H oxidative addition to Pt(II) has proved to be elusive since there are few examples of Pt(II) complexes that oxidatively add C-H bonds to form stable Pt(IV) alkyl hydride (32) or aryl hydride (56) complexes. The mechanism by which C-H bond formation occurs during reductive elimination provides a mirror for revealing mechanistic information about C-H bond activation. Our results with Pt(IV) alkyl hydride reagents support the involvement of a Pt(II) σ-alkane intermediate which interconverts rapidly with a five-coordinate Pt(IV) oxidative addition product prior to methane elimination.

Five-Coordinate Pt(IV) Intermediates

The first step in alkane reductive elimination from many Pt(IV) complexes has been shown to involve dissociation of a labile ligand to generate a 16-electron platinum(IV) intermediate (20,28,40,57-70). Isolation of related five-coordinate Pt(IV) complexes resembling the proposed intermediates discussed in previous sections was reported in 2001 (71,72). Prior attempts to prepare five-coordinate complexes resulted either in weakly coordinating anions occupying the open coordination site or in immediate reductive elimination of alkane (64,73). In our laboratory, the use of hydride and silyl ligands which are known to stabilize high-oxidation state metals (74) coupled with the bulky Tp' ligand allowed the synthesis, isolation, and structural characterize of the five-coordinate [κ^2-(HTp')Pt(H)$_2$(SiEt$_3$)][BAr'$_4$] (**8**) complex (72). Simultaneously, Goldberg and co-workers reported isolation and structural characterization of the neutral five-coordinate complex [(nacnac)PtMe$_3$] [nacnac = {(2,6-iPr$_2$C$_6$H$_3$)NCMe}$_2$CH] (71).

The structure of Tp'Pt(SiEt$_3$)(H)$_2$ revealed an anomalously long Pt-N bond of 2.37 Å trans to the silyl group. Could protonation at nitrogen simply remove this pyrazolyl ring from the platinum coordination sphere and not induce reductive elimination of HSiEt$_3$? Low temperature protonation of the Pt(IV) precursor Tp'Pt(SiR$_3$)(H)$_2$ with [H(OEt$_2$)$_2$][BAr'$_4$] indeed afforded five-coordinate complexes [κ^2-(HTp')Pt(H)$_2$(SiR$_3$)][BAr'$_4$] [R$_3$ = Et$_3$ (**8**), Ph$_3$ (**9**), and Ph$_2$H (**10**)] (eq 5) (*72*). The five-coordinate geometry of these complexes was assigned based on a variety of NMR experiments and one X-ray structural analysis. It was necessary to consider a formulation of complex **8** as a Pt(II) σ-silane adduct, [(HTp')Pt(H)(H-SiEt$_3$)][BAr'$_4$], with rapid exchange of the Pt-H and Si-H to equilibrate them on the NMR timescale since only one upfield ^{1}H NMR signal integrating for 2H was observed. The absence of ^{29}Si coupling to the Pt-H resonance and the fact that addition of excess acetonitrile to complex **8** does not displace the silane are pieces of evidence against assignment of **8** as a Pt(II) silane adduct. Most compelling is the absence of a significant chemical shift jump when the partially deuterated complex was prepared since a significant zero-point energy difference between Pt-H and (σ-R$_3$Si-H)Pt should produce a large ^{1}H chemical shift difference upon partial deuterium incorporation (*75*). Additionally, the ^{1}H NMR spectrum for complex **10**, [(HTp')Pt(H)$_2$(SiPh$_2$H)][BAr'$_4$], shows distinct resonances for the platinum hydrides and the silicon-bound hydrogen which do not exchange on the NMR timescale. X-ray analysis of complex **8** indicated a square-pyramidal geometry for the dihydridosilyl platinum(IV) complex, and no solvent molecules were found near the open coordination site on platinum. All evidence is consistent with the assignment of complexes **8-10** as five-coordinate dihydridosilyl platinum(IV) complexes.

Complexes **8** - **10** can also be synthesized directly by protonation of the methyldihydride complex **3** to generate the cationic platinum(II) solvent complex, [(HTp')Pt(H)(solv)][BAr'$_4$] [solv = CD$_2$Cl$_2$ or Et$_2$O] (**11**), which can then be trapped by added silane (eq 5) (*72*). Immediate oxidative addition of the silane Si-H bond occurs to give the five-coordinate complex; no evidence for a σ-silane adduct was observed. Exclusive methane elimination (i.e. no dihydrogen) was observed from complex **3**. This is similar to the results seen by Vedernikov and Caulton both experimentally and by DFT calculations in the thermolysis of [LPtMe(H)$_2$] [L = [2.1.1]-(2,6)-pyridinophane] to give exclusive methane loss (76). Protonation of complex **3** induces reductive methane elimination as was seen by thermolysis and protonation of complex **1** (vide supra).

Stable Platinum(IV) Aryl Hydride Complexes

Thermolysis of complex **1** in benzene (or benzene-d_6) leads to complete conversion to Tp'PtPh$_2$H (**5**) {or Tp'Pt(C$_6$D$_6$)$_2$D (**5-d**)} in a few hours (38). Similar results were observed in seconds at room temperature with the addition of substoichiometric amounts of B(C$_6$F$_5$)$_3$ as a catalyst. The observation that Lewis acids promote reductive methane elimination at ambient temperatures, in contrast to thermolysis at high temperatures, prompted exploration of these milder conditions as a route to new platinum(IV) compounds. Several platinum(IV) aryl dihydrides were synthesized from Tp'PtMeH$_2$ (**3**) via Lewis acid induced methane elimination and subsequent aromatic C-H activation (56).

Generation of these platinum(IV) aryl dihydride complexes was achieved by the reaction of complex **3** with substoichiometric B(C$_6$F$_5$)$_3$ in various aromatic solvents to give the C-H activation products Tp'Pt(Ar)(H)$_2$ [Ar = Ph (**12**), tolyl (**13**), o-xylyl (**14**), m-xylyl (**15**), and p-xylyl (**16**)] (eq 6) (56). The mechanism of the Lewis acid induced methane elimination from **3** is currently unknown, but it may involve reversible attack of the Lewis acid on a pyrazole nitrogen atom to remove one arm of the chelate from the metal center. This would generate a reactive five-coordinate intermediate from which methane elimination followed by C-H activation would give the final product after borane dissociation. Further support for this mechanism comes from the evolution of methane (evident in NMR scale reactions) and the observation that only substoichiometric amounts of borane are required. In contrast to the reaction of [K][Tp'PtMe$_2$] with B(C$_6$F$_5$)$_3$ (32), methide abstraction by the Lewis acid [Ph$_3$C][BAr'$_4$] (77) was not observed with complex **3**: no signals for Ph$_3$C-CH$_3$ were observed in the ^{1}H NMR.

Pt(II) η²-Arene Intermediates

Just as σ-bound alkane complexes are putative intermediates in alkane activation, η²-arene complexes are key intermediates in aromatic C-H bond activation (7,20,24,32,57,59,78-82). Although formation of a σ-bound arene complex is thought to precede arene C-H activation, a C-H σ-arene complex is presumably unstable relative to the π-complex (59). NMR characterization at low temperature of the first Pt(II)-η²-benzene adduct was recently reported (59). In our Tp'Pt system, acid induced reductive elimination of benzene from the phenyl dihydride complex **12** afforded a surprisingly stable Pt(II)-η²-benzene adduct [κ²-(HTp')Pt(C,C-η²-C₆H₆)(H)][BAr'₄] (**17**) which was crystallized and examined by X-ray crystallography (83).

Low temperature protonation of Tp'Pt(Ar)(R)(H) complexes led to the isolation of platinum(II) η²-arene adducts [κ²-(HTp')Pt(C,C-η²-Ar-H)(R)][BAr'₄] [Ar = Ph, R = H (**17**), Ar = tolyl, R = H (**18**); Ar = *p*-xylyl, R = H (**19**); Ar = R = Ph (**20**)] (eq 7) (83,84). Protonation occurs exclusively at the pyrazole nitrogen to generate a five-coordinate intermediate which reductively eliminates arene to generate an η²-arene adduct of Pt(II), in contrast to the silyl dihydride complexes where the five-coordinate Pt(IV) complex is the ground state structure (72). The η²-arene does not dissociate from the Pt(II) metal center at low temperatures, in contrast to the results of protonation of complexes **1** and **3** where methane dissociation occurs readily (35).

$$\text{(eq 7)}$$

These Pt(II) η²-arene adducts were probed with NMR techniques, and for the benzene hydride adduct **17** (83) and the *p*-xylene adduct **19** (84) single crystal X-ray structural analysis was accomplished. ¹H NMR spectra for both the benzene hydride and the benzene phenyl adducts exhibit a singlet (6.9 ppm) upfield from free benzene (7.2 ppm) which integrates for six hydrogens and corresponds to the hydrogens of the benzene ring bound to platinum. The equivalency of all six hydrogens on the NMR timescale could be attributed either to migration of the platinum around all six positions on the same face of the arene or to a combination of face flipping and ring rotation. Migration of the platinum around the ring of the toluene adduct **18** is not an NMR observable process, since all of the aromatic hydrogens are inequivalent, a fact that is independent of ring rotation. To address the question of platinum migration, we turned to the *p*-xylene adduct **19**. At low temperatures, distinct resonances are observed in the ¹H NMR spectrum for all four aromatic hydrogens and for both aryl methyl groups. As the temperature is raised, coalescence of the two aryl methyl groups is observed, a definitive indication of platinum migration on the same face of the arene. In contrast, a face flipping dynamic process would not exchange the two aryl methyl groups. The observed coalescence of the two methyl groups led to a calculated barrier ($\Delta G^{\ddagger}$) of 9.2 kcal/mol for platinum migration around the ring at $T_c = 209$ K.

The X-ray structures of the benzene hydride adduct **17** (*83*) and the *p*-xylene adduct **19** (*84*) confirmed η^2-coordination of the arene to the square-planar platinum(II) center. The average C-C bond lengths of the coordinated benzene (1.36 Å) are comparable to those of free benzene (1.39 Å) (*85*); in the *p*-xylene adduct the average C-C bond lengths (1.45 Å) are slightly longer than that for free *p*-xylene (1.39 Å) (*86*). The orientation of the arene plane relative to the square-plane of the metal is almost perdendicular in both complexes (**17**: 85.0° and **19**: 77.9°).

The presence of a hydride ligand cis to the η^2-arene ligand provides an important spectroscopic probe for assessing the barrier to C-H oxidative addition. These η^2-arene platinum(II) adducts exhibit dynamic ^{1}H NMR spectra which can be attributed to equilibration of the η^2-arene adduct with a five-coordinate platinum aryl dihydride intermediate via arene C-H oxidative addition (eq 8) (*83,84*). Scrambling of the two hydride ligands in the five-coordinate intermediate allows for exchange to occur. Variable temperature NMR experiments allowed us to quantitatively measure the barrier to oxidative addition of arene C-H bonds to Pt(II) by monitoring exchange of the hydride and arene signals. The barriers ($\Delta G^{\ddagger}$) to arene C-H oxidative addition for the benzene hydride adduct **17** (12.7 kcal/mol), the toluene adduct **18** (*m*-H: 13.6 kcal/mol; *p*-H: 13.3 kcal/mol) and the *p*-xylene adduct **19** (14.2 kcal/mol) were measured using this technique. The temperature dependence for intramolecular C-H oxidative addition of benzene in the benzene hydride adduct **17** was monitored by ^{1}H NMR to give activation parameters of $\Delta H^{\ddagger} = 11.7 \pm 0.5$ kcal/mol and $\Delta S^{\ddagger} = -3.8 \pm 2$ cal/(mol K). The entropy of activation is near zero, as one might expect for an intramolecular process.

$$\begin{array}{c}\text{H}_B\overset{\frown}{\underset{N}{\text{N}}}\overset{\text{NH}}{\underset{}{|}}\text{Pt}\overset{\text{H}}{\underset{}{\cdot}}\\\text{(CH}_3)_n\\\textbf{17 - 19}\end{array}\quad\rightleftharpoons\quad\left[\begin{array}{c}\text{H}_B\overset{\frown}{\underset{N}{\text{N}}}\overset{\text{NH}}{\underset{}{|}}\text{Pt}\overset{\text{H}}{\underset{}{\cdot}}\text{H}\\\text{(CH}_3)_n\\\textbf{17a - 19a}\end{array}\right]^{\text{BAr}'_4} \quad (8)$$

Unlike the benzene hydride complex **17**, the benzene phenyl adduct **20** had no adjacent hydride ligand to exchange with the platinum bound arene aromatic hydrogens. However, oxidative addition of the η^2-benzene was reflected by averaging of two Tp' arms as complex **20**, of C_1 symmetry, equilibrated with the oxidative addition product, a five-coordinate platinum(IV) diphenyl hydride intermediate (**20a**) with effective C_s symmetry on the NMR timescale (eq 9). The barrier for exchange between the η^2-benzene and σ-phenyl rings in **20**, assumed to reflect the oxidative addition barrier, was calculated to be 12.9 kcal/mol.

$$\begin{array}{c}\text{NH}\\\text{HB}\\\text{N-Pt-Ph}\\\text{N}\end{array}\left]\text{BAr}'_4 \longrightarrow \left[\begin{array}{c}\text{NH}\\\text{HB} \quad \text{H}\\\text{N-Pt-Ph}\\\text{N} \quad \text{Ph}\end{array}\right]\text{BAr}'_4 \quad (9)$$

20　　　　　　　**20a**

Kinetic isotope effects have been measured previously for arene C-H bond activation to probe the transition state (*81,87-91*). The method used by Jones and Feher in the Cp*Rh(PMe$_3$)(Ph)H system involved examining the bimolecular reaction of arene complexation to the metal center followed by oxidative cleavage (*88*). In cleverly designed experiments, a 1:1 mixture of C$_6$H$_6$/C$_6$D$_6$ was used to assess the equilibrium isotope effect for arene complexation and the kinetic isotope effect for the oxidative cleavage step was determined using 1,3,5-C$_6$D$_3$H$_3$. Since the arene ring is already complexed to the metal center in our system, the kinetic isotope effect measured reflects arene C-H(D) bond activation step only. Isolation of the perdeuterio benzene hydride adduct (**17-*d***) and benzene phenyl adduct (**20-*d***) permitted direct measurement of the kinetic isotope effects for intramolecular arene C-H(D) bond activation. The k_H/k_D value for complexes **17** and **20** were 3.0 at 259 K and 4.7 at 241 K, respectively, consistent with significant C-H(D) bond cleavage to reach the transition state (*84*).

As discussed previously, the mechanism for C-H reductive elimination from Pt(IV) alkyl hydride complexes is proposed to involve ligand dissociation to form a five-coordinate Pt(IV) intermediate which interconverts rapidly with a Pt(II) σ-alkane intermediate. According to the principle of microscopic reversibility, the reaction coordinate for C-H activation must be the same. Isolation of Pt(II)-η^2-arene adducts with adjacent hydride or phenyl ligands has allowed us to measure the barrier to interconversion of these two intermediates for arene C-H bond activation and thereby provided insight into the energetics of arene C-H bond activation.

Conclusion

Considerable evidence supports the hypothesis that the first step in alkane C-H activation by platinum(II) reagents involves coordination of the alkane to a coordinatively unsaturated metal center to form a σ-bound alkane intermediate Pt^{II}(R-H) which is thought to interconvert rapidly with a Pt^{IV}(R)(H) oxidative addition product (*20,28,39*). The existence of these intermediates has only been inferred from kinetic data; neither one has been observed directly. In an effort to map out the reaction coordinate, analogs of these intermediates have been isolated and structurally characterized for the Tp'PtR(H)$_2$ system (Scheme 2). For R = SiR$_3$, five-coordinate Pt(IV) complexes of the type [κ^2-(HTp')Pt(H)$_2$(SiR$_3$)][BAr'$_4$] have been isolated (*72*). In addition, isolation of a Pt(II) arene adduct prior to replacement of the Ar-H ligand was achieved in the form of Pt(II) η^2-arene adduct (*83,84*).

Scheme 2

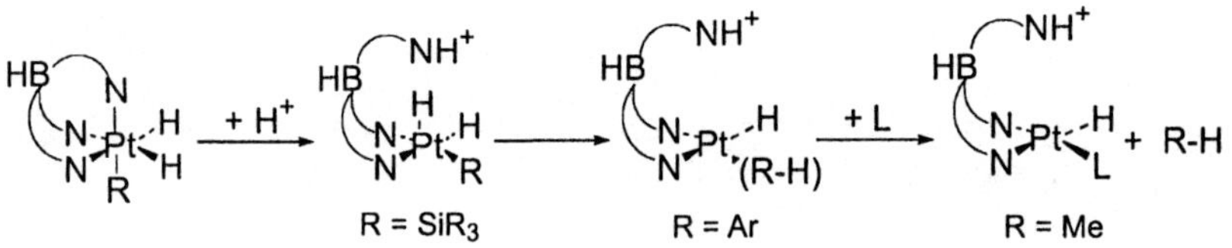

By utilizing R = SiR$_3$ and C$_6$R$_5$ rather than R = alkyl in the low temperature protonation of Tp'PtR(H)$_2$, isolation of analogs of proposed intermediates in C-H bond activation by Pt(II) complexes was possible (Scheme 3). Protonation of Tp'PtR(H)$_2$ occurs readily at 193 K and provides access to the manifold including the Pt(IV) five-coordinate complex and the Pt^{II}(Ar-H) adduct. Changing the R group from silyl to aryl shifts the ground state structure from the five-coordinate complex **B** to the arene adduct **C**, respectively. The barrier (ΔG‡) to interconversion between **C** and **B** was calculated for R = aryl and varied between 12.7 and 14.2 kcal/mol depending on the arene. For R = silyl, this energy barrier is unknown, but the σ-silane adduct **C** is higher in energy than the five-coordinate species **B**. In effect, the relative energies of complexes **B** and **C** can be reversed by varying the R group in the Tp'PtR(H)$_2$ reagent prior to protonation. Variation of R then allows for isolation of complexes resembling key intermediates in C-H bond activation.

Although the ultimate goal of selective hydrocarbon functionalization still remains as a target on the horizon, significant advances have been made towards understanding factors that govern C-H activation at platinum(II) centers. By undertaking systematic studies with Pt(IV) alkyl, silyl, and aryl hydride reagents, reductive elimination has provided analogs of key Pt(II) intermediates.

Acknowledgment

We gratefully acknowledge the National Science Foundation (Grant No. 0109655) for support of this research.

References

1. Crabtree, R. H. *Chem. Rev.* **1985**, *85*, 245.
2. Hill, C. L. *Activation and Functionalization of Alkanes*; John Wiley & Sons: New York, 1989.

Scheme 3

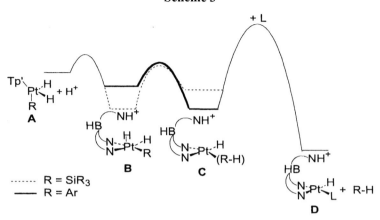

3. Davies, J. A.; Watson, P. L.; Liebman, J. F.; Greenberg, A. *Selective Hydrocarbon Activation: Principles and Progress*; VCH Publishers: New York, 1990.
4. Crabtree, R. H. *Chem. Rev.* **1995**, *95*, 987.
5. Arndtsen, B. A.; Bergman, R. G.; Mobley, T. A.; Peterson, T. H. *Acc. Chem. Res.* **1995**, *28*, 154.
6. Shilov, A. E.; Shul'pin, G. B. *Chem. Rev.* **1997**, *97*, 2879.
7. Stahl, S. S.; Labinger, J. A.; Bercaw, J. E. *Angew. Chem. Int. Ed.* **1998**, *37*, 2180.
8. Periana, R. A.; Taube, D. J.; Gamble, S.; Taube, H.; Satoh, T.; Fujii, H. *Science* **1998**, *280*, 560.
9. Sen, A. *Acc. Chem. Res.* **1998**, *31*, 550.
10. Dyker, G. *Angew. Chem. Int. Ed.* **1999**, *38*, 1698.
11. Shilov, A. E.; Shul'pin, G. B. *Activation and Catalytic Reactions of Saturated Hydrocarbons in the Presence of Metal Complexes*; Kluwer Academic Publishers: Boston, 2000.
12. Jones, W. D. *Science* **2000**, *287*, 1942.
13. Jia, C.; Kitamura, T.; Fujiwara, Y. *Acc. Chem. Res.* **2001**, *34*, 633.
14. Crabtree, R. H. *J. Chem. Soc., Dalton Trans.* **2001**, 2437.
15. Labinger, J. A.; Bercaw, J. E. *Nature* **2002**, *417*, 507.
16. Gol'dshleger, N. F.; Es'kova, V. V.; Shilov, A. E.; Shteinman, A. A. *Russ. J. Phys. Chem.* **1972**, *46*, 785 (English Translation).
17. Lin, M.; Shen, C.; Garcia-Zayas, E. A.; Sen, A. *J. Am. Chem. Soc.* **2001**, *123*, 1000.
18. Geletti, Y. V.; Shilov, A. E. *Kinetics Catal. Engl. Trans.* **1983**, *24*, 413.
19. Luinstra, G. A.; Wang, L.; Stahl, S. S.; Labinger, J. A.; Bercaw, J. E. *J. Organomet. Chem.* **1995**, *504*, 75.
20. Stahl, S. S.; Labinger, J. A.; Bercaw, J. E. *J. Am. Chem. Soc.* **1996**, *118*, 5961.

21. Holtcamp, M. W.; Labinger, J. A.; Bercaw, J. E. *Inorg. Chim. Acta* **1997**, *265*, 117.
22. Wang, L.; Stahl, S. S.; Labinger, J. A.; Bercaw, J. E. *J. Mol. Catal.* **1997**, *116*, 269.
23. Holtcamp, M. W.; Henling, L. M.; Day, M. W.; Labinger, J. A.; Bercaw, J. E. *Inorg. Chim. Acta* **1998**, *270*, 467.
24. Holtcamp, M. W.; Labinger, J. A.; Bercaw, J. E. *J. Am. Chem. Soc.* **1997**, *119*, 848.
25. Puddephatt, R. J. *Coord. Chem. Rev.* **2001**, *219-221*, 157.
26. de Felice, V.; de Renzi, A.; Panunzi, A.; Tesauro, D. *J. Organomet. Chem.* **1995**, *488*, C13.
27. Hill, G. S.; Rendina, L. M.; Puddephatt, R. J. *Organometallics* **1995**, *14*, 4966.
28. Stahl, S. S.; Labinger, J. A.; Bercaw, J. E. *J. Am. Chem. Soc.* **1995**, *117*, 9371.
29. Trofimenko, S. *Scorpionates-The Coordination Chemistry of Polypyrazolylborate Ligands*; Imperial College Press: London, 1999.
30. O'Reilly, S. A.; White, P. S.; Templeton, J. L. *J. Am. Chem. Soc.* **1996**, *118*, 5684.
31. Canty, A. J.; Dedieu, A.; Jin, H.; Milet, A.; Richmond, M. K. *Organometallics* **1996**, *15*, 2845.
32. Wick, D.; Goldberg, K. I. *J. Am. Chem. Soc.* **1997**, *119*, 10235.
33. Chen, E. Y.; Marks, T. J. *Chem. Rev.* **2000**, *100*, 1391.
34. Reinartz, S.; Baik, M. H.; White, P. S.; Brookhart, M.; Templeton, J. L. *Inorg. Chem.* **2001**, *40*, 4726.
35. Reinartz, S.; White, P.; Brookhart, M.; Templeton, J. L. *Organometallics* **2000**, *19*, 3854.
36. Haskel, A.; Keinan, E. *Organometallics* **1999**, *18*, 4677.
37. Reinartz, S.; White, P. S.; Brookhart, M.; Templeton, J. L. *Organometallics* **2000**, *19*, 3748.
38. Jensen, M. P.; Wick, D. D.; Reinartz, S.; White, P. S.; Templeton, J. L.; Goldberg, K. I. *J. Am. Chem. Soc.* **2003**, *125*, 8614.
39. Puddephatt, R. J. *Angew. Chem. Int. Ed.* **2002**, *41*, 261.
40. Fekl, U.; Zahl, A.; van Eldik, R. *Organometallics* **1999**, *18*, 4156.
41. Buchanan, J. M.; Stryker, J. M.; Bergman, R. G. *J. Am. Chem. Soc.* **1986**, *108*, 1537.
42. Periana, R. A.; Bergman, R. G. *J. Am. Chem. Soc.* **1986**, *108*, 7332.
43. Bullock, R. M.; Headford, C. E.; Hennessy, K. M.; Kegley, S. E.; Norton, J. R. *J. Am. Chem. Soc.* **1989**, *111*, 3897.
44. Parkin, G.; Bercaw, J. E. *Organometallics* **1989**, *8*, 1172.
45. Gould, G. L.; Heinekey, D. M. *J. Am. Chem. Soc.* **1989**, *111*, 5502.
46. Wang, C.; Ziller, J. W.; Flood, T. C. *J. Am. Chem. Soc.* **1995**, *117*, 1647.
47. Mobley, T. A.; Schade, C.; Bergman, R. G. *J. Am. Chem. Soc.* **1995**, *117*, 7822.
48. Chernaga, A.; Cook, J.; Green, M. L. H.; Labella, L.; Simpson, S. J.; Souter, J.; Stephens, A. H. H. *J. Chem. Soc., Dalton Trans.* **1997**, 3225.
49. Gross, C. L.; Girolami, G. S. *J. Am. Chem. Soc.* **1998**, *120*, 6605.

50. Wick, D. D.; Reynolds, K. A.; Jones, W. D. *J. Am. Chem. Soc.* **1999**, *121*, 3974.
51. Northcutt, T. O.; Wick, D. D.; Vetter, A. J.; Jones, W. D. *J. Am. Chem. Soc.* **2001**, *123*, 7257.
52. Lo, H. C.; Haskel, A.; Kapon, M.; Keinan, E. *J. Am. Chem. Soc.* **2002**, *124*, 3226.
53. Iron, M. A.; Lo, H. C.; Martin, J. M. L.; Keinan, E. *J. Am. Chem. Soc.* **2002**, *124*, 7041.
54. Lo, H. C.; Haskel, A.; Kapon, M.; Keinan, E. *J. Am. Chem. Soc.* **2002**, *124*, 12626.
55. Brookhart, M.; Grant, B.; Volpe, Jr., A. F. *Organometallics* **1992**, *11*, 3920.
56. Reinartz, S.; White, P. S.; Brookhart, M.; Templeton, J. L. *Organometallics* **2001**, *20*, 1709.
57. Johansson, L.; Tilset, M. *J. Am. Chem. Soc.* **2001**, *123*, 739.
58. Heiberg, H.; Johansson, L.; Gropen, O.; Ryan, O. B.; Swang, O.; Tilset, M. *J. Am. Chem. Soc.* **2000**, *122*, 10831.
59. Johansson, L.; Tilset, M.; Labinger, J. A.; Bercaw, J. E. *J. Am. Chem. Soc.* **2000**, *122*, 10846.
60. Bartlett, K. L.; Goldberg, K. I.; Borden, W. T. *J. Am. Chem. Soc.* **2000**, *122*, 1456.
61. Crumpton, D. M.; Goldberg, K. I. *J. Am. Chem. Soc.* **2000**, *122*, 962.
62. Williams, B. S.; Holland, A. W.; Goldberg, K. I. *J. Am. Chem. Soc.* **1999**, *121*, 252.
63. Goldberg, K. I.; Y., Y. J.; Breitung, E. M. *J. Am. Chem. Soc.* **1995**, *117*, 6889.
64. Hill, G. S.; Yap, G. P. A.; Puddephatt, R. J. *Organometallics* **1999**, *18*, 1408.
65. Hill, G. S.; Puddephatt, R. J. *Organometallics* **1998**, *17*, 1478.
66. Hill, G. S.; Puddephatt, R. J. *Organometallics* **1997**, *16*, 4522.
67. Jenkins, H. A.; P.A., Y. G.; Puddephatt, R. J. *Organometallics* **1997**, *16*, 1946.
68. Roy, S.; Puddephatt, R. J.; Scott, J. D. *J. Chem. Soc., Dalton Trans.* **1989**, 2121.
69. Brown, M. P.; Puddephatt, R. J.; Upton, C. E. E. *J. Chem. Soc., Dalton Trans.* **1974**, 2457.
70. Canty, A. J. *Acc. Chem. Res.* **1992**, *25*, 83.
71. Fekl, U.; Kaminsky, W.; Goldberg, K. I. *J. Am. Chem. Soc.* **2001**, *123*, 6423.
72. Reinartz, S.; White, P. S.; Brookhart, M.; Templeton, J. L. *J. Am. Chem. Soc.* **2001**, *123*, 6425.
73. Asselt, R.; Rijnberg, E.; Elsevier, C. J. *Organometallics* **1994**, *13*, 706.
74. Corey, J. Y.; Braddock-Wilking, J. *Chem. Rev.* **1999**, *99*, 175.
75. Calvert, R. B.; Shapley, J. R. *J. Am. Chem. Soc.* **1978**, *100*, 7726.
76. Vedernikov, A. N.; Caulton, K. G. *Angew. Chem. Int. Ed.* **2002**, *41*, 4102.
77. Bahr, S. R.; Boudjouk, P. *J. Org. Chem.* **1992**, *57*, 5545.
78. Johansson, L.; Ryan, O. B.; Tilset, M. *J. Am. Chem. Soc.* **1999**, *121*, 1974.
79. Johansson, L.; Ryan, O. B.; Romming, C.; Tilset, M. *J. Am. Chem. Soc.* **2001**, *123*, 6579.

80. Thomas, J. C.; Peters, J. C. *J. Am. Chem. Soc.* **2001**, *123*, 5100.
81. Jones, W. D.; Feher, F. J. *Acc. Chem. Res.* **1989**, *22*, 91.
82. Iverson, C. N.; Lachicotte, R. J.; Muller, C.; Jones, W. D. *Organometallics* **2002**, *21*, 5320.
83. Reinartz, S.; White, P. S.; Brookhart, M.; Templeton, J. L. *J. Am. Chem. Soc.* **2001**, *123*, 12724.
84. Norris, C. M.; Reinartz, S.; White, P. S.; Templeton, J. L. *Organometallics* **2002**, *21*, 5649.
85. Meiere, S. H.; Brooks, B. C.; Gunnoe, T. B.; Sabat, M.; Harman, W. D. *Organometallics* **2001**, *20*, 1038.
86. van Koningsveld, H.; van den Berg, A. J.; Jansen, J. C.; de Goede, R. *Acta Cryst.* **1986**, *B42*, 491.
87. Brown, S. N.; Myers, A. W.; Fulton, J. R.; Mayer, J. M. *Organometallics* **1998**, *17*, 3364.
88. Jones, W. D.; Feher, F. J. *J. Am. Chem. Soc.* **1986**, *108*, 4814.
89. Yandulov, D. V.; Caulton, K. G. *New. J. Chem.* **2002**, *26*, 498.
90. Sadow, A. D.; Tilley, T. D. *J. Am. Chem. Soc.* **2002**, *124*, 6814.
91. Jones, W. D. *Acc. Chem. Res.* **2003**, *36*, 140.

Chapter 19

C–H Bond Activation with Neutral Platinum Methyl Complexes

Carl N. Iverson[1], Charles A. G. Carter[1], John D. Scollard[2], Melanie A. Pribisko[1], Kevin D. John[1], Brian L. Scott[1], R. Tom Baker[1,*], John E. Bercaw[2,*], and Jay A. Labinger[2,*]

[1]Los Alamos Catalysis Initiative, Chemistry Division, MS J514, Los Alamos National Laboratory, Los Alamos, NM 87545
[2]Department of Chemistry, California Institute of Technology, Pasadena, CA 91125

The selective metal-catalyzed oxidation of alkanes to alcohols offers immense opportunities for saving energy and reducing waste in the petroleum and chemical manufacturing industries. Robust transition metal complexes with chelating nitrogen ligands show great promise as homogeneous catalysts for this reaction, but a practical system has yet to be identified. In this work a variety of neutral platinum methyl complexes with bidentate anionic *N,N*- and *N,C*-donor ligands were prepared. Ligand backbones included one, two, and three-atom bridges between the donor atoms. Redox properties of the new complexes were investigated using competition studies with I_2 and evaluating equilibrium constants between divalent methyl complexes and their tetravalent diiodides. The ease of oxidation was amidinate > β-diketiminate > iminopyrrolide. Studies of benzene C-H bond activation using iminopyrrolide platinum complexes were consistent with rate determining benzene association and revealed a novel geometric effect on the rate for iminopyrrolide ligands.

© 2004 American Chemical Society

Introduction

Shilov et al. first demonstrated Pt(II)-catalyzed oxidation of alkanes to alcohols and alkyl chlorides some 30 years ago with simple platinum salts (*1*). In spite of the limitations imposed by the Pt(IV) stoichiometric oxidant and poor catalyst lifetime due to Pt metal precipitation, this mild (120°C) alkane functionalization is notable for its selectivity (1° > 2° > 3°) that is opposite to that of radical processes. Much effort has been expended to elucidate the mechanism of this transformation (*2,3*). The first step is proposed to involve electrophilic activation of an alkane C–H bond by a Pt(II) salt to form a metal alkyl complex. Subsequent oxidation of the Pt(II) alkyl to Pt(IV) is followed by nucleophilic attack at the Pt(IV) alkyl by water or chloride to form the alcohol or alkyl halide and regenerate the starting Pt(II) salt.

More recently, Periana et al. reported that solutions of bipyrimidine platinum dichloride in boiling sulfuric acid selectively oxidize methane to methylbisulfate in high one-pass yields (*4,5*). Subsequent studies showed that nitrogen-ligated, cationic platinum methyl complexes with diamine or diimine ligands undergo facile C-H bond activation reactions in non-coordinating solvents (*6,7*). While neutral platinum dimethyl complexes with these ligands could also be oxidized with dioxygen in methanol to give Pt(IV) alkyl products (*8*), cationic Pt methyl complexes derived from methane C-H bond activation proved more difficult to oxidize (*9*). As a result, platinum methyl complexes with monoanionic nitrogen-based chelates were investigated for their ability to activate C-H bonds and undergo facile oxidation reactions (*10*). The first example, reported by Goldberg et al., involved a tris(3,5-dimethylpyrazolyl)borate (Tp') platinum complex that oxidatively adds alkane C–H bonds to afford a Pt(IV) methyl alkyl hydride complex stabilized in a six-coordinate geometry by the tridentate Tp' ligand (*11*). More recently, the same group used a bulky β-diketiminate ligand to support the first five-coordinate Pt(IV) methyl compound (*12*), and further work with Tp ligands by the groups of Templeton and Keinan afforded a rare η^2-benzene complex and showed reversible H/D exchange of methyl and hydride positions with alcohol solvent, respectively (*13,14*). Finally, Peters et al. have developed a bis(phosphinoalkyl)borate ligand that has also shown C–H bond activation capabilities when ligated to platinum (*15*).

In this work a variety of neutral platinum methyl complexes ligated with bidentate anionic *N,N*- and *N,C*-donor ligands were prepared. Ligand backbones included one,- two,- and three-atom bridges between the donor atoms. Several of the resulting complexes were then compared for their ability to activate C-H bonds and their ease of oxidation to Pt(IV) methyl products.

Preparation and Characterization of Platinum Methyl Complexes with Anionic Bidentate Ligands

The ligands **1-5** used in this study were prepared by standard literature procedures (*16-20*). Reactions of PtMeCl(cod) (**6**, cod = 1,5-cyclooctadiene)

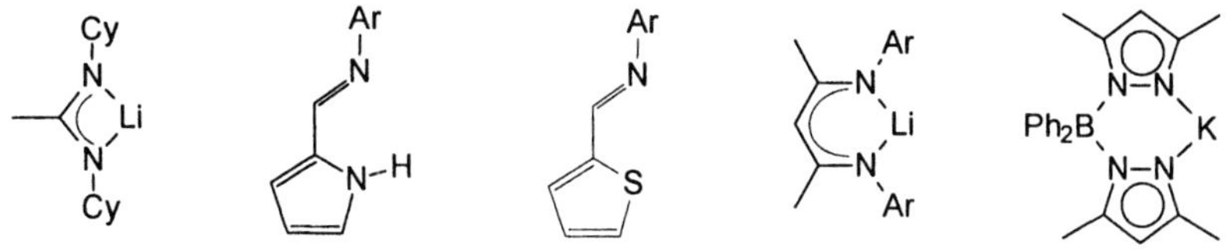

1

2a: R = 4-MeO-Ph
2b: R = 4-CF$_3$-Ph
2c: R = 3,5-(CF$_3$-)$_2$-Ph
2d: R = 2,6-^{i}Pr$_2$-Ph
2e: R = 4-Me-Ph

3: R = 2,6-^{i}Pr$_2$-Ph

4a: R = 4-MeO-Ph
4b: R = 4-Cl-Ph

5

with ligands **1**, **5** or deprotonated **2** afforded five-coordinate platinum methyl complexes **7-9** (Eqs. 1-3).

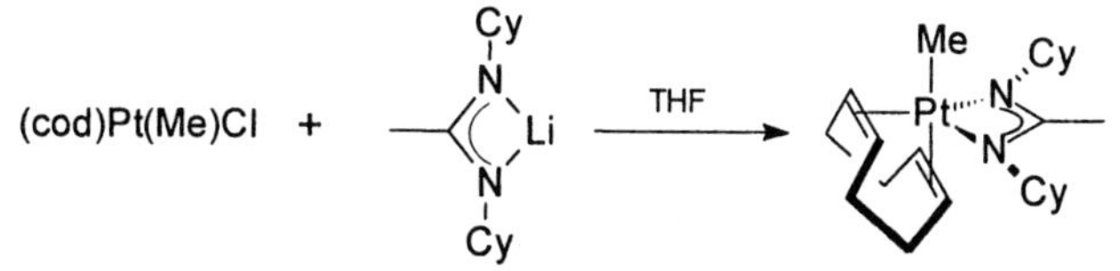

(cod)Pt(Me)Cl + [1] →(THF) **7**   (1)

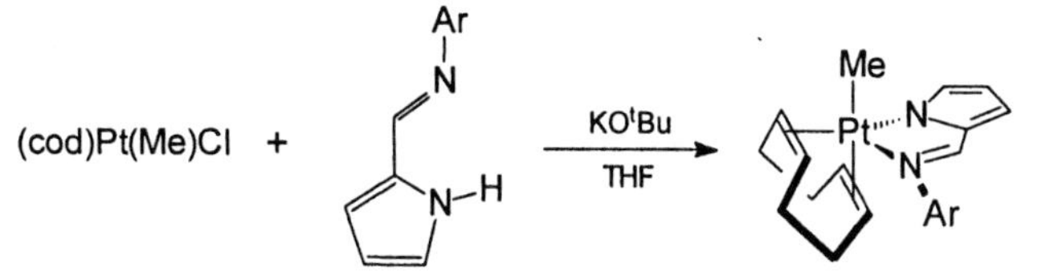

(cod)Pt(Me)Cl + [2a] →(KO^{t}Bu, THF) **8a**: Ar = 4-MeO-Ph   (2)

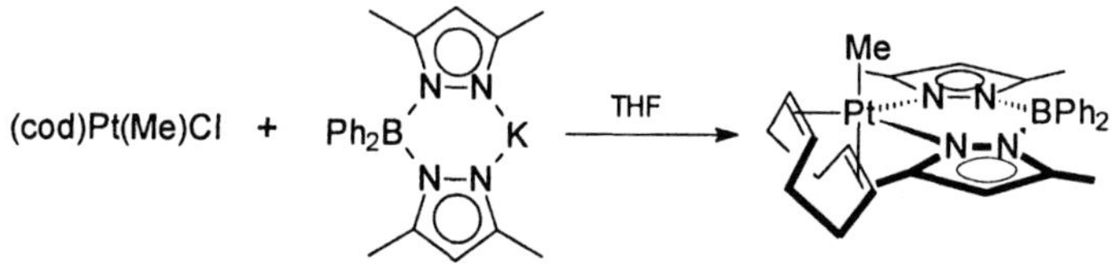

(cod)Pt(Me)Cl + [5] →(THF) **9**   (3)

These complexes are all proposed to have a pseudo trigonal bipyramidal geometry with an apical methyl group, based on the similar values of J_{HPt} (73, 78 Hz; Table I) and the molecular structure of **7** as determined by X-ray diffraction (Figure 1).

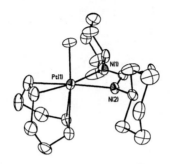

Figure 1. Molecular Structure of PtMe(cod)(CyNCMeNCy), 7

Table I. ^{1}H NMR chemical shifts and Pt-H Coupling Constants (Hz) of Divalent Platinum Methyl Complexes

Complex	Solvent	$\delta\,Pt\text{-}Me,\,^2J_{HPt}$	$\delta\,Pt\text{-}SMe_2,\,^3J_{HPt}$
7	M	0.70, 73	-
8a	C	0.63, 78	-
9	A	0.76, 73	-
13	B	1.13, 83	1.73, 52
14a	B	1.31, 79	1.52, 51
14b	B	1.26, 80	1.43, 51
14c	B	1.22, 81	1.45, 52
15a	B	0.74, 73	1.73, 57
15b	B	0.58, 73	1.68, 58
15d	B	0.42, 73	1.71, 58
16	B	1.72, 84	1.58, 29
17	B	1.30, 79	-
18a	B	0.33, 72	1.48, 50
18b	M	-0.20, 78	2.01, 46
19	B	0.38, 75	2.05, 54

A = acetone-d_6, B = C_6D_6, C = $CDCl_3$, M = CD_2Cl_2

Four coordinate platinum methyl complexes 13-19 were prepared by a similar salt metathesis route from PtMeCl(SMe$_2$)$_2$ 10 or via protonolysis of a methyl group in [PtMe$_2$(μ-SR$_2$)]$_2$ (11, R = Me; 12, R= Et) by the N-H bond of the neutral ligand (Eqs. 4-10). With the exception of the bulky 2,6-diisopropylphenyl-substituted example, the unsymmetrical iminopyrrolide ligands 2 gave both stereoisomers from the metathesis route, but a single isomer from the protonation route. Complexes 14 and 15 are labeled *cis* and *trans*, respectively, according to the relationship of the anionic methyl and pyrrolide ligands. Reaction of 11 with ligand 3 yielded a methyl complex with an *N,C*-chelate arising from orthometallation of the β–C-H bond of the thiophene ring (*21*). The analogous diethylsulfide dimer, 12, also reacts with the neutral iminopyrollides via deprotonation (Eq. 8), albeit at a much slower rate, to produce the *cis* product 17.

Characterization of complexes 14 and 15 was facilitated by the magnitude of J_{HPt} for both Pt-Me and SMe$_2$ groups (Table I) that reflected the donor nature of the *trans* ligand. The two-bond H-Pt coupling constant for the Pt-Me group in 14 (*trans* to the weaker σ-donor imine) averaged 80 Hz vs. 73 Hz for 15 with Pt-Me *trans* to pyrrolide. Likewise, the coupling constants for the SMe$_2$ resonances were 58 Hz in 14 vs. 51 Hz for 15. Although comparisons with other complexes were complicated by the different backbone lengths, the coupling constant differences between the symmetrical, three atom bridge β-diketiminate and bis(pyrazolyl)borate ligands were consistent with the higher donor power of the former.

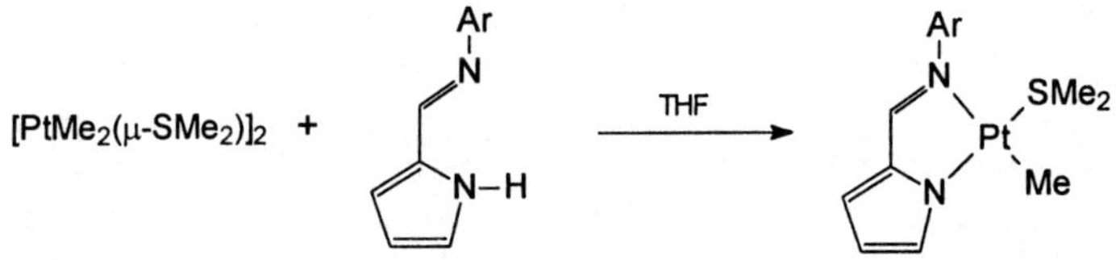

$$[PtMe_2(\mu\text{-}SMe_2)]_2 \; + \; \underset{\textbf{2a,c}}{\text{Ar-N=CH-pyrrole-NH}} \xrightarrow{\text{THF}} \underset{\textbf{14a: Ar = 4-MeO-Ph} \atop \textbf{14c: Ar = 4-CF}_3\text{-Ph}}{\text{complex}} \quad (6)$$

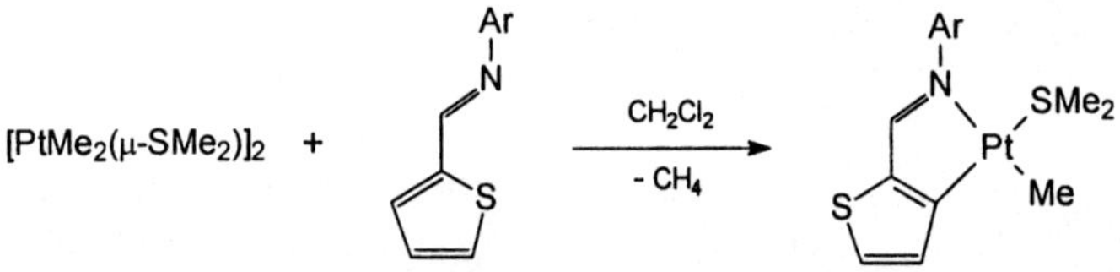

$$[PtMe_2(\mu\text{-}SMe_2)]_2 \; + \; \underset{\textbf{3}}{\text{Ar-N=CH-thiophene}} \xrightarrow[-CH_4]{CH_2Cl_2} \underset{\textbf{16: Ar = 2,6-}^i\text{Pr}_2\text{-Ph}}{\text{complex}} \quad (7)$$

11   **3**

$$[PtMe_2(\mu\text{-}SEt_2)]_2 \; + \; \underset{\textbf{2a}}{\text{Ar-N=CH-pyrrole-NH}} \xrightarrow[-CH_4]{C_6H_6} \underset{\textbf{17: Ar = 4-MeO-Ph}}{\text{complex}} \quad (8)$$

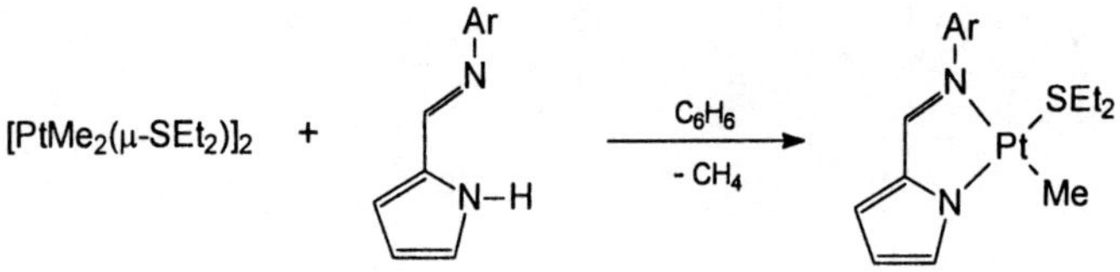

12

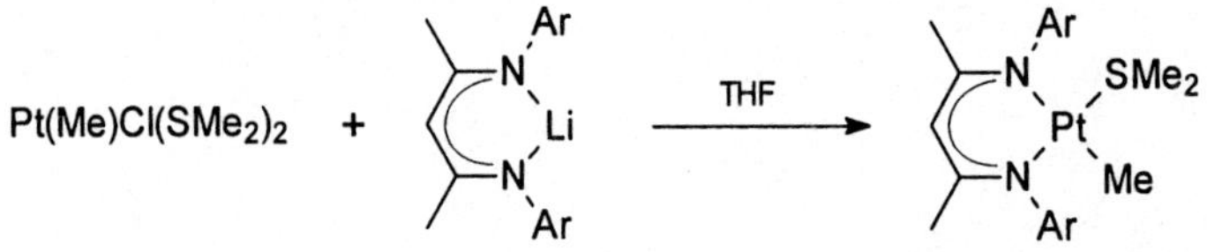

$$\text{Pt(Me)Cl(SMe}_2)_2 \; + \; \underset{\textbf{4}}{\text{Li-diketiminate}} \xrightarrow{\text{THF}} \underset{\textbf{18a: Ar = 4-MeO-Ph} \atop \textbf{18b: Ar = 4-Cl-Ph}}{\text{complex}} \quad (9)$$

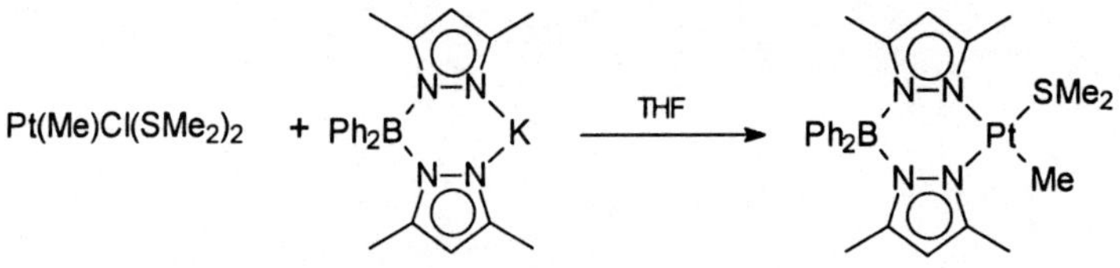

10

$$\text{Pt(Me)Cl(SMe}_2)_2 \; + \; \text{Ph}_2\text{B(pz}^{Me_2})_2\text{K} \xrightarrow{\text{THF}} \textbf{19} \quad (10)$$

10   **5**

Complexes **13**, **16** and **17** were also characterized by X-ray diffraction (Figure 2) and the effect of different chelate ligands on geometric parameters of the Pt-Me and Pt-SMe$_2$ moieties was examined. The Pt-C bond distances (Table II) are nearly invariable among the complexes; however, the Pt-S bond in **16** is ~0.1 Å longer than in either **13** or **17**. A related structure reported by Templeton has bond lengths similar to those observed in **13** and **17** (*22*). The Pt-C and Pt-S bonds in Tp'PtMe(SMe$_2$) are 2.049 and 2.249Å, respectively. Steric interactions between the 2,6-diisopropylphenyl substituent of the imine and the dimethylsulfide ligand, due to their mutual *cis* arrangement, is a likely reason for this lengthening. The longer bond is also reflected in the $^3J_{HPt}$ value of 29 Hz for **16** in comparison to ~51 Hz for complexes **14**.

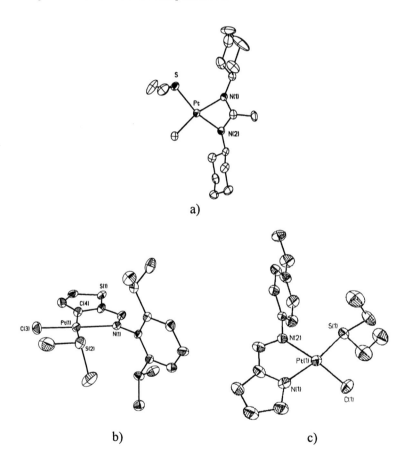

*Figure 2. Molecular Structures of a) PtMe(SMe$_2$)(CyNCMeNCy), **13**, b) PtMe(SMe$_2$)[2-(N-[2,6-(i-Pr)$_2$-Ph]imino)thiophenide], **16**, and c) PtMe(SEt$_2$)[2-(N-[4-MeO-Ph]imino)pyrollide], **17**.*

Table II. Selected bond lengths (Å) for complexes 13,16 and 17

Complex	Pt-Me	Pt-SMe$_2$
13	2.048(6)	2.238(2)
16	2.041(6)	2.352(2)
17	2.077(9)	2.249(2)

Redox Properties of Neutral Platinum Methyl Complexes

In previous work we determined the ease of oxidation of nitrogen-ligated platinum complexes from competition reactions with I_2 and also evaluated equilibrium constants between divalent methyl complexes and their tetravalent diiodides (9). Complexes **13-15** and **18** all reacted cleanly with iodine to afford the *trans* diiodides **20-23** (Eq. 11, complexes **21** and **22** are produced from **14** and **15**, respectively). Complex **19** did not react with iodine at room temperature over several days. Presumably, the steric bulk from the two phenyl rings attached to boron in addition to dimethyl substitution of the pyrazole rings prevents I_2 from reacting with the platinum metal center (23). The reduction of electron density in the tetravalent diiodides (vs. the Pt(II) starting compounds) is reflected in the greatly reduced values of J_{HPt} for the SMe$_2$ resonances (Table III) (24). Competition reactions with pairs of complexes indicated that the ease of oxidation of the divalent platinum methyl complexes is **13 > 18 > 15 > 14**.

This ordering was also reflected in estimates of the equilibrium constants between pairs of divalent methyl complexes and the corresponding tetravalent diiodides (Eq. 12). Representative equilibrium constants are given in Table IV. The reaction of **13** with **23**, for example, shows the equilibrium greatly favors

Table III. ^{1}H NMR chemical shifts and Pt-H Coupling Constants (Hz) of Tetravalent Methylplatinum Diiodides in C_6D_6

Complex	δ Pt-Me, $^2J_{HPt}$	δ Pt-SMe$_2$, $^3J_{HPt}$
20	2.65, 70	2.05, 31
21a	2.79, 70	1.72, 32
21b	2.74, 71	1.68, 33
22a	2.30, 67	2.05, 36
22b	2.14, 67	1.99, 38
23a	1.87, 67	1.66, 32

the amidinate ligated Pt(IV) complex as **23** converts completely to complex **18b** as judged by ^{1}H NMR spectroscopy. In accordance with electronic substitution effects of the aryl group, the more nucleophilic complex **14a** (*p*-OMe substituted) favors oxidation over **14b** (*p*-CF$_3$ substituted). Also, from these equilibrium oxidation studies it would appear that the *trans* products (**15**) of the iminopyrrolide ligated complexes are more electron rich than the corresponding *cis* products (**14**).

$$\begin{array}{c}\text{N'} \quad \text{SMe}_2\\ \text{Pt}\\ \text{N'} \quad \text{Me}\end{array} + \begin{array}{c}\text{N} \quad \text{I} \quad \text{SMe}_2\\ \text{Pt}\\ \text{N} \quad \text{I} \quad \text{Me}\end{array} \xrightleftharpoons{C_6D_6} \begin{array}{c}\text{N'} \quad \text{I} \quad \text{SMe}_2\\ \text{Pt}\\ \text{N'} \quad \text{I} \quad \text{Me}\end{array} + \begin{array}{c}\text{N} \quad \text{SMe}_2\\ \text{Pt}\\ \text{N} \quad \text{Me}\end{array} \quad (12)$$

Table IV. Equilibrium Constants for Reactions of Pt(II) with Pt(IV) Complexes in C_6D_6

Equilibrium system	K
[20][18a]/[13][23a]	≥ 13
[20][14a]/[13][21a]	≥ 400
[23a][14a]/[18a][21a]	3.0
[21a][14b]/[14a][21b]	2.8
[21b][15b]/[14b][22b]	2.9

Electrophilic Activation of Benzene C-H Bonds with Neutral Platinum Methyl Complexes

The reactivity of platinum methyl complexes **13-15, 18** and **19** with benzene at 85°C was monitored by ^{1}H NMR spectroscopy (Eq. 13). While most of the complexes reacted sluggishly (**13, 14a,b, 18b**) or not at all (**19**), iminopyrrolide complexes **15a,b,d** activated solvent C-D bonds to form a platinum phenyl product with concomitant formation of CH_3D within hours.

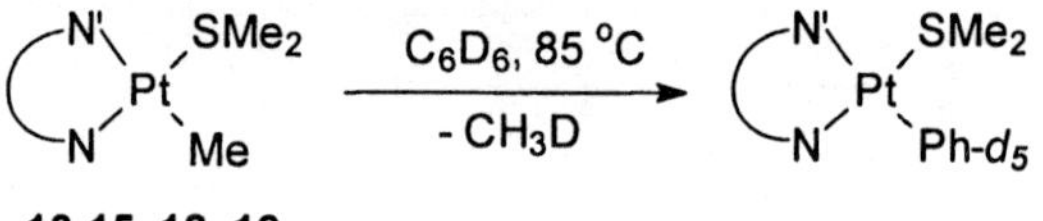
N'
SMe2
Pt
N
Me
C6D6, 85 °C
- CH3D
N'
SMe2
Pt
N
Ph-d5

13-15, 18, 19

(13)

Electronic effects on the rate of benzene C-H bond activation (cf. **15b** is ca. four times faster than **15a**) were consistent with an electrophilic reaction. More surprising was the ca. 80-fold faster reactions observed for **15** in which the Pt-Me is *trans* to the pyrrolide nitrogen, vs. isomeric **14** with Pt-Me *trans* to the imine nitrogen (Scheme 1). We suggest that rate-determining associative displacement of SMe_2 by benzene accounts for these differences (*25*) and the greater electronegativity of the ligand in **14b** and **15b** would tend to favor associative displacement.

Scheme 1

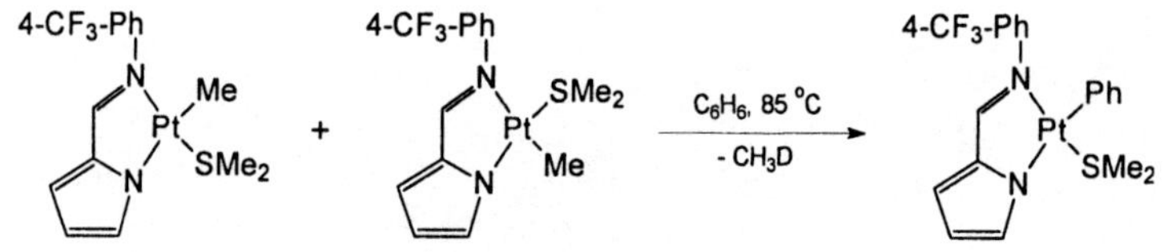
4-CF3-Ph
N
Me
Pt
SMe2
N
+
4-CF3-Ph
N
SMe2
Pt
Me
N
C6H6, 85 °C
- CH3D
4-CF3-Ph
N
Ph
Pt
SMe2
N

$t_{1/2}$ **15b:** 1 hr **14b:** 14 days

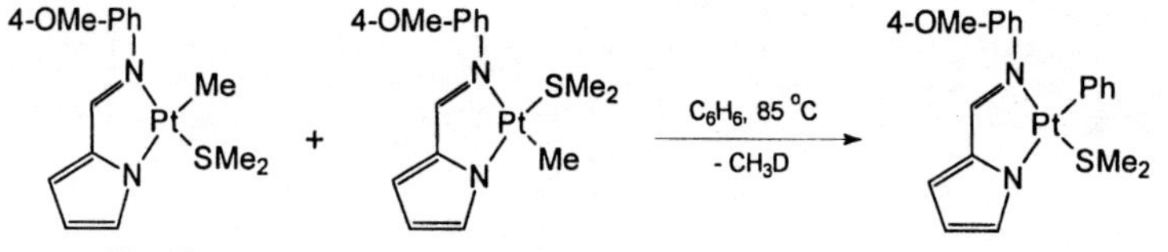
4-OMe-Ph
N
Me
Pt
SMe2
N
+
4-OMe-Ph
N
SMe2
Pt
Me
N
C6H6, 85 °C
- CH3D
4-OMe-Ph
N
Ph
Pt
SMe2
N

$t_{1/2}$ **15a**:4 hr **14a:** 3 days

An associative mechanism is favored in systems with little steric hindrance and also where the leaving group is *trans* to a poor donor (*26*). However, the trigonal bipyramidal intermediate for such displacement in *trans* complexes **15** would have the better π-acceptor imine ligand equatorial and the better σ-donor pyrrolide axial, the strongly preferred conformation (Scheme 2). Slower displacement in the *cis* isomer must pass through a higher-energy intermediate and is accompanied by isomerization to the thermodynamically preferred *cis* isomer of the platinum phenyl product.

Scheme 2

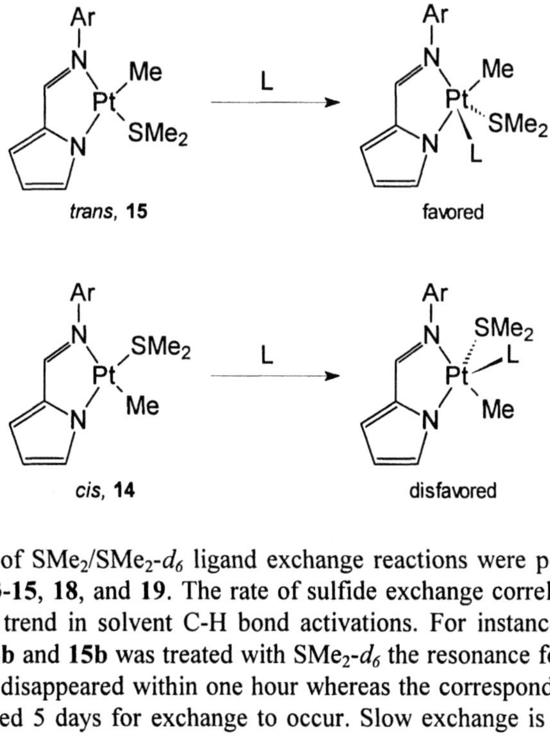

A series of SMe_2/SMe_2-d_6 ligand exchange reactions were performed with complexes **13-15**, **18**, and **19**. The rate of sulfide exchange correlates well with the observed trend in solvent C-H bond activations. For instance, when a 1:1 mixture of **14b** and **15b** was treated with SMe_2-d_6 the resonance for coordinated SMe_2 in **15b** disappeared within one hour whereas the corresponding resonance in **14b** required 5 days for exchange to occur. Slow exchange is also observed for complexes **13** and **18a**. The borate complex **19** is again the slowest in terms of reactivity, not showing any appreciable exchange at room temperature after several days. In addition, the presence of excess sulfide isomerizes **15b** to the more stable thermodynamic complex **14b**.

Sulfide exchange is also observed between complexes in a crossover experiment involving co-thermolysis of **14e** and **17** in C_6D_6 (Eq. 14); however, the process is very sluggish with respect to the complex/ligand exchange reactions. The mixture requires heating at 85 °C for 48 hours before significant

(~ 35%) sulfide exchange is observed. C–D bond activation of the solvent was also observed but it occurs at an even slower rate than the ligand crossover.

$$\text{(14)}$$

In contrast to these sluggish reactions, reaction of **11** with bulky iminopyrrole **2d** in perdeuterobenzene at 25°C over several hours afforded two equivalents of methane isotopomers and *trans*-phenyl complex **24** (Eq. 15). Ca. 10% of *trans* methyl complex **15e** was also formed and methyl groups of the isopropyl substituents were partially deuterated (*27*). In this reaction we propose

$$1/2\ [PtMe_2(\mu\text{-}SMe_2)]_2\ +\ \mathbf{2d}\ \xrightarrow{C_6D_6}\ \mathbf{24d} + 2\ CH_{4-x}D_x \quad (15)$$

24d: Ar = 2,6-^{i}Pr$_2$Ph
+ 10% **15d**

that benzene C-H bond activation must *precede* chelation since preformed complexes **15** only react with benzene at elevated temperature. The fact that only *trans* products are observed in equation 15 is presumably a consequence of steric interactions between the SMe$_2$ ligand and the *ortho*-isopropyl substituents.

Conclusions

Neutral platinum methyl complexes have been shown to be easily oxidized to tetravalent complexes and to activate benzene C-H bonds under mild conditions. In this study, the inherent reactivity of *N,N*-chelated platinum methyl complexes toward hydrocarbon C-H bonds was masked by the need to substitute strongly coordinating dialkylsulfide ligands. Use of unsymmetrical iminopyrroles and iminopyrrolide ligands shed some light on this process and further work is in progress (*28*) to identify nitrogen-ligated systems for alkane functionalization outside of strong acid media.

Acknowledgements

Akzo Nobel, Inc., the US Department of Energy's Office of Industrial Technologies, the Laboratory-Directed Research and Development program at Los Alamos, and BP are gratefully acknowledged for support and CNI thanks Los Alamos for a Director's Funded Postdoctoral Fellowship.

References

1. Gol'dshleger, N. F.; Es'kova, V. V.; Shilov, A. E.; Shteinman, A. A. *Zh. Fiz. Khim. (Engl. Transl.)* **1972**, *46*, 785.
2. a) Kushch, L. A.; Lavrushko, V. V.; Misharin, Y. S.; Moravsky, A. P.; Shilov, A. E. *Nouv. J. Chem.* **1983**, *7*, 729. b) Horvath, I. T.; Cook, R. A.; Millar, J. M.; Kiss, G. *Organometallics* **1993**, *12*, 8. c) Hutson, A. C.; Lin, M. R.; Basickes, N.; Sen, A. *J. Organomet. Chem.* **1995**, *504*, 69-74. d) Zamashchikov, V. V.; Popov, V. G.; Rudakov, E. S.; Mitchenko, S. A. *Dokl. Akad. Nauk SSSR* **1993**, *333*, 34.
3. a) Labinger, J. A.; Herring, A. M.; Lyon, D. K.; Luinstra, G. A.; Bercaw, J. E.; Horvath, I. T.; Eller, K. *Organometallics* **1993**, *12*, 895. b) Luinstra, G. A.; Labinger, J. A.; Bercaw, J. E. *J. Am. Chem. Soc.* **1993**, *115*, 3004. c) Luinstra, G. A.; Wang, L.; Stahl, S. S.; Labinger, J. A.; Bercaw, J. E. *Organometallics* **1994**, *13*, 755. d) Luinstra, G. A.; Wang, L.; Stahl, S. S.; Labinger, J. A.; Bercaw, J. E. *J. Organomet. Chem.* **1995**, *504*, 75. e) Stahl, S. S.; Labinger, J. A.; Bercaw, J. E. *J. Am. Chem. Soc.* **1995**, *117*, 9371-9372. f) Stahl, S. S.; Labinger, J. A.; Bercaw, J. E. *J. Am. Chem. Soc.* **1996**, *118*, 5961-5976.
4. Periana, R. A.; Taube, D. J.; Gamble S.; Taube, H.; Satoh, T.; Fujii, H. *Science* **1998**, *280*, 560-564.
5. Periana's catalytic system of has been examined in detail *via* theoretical methods. See: a) Kua, J.; Xu, X.; Periana, R. A.; Goddard, III, W. A. *Organometallics* **2002**, *21*, 511-525. b) Gilbert, T. M.; Hristov, I.; Ziegler, T. *Organometallics* **2001**, *20*, 1183-1189. c) Mylvaganam, K.; Bacskay, G. B.; Hush, N. S. *J. Am. Chem. Soc.* **2000**, *122*, 2041-2052.
6. a) Holtcamp, M. H.; Labinger, J. A.; Bercaw, J. E. *J. Am. Chem. Soc.* **1997**, *119*, 848-849. b) Johansson, L.; Tilset, M.; Labinger, J. A.; Bercaw, J. E *J. Am. Chem. Soc.* **2000**, *122*, 10846-10855. c) Zhong, H. A.; Labinger, J. A.; Bercaw, J. E. *J. Am. Chem. Soc.* **2002**, 124, 1378-1399.

7. a) Johansson, L.; Ryan, O. B.; Tilset, M. *J. Am. Chem. Soc.* **1999**, *121*, 1974-1975. b) Johansson, L.; Ryan, O. B.; Romming, C.; Tilset, M. *J. Am. Chem. Soc.* **2001**, *123*, 6579-6590.
8. a) Rostovtsev, V. V.; Labinger, J. A.; Bercaw, J. E.; Lasseter, T. L.; Goldberg, K. I. *Organometallics* **1998**, *17*, 4530-4531. b) Rostovtsev, V. V.; Henling, L. M.; Labinger, J. A.; Bercaw, J. E. *Inorg. Chem.* **2002**, *41*, 3608-3619.
9. Scollard, J. D.; Day, M.; Labinger, J. A.; Bercaw, J. E. *Helv. Chim. Acta* **2001**, *84*, 3247-3268.
10. a) Baker, R. T.; Watkin, J.; Carter, C. A. G.; Bercaw, J. E.; Labinger, J. A.; Whitwell, G. E., Abstr. I&EC 7, 218th ACS Meeting, New Orleans, LA, August 22, **1999**. b) Carter, C. A. G.; Day, M. W.; John, K. D.; Scollard J. D.; Baker, R. T.; Bercaw, J. E.; Labinger, J. A., Abstr. INOR 161 , 223rd ACS Meeting, Orlando, FL, April 8, **2002**.
11. Wick, D. D.; Goldberg, K. I. *J. Am. Chem. Soc.* **1997**, *119*, 10235-10236.
12. Fekl, U.; Kaminsky, W.; Goldberg, K. I. *J. Am. Chem. Soc.* **2001**, *123*, 6579-6590.
13. a) Reinartz, S.; White, P. S.; Brookhart, M.; Templeton, J. L. *J. Am. Chem. Soc.* **2001**, *123*, 12724-12725. b) Reinartz, S.; White, P. S.; Brookhart, M.; Templeton, J. L. *Organometallics*, **2001**, *20*, 1709-1712.
14. a) Lo, H. C.; Haskel, A.; Kapon, M.; Keinan, E. *J. Am. Chem. Soc.* **2002**, *124*, 3226-3228. b) Iron, M. A.; Lo, H. C.; Martin, J. M. L.; Keinan, E. *J. Am. Chem. Soc.* **2002**, *124*, 7041-7054.
15. Thomas, J. C.; Peters, J. C. *J. Am. Chem. Soc.* **2001**, *123*, 5100-5101.
16. Reviews on amidinate chemistry: a) Edelman, F. T. *Coord. Chem. Rev.* **1994**, *137*, 403. b) Barker, J.; Kilner, M. *Coord. Chem. Rev.* **1994**, *133*, 219.
17. iminopyrrole ligands: Tanaka, T.; Tamauchi, O. *Chem. Pharm. Bull.* **1961**, *9*, 588.
18. iminothiophene ligand: Drisko, R. W.; McKennis, Jr., H. *J. Am. Chem. Soc.* **1952**, *74*, 2626-2628.
19. β-diketiminate (nacnac) ligands: Parks, J. E.; Holm, R. H. *Inorg. Chem.* **1968**, *7*, 1408-1416. b) McGeachin, S. G. *J. Can. Chem.* **1968**, *46*, 1903-1912.
20. Komorwski, L.; Maringgele, W.; Meller, A.; Niedenzu, K.; Serwatowski, J. *Inorg. Chem.* **1990**, *29*, 3845-3849.
21. β-metalation of iminothiophene by dinuclear metal complexes has been reported previously: Wang, D.-L.; Hwang, W.-S.; Liang, L.-C.; Wang, L.-I.; Lee, L.; Chiang, M. Y. *Organometallics.* **1997**, *16*, 3109-3113; Imhof, W. *J. Organomet. Chem.* **1997**, *533*, 31-43.
22. Reinartz, S. White, P. S.; Brookhart, M.; Templeton, J. L. *Organometallics* **2000**, *19*, 3854-3866.

23. Oxidative addition of methyl iodide to a similar square planar Pt(II) complex supported by the less sterically demanding bis(pyrazolyl)ethane ligand has been reported. Byers, P. K.; Canty, A. J.; Honeyman, R. T.; Skelton, B. W.; White, A. H. *J. Organomet. Chem.* **1992**, *433*, 223-229.
24. Only a small reduction in $^2J_{HPt}$ values is seen for the Pt-Me group protons of complexes **20-23**. These values, however, are typical for methyl groups *trans* to nitrogen ligands in Pt(IV) complexes. For example, see refs. 11, 12, 20, and: a) Canty, A. J.; Honeyman, R. T. *J. Organomet. Chem.* **1990**, *387*, 247-263. b) Prokopchuk, E. M.; Puddephatt, R. J. *Organometallics* **2003**, *22*, 563-566. c) Hinman, J. G.; Baar, C.R.; Jennings, M. C.; Puddephatt, R. J. *Organometallics* **2000**, *19*, 563-570. d) Hill, G. S.; Puddephatt, R. J. *J. Am. Chem. Soc.* **1996**, *118*, 8745-8746
25. A recent paper detailed volume of activation studies with respect to the mechanism of benzene C–H bond activation by related cationic platinum-methyl complexes. This report supports a benzene based associative mechanism. Procelewska, J.; Zahl, A.; van Eldik, R.; Zhong, H. A.; Labinger, J. A.; Bercaw, J. E. *Inorg. Chem.* **2002**, *41*, 2808-2810.
26. a) Frey, U.; Helm, L.; Merbach, A. E.; Romeo, R. *J. Am. Chem. Soc.* **1989**, *111*, 8161-8165. b) Pienaar, J. J.; Kotowski, M.; van Eldik, R. *Inorg. Chem.* **1989**, *28*, 373-375. c) Cooper, J.; Ziegler, T. *Inorg. Chem.* **2002**, *41*, 6614-6622.
27. Similar exchange has been observed in a related system: Fekl, U.; Goldberg, K. I. *J. Am. Chem. Soc.* **2002**, *124*, 6804-6805.
28. Iverson, C. N.; Carter, C. A. G.; Baker, R. T.; Scollard, J. D.; Labinger, J. A.; Bercaw, J. E., *J. Am. Chem. Soc.*, **2003**, *125*, 12674-12675.

Chapter 20

Issues Relevant to C–H Activation at Platinum(II): Comparative Studies between Cationic, Zwitterionic, and Neutral Platinum(II) Compounds in Benzene Solution

Jonas C. Peters, J. Christopher Thomas, Christine M. Thomas, and Theodore A. Betley

Division of Chemistry and Chemical Engineering, California Institute of Technology, Pasadena, CA 91125

Cationic late metal systems are being highly scrutinized due to their propensity to mediate so-called *electrophilic* C-H activation reactions. This contribution compares the reactivity of highly reactive cationic platinum(II) systems with structurally related but neutral species. Our experimental design exploits isostructural neutral and cationic complexes supported by bis(phosphine) ligands amenable to mechanistic examination in benzene solution. The data presented herein collectively suggests that neutral platinum complexes can be equally if not more reactive towards benzene than their cationic counter-parts. Moreover, a number of unexpected mechanistic distinctions between the two systems arise that help to explain their respective reactivity.

Introduction

Recent decades have seen a flurry of interest in homogeneous organometallic species capable of mediating C-H bond activation processes (*1*). To develop inorganic systems for productive alkane or arene functionalization chemistry, an intimate understanding of the dominant factors controlling C-H activation processes is paramount. Mechanistic model studies that expose key factors controlling rates of substrate coordination, C-H bond activation, and kinetic selectivity are critical to the systematic design and re-design of potential catalyst systems.

As of late, much effort has focused on developing highly electrophilic, late metal (e.g., Rh, Ir, Pt, Hg) C-H activation systems, the presumption being that "electrophilic" systems efficiently coordinate alkanes (σ-complexes) and mediate subsequent C-H bond activation processes (*1-7*). This has been particularly true of cationic platinum(II) systems. Despite this fact, a collection of recent papers has put into question whether more electrophilic systems are more active for C-H activation (*3*). Indeed, several neutral platinum(II) systems are now known that efficiently mediate intermolecular C-H activation processes. The first such system was reported by Goldberg and coworkers in 1997 (*5*). In a key experiment, they showed that the anionic complex K[Tp'PtMe$_2$] reacts with B(C$_6$F$_5$)$_3$ to generate an unobserved, presumed 3-coordinate species, "[η^2-Tp']PtMe", that undergoes oxidative C-H bond addition to afford octahedral platinum(IV) alkyl hydride products, [η^3-Tp']Pt(Me)(R)H. Building on this report, our group later identified a well-defined, neutral platinum(II) complex, [Ph$_2$BP$_2$]Pt(Me)(THF) ([Ph$_2$BP$_2$] = [Ph$_2$B(CH$_2$PPh$_2$)$_2$]), that underwent aromatic C-H bond activation processes to afford platinum(II) byproducts (e.g., [Ph$_2$BP$_2$]Pt(Ph)(THF) in benzene) (*4*). We highlighted the zwitterionic nature of this neutral complex, owing to the fact that its anionic [Ph$_2$BP$_2$] ligand contains a partially insulated borate counter-anion. The C-H activation reaction exhibited by [Ph$_2$BP$_2$]Pt(Me)(THF) is hence reminiscent of its cationic platinum(II) relatives.

Several intriguing questions arise from our neutral [Ph$_2$BP$_2$]Pt(II) system. (1) How important/appropriate is a zwitterionic resonance contributor in describing the electronic distribution of [Ph$_2$BP$_2$]Pt(X)(L) systems? (2) Would an isostructural but formally cationic complex, (P-P)Pt(X)(L)$^+$, exhibit similar C-H bond activation chemistry? (3) If so, how would the mechanisms by which the neutral and cationic complexes mediate such processes compare? Which system would react faster? What factors would govern their relative C-H bond activation rates? (4) How would the reactivity of neutral platinum(II) complexes that feature other donor ligands (e.g., N-donor ligands) compare?

Each of the above issues is likely to be system dependent. We therefore identified a need for a study that would cross-compare the reactivity of several structurally related, neutral and cationic platinum(II) systems as shown in Figure 1. In this contribution, we discuss our recent progress in this undertaking (8).

Ligand systems

The top row displayed in Figure 1 shows three neutral platinum(II) systems (A, C, and E) that are supported by a bidentate, uni-negative donor ligand containing a diphenylborate unit tethered to the metal via its donor arms. In this contribution, X represents a methyl ligand, and L represents a neutral donor ligand. While most of our work to date has focused on system A, we have more recently begun to examine systems of the type C and E for comparison. Aside from the obvious change in the donor group of the three systems, the possible degree of borate charge insulation is likely different. For example, in system C convenient resonance contributors are available to distribute the anionic charge throughout the ligand. In systems A and E, on the other hand, aromatic delocalization of the borate charge is not available and the zwitterionic resonance depiction is *perhaps* more appropriate.

While bis(pyrazolyl)borate ligands have been known for decades (9), bis(phosphino)- and bis(amino)borate ligands were unknown at the outset of our study. It was therefore necessary to tackle their syntheses first. The preparation of a variety of phenyl-substituted phosphine systems, [R$_2$BPR'$_2$] (10), and a methyl-substituted amine system, [Ph$_2$BN^{Me}$_2$] (11), has been reported.

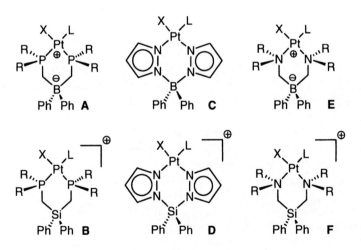

Figure 1. Design scheme to examine factors that influence C-H activation reactivity at platinum(II) centers.

{[Ph$_2$BL$_2$]PtMe$_2$}⁻ Complexes

While rapid displacement of COD from (COD)PtMe$_2$ by [R$_2$BPR'$_2$][NR$_4$] salts in general occurred at 22 °C, ammonium salts of [Ph$_2$B(pz)$_2$] and [Ph$_2$BN^{Me}$_2$] were much less reactive. Dimeric [Me$_2$Pt(SMe$_2$)]$_2$ reacted more rapidly with both N-donor borate ligands, affording [[Ph$_2$B(pz)$_2$]PtMe$_2$][NBu$_4$] and [[Ph$_2$BN^{Me}$_2$]PtMe$_2$][NEt$_4$], respectively (ASN = 5-azonia-spiro[4.4]nonane; pz = pyrazolyl). The complexes [[Ph$_2$BP$_2$]PtMe$_2$][ASN] and [[Ph$_2$B(pz)$_2$]PtMe$_2$][NBu$_4$] have been structurally characterized, and the $^2J_{PtH}$ NMR coupling constant for each complex has been measured (see Figure 2). We presume the conformation of the bis(amino)borate complex generally mimics that observed for the bis(phosphino)borate derivatives, by analogy to structurally characterized square planar [Ph$_2$BN^{Me}$_2$]Rh complexes (*11*). The borate anion in each system lies along a vector that bisects the midpoint of a square planar complex, spatially well separated from the coordinated metal center. For platinum bis(phosphino)borate complexes, the M-B distance is typically between 4.0 Å and 4.2 Å. For bis(amino)borate complexes, we estimate this distance to be between 3.65 and 3.68 Å. The conformation of square planar bis(pyrazolyl)borate complexes, in contrast, allows the borate unit to slide closer to the coordinated metal center in the solid-state. The range of M-B distances observed is quite broad (M = Rh, Pd, Pt), between 3.0 and 3.5 Å, attributable to a flexible degree of puckering exhibited by the bidentate ligand. The $^2J_{PtH}$ NMR coupling constants measured for the anionic dimethyl complexes suggest that the (phosphino)borate exhibits a stronger trans influence than the comparable bis(pyrazolyl)- and bis(amino)borate derivatives. The relative $^2J_{PtH}$ coupling constants of [[Ph$_2$BP$_2$]PtMe$_2$][ASN] and [[Ph$_2$B(pz)$_2$]PtMe$_2$][NBu$_4$] track well with their observed Pt-Me distances, the average distance in the (phosphino)borate derivative being ca. 0.1 Å longer than the (pyrazolyl)borate system.

[Ph$_2$BP$_2$]PtMe(L) and [Ph$_2$B(pz)$_2$]PtMe(L) complexes

The anionic complexes [[Ph$_2$B(pz)$_2$]PtMe$_2$][NBu$_4$] and [[Ph$_2$BP$_2$]PtMe$_2$][ASN] are easily protonated by weak acids in moderately polar solvents such as THF. Their basicity provides straightforward access to complexes of the type [Ph$_2$BL$_2$]PtMeL' by judicious choice of a bulky ammonium salt whose conjugate base is non-coordinating (e.g., HN^{i}Pr$_2$Et$^+$). This allows selective delivery of an L' donor ligand. For the bis(phosphino)borate system, a wide variety of L' ligands have been canvassed (e.g., CO, PMe$_3$, CH$_3$CN, SMe$_2$). For the C-H activation studies relevant to this contribution, we focus only on the case where L' is THF. The synthesis and structural characterization of the feature complex [Ph$_2$BP$_2$]Pt(Me)(THF) has been described (4,8). The bis(pyrazolyl)borate complexes [Ph$_2$B(pz)$_2$]PtMeL' (L' = CO, PMe$_3$, CH$_3$CN, SMe$_2$) are similarly available, though the THF-adduct complex is too reactive to be isolated and rigorously characterized. The bis(amino)borate precursors, [Ph$_2$BN^{Me}$_2$]PtMeL', have yet to be systematically developed.

Figure 2. Noteworthy NMR and structural parameters collected for the anionic dimethyl complexes [[Ph$_2$B(pz)$_2$]PtMe$_2$][NBu$_4$], [[Ph$_2$BP$_2$]PtMe$_2$][ASN], and [[Ph$_2$BN^{Me}$_2$]PtMe$_2$][NEt$_4$].

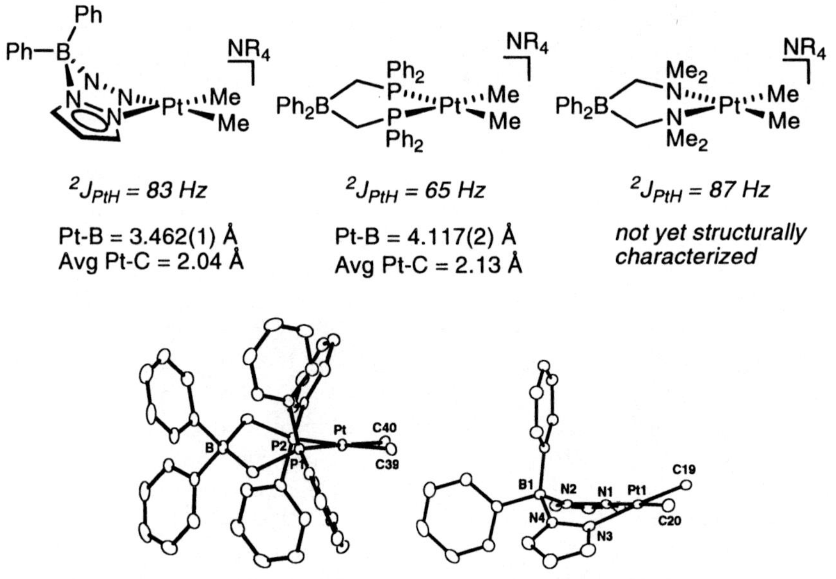

Electronic information from carbonyl model complexes

To gain an appreciation of the relative electron-releasing character of the various [Ph$_2$BL$_2$] ligands, it is instructive to consider relative CO stretching frequencies for a structurally comparable set of square planar carbonyl complexes. For the three borate ligand types under consideration, we have obtained pertinent data using rhodium(I) dicarbonyl and platinum(II) monocarbonyl platforms. We have also collected infrared data for several isostructural but formally cationic complexes. This data is summarized in Figure

3. Several conclusions can be drawn from the data. First, despite the zwitterionic resonance contributor we have emphasized, neutral complexes supported by either the [Ph$_2$BP$_2$] ligand or the [Ph$_2$BN$^{Me}{}_2$] ligand are appreciably less electrophilic than their cationic counter-parts in which a diphenylsilane unit substitutes the diphenylborate unit of the ligand backbone. For the platinum system in particular, the cationic complex [(Ph$_2$SiP$_2$)Pt(Me)(CO)][B(C$_6$F$_5$)$_4$] provides a CO stretching vibration that is 24 cm^{-1} higher in energy than its neutral derivative [Ph$_2$BP$_2$]Pt(Me)(CO) (Ph$_2$SiP$_2$ refers to the neutral ligand Ph$_2$Si(CH$_2$PPh$_2$)$_2$). This implies that the cationic complex is significantly more electrophilic than its neutral congener, a conclusion that would seem to suggest there is a significant degree of electronic communication between the borate anion and the platinum center in the neutral complexes. However, examination of a series of *para*-substituted aryl bis(phosphino)borates shows that substitution on the aryl borate has almost no effect on $v_{(CO)}$, while *para*-substitution of the aryl phosphine shows a stronger effect (Figure 3) (*10*). Considered independently, the *para*-substitution data is consistent with a localized resonance contributor for anionic bis(phosphino)borate ligands (resonance form **A** in Figure 4), whereas the cationic versus neutral comparative data is perhaps more consistent with resonance contributors that emphasize an ylide form of the bis(phosphino)borate ligands (e.g., resonance forms **B** and **C** in Figure 4).

To reconcile the two seemingly disparate conclusions, it needs to be underscored that the absolute magnitude in $\Delta v_{(CO)}$ measured between a zwitterionic complex (e.g., [Ph$_2$BP$_2$]Pt(Me)(CO)) and a cationic complex (e.g., [(Ph$_2$SiP$_2$)Pt(Me)(CO)][BAr$_4$]) is somewhat misleading for the following reason. In cationic late metal carbonyls, where π-backbonding is relatively weak, strong polarization of the CO σ-bond by the cationic complex is anticipated (*12*). This raises the energy of the force constant F_{CO} dramatically, and in extreme cases, polarization can dominate the observed F_{CO}. In this context, cationic complexes such as [(Ph$_2$SiP$_2$)PtMe]$^+$ are expected to have characteristically high force constants F_{CO} due to a strong polarization effect. This effect will be much reduced for CO coordinated to a neutral [Ph$_2$BP$_2$]PtMe fragment, regardless of whether or not its charge is distributed asymmetrically due to a zwitterionic resonance contributor. We therefore caution that the absolute magnitude of $\Delta v_{(CO)}$ is not so reliable a gauge of relative backbonding ability between complexes that are *formally* cationic and complexes that are *formally* neutral. Large differences in polarization between isostructural cationic and neutral complexes likely compete with the electronic contributors of sigma donation, π-backbonding, and/or π-acceptor character that we typically rely upon to correlate measured $\Delta v_{(CO)}$ values to the electron-releasing character of a ligand. The dilemma of variable polarization is avoided by examining a contiguous series of neutral platinum carbonyl complexes (*10*). The relative differences in $v_{(CO)}$ values recorded in such a series are likely more reflective of the relative "electron-releasing" character of the donor ligand, a point we have discussed elsewhere (10).

One other interesting point to note from these infrared data is that the complex [Ph$_2$B(pz)$_2$]Pt(Me)(CO) provides a CO stretch 7 cm^{-1} lower in energy than in [Ph$_2$BP$_2$]Pt(Me)(CO), suggesting that the former complex, at least according to this gauge, is the more electron-rich of the two.

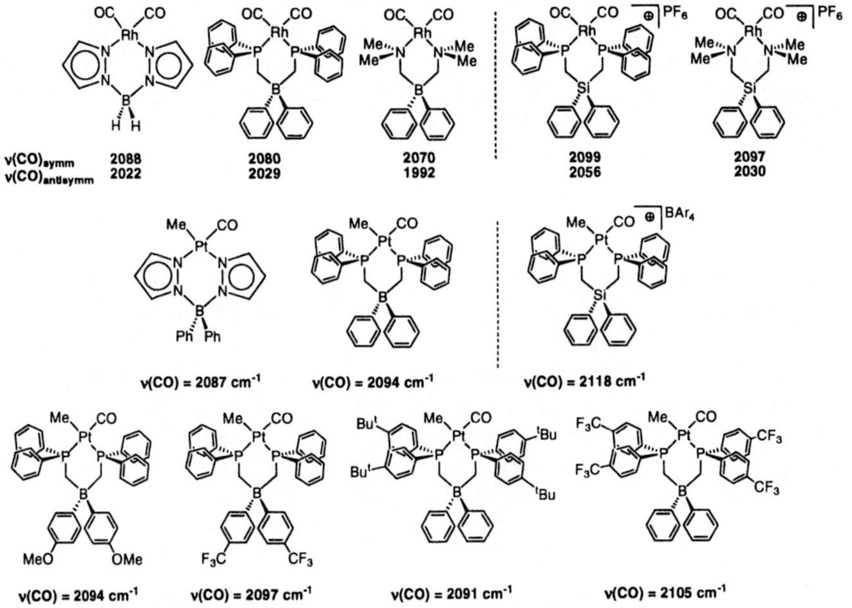

v(CO)$_{symm}$	2088	2080	2070	2099	2097
v(CO)$_{antisymm}$	2022	2029	1992	2056	2030

v(CO) = 2087 cm^{-1} v(CO) = 2094 cm^{-1} v(CO) = 2118 cm^{-1}

v(CO) = 2094 cm^{-1} v(CO) = 2097 cm^{-1} v(CO) = 2091 cm^{-1} v(CO) = 2105 cm^{-1}

Figure 3. Infrared data collected for a series of rhodium and platinum carbonyl complexes. All data are reported in cm^{-1} obtained by FTIR in CH$_2$Cl$_2$/KBr (BAr$_4$ = B(C$_6$F$_5$)$_4$$^-$). The complex [H$_2$B(pz)$_2$]Rh(CO)$_2$ was reported previously by Flavio et. al. (13).

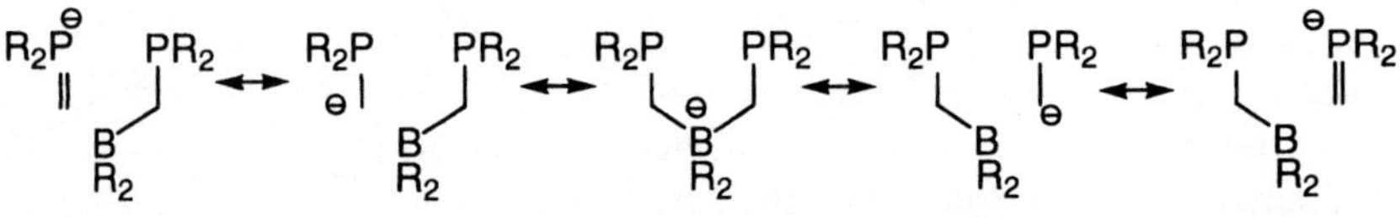

Figure 4. Selected resonance contributors for bis(phosphino)borate ligands.

Mechanistic Comparisons Between Isostructural Neutral and Cationic Complexes

In this section we discuss the reaction profiles of the structurally related neutral and cationic platinum(II) systems, [Ph$_2$BP$_2$]Pt(Me)(THF) and [(Ph$_2$SiP$_2$)Pt(Me)(THF)][B(C$_6$F$_5$)$_4$], each of which is capable of mediating an elementary C-H bond activation process in benzene solution. Several important and unexpected mechanistic distinctions between the two systems are revealed.

Both [Ph$_2$BP$_2$]Pt(Me)(THF) and [(Ph$_2$SiP$_2$)Pt(Me)(THF)][B(C$_6$F$_5$)$_4$] are appreciably soluble in benzene solution. This fact allowed us to study their respective benzene solution chemistries by NMR spectroscopy under comparable reaction conditions. [Ph$_2$BP$_2$]Pt(Me)(THF) reacts in benzene solution at 50 °C over a period of several hours to form [Ph$_2$BP$_2$]Pt(Ph)(THF) as the major product (~ 80%) with concomitant liberation of methane. The cationic complex [(Ph$_2$SiP$_2$)Pt(Me)(THF)]$^+$ reacted similarly, liberating methane to produce the corresponding phenyl derivative [(Ph$_2$SiP$_2$)Pt(Ph)(THF)]$^+$. Neither of the phenyl products proved stable to extended thermolysis; however, the addition of excess THF significantly inhibited the degradation of both [Ph$_2$BP$_2$]Pt(Ph)(THF) and [(Ph$_2$SiP$_2$)Pt(Ph)(THF)]$^+$. To probe these downstream reaction processes, the independently prepared phenyl complexes were warmed and monitored in benzene solution. Prolonged thermolysis of [Ph$_2$BP$_2$]Pt(Ph)(THF) resulted in two apparent reaction pathways. The dominant pathway was that of disproportionation to generate the colorless molecular salt {[Ph$_2$BP$_2$]Pt(Ph)$_2$}$^-${[Ph$_2$BP$_2$]Pt(THF)$_2$}$^+$. Formation of this cation/anion pair was suggested by the appearance of two singlets (^{31}P{^{1}H} NMR) in a 1:1 ratio, and by a positive identification of each ion by electrospray mass spectroscopy. Additionally, the species [[Ph$_2$BP$_2$]Pt(Ph)$_2$][ASN] was independently prepared to corroborate its spectroscopic and electrospray mass spectral data. A small amount of a presumed bridged-biphenyl species was also evident by ^{31}P{^{1}H} NMR. This minor species was isolated in very low yield (~ 10%) by fractional crystallization, and an independent XRD study on crystals of this minor species established a connectivity consistent with the dinuclear complex {[Ph$_2$BP$_2$]Pt}$_2$(μ-η^3:η^3-biphenyl). A similar bridged-biphenyl species was the only product observed in the cationic system. Thus, both spectroscopic and structural evidence established that [(Ph$_2$SiP$_2$)Pt(Ph)(THF)][B(C$_6$F$_5$)$_4$] decayed quantitatively to the orange dinuclear species, [{(Ph$_2$SiP$_2$)Pt}$_2$(μ-η^3:η^3-biphenyl)][B(C$_6$F$_5$)$_4$]$_2$. During the course of this study, Konze, Scott, and Kubas reported a related coupling reaction for cationic complexes of the type [(R$_2$PC$_2$H$_4$PR$_2$)PtMe(OEt$_2$)][B(3,5-(CF$_3$)$_2$C$_6$H$_3$)$_4$] *(14)*.

Fortunately, the second, bimolecular reaction is in each case slow. Clean first order kinetics could therefore be obtained for the decay of [Ph$_2$BP$_2$]Pt(Me)(THF) and [(Ph$_2$SiP$_2$)Pt(Me)(THF)][B(C$_6$F$_5$)$_4$], respectively. To our initial surprise, the rate of decay of the neutral system was significantly faster. At 55 °C the rate of decay of [(Ph$_2$SiP$_2$)Pt(Me)(THF)][B(C$_6$F$_5$)$_4$] was 1.80(6) x 10^{-4} s^{-1} (t$_{1/2}$ = 64 min), while the rate of decay of [Ph$_2$BP$_2$]Pt(Me)(THF) was too fast to be accurately measured (^{31}P NMR). The decay rate of [Ph$_2$BP$_2$]Pt(Me)(THF) at 45 °C was 1.42(5) x 10^{-4} s^{-1} (t$_{1/2}$ = 81 min). The difference in rates was much more pronounced in C$_6$D$_6$ due to a large kinetic deuterium isotope effect in the cationic system (k$_H$/k$_D$ = 6.52) and a negligible isotope effect in the neutral system (k$_H$/k$_D$ = 1.26). These data are shown graphically in Figures 5(A), (B), and (C).

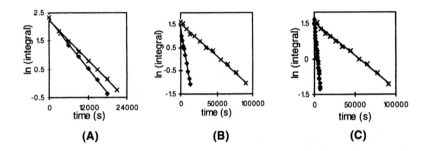

Figure 5. Representative plots of: (A) [Ph$_2$BP$_2$]Pt(Me)(THF) in C$_6$H$_6$ (♦) and in C$_6$D$_6$ (×) acquired at 45 °C; (B) [(Ph$_2$SiP$_2$)Pt(Me)(THF)][B(C$_6$F$_5$)$_4$] in C$_6$H$_6$ (♦) and in C$_6$D$_6$ (×) acquired at 55 °C; (C) [Ph$_2$BP$_2$]Pt(Me)(THF) (♦) and [(Ph$_2$SiP$_2$)Pt(Me)(THF)][B(C$_6$F$_5$)$_4$] (×) in C$_6$D$_6$ at 55 °C.

Also curious were the methane byproducts for each of these thermolysis reactions. When [(Ph$_2$SiP$_2$)Pt(Me)(THF)][B(C$_6$F$_5$)$_4$] was incubated in C$_6$D$_6$, the methane byproduct was predominantly CH$_3$D, but a small amount of CH$_4$ was also observed (*CH$_4$: CH$_3$D = 1 : 7.6*). When neutral [Ph$_2$BP$_2$]Pt(Me)(THF) was similarly incubated, CH$_4$ was the dominant byproduct (*CH$_4$: CH$_3$D = 3 : 1*). The ratio of methane byproducts for the latter system indicates that protons from either the [Ph$_2$BP$_2$] ligand, or from THF, are incorporated into the methane byproduct to a large extent. Control experiments have established that these protons arise from the [Ph$_2$BP$_2$] ligand itself due to reversible ligand metalation processes. For example, we prepared the d_{20}-labeled [Ph$_2$BP$_2$] ligand [Ph$_2$B(CH$_2$P(C$_6$D$_5$)$_2$)$_2$] and examined the solution chemistry of [Ph$_2$B(CH$_2$P(C$_6$D$_5$)$_2$)$_2$]Pt(Me)(THF) in C$_6$D$_6$. For this complex, thermolysis provided CH$_3$D as the dominant methane byproduct (*CH$_4$: CH$_3$D = 1.0 : 7.3*). This and other mechanistically informative labeling experiments are discussed more thoroughly elsewhere (*8*). *For the present contribution, we underscore the* *likelihood that reversible metalation at a phenylphosphine arm of [Ph₂BP₂]Pt(Me)(THF) is operative, and a contributing factor to the overall rate of intermolecular benzene C-H bond activation chemistry. The degree of intramolecular metalation processes operative in the cationic system appears to be much less prevalent.*

THF Ligand Self-exchange in Benzene

An interesting question concerns the mechanism by which C_6H_6 (or C_6D_6) enters the coordination sphere in a step preceding C-H (or C-D) activation in the two systems. Benzene was chosen as the comparative solvent of choice for the two systems because both systems are appreciably soluble in benzene. Moreover, each complex is *relatively* stable at room temperature in benzene. This was not true of other solvents. For example, [Ph₂BP₂]Pt(Me)(THF) degrades rapidly at 22 °C in dichloromethane, whereas both systems have very limited solubility in less reactive hydrocarbon solvents (e.g., pentane, methylcyclohexane). Under a comparative set of reaction conditions, it was consequently impractical to examine C-H activation rates and ligand exchange processes as a function of benzene concentration. We therefore chose to study THF self-exchange in benzene solution as a model for benzene replacement of THF. Once again, the cationic and neutral systems yielded disparate results.

The dependence of the observed rate constant for self-exchange, k_{ex} (determined by magnetization transfer) (*8*), on the concentration of THF was strikingly different between the neutral and cationic platinum systems. For neutral [Ph₂BP₂]Pt(Me)(THF), k_{ex} showed no [THF] dependence. In sharp contrast, cationic [(Ph₂SiP₂)Pt(Me)(THF)][B(C₆F₅)₄] showed a first-order dependence on [THF] for the observed rate constant. The extrapolated intercept for the plot of k_{ex} versus THF equivalents for [(Ph₂SiP₂)Pt(Me)(THF)]⁺ intersects at the origin and thereby suggests negligible mechanistic dependence on the solvent (benzene) and/or the [B(C₆F₅)₄] anion (*15*). The absolute difference in the rate constant of THF self-exchange (k_{ex}) at a given temperature between the two systems was only modest. For example, at 25 °C, the rate constant for [Ph₂BP₂]Pt(Me)(THF) ($k_{ex(298K)}$ = 12.0 s⁻¹) is ~ one third as large as that for [(Ph₂SiP₂)Pt(Me)(THF)]⁺ ($k_{ex(298K)}$ = 38.5 M⁻¹ s⁻¹). More interesting was that the temperature dependence of k_{ex} varied dramatically between the two systems. The rate constant k_{ex} of [Ph₂BP₂]Pt(Me)(THF) was examined over the temperature range 11.2 - 48.9 °C and provided an entropy and enthalpy of activation as follows: $\Delta S^‡$ = 0.1 ± 5.4 e.u.; $\Delta H^‡$ = 16.0 ± 1.6 kcal/mol. Analogous data collected for cationic [(Ph₂SiP₂)Pt(Me)(THF)]⁺ over the temperature range 16.0 - 44.6 °C provided distinctly different values: $\Delta S^‡$ = -30.2 ± 5.2 e.u. and $\Delta H^‡$ = 1.9 ± 0.5 kcal/mol.

The activation parameters obtained for [(Ph$_2$SiP$_2$)Pt(Me)(THF)][B(C$_6$F$_5$)$_4$] are consistent with a classic associative mechanism of ligand exchange, in accord with the linear dependence of the exchange rate constant on THF concentration (15). Associative ligand exchange is commonplace for ligand substitution in square planar platinum(II) systems and is the mechanism we had anticipated. Interpreting the activation parameters and lack of THF dependence on k$_{ex}$ of [Ph$_2$BP$_2$]Pt(Me)(THF) is less straightforward. Perhaps the simplest scenario to put forward for ligand exchange is therefore that of a purely dissociative mechanism which proceeds via a neutral 3-coordinate intermediate, "[Ph$_2$BP$_2$]Pt(Me)". Given the increased electron-richness of the platinum center in the neutral system, dissociation of a σ-donor ligand might be expected to be more favorable. However, simple dissociative exchange mechanisms are rarely observed in platinum(II) substitution chemistry. Even in cases where they have been reported, such as the systems elegantly put forth by Romeo (16), certain assumptions are required to suggest the presence of a truly 3-coordinate intermediate species. Therefore, two additional mechanisms for THF exchange in [Ph$_2$BP$_2$]Pt(Me)(THF) need also to be considered: solvolytic displacement of the bound THF by benzene itself, and an anchimeric mechanism whereby a bond pair from the ancillary [Ph$_2$BP$_2$] ligand coordinates the platinum center prior to appreciable Pt-O bond breaking. These latter two possibilities constitute associative interchange mechanisms involving discrete 5-coordinate, rather than 3-coordinate, intermediates. While we cannot distinguish between dissociative, solvent-assisted, or ligand-assisted exchange pathways from the THF exchange data alone, solution NMR data obtained for [Ph$_2$BP$_2$]Pt(Me)(THF) (8) encourages us to suggest that an anchimeric pathway for ligand exchange is most likely operative for the neutral complex [Ph$_2$BP$_2$]Pt(Me)(THF) in benzene (Figure 6). We speculate that a similar mechanism is operative for exchange of THF by benzene in the overall C-H activation process in the neutral system.

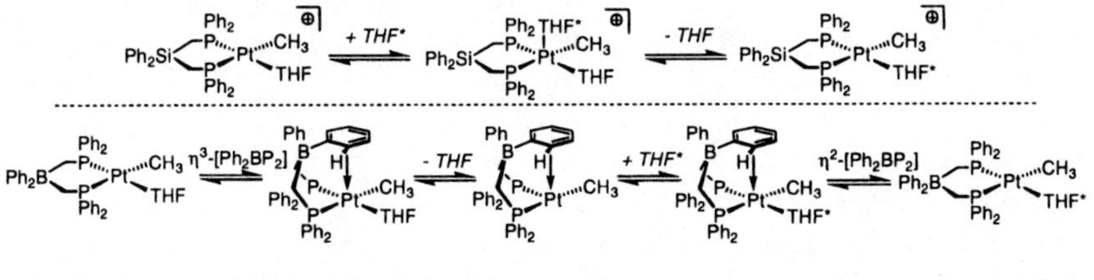

Figure 6. Possible mechanisms for THF ligand self-exchange in benzene for the complexes [(Ph$_2$SiP$_2$)Pt(Me)(THF)][B(C$_6$F$_5$)$_4$] (top) and [Ph$_2$BP$_2$]Pt(Me)(THF) (bottom). Note: An anchimeric exchange pathway for [Ph$_2$BP$_2$]Pt(Me)(THF) involving the phenyl groups of the phosphine donor arms is also likely, but is not shown explicitly.

Direct Evidence for Reversible Metalation Processes in [Ph$_2$BP$_2$]Pt(Me)(THF) and for a Pt(IV) Hydride Intermediate

When a sample of [Ph$_2$BP$_2$]Pt(Me)(THF) slightly wetted with excess THF was dissolved in benzene-d_6, its ^{1}H and ^{31}P{^{1}H} NMR spectra were unremarkable. This was also true of related spectra for an analytically pure sample of cationic [(Ph$_2$SiP$_2$)Pt(Me)(THF)][B(C$_6$F$_5$)$_4$]. However, when analytically pure [Ph$_2$BP$_2$]Pt(Me)(THF), obtained by careful drying under an argon purge to remove residual THF, was dissolved in benzene-d_6 and examined at 25 °C by ^{31}P{^{1}H} NMR spectroscopy, additional signals were observed that clearly indicated the presence of species distinct from [Ph$_2$BP$_2$]Pt(Me)(THF), and distinct from its known products of benzene C-H activation. A series of NMR experiments (^{1}H, ^{1}H{^{31}P}, ^{31}P{^{1}H}, and ^{13}C{^{1}H}) were performed on [Ph$_2$BP$_2$]Pt(Me)(THF), the d_{20}-labeled derivative [Ph$_2$B(CH$_2$P(C$_6$D$_5$)$_2$)$_2$]Pt(Me)(THF), and the ^{13}C-labeled derivative [Ph$_2$BP$_2$]Pt(^{13}CH$_3$)(THF) to assign the number and nature of the new species present. The interested reader is referred to the full paper for complete details (8). The collective NMR data allows us to draw three important conclusions. First, when [Ph$_2$BP$_2$]Pt(Me)(THF) is dissolved in C$_6$D$_6$ at 25 °C, three distinct, methyl-containing secondary species can be detected, in addition to the expected and dominant complex, [Ph$_2$BP$_2$]Pt(Me)(THF). This is most evident from a ^{13}C{^{1}H} NMR spectrum of [Ph$_2$BP$_2$]Pt(^{13}CH$_3$)(THF). Second, the dominant secondary complex contains a hydride ligand whose signal persists in the spectrum of d_{20}-labeled [Ph$_2$B(CH$_2$P(C$_6$D$_5$)$_2$)$_2$]Pt(Me)(THF). This secondary species appears to be a platinum(IV) hydride that is generated by *ortho*-metalation of the diphenylborate subunit. Possible structures of the spectroscopically observable platinum(IV) intermediate are proposed below in Figure 7. As a final point, we emphasize that the starting complex [Ph$_2$BP$_2$]Pt(Me)(THF), as well as the secondary species observed at 25 °C, are all consumed and ultimately funneled along the intermolecular benzene C-H activation pathway over a period of several hours at 45 °C. Important data detailed above for the neutral and cationic platinum(II) systems are summarized in Table 1.

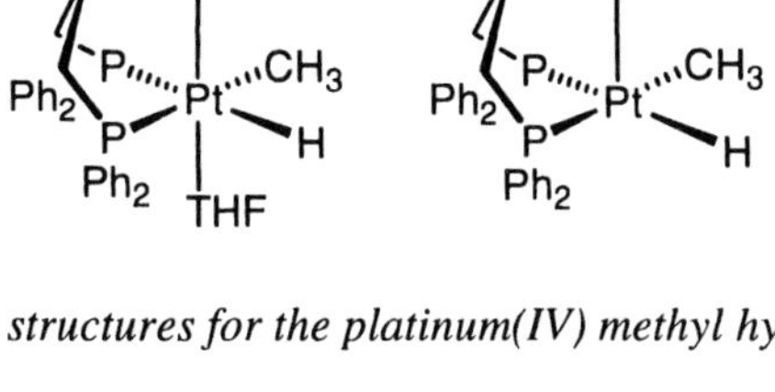

Figure 7. Possible structures for the platinum(IV) methyl hydride intermediate.

Table 1. Summary of key comparative data relevant to the overall mechanism of benzene C-H activation between the cationic and neutral complexes [(Ph$_2$SiP$_2$)Pt(Me)(THF)][B(C$_6$F$_5$)$_4$] and [Ph$_2$BP$_2$]Pt(Me)(THF).

[(Ph$_2$SiP$_2$)Pt(Me)(THF)][B(C$_6$F$_5$)$_4$] scheme	[Ph$_2$BP$_2$]Pt(Me)(THF) scheme
• ν(CO) of [(Ph$_2$SiP$_2$)Pt(Me)(CO)][B(C$_6$F$_5$)$_4$] = 2118 cm^{-1}	• ν(CO) of [Ph$_2$BP$_2$]Pt(Me)(CO) = 2094 cm^{-1}
• Rate of C-H activation at 55 °C: k = 1.80(6) x 10^{-4} s^{-1} • Rate of C-D activation at 55 °C: k = 2.76(7) x 10^{-5} s^{-1} k$_H$/k$_D$ = 6.52	• Rate of C-H activation at 45 °C: k = 1.42(5) x 10^{-4} s^{-1} • Rate of C-D activation at 45 °C: k = 1.13(3) x 10^{-4} s^{-1} k$_H$/k$_D$ = 1.26
• Mechanism of THF self-exchange is associative. Activation parameters: $\Delta S^\ddagger$ = -30.2 ± 5.2 e.u. and $\Delta H^\ddagger$ = 1.9 ± 0.5 kcal/mol	• Mechanism of THF self-exchange is ligand-assisted (or dissociative). Activation parameters: $\Delta S^\ddagger$ = -0.1 ± 5.4 e.u. and $\Delta H^\ddagger$ = 16.0 ± 1.6 kcal/mol
• CH$_4$: CH$_3$D ratio in methane byproduct after thermolysis in C$_6$D$_6$: 1.0 : 7.6	• CH$_4$: CH$_3$D ratio in methane byproduct after thermolysis in C$_6$D$_6$: 3.0 : 1.0
• Negligible deuterium incorporation into the (Ph$_2$SiP$_2$) ligand after thermolysis in benzene-d$_6$.	• Significant deuterium incorporation into the [Ph$_2$BP$_2$] ligand after thermolysis in benzene-d$_6$.
• (Ph$_2$SiP$_2$) metalation in benzene solution is kinetically non-competitive with benzene C-H activation processes.	• [Ph$_2$BP$_2$] metalation in benzene solution are kinetically very competitive with benzene C-H activation processes.

Discussion of the Mechanisms of Benzene C-H Activation

The benzene solution chemistry we have observed for the neutral complex [Ph$_2$BP$_2$]Pt(Me)(THF) is generally comparable to that observed for its isostructural but cationic relative [(Ph$_2$SiP$_2$)Pt(Me)(THF)][B(C$_6$F$_5$)$_4$]. The zwitterionic descriptor offered for [Ph$_2$BP$_2$]Pt(Me)(THF) is therefore useful in that it predicts its overall C-H activation reactivity. However, important mechanistic differences exist that can be attributed to the role that the bis(phosphine)-ligand auxiliary plays in each respective system. *These mechanistic distinctions most likely reflect electronic rather than steric differences.*

In Figure 8, we outline the simplest plausible mechanism (**Path A**) by which cationic (Ph$_2$SiP$_2$)Pt(Me)(THF)$^+$ undergoes intermolecular benzene activation.

The outlined mechanism is consistent with all of our data, and is generally similar to that proposed for other L$_2$Pt(Me)$^+$ systems that have been thoroughly described elsewhere. *Key points to note in* **Path A** *are that benzene coordination to the cationic platinum center is likely an associative process, and the benzene activation step is likely to be rate-determining, intimated by the large primary kinetic isotope effect that was observed for (Ph$_2$SiP$_2$)Pt(Me)(THF)$^+$ (k$_H$/k$_D$ ~ 6.52).* While we favor a benzene C-H activation step for the cationic system that occurs by oxidative addition from platinum(II) to give a platinum(IV) phenyl hydride, our data offers no direct evidence for this hypothesis.

Neutral [Ph$_2$BP$_2$]Pt(Me)(THF) differs from (Ph$_2$SiP$_2$)Pt(Me)(THF)$^+$ in that the bis(phosphine) auxiliary is intimately involved in both ligand exchange, and in C-H activation processes operative in benzene solution. The zero-order dependence in THF for THF self-exchange reflects the ability of the [Ph$_2$BP$_2$] ligand to assist in ligand exchange by an η^3-binding mode, an intramolecular process akin to a solvent-assisted ligand substitution process. While THF loss might also be dissociative based upon our exchange data, the prevailing ligand metalation chemistry of [Ph$_2$BP$_2$]Pt(Me)(THF) persuades us to discount this latter possibility. A propensity for the [Ph$_2$BP$_2$] ligand to achieve an η^3-binding mode dramatically impacts the nature of the C-H activation processes that are observed in benzene solution.

In Figure 8, we outline three mechanistic pathways to account for the solution chemistry of [Ph$_2$BP$_2$]Pt(Me)(THF). These are labeled **Path B**, **Path C**, and **Path D**, respectively. Association of an aryl ring from the diphenylborate unit of [Ph$_2$BP$_2$]Pt(Me)(THF) leads down **Path B** to a metalation process that generates a platinum(IV) methyl hydride complex (product **B**), an intermediate that can be spectroscopically detected. We do not think product **B** precedes an intermolecular benzene C-H activation step. Rather, we think that metalation at the diphenylborate unit is reversible, and that product **B** is ultimately funneled along **Paths C** and **D**. Common to **Paths C** and **D** is an η^3-binding mode for the [Ph$_2$BP$_2$] auxiliary that involves the arylphosphine donor rather than the diphenylborate unit. **Path C** proceeds along a simpler scenario that invokes a [Ph$_2$BP$_2$]-assisted benzene-d_6 substitution for THF, followed by oxidative addition of benzene-d_6 and reductive elimination of CH$_3$D, the methane byproduct expected. *The key distinction between* **Path C** *and* **Path A** *is the mechanism by which benzene enters the platinum coordination sphere. Our intuition is to suggest that the rate-determining step along* **Path C** *is the C-H activation step, and that the negligible primary kinetic isotope effect that was measured for [Ph$_2$BP$_2$]Pt(Me)(THF) (k$_H$/k$_D$ ~ 1.26) was due to the kinetic dominance of the fourth path,* **Path D**. In this last pathway, arylphosphine ligand metalation processes occur that produce platinum(IV) methyl hydride-species distinct from product **B** (shown in **Path B**). Subsequent to ligand metalation, benzene-d_6 enters the platinum coordination sphere at one of several

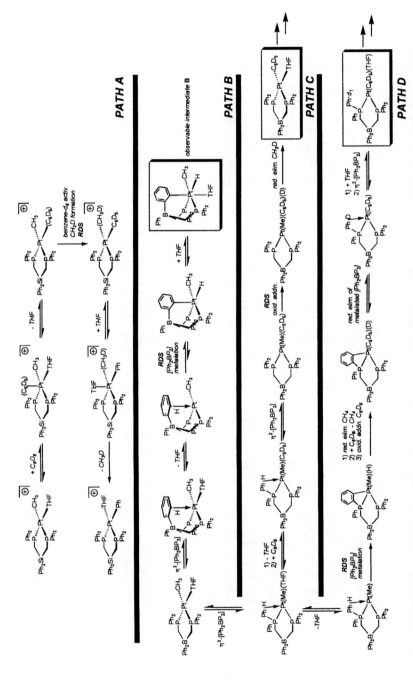

Figure 8. Postulated mechanisms for the intermolecular C–H activation of benzene for cationic [(Ph$_2$SiP$_2$)Pt(Me)(THF)][B(C$_6$F$_5$)$_4$] (Path A), and neutral [Ph$_2$BP$_2$]Pt(Me)(THF) (Paths B, C, and D). RDS refers to the suggested rate-determining step along each path.

indistinguishable stages, each of which involves the reductive elimination of CH_4 (for simplicity only one scenario is presented in Figure 8 explicitly). C-D activation of benzene-d_6, followed by a reverse metalation process that transfers deuteride into the [Ph$_2$BP$_2$] ligand, ultimately leads to the phenyl platinum complex. **Path D** thus accounts for the high degree of CH$_4$ released by [Ph$_2$BP$_2$]Pt(Me)(THF) in benzene-d_6, and the incorporation of deuteride into the [Ph$_2$BP$_2$] ligand. We are comfortable explicitly invoking platinum(IV) intermediates along both **Paths C** and **D** that arise from oxidative addition of benzene-d_6 because of our spectroscopic evidence for a platinum(IV) species resulting from [Ph$_2$BP$_2$] metalation (product **B**, **Path B**) (8). Also, we emphasize that our inability to detect the platinum(IV) hydride species produced by [Ph$_2$BP$_2$] metalation along **Path D** is because the ligand metalation process is itself rate-determining. A key piece of evidence that supports this assertion is that in both benzene and benzene-d_6, *the rate of decay of the d_{20}-labeled derivative [Ph$_2$B(CH$_2$P(C$_6$D$_5$)$_2$)$_2$]Pt(Me)(THF) is significantly slower than that of [Ph$_2$BP$_2$]Pt(Me)(THF) itself* ($k_{H\text{-}20}/k_{d\text{-}20}$ ~ 3 in benzene-d_6). Under conditions in which **Path D** dominates and ligand metalation is rate-determining, this is just what we expect.

The observation that the rate of decay of d_{20}-[Ph$_2$BP$_2$]Pt(Me)(THF) is modestly slower in benzene-d_6 than in protio benzene (k_H/k_D ~ 1.8) is perhaps more curious, but is conveniently explained as follows: deuteration of the aryl positions of the [Ph$_2$BP$_2$] ligand slows the rate of ligand metalation, and thereby attenuates the overall rate by which d_{20}-[Ph$_2$BP$_2$]Pt(Me)(THF) traverses down **Path D**. This in turn funnels more of the system down **Path C**, a path insensitive to arylphosphine deuteration. In this manner, a pre-equilibrium shift in benzene-d_6 serves to amplify the primary kinetic isotope effect of **Path C**, and thereby exposes C-H activation as rate-determining along this path as well. *We can therefore suggest that a C-H activation process of some sort is rate-determining for each of the four distinct pathways that are outlined in Figure 8.*

Comparing the [Ph$_2$BP$_2$] and [Ph$_2$B(pz)$_2$] Systems with Respect to C-H Activation

We had also hoped to compare the benzene C-H activation chemistry of [Ph$_2$BP$_2$]Pt(Me)(THF) with the bis(pyrazolyl)borate complex [Ph$_2$B(pz)$_2$]Pt(Me)(THF). Unfortunately, the latter complex proved too reactive to isolate or even spectroscopically identify in benzene solution. We therefore canvassed the relative reactivities of the anionic dimethyl complexes, [[Ph$_2$BL$_2$]PtMe$_2$][NR$_4$], in benzene solution upon addition of [HN^{i}Pr$_2$Et][BPh$_4$] or B(C$_6$F$_5$)$_3$ to expose a coordination site in situ. Several qualititative differences in the reactivity of the two systems were immediately apparent. When a slurry of

[[Ph$_2$BP$_2$]PtMe$_2$][ASN] was exposed to one equiv of either B(C$_6$F$_5$)$_3$ or HN^{i}Pr$_2$Et$^+$ in C$_6$D$_6$, ill-defined chemistry occurred at a very slow rate. The starting material persisted for hours, and the only identifiable byproducts were CH$_4$ and CH$_3$D. Consumption of the starting material proceeded much more rapidly on addition of either reagent in the presence of several equiv of THF. Under these conditions, [Ph$_2$BP$_2$]Pt(Me)(THF) is generated immediately. In contrast, the addition of B(C$_6$F$_5$)$_3$ or [HN^{i}Pr$_2$Et][NBu$_4$] at 22 °C to a slurry of [[Ph$_2$B(pz)$_2$]PtMe$_2$][NR$_4$] in benzene rapidly (< 5 min) and cleanly produced the diphenyl complex, [[Ph$_2$B(pz)$_2$]PtPh$_2$][NBu$_4$], with the concomitant evolution of each of the possible methane isotopomers (CH$_4$, CH$_3$D, CH$_2$D$_2$, CHD$_3$, CD$_4$). This latter reaction occurred at the same rate in both C$_6$H$_6$ and C$_6$D$_6$: a competition experiment using a 1:1 mixture of C$_6$H$_6$:C$_6$D$_6$ and subsequent examination of the products by ^{1}H NMR demonstrated that approximately 50% of the platinum-phenyl groups were Pt-C$_6$H$_5$, consistent with no significant kinetic isotope effects. It appears that methane loss is rate limiting in the (pyrazolyl)borate system, but not in the bis(phosphino)borate system. Even more interesting is that catalytic quantities (0.1 – 0.25 equiv) of [HN^{i}Pr$_2$Et][BPh$_4$] and B(C$_6$F$_5$)$_3$ drive the conversion of [{Ph$_2$B(pz)$_2$}PtMe$_2$][NR$_4$] to [{Ph$_2$B(pz)$_2$}Pt(C$_6$D$_6$)$_2$][NR$_4$]. Addition of 0.1 equiv of B(C$_6$F$_5$)$_3$ mediates the conversion rapidly (< 5 min). Currently, we cannot distinguish whether methide abstraction initiates subsequent arene activation, or whether trace "HOB(C$_6$F$_5$)$_3$" initiates the activation chemistry by protonation at platinum followed by methane loss. In the case of [HN^{i}Pr$_2$Et][BPh$_4$], there is little doubt that protonation (presumably at platinum) is the first step that exposes a site by methane loss. An additional point is that the activation chemistry will occur below −20 °C in toluene-d_8. In a VT experiment carried out from −60 °C to 0 °C, a stoichiometric mixture of [HN^{i}Pr$_2$Et][BPh$_4$] and [[Ph$_2$B(pz)$_2$]PtMe$_2$][NBu$_4$] began to produce deuterated methane isotopomers as low as −40 °C. A mixture of platinum products resulted consistent with toluene-d_8 C-D activation in various positions. ^{1}H NMR data verified a product consistent with benzylic activation. We also canvassed the reaction of [HN^{i}Pr$_2$Et][BPh$_4$] and [[Ph$_2$B(pz)$_2$]PtMe$_2$][NBu$_4$] in mesitylene. In this case, we were able to isolate and structurally characterize a product establishing benzylic activation. Unfortunately, the pyrazole ligand occupying the fourth coordination site shows that some [Ph$_2$B(pz)$_2$] ligand degradation occurred (Figure 9). While we have yet to probe the mechanism of the double C-H bond activation reaction, we suggest a plausible mechanism that might account for the ability of catalytic [HN^{i}Pr$_2$Et][BPh$_4$] to mediate the overall process (Figure 10) (*17*).

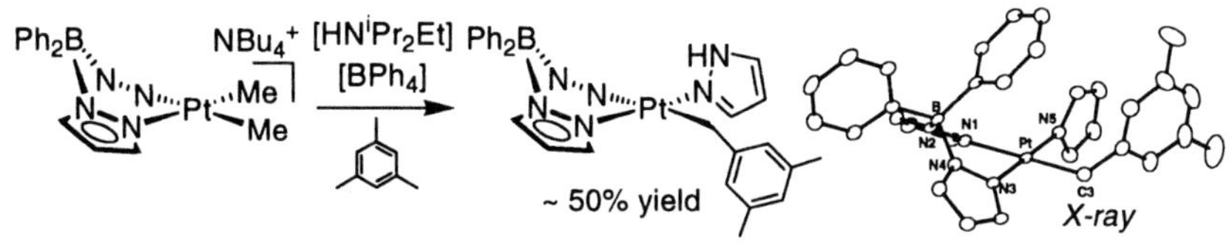

Figure 9. C-H bond activation of mesitylene with concomitant ligand degradation. Displacement ellipsoid representation of [Ph$_2$B(pz)$_2$]Pt(CH$_2$-3,5-Me$_2$Ph)(pyrazole) is also shown.

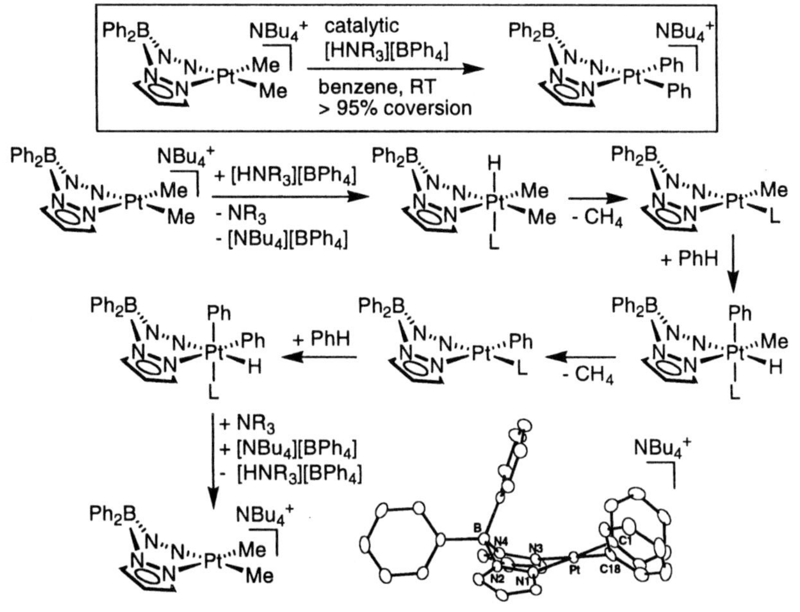

Figure 10. Mechanistic scenario that might account for the ability of [HN^{i}Pr$_2$Et][BPh$_4$] to catalyze the conversion of [[Ph$_2$B(pz)$_2$]PtMe$_2$][NBu$_4$] to [[Ph$_2$B(pz)$_2$]PtPh$_2$][NBu$_4$] in benzene solution. A displacement ellipsoid representation of the anion in [[Ph$_2$B(pz)$_2$]PtPh$_2$][NBu$_4$] is also shown.

Summary

This contribution has emphasized the reactivity of several platinum(II) complexes with respect to arene C-H bond activation. Most noteworthy are the results that compare the benzene solution chemistry of a neutral and cationic platinum(II) complex in which the platinum centers reside in an isostructural environment. Our data reveals that, even for very similar complexes, the mechanism of a C-H bond activation process can be dramatically different for two complexes capable of mediating the same overall reaction. Key differences described herein concern respective mechanisms of ligand exchange, and the relative propensity for inter- versus intramolecular arene C-H bond activation processes. Two surprising results uncovered within the neutral system [Ph$_2$BP$_2$]Pt(Me)(THF) concern the (i) lack of THF dependence on the rate of THF self-exchange in benzene, and (ii) the presence of a spectroscopically observable platinum(IV) methyl hydride complex that is presumed to be in equilibrium with the [Ph$_2$BP$_2$]Pt(Me)(THF) in benzene solution.

We had hoped to correlate key mechanistic differences between the cationic complex [(Ph$_2$SiP$_2$)Pt(Me)(THF)][B(C$_6$F$_5$)$_4$] and the neutral complex [Ph$_2$BP$_2$]Pt(Me)(THF) to electronic factors, the neutral system being the more electron-rich of the two despite the fact that it can be represented as a zwitterion. A key question with which we continue to struggle is the degree to which measured CO stretching frequencies can be used to correlate the relative electrophilicity of two isostructural complexes where one is cationic and one is neutral. The increased polarizability of CO when bound to a cationic platinum center complicates this issue. While comparisons that rely on such data are in and of themselves limited in a quantitative sense, our intuition is still to suggest the neutral zwitterion [Ph$_2$BP$_2$]Pt(Me)(THF) is appreciably more electron rich at platinum than the cation [(Ph$_2$SiP$_2$)Pt(Me)(THF)][B(C$_6$F$_5$)$_4$].

The final section in this contribution briefly compared the reactivity of a bis(pyrazolyl)borate platinum(II) system with the bis(phosphino)borate system. While these two systems are similar, the nature of the donor atoms coordinating platinum are distinct. Marked differences in reactivity were discerned, the bis(pyrazolyl)borate being the more reactive of the two *under the conditions discussed* (*18*). An intriguing double C-H bond activation reaction was identified that converted [[Ph$_2$B(pz)$_2$]PtMe$_2$][NBu$_4$] directly to [[Ph$_2$B(pz)$_2$]PtPh$_2$][NBu$_4$] rapidly in benzene at room temperature and below. The reaction is initiated by both stoichiometric and catalytic quantities of either B(C$_6$F$_5$)$_3$ or [HN^{i}Pr$_2$Et][BPh$_4$]. It appears that benzene C-H activation at [Ph$_2$B(pz)$_2$]Pt(II) can be very facile, occurring at temperatures as low as -40 °C, in cases where coordinating donor ligands other than benzene solvent are removed from the reaction system. In this regard, generating neutral platinum species in situ may afford highly productive C-H activation systems, but their activation chemistry needs to be controlled to produce high selectivity.

Acknowledgements

Financial support was provided by the Department of Energy (PECASE), the NSF (CHE-0132216), BP, and the ACS PRF. Larry Henling provided assistance with crystallographic studies.

References

1. Recent summaries and examples of cationic metal-mediated C-H bond activation and functionalization include: a) Arndtsen, B. A.; Bergman, R. G.; Mobley, T. A.; Peterson, T. H. *Acc. Chem. Res.* **1995**, *28*, 154. b) Lohrenz, J. C. W.; Jacobsen, H. *Angew. Chem., Int. Ed. Engl.* **1996**, *35*, 1305. c) Sen, A. *Acc. Chem. Res.* **1998**, *31*, 550. d) Stahl, S. S.; Labinger, J. A.; Bercaw, J. E. *Angew. Chem., Int. Edit.* **1998**, *37*, 2181. e) Crabtree, R. H. *J. Chem. Soc., Dalton Trans.* **2001**, 2437. f) Shilov, A. E.; Shul'pin, G. B. *Activation and Catalytic Reactions of Saturated Hydrocarbons in the Presence of Metal Complexes*; Klower: Boston, **2000**. g) Cho, J.-Y.; Iverson, C. N.; Smith, M. R. *J. Am. Chem. Soc.* **2000**, *122*, 12868. h) Chen, H. Y.; Schlecht, S.; Semple, T. C.; Hartwig, J. F. *Science* **2000**, *287*, 1995.
2. a) Stahl, S. S.; Labinger, J. A.; Bercaw, J. E. *J. Am. Chem. Soc.* **1995**, *117*, 9371. b) Holtcamp, M. W.; Labinger, J. A.; Bercaw, J. E. *J. Am. Chem. Soc.* **1997**, *119*, 848. c) Periana, R. A.; Taube, D. J.; Gamble, S.; Taube, H.; Satoh, T.; Fujii, H. *Science* **1998**, *280*, 560. d) Vedernikov, A. N.; Caulton, K. G. *Angew. Chem. Int. Ed.* **2002**, *41*, 4102. e) Johansson, L.; Ryan, O. B.; Tilset, M. *J. Am. Chem. Soc.* **1999**, *121*, 1974.
3. a) Zhong, H. A.; Labinger, J. A.; Bercaw, J. E. *J. Am. Chem. Soc.* **2002**, *124*, 1378. b) Tellers, D. M.; Yung, C. M.; Arndtsen, B. A.; Adamson, D. R.; Bergman, R. G. *J. Am. Chem. Soc.* **2002**, *124*, 1400.
4. Thomas, J. C.; Peters, J. C. *J. Am. Chem. Soc.* **2001**, *123*, 5100.
5. Wick, D. D.; Goldberg, K. I. *J. Am. Chem. Soc.* **1997**, *119*, 10235.
6. Iverson, C. N.; Carter, C. A. G.; Baker, R. T.; Scollard, J. D.; Labinger, J. A.; Bercaw, J. E. *J. Am. Chem. Soc.* **2003**, *125*, 12674.
7. Reinartz, S.; White, P. S.; Brookhart, M.; Templeton, J. L. *Organometallics* **2001**, *20*, 1709.
8. Thomas, J. C.; Peters, J. C. *J. Am. Chem. Soc.* **2003**, *125*, 8870.
9. a) Trofimenko, S. *J. Am. Chem. Soc.* **1966**, *88*, 1842. b) Trofimenko, S. *J. Am. Chem. Soc.* **1967**, *89*, 3170.
10. Thomas, J. C.; Peters, J. C. *Inorg. Chem.* **2003**, *42*, 5055.
11. Betley, T. A.; Peters, J. C. *Inorg. Chem.* **2002**, *41*, 6541.

12. a) Goldman, A. S.; Krogh-Jespersen, K. *J. Am. Chem. Soc.* **1996**, *118*, 12159. b) Caulton, K. G.; Fenske, R. F. *Inorg. Chem.* **1968**, *7*, 1273. c) Hush, N. S.; Williams, M. I. *J. Mol. Spectrosc.* **1974**, *50*, 349.
13. Flavio, B.; Minghetti, G.; Banditelli, G. *J. Organomet. Chem.* **1975**, *87*, 365.
14. Konze, W. V.; Scott, B. L.; Kubas, G. J. *J. Am. Chem. Soc.* **2002**, *124*, 12550.
15. Langford, C. H.; Gray, H. B. *Ligand Substitution Processes*; Benjamin: New York, 1966; pp 18-54.
16. a) Romeo, R. *Comments Inorg. Chem.* **1990**, *11*, 21. b) Romeo, R.; Scolaro, L. M.; Nastasi, N.; Arena, G. *Inorg. Chem.* **1996**, *35*, 5087. c) Romeo, R.; Alibrandi, G. *Inorg. Chem.* **1997**, *36*, 4822. d) Romeo, R.; Plutino, M. R.; Elding, L. I. *Inorg. Chem.* **1997**, *36*, 5909. e) Plutino, M. R.; Scolaro, L. M.; Romeo, R.; Grassi, A. *Inorg. Chem.* **2000**, *39*, 2712.
17. The mechanistic scheme shown in figure 10 does not consider protonation at one of the pyrazolyl arms as a means to expose a coordination site for arene coordination/activation. Such a scenario would also be reasonable and could account for the observed diphenyl product. For a reference that pertains to the protonation of (pyrazolyl)borate ligands in a related context see: Reinartz, S.; White, P. S.; Brookhart, M.; Templeton, J. L. *Organometallics* **2000**, *19*, 3854.
18. In separate studies, we have found examples whereby [Ph$_2$BP$_2$]Pt(Me)(L) complexes undergo benzene C-H activation more readily than [Ph$_2$B(pz)$_2$]Pt(Me)(L) complexes. For example, the complex [Ph$_2$BP$_2$]Pt(Me){P(C$_6$F$_5$)$_3$} degrades quantitatively to [Ph$_2$BP$_2$]Pt(Ph){P(C$_6$F$_5$)$_3$} in benzene at 80 °C, whereas [Ph$_2$BP$_2$(pz)$_2$]Pt(Me){P(C$_6$F$_5$)$_3$} is stable for prolonged periods at temperatures above 100 °C in benzene (see ref 8).

Oxidations

Chapter 21

Non-Organometallic Mechanisms for C–H Bond Oxidation: Hydrogen Atom versus Electron versus Hydride Transfer

James M. Mayer, Anna S. Larsen, Jasmine R. Bryant, Kun Wang, Mark Lockwood, Gordon Rice, and Tae-Jin Won

Department of Chemistry, University of Washington, Seattle WA 98195-1700

Oxidations of C–H bonds in alkylaromatic substrates can occur by initial electron, hydrogen atom, or hydride transfer. Presented here are oxidations by [(phen)$_2$Mn(μ-O)$_2$Mn(phen)$_2$]$^{n+}$ (n = 3, 4) and [(bpy)$_2$(py)Ru(O)]$^{2+}$. All three pathways are observed; which mechanism is preferred depends on the thermochemistry and the intrinsic barriers.

The activation of C–H bonds by metal complexes via organometallic mechanisms has been a major focus of research for some time, and is the dominant theme of this volume. Organometallic mechanisms for C–H activation include oxidative addition, formation of σ-complexes, σ-bond metathesis, and [2+2] additions to metal-ligand multiple bonds. These hold great promise for the functionalization of alkanes because all of these processes are selective for the least hindered C–H bond, such as the terminal methyl group of a chain. The promise of organometallic C–H activation has, however, been slowed by the difficulty in functionalizing the carbon after the activation step.

The organometallic approach should be contrasted with non-organometallic metal-mediated oxidations of hydrocarbons. In these mechanisms, the hydrocarbon substrate does not bind to the metal. Instead, bond cleavage and bond formation occur at a metal-bound ligand, such as an oxo group. Such mechanisms are in general selective for the weakest bond in the molecule, typically activating tertiary or benzylic C–H bonds. Because of this selectivity, reactions that follow these mechanisms are unlikely to solve the challenge (the "holy grail" (1)) of selectively converting methane to methanol or linear hydrocarbons to terminally functionalized products. Still, it should be emphasized that essentially all commercial metal-mediated C–H bond oxidations proceed by non-organometallic pathways (including metal-mediated radical processes in which the metal is not involved in C–H activation) (2). This includes both homogeneous oxidations, such as *para*-xylene to terephthalic acid (for polyethylene terephthalate [polyester]) and heterogeneous processes over metal oxides such as butane to maleic anhydride (for THF).

It is interesting to compare the terminology of the two approaches. The organometallic approach is termed "activation and functionalization" because these are separate processes. This distinction is not made in the non-organometallic processes, typically called "selective oxidations," because in these processes functionalization almost inevitably follows C–H bond cleavage. The difference in large part follows from the properties of the metal complexes involved. The active species or reagents in the non-organometallic approaches are oxidants, and often strong ones. In the organometallic reactions, however, the metal center that activates the C–H bond is typically reducing (as required for C–H bond oxidative addition) or redox-neutral (as in the formation of σ-complexes, σ-bond metatheses, and [2+2] additions). Since the "functionalization" desired is typically oxidative, it is not surprising that the organometallic compounds that activate C–H bonds do not functionalize them. In general, functionalization of the metal-bound C–H activated species requires a subsequent oxidation step. It is interesting that an exception to this pattern, the catalytic alkane borylation developed by Hartwig *et al.*, is formally not an oxidation of carbon ($C^{\delta-}-H^{\delta+} \rightarrow C^{\delta-}-B^{\delta+}$) (3).

Work in the Mayer labs has focused on non-organometallic oxidations, and hydrogen atom transfer reactions in particular. Rates of H-atom transfer to a transition metal oxidant depend not on the radical character of the oxidant, but on the thermochemical affinity of the oxidant for H˙ (4) and its intrinsic barrier toward hydrogen atom transfer (5). In this chapter we discuss the factors that influence the choice of non-organometallic mechanism for C–H bond activation, between initial H-atom transfer, electron transfer, and hydride transfer.

Mechanistic overview

Common pathways in non-organometallic metal-mediated oxidations of hydrocarbons are summarized in Scheme 1a, using toluene as a typical substrate. Electron transfer oxidation of alkanes is quite difficult because of their high ionization energies, but electron transfer from aromatic compounds is frequently

Scheme 1. (Adapted in part from reference (*6*) with permission.)

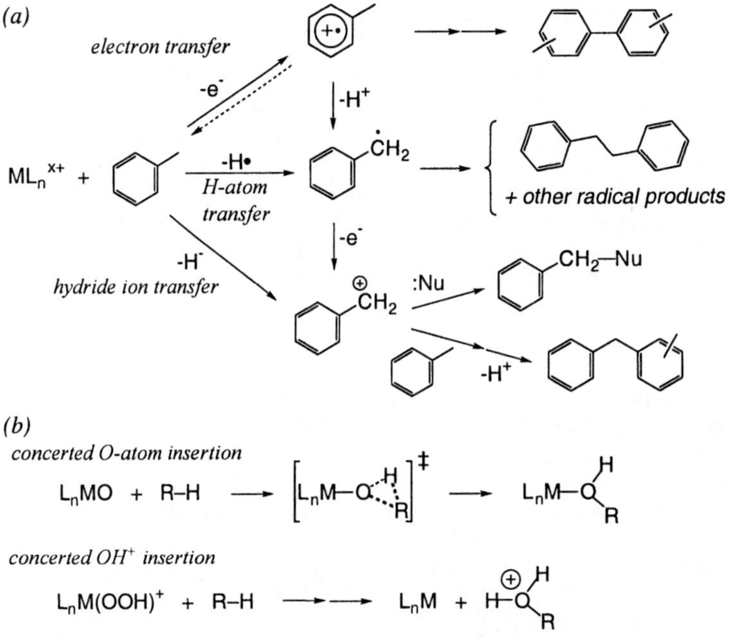

(a)

(b)

concerted O-atom insertion

$$L_nMO + R-H \longrightarrow [L_nM-O\cdots{}^H_R]^\ddagger \longrightarrow L_nM-O{}^H_R$$

concerted OH⁺ insertion

$$L_nM(OOH)^+ + R-H \longrightarrow \longrightarrow L_nM + H-\overset{\oplus}{O}{}^H_R$$

observed and often leads to biaryl products. Hydrogen atom transfer yields carbon radicals, which are sometimes indicated by radical coupling or rearrangement products. Hydride transfer gives carbocations which can be trapped by nucleophiles, including the aromatic substrate itself. Carbon radicals can also be formed by deprotonation of organic radical cations (which are often quite acidic) and similarly, carbocations can be formed by oxidation of radicals.

Concerted oxygen atom insertion into C–H bonds has also been discussed (Scheme 1b), for instance by the oxo complexes [(bpy)₂(py)Ru(O)]²⁺ (*7*) and [(TPA)Fe^V=O] (TPA = tris(2-pyridylmethyl)amine and its derivatives (*8*)). Newcomb and co-workers have also suggested a mechanism of concerted OH⁺ insertion into a C–H bond (*9*). This mechanism (with subsequent water loss) is used to explain the observed carbocation intermediates in C–H bond oxidations by cytochrome P450 and methane monooxygenase enzymes. These direct insertions in C–H bonds are analogous to the proposed reactivity of singlet methylene ($^1CH_2$) (*10a*). Concertedness is usually indicated by observing only the products of direct insertion, without any rearrangement, racemization, or inversion of an intermediate radical or carbocation. It should be noted, however, that there is a mechanistic continuum between concerted insertion and hydrogen removal (as H• or H⁻). Computational studies suggest that O–H bond formation is typically more advanced than C–O bond formation (*11*), as indicated in Scheme 1b. If C–O bond formation is sufficiently rapid or if the C–O bond is formed to only a small extent, then reactions may appear concerted while the transition state closely resembles a hydrogen atom or hydride transfer.

Thermochemistry and its connection to kinetic barriers

The thermodynamic properties of the reactants and products are very valuable in understanding a chemical reaction and the pathway(s) it follows. For each of the mechanisms in Scheme 1, the thermochemistry varies significantly with the nature of the C–H bond being oxidized. For instance, oxygen atom insertion into the tertiary C–H bond of isobutane is 12 kcal mol^{-1} more favorable than the analogous conversion of methane to methanol. Such values can be calculated from standard Tables (and are given in reference (12)). If C–H bond strengths are needed, the most current values (13) should be used if possible as the "accepted" values have risen 1-4 kcal mol^{-1} over the last few decades (13). A valuable advance in this area was the development by Bordwell and coworkers (based on precedents) of a simple thermochemical cycle to determine X–H homolytic bond strengths from redox potentials ($E°$) and acidities (pK_a values) (14). A similar approach to hydride affinities in MeCN has been described by Parker (15).

This chapter describes oxidations by two metal-containing oxidants, manganese-oxo-phenanthroline dimers [(phen)$_2$Mn(μ-O)$_2$Mn(phen)$_2$]$^{n+}$ (n = 3, 4) and the ruthenium-oxo-polypyridyl complex [(bpy)$_2$(py)Ru(O)]$^{2+}$. The abbreviations used here omit the pyridyl ligands, **Mn$_2$O$_2^{n+}$** and **RuO^{2+}**. In both cases, addition of electrons and protons generates reduced hydroxo (or aquo) species. Measured redox potentials and pK_a values are presented in Schemes 2 and 3, which are organized such that one row differs from the next by one electron, and one column differs from the next by one proton. The affinities of the oxidants for H· or H⁻ are calculated using thermochemical cycles (12,15).

It is informative to compare values for $\Delta G°$ for a series of related reactions with the activation free energies $\Delta G^{\ddagger}$ (a linear free energy relationship). The ratio of the changes in these quantities is usually referred to as α (eq 1), from the Brønsted relation for proton transfer reactions (16). For most reactions, linear

$$\alpha = \frac{\Delta\Delta G^{\ddagger}}{\Delta\Delta G°} = \frac{\Delta\log(k)}{\Delta\log(K_{eq})} \tag{1}$$

free energy relationships hold only within a closely related series of substrates over a narrow range of driving force. For a few classes of reactions, rate constants correlate with driving force over a wider range of reactants, the most established example being electron transfer. By Marcus Theory, barriers for a series of reactions with similar intrinsic barriers will have a quadratic dependence on $\Delta G°$ – linear over a small range of $\Delta G°$ (17). Hydrogen atom transfer reactions of organic and main group radicals have long been known to have a similar correlation of barrier with driving force, typically stated as an

Scheme 2. Thermodynamic data for [(phen)$_2$Mn(OH$_x$)$_2$Mn(phen)$_2$]$^{n+}$ compounds (values in MeCN, potentials vs. Cp$_2$Fe$^{+/0}$; adapted from (6) with permission.)

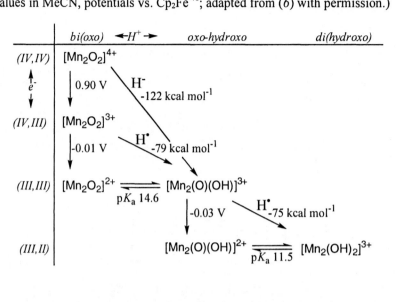

Scheme 3. Thermodynamic data for [(bpy)$_2$(py)Ru(OH$_x$)]$^{n+}$ compounds (in H$_2$O, potentials at pH 0 vs. NHE from (18); adapted from (19) with permission.)

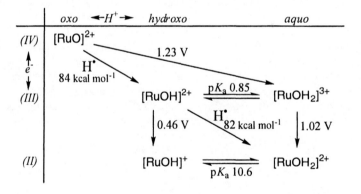

enthalpy relationship (E_a vs. bond strength) (20). This relationship holds well as long as "similar" radicals and substrates are compared. We have shown that such correlations also apply to hydrogen atom transfer reactions of transition metal complexes (4). This is one component of a more complete analysis, based on a Marcus-type approach using both driving force and intrinsic barriers (5). A Marcus treatment has also been applied to hydride transfer reactions (21).

Oxidations of alkylaromatic compounds by $[(phen)_2Mn(\mu-O)_2Mn(phen)_2]^{n+}$.

A full report of oxidations by $Mn_2O_2^{4+}$ and $Mn_2O_2^{3+}$ has appeared (6) and only the conclusions will be described here. $Mn_2O_2^{4+}$ is a quite reactive species, oxidizing all of the aromatic compounds we have examined except nitrotoluene. Organic compounds with relatively low redox potentials, such as naphthalene and *p*-methoxytoluene, are oxidized predominantly to biaryl compounds (*e.g.*, eq 2). As shown in Scheme 1a, biaryl products indicate a

mechanism of initial electron transfer. This conclusion is supported by kinetic studies showing inhibition of the reactions by added $Mn_2O_2^{3+}$, the reduced form of the oxidant. Such inhibition is a clear marker for a pre-equilibrium electron transfer mechanism (eq 3).

$$[ML_n]^{x+} + RH \rightleftarrows [ML_n]^{(x-1)+} + [RH^{\cdot+}] \rightarrow \rightarrow \quad (3)$$

The $Mn_2O_2^{4+}$ oxidation of toluene gives predominantly benzyl-toluene products from Friedel-Crafts addition of benzyl cation to toluene (along with small amount of the further oxidized products, eq 4). *p*-Methoxytoluene oxidation, which is 4×10^6 times faster than oxidation of toluene, also gives a trace of Friedel-Crafts products (eq 2). These products indicate the intermediacy of benzyl cations (Scheme 1a). The benzyl cations could be formed by initial electron transfer followed by deprotonation and oxidation, or by direct hydride transfer. A hydrogen atom transfer mechanism is ruled out by the very large substituent effects observed. For instance, p-xylene is 68 times more reactive than toluene and p-nitrotoluene is unreactive.

In contrast to substrates with low oxidation potentials, there are two indications that the oxidation of toluene by $Mn_2O_2^{4+}$ does not occur by electron transfer. First, addition of $Mn_2O_2^{3+}$ does not inhibit the reaction. While this could be consistent with rate-limiting (rather than preequilibrium) electron transfer, that mechanism is ruled out by the primary isotope effect of 4.3 ± 0.8 observed upon oxidation of a mixture of $C_6H_5CH_3$ and $C_6H_5CD_3$. Second, electron transfer from toluene is quite difficult because of its very high redox potential, 2.6 V vs. NHE in MeCN (0.5 V higher than naphthalene) (22). Based on this potential and the value in Scheme 2, electron transfer from toluene to $Mn_2O_2^{4+}$ is 1.2 V or 27 kcal mol^{-1} endergonic, larger than the observed $\Delta G^{\ddagger} = 23.6$ kcal mol^{-1}. [$\Delta G^{\circ\prime}$, the corrected potential within the precursor complex (22), is even more unfavorable because of coulombic repulsion in the successor complex $Mn_2O_2^{3+}\|C_7H_8^{\bullet+}$.] Thus electron transfer is not kinetically competent to account for the observed rate of oxidation, and the oxidation of toluene by $Mn_2O_2^{4+}$ occurs via initial hydride transfer.

A note of caution should be added here, that redox potentials for hard to oxidize substrates such as toluene are very difficult to measure. Highly irreversible electrochemical behavior is often observed, and the radical cations are very reactive with trace impurities in the solvent. One often finds a number of different values in the literature for the redox potential of a substrate such as toluene. Therefore arguments solely based on redox potentials – particularly relative potentials from different sources – should be viewed with caution. We have yet to find a single source that gives a consistent list for all the substrates of interest; recommended references are (22) and (23). In the case at hand, the toluene redox potential would have to be lower by more than 0.25 V lower than the value given by Eberson (22) in order for $\Delta G(Mn_2O_2^{4+}$ + toluene) to even be equal to $\Delta G^{\ddagger}$. In addition, the primary kinetic isotope effect and lack of inhibition by $Mn_2O_2^{3+}$ provide a second argument against electron transfer.

$Mn_2O_2^{3+}$ is a much weaker oxidant than $Mn_2O_2^{4+}$, only oxidizing compounds with very weak C–H bonds such as xanthene and dihydroanthracene (DHA) (6). Xanthene is oxidized to a mixture of xanthone and bixanthenyl (eq 5). Bixanthenyl is the radical coupling product, indicating that carbon radicals are present. The ratio of xanthone to bixanthenyl depends on the concentrations of $Mn_2O_2^{3+}$ and xanthene in a way that is consistent with the entire reaction proceeding via the xanthenyl radical, which can dimerize or be trapped by $Mn_2O_2^{3+}$ at ~2 × 10^5 M^{-1} s^{-1} (or by $Mn_2O(OH)^{3+}$ at ~2 × 10^4 M^{-1} s^{-1}). Consistent with a radical process, the relative rates inversely correlate with the substrate C–H bond strengths ($k_{xanthene} > k_{DHA} > k_{fluorene}$).

So why does $Mn_2O_2^{3+}$ react by hydrogen atom abstraction while $Mn_2O_2^{4+}$ reacts by electron transfer and hydride transfer? The first step in answering such questions is to look at the energetics of each possible rate limiting step. $Mn_2O_2^{3+}$ is thermodynamically a reasonable hydrogen atom abstractor because it can form a 79 kcal mol^{-1} bond to H$^\bullet$ (Scheme 2). Thus reactions with xanthene, DHA, and fluorene have $\Delta H° \cong$ -3.5, -1, and +1 kcal mol^{-1} (*cf.*, eq 6; BDE = bond dissociation energy). The affinity of $Mn_2O_2^{3+}$ for H$^\bullet$ is due to its being a moderate outer-sphere oxidant ($E_{1/2}$ = -0.01 V in MeCN vs. $Cp_2Fe^{+/0}$) and the

$$\Delta H°(Mn_2O_2^{3+} + XH \rightarrow Mn_2O(OH)^{3+} + X^\bullet) = BDE(X-H) - BDE[Mn_2O(O-H)^{3+}] \quad (6)$$

reduced form being a good base (pK_a 14.6 in MeCN). In contrast, $Mn_2O_2^{4+}$ is not a good H-atom abstractor because its reduced form, $Mn_2O_2^{3+}$, does not have significant basicity. Thermodynamically, it is most favorable for $Mn_2O_2^{4+}$ to accept an electron or a hydride, forming stable $Mn_2O_2^{3+}$ or $Mn_2O(OH)^{3+}$.

Comparing the two possible pathways for toluene oxidation by $Mn_2O_2^{4+}$, electron transfer is uphill by $\Delta G°$ = ~27 kcal mol^{-1} while hydride transfer has $\Delta G°$ = ~-4 kcal mol^{-1} (eqs 7, 8). The latter value results from the hydride affinity of $Mn_2O_2^{4+}$ being slightly larger than that of benzyl cation (122 vs. 118 kcal mol^{-1} (*15*)). The hydride affinities are derived with Parker's thermochemical

$$Mn_2O_2^{4+} + PhCH_3 \rightarrow Mn_2O_2^{3+} + PhCH_3^{\bullet+} \qquad \Delta G° = \sim+27 \text{ kcal mol}^{-1} \quad (7)$$

$$Mn_2O_2^{4+} + PhCH_3 \rightarrow Mn_2O(OH)^{3+} + PhCH_2^+ \qquad \Delta G° = \sim-4 \text{ kcal mol}^{-1} \quad (8)$$

cycle and with potentials and pK_a's in MeCN, so they should be directly comparable. There is thus a 31 kcal mol^{-1} thermodynamic bias for toluene oxidation to occur by hydride rather than electron transfer. This bias is large enough for the system to prefer the hydride transfer path.

For *p*-methoxytoluene, electron transfer has $\Delta G°$ = 9 kcal mol^{-1} and hydride transfer has $\Delta G°$ = -15 kcal mol^{-1}. Again there is a large thermochemical preference for hydride transfer pathway, in this case 24 kcal mol^{-1}. *In spite of this 24 kcal mol^{-1} bias, the oxidation of p-methoxytoluene occurs by initial electron transfer.* This is indicated by the predominant formation of biaryl products and the inhibition by added $Mn_2O_2^{3+}$. Hydride transfer is intrinsically more difficult than electron transfer in this system. If this is generally the case, hydride transfer will occur as a one step process only when it has a very large thermodynamic bias over electron transfer.

C–H bond oxidations by [(bpy)$_2$(py)Ru(O)]$^{2+}$.

In 1982, Thompson and Meyer described oxidations of alkylaromatic compounds by **RuO^{2+}** in both water and acetonitrile solvents (24). They proposed a new mechanism for a C–H bond oxidation, hydride abstraction with assistance from an entering nucleophile (eq 9). The compelling evidence for this pathway was a termolecular rate law, first order each in **RuO^{2+}**, substrate

$$\text{RuO}^{2+} + \underset{\text{}}{\text{[cumene]}} + :\text{Nu} \rightarrow \left[\text{RuO}\cdots\text{H}\cdots\text{C}-\text{Nu} \right]^{2+\ddagger} \rightarrow \text{RuOH}^+ + \underset{\text{}}{\text{[C-Nu]}^+} \quad (9)$$

(cumene), and the entering nucleophile (H$_2$O, ^{t}BuOH, or Br$^-$). We have been particularly interested in this pathway, and we invoked it in a permanganate oxidation (25). It is a type of S$_N$2 reaction and therefore could show primary > secondary > tertiary selectivity. However, it is quite unusual for an S$_N$2-type process to occur at a tertiary center (as in eq 9), and unusual that the relative rates, H$_2$O > ^{t}BuOH > Br$^-$, were opposite to typical trends in nucleophilicity (10b).

We and the Meyer group have independently re-examined the oxidation of cumene by **RuO^{2+}** (26). Both groups have found that the rate in MeCN is independent of the concentration of added water, contrary to the previous report. We have found no dependence on added ^{t}BuOH, and that **RuO^{2+}** is reduced by Br$^-$ in MeCN even in the absence of substrates. Without a third-order rate law, the mechanism in eq 9 must unfortunately be put aside. We have therefore looked more broadly at oxidations of alkylaromatics by **RuO^{2+}** (19).

DHA is rapidly oxidized by **RuO^{2+}** to give a mixture of anthracene, anthrone, and anthraquinone (eq 10 (19)). Oxidation of xanthene gives predominantly xanthone, with a small amount of bixanthenyl (eq 11); the

$$\text{[DHA]} \xrightarrow{\text{RuO}^{2+}} \text{[anthracene]} + \text{[anthrone]} + \text{[anthraquinone]} \quad (10)$$

$$\text{[xanthene]} \xrightarrow{\text{RuO}^{2+}} \text{[xanthone]} + \text{[bixanthenyl]} \quad (11)$$

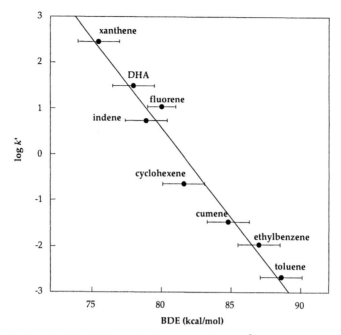

Figure 1. Graph of log k' (rate of abstraction by RuO^{2+}, per hydrogen) versus the bond dissociation energy for the substrate.

fluorene reaction is similar. Formation of bixanthenyl and bifluorenyl indicates the presence of xanthenyl and fluorenyl radicals in sufficient concentrations to couple. Radical coupling is competitive with trapping of the radicals by the oxidants present (RuO^{2+} and $RuOH^{2+}$) which leads to the oxygenated products. Consistent with this model, the yields of coupled products decrease with increasing RuO^{2+} starting concentrations. The chemistry is similar to that found for $Mn_2O_2^{3+}$ with these substrates (see above). The carbon radicals must be made by direct hydrogen atom transfer because RuO^{2+} is a very poor outer-sphere (electron transfer) oxidant (27).

The kinetics of RuO^{2+} oxidations, as shown by Meyer and co-workers (28), are complicated by the presence of multiple ruthenium species: RuO^{2+}, $RuOH^{2+}$, $Ru(H_2O)^{2+}$, and $Ru(NCMe)^{2+}$ [Ru = $Ru(bpy)_2py$]. In addition, there is a conproportionation/disproportionation equilibrium among the oxygenated species (eq 12), with K_{eq} = 50. The DHA, xanthene and fluorene oxidations have been monitored with a stopped-flow spectrophotometer, and the optical spectra have been fit to an A→B→C model with the SPECFIT™ global analysis software. In the oxidation of 40 mM DHA by 0.2 mM RuO^{2+}, $t_{1/2}$ for the A→B phase of the reaction is 0.13 s. After this first phase, the spectra indicate most of

$$RuO^{2+} + Ru(H_2O)^{2+} \rightleftarrows 2\ RuOH^{2+} \qquad (12)$$

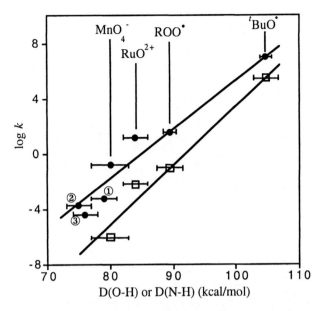

Figure 2. Plot of the log of the rate constant for H-atom abstraction for DHA (●, upper line) and toluene (□, lower line) vs. the strength of the X-H bond formed by the oxidant. In both cases, a straight line is drawn through the two oxygen radical points.
Key for numbered points: ①, [(phen)₂Mn(O)₂Mn(phen)₂]³⁺; ②, [(phen)₂Mn(O)(OH)Mn(phen)₂]³⁺; and ③, [Fe(Hbim)(H₂bim)₂]²⁺ (6,29)

the ruthenium is **RuOH^{2+}** or a related Ru^{III} complex. The Ru^{III} is *not* made by conproportionation (eq 12), since under these conditions the half life for **Ru(H$_2$O)$^{2+}$** reacting to form **RuOH^{2+}** is ≥0.83 s. Thus the observation of **RuOH^{2+}** indicates that **RuO^{2+}** is acting as a one-electron/one-proton oxidant and supports a mechanism of initial hydrogen atom transfer.

Oxidation of a number of alkylaromatic compounds by **RuO^{2+}** have been examined. Rate constants for the A→B step are taken as the rate of hydrogen abstraction k_H (*30*). These k_H values (divided by the number of reactive hydrogens in the substrate) correlate well with the C–H bond strength of the substrate (Figure 1). This linear correlation is confirmation of the hydrogen abstraction mechanism. In addition, we have often found that rate constants for hydrogen abstraction by metal-containing oxidants correlate with those for oxygen radicals (*4,6*). Figure 2 shows such a correlation for the oxidations of both DHA and toluene. Also included on the DHA line are rate constants for oxidations by **Mn$_2$O$_2^{3+}$**, **Mn$_2$O(OH)$^{3+}$**, and an iron(III) bi-imidazoline complex.

To further probe the hydrogen atom-transfer reactivity of **RuO^{2+}**, we have measured the rate of its H-atom self exchange with **RuOH^{2+}**: $k_{Ru/Ru} = 7.6 \times 10^4$ M^{-1} s^{-1} (*19*). Rate constants were determined by ^{1}H NMR line broadening in

MeCN solutions containing **RuO^{2+}** and **RuOH^{2+}**. The broadening was followed as a function of the concentration of **RuOH^{2+}**, which changes over time in these solutions because of the disproportionation of **RuOH^{2+}** (eq 12) and the solvolysis of the formed **RuOH$_2^{2+}$**. This self exchange rate constant can then be used in the Marcus cross relation (eq 13 (*17*)) to predict the rate constant of an oxidation reaction. Using the reported benzyl radical/toluene H-atom self

$$k_{calc}(\mathbf{RuO^{2+}} + \text{PhCH}_3) = \sqrt{k_{Ru/Ru}\, k_{C/CH}\, K_{Ru/CH}\, f} \qquad (13)$$

exchange rate constant ($k_{C/CH}$ = ~1 × 10^{-5} M^{-1} s^{-1}, per hydrogen) and the equilibrium constant $K_{Ru/CH}$ (2 × 10^{-4} from the bond strengths assuming $\Delta S°$ = 0), eq 13 gives k_{calc}(**RuO^{2+}** + PhCH$_3$) = 1.4 × 10^{-2} M^{-1} s^{-1} (*19*). This is within a factor of seven of the experimental rate constant (2.1 × 10^{-3} M^{-1} s^{-1}, per hydrogen). The good agreement provides another example of the applicability of the Marcus cross relation to hydrogen atom transfer reactions.

Conclusions

Many industrial and enzymatic oxidations of C–H bonds occur by non-organometallic mechanisms, without formation of M–H or M–C bonds. Mechanistic studies of two model systems are described here, to probe why a particular C–H bond oxidation follows one non-organometallic pathway versus another.

The manganese dimers [(phen)$_2$Mn(μ-O)$_2$Mn(phen)$_2$]$^{n+}$ (**Mn$_2$O$_2^{n+}$**) oxidize alkylaromatic compounds by electron transfer, hydrogen atom transfer, or hydride transfer. **Mn$_2$O$_2^{3+}$** acts as an H-atom abstractor in large part because it can form a 79 kcal mol^{-1} bond to a hydrogen atom (Scheme 2). **Mn$_2$O$_2^{4+}$** is a much more potent outer-sphere electron transfer oxidant, and it oxidizes substrates by electron transfer or hydride transfer. It does not react by H-atom transfer because it does not form a strong bond to H•, due to the low basicity of **Mn$_2$O$_2^{3+}$**. Oxidation of *p*-methoxytoluene by **Mn$_2$O$_2^{4+}$** occurs by electron transfer, in a step that is uphill by $\Delta G°$ = 9 kcal mol^{-1}. It is remarkable that electron transfer is favored over direct hydride transfer from *p*-MeOC$_6$H$_4$CH$_3$ to **Mn$_2$O$_2^{4+}$**, since the latter is *downhill* by $\Delta G°$ = -15 kcal mol^{-1}. This shows that electron transfer is intrinsically more facile than hydride transfer.

The ruthenium-oxo-polypyridyl complex [(bpy)$_2$(py)Ru(O)]$^{2+}$ (**RuO^{2+}**) does not oxidize cumene by the previously suggested mechanism of nucleophile-assisted hydride transfer. Instead, a common mechanism of hydrogen atom abstraction is indicated for a range of alkylaromatic compounds. One piece of evidence is the good correlation of log(k_H) with the strength of the C–H bond being cleaved (Figure 1). **RuO^{2+}** is not a good outersphere oxidant, but it does have a large affinity for a hydrogen atom (84 kcal mol^{-1}). The rate constant for toluene oxidation by **RuO^{2+}** is predicted within a factor of 7 by the Marcus cross relation (eq 13). These results and previous studies suggest that electron transfer and hydrogen atom transfer are the two most facile mechanisms for non-organometallic metal mediated oxidations. The choice between these pathways for a particular combination of oxidant and substrate is in large part determined by the thermochemistry of these reaction steps.

Acknowledgements

We are grateful for the financial support of the National Institutes of Health to JMM and to JCY, for a UW PRIME Fellowship to JRB, and to the National Science Foundation for support of an NMR spectrometer upgrade (CHE9710008) and for the purchase of a mass spectrometer (CHE9807748).

References

1 Arndtsen, B. A.; Bergman, R. G.; Mobley, T. A.; Peterson, T. A. *Acc. Chem. Res.;* **1995**, *28*, 154-162.
2. Olah, G. A.; Molnár, Á. *Hydrocarbon Chemistry* Wiley, New York, 1995.
3 Chen, H.; Schlecht, S.; Semple, T. C.; Hartwig, J. F. *Science* **2000**, *287*,1995.
4. Mayer, J. M. *Acc. Chem. Res.* **1998**, *31*, 441-450.
5. Roth, J. P.; Yoder, J. C.; Won, T.-J.; Mayer, J. M. *Science* **2001**, *294*, 2524.
6. Larsen, A. S.; Wang, K.; Lockwood, M. A.; Rice, G. L.; Won, T.-J.; Lovell, S.; Sadílek, M.; Tureček, F.; Mayer, J. M. *J. Am. Chem. Soc.* **2002**, *124*, 10112-10123.
7. Stultz, L. K.; Huynh, M. H. V.; Binstead, R. A.; Curry, M; Meyer, T. J. *J. Am. Chem. Soc.* **2000**, *122*, 5984-5996.
8 Chen, K.; Que, L., Jr. *J. Am. Chem. Soc.* **2001**, *123*, 6327-6337.
9 (a) Newcomb, M.; Toy, P. H. *Acc. Chem. Res.* **2000**, *33*, 449-455. (b) Choi, S.-Y.; Eaton, P. E.; Kopp, D. A.; Lippard, S. J.; Newcomb, M.; Shen, R. *J. Am. Chem. Soc.* **1999**, *121*, 12198-12199. b) Newcomb, M.; Shen, R.; Choi, S. Y.; Toy, P. H.; Hollenberg, P. F.; Vaz, A. D. N.; Coon, M. J.; *J. Am. Chem. Soc.* **2000**, *122*, 2677-2686.
10 March, J. *Advanced Organic Chemistry* 3rd ed.; Wiley-Interscience: New York, 1985 (a) pp. 604-5; (b) pp. 348-352.
11 See, for instance: (a) Ogliaro, F.; Harris, N.; Cohen, S.; Filatov, M.; de Visser, S. P.; Shaik, S. *J. Am. Chem. Soc.* **2000**, *122*, 8977-8989. (b) Hata, M.; Hirano, Y.; Hoshino, T.; Tsuda, M.; *J. Am. Chem. Soc.* **2001**, *123*, 6410-6416. (c) Baik, M.-H.; Gherman, B. F.; Friesner, R. A.; Lippard, S. J. *J. Am. Chem. Soc.* **2002**, *124*, 14608-14615. (d) Strassner, T.; Houk, K. N.; *J. Am. Chem. Soc.* **2000**, *122*, 7821-7822.
12. Mayer, J. M. Chapter 1 in *Biomimetic Oxidations Catalyzed by Transition Metal Complexes*, B. Meunier, Ed., Imperial College Press, **2000**.

13. (a) Berkowitz, J.; Ellison, G. B.; Gutman, D. *J. Phys. Chem.* **1994**, *98*, 2744-2765. (b) Blanksby, S. J.; Berkowitz, J. *Acc. Chem. Res.* **2003**, *36*, 255-263.
14. a) Bordwell, F. G.; et al. *J. Am. Chem. Soc.* **1991**, *113*, 9790; **1996**, *188*, 8777. (b) Jaun, B.; Schwarz, J.; Breslow, R. *J. Am. Chem. Soc.* **1980**, *102*, 5741-5748. (c) See also reference 12.
15. Cheng, J. -P.; Handoo, K. L.; Parker, V. D. *J. Am. Chem. Soc.* **1993**, *115*, 2655-2660.
16. Lowry, T. H.; Richardson. K. S. *Mechanism and Theory in Organic Chemistry* Harper & Row, NY 1976, pp. 141ff and 406-416.
17. (a) Marcus, R. A.; Sutin, N. *Biochim. Biophys. Acta* **1985**, *811*, 265-322. (b) Sutin, N. *Prog. Inorg. Chem.* **1983**, *30*, 441-499. (c) $\ln f_{xy} = [1/4(\ln K_{xy})^2]/[\ln(k_{xx}k_{yy}/Z^2)]$.
18. LeBeau, E. L.; Binstead, R. A.; Meyer, T. J. *J. Am. Chem. Soc.* **2001**, *123*, 10535-10544.
19. Bryant, J. R.; Mayer, J. M. *J. Am. Chem. Soc.* **2003**, *125*, 10351-10361.
20. (a) Ingold, K. U., Chapter 2, pp. 69ff and Russell, G. A. Chapter 7, pp. 283-293 in *Free Radicals* Kochi, J. K., Ed.; Wiley: New York, 1973. (b) Tedder, J. M. *Angew. Chem., Int. Ed. Engl.* **1982**, *21*, 401. (c) Knox, J. H. in *Oxidation of Organic Compounds, Volume II*, F. R. Mayo, Ed., *Adv. Chem. Series* **1968**, *76*, 11.
21. (a) Roberts, R. M. G.; Ostovíc, D.; Kreevoy, M. M. *Faraday Disc. Chem. Soc.* **1982**, *74*, 257-265; Lee, I.-S. H.; Jeoung, E. H.; Kreevoy, M. M. *J. Am. Chem. Soc.* **1997**, *119*, 2722-2728 and *ibid.* **2001**, *123*, 7492-7496; and references therin. (b) Schindele, C.; Houk, K. N.; Mayr, H. *J. Am. Chem. Soc.* **2002**, *124*, 11208-11214.
22. Eberson, L. *Electron Transfer Reactions in Organic Chemistry*, Springer-Verlag, Berlin, 1987 (redox potentials, p. 44).
23. Kochi, J. K., in *Comprehensive Organic Synthesis*, Vol. 7 (Oxidation); Trost, B. M., Ed.; Pergamon: New York, 1991 p. 849-855.
24. Thompson, M. S.; Meyer, T. J. *J. Am. Chem. Soc.* **1982**, *104*, 5070-5076.
25. Gardner, K. A.; Mayer, J. M. *Science* **1995**, *269*, 1849-1851.
26. Curry, M.; Huynh, M. H. V.; Stultz, L. K.; Binstead, R. A.; Meyer, T. J.; Bryant, J. R.; Mayer, J. M. manuscript in preparation.
27. Lebeau, E. L.; Binstead, R. A.; Meyer, T. J. *J. Am. Chem. Soc.* **2001**, *123*, 10535-10544.
28. a) Binstead, R. A.; Stultz, L. K.; Meyer, T. J. *Inorg. Chem.* **1995**, *34*, 546-551. b) Lebeau, E. L.; Binstead, R. A.; Meyer, T. J. *J. Am. Chem. Soc.* **2001**, *123*, 10535-10544.
29 Roth, J. P.; Mayer, J. M. *Inorg. Chem.* **1999**, *38*, 2760-2761.
30. While $k_{A \to B}$ should be corrected for the reaction stoichiometry to give k_H, this is not possible because of the complexity of the kinetics and the product mixtures. *Cf.*, reference (*19*) and Gardner, K. A.; Kuehnert, L. L.; Mayer, J. M. *Inorg. Chem.* **1997**, *36*, 2069-2078.

Chapter 22

Oxidation of Organic Molecules with Molecular Oxygen Catalyzed by Heterometallic Complexes

Patricia A. Shapley

Department of Chemistry, University of Illinois, Urbana, IL 61801

The reactions between $[N(n\text{-Bu})_4][M(N)R_2Cl_2]$ (where M= Ru, Os; R= CH_3, CH_2SiMe_3, Ph) and chromate or other oxyanions (WO_4^{2-}, ReO_4^-) produce stable but coordinatively unsaturated heterometallic complexes. The Os-Cr and Ru-Cr heterobimetallic complexes $[N(n\text{-Bu})_4][M(N)R_2(\mu\text{-O})_2CrO_2]$ are selective catalysts for the oxidation of alcohols with molecular oxygen. Heterotrimetallic complexes $\{M(N)R_2\}_2(\mu_3\text{-S})_2M'(dppe)$ (where M'= Pt, Pd, Ni) result from the reactions between $[N(n\text{-Bu})_4][Ru(N)R_2Cl_2]$ or $[Os(N)R_2(py)_2][BF_4]$ and $M'(SSiMe_3)_2(dppe)$. The Ru-Pt complex $Pt(dppe)(\mu_3\text{-S})_2\{Ru(N)Me_2\}_2$ is a catalyst for the oxidation of alkenes and alcohols with O_2.

Oxidation reactions are important in the synthesis of organic compounds because these reactions create new functional groups or modify existing functional groups in a molecules. However, autooxidations and other reactions with radical intermediates frequently exhibit low chemo- and regioselectivity. Stoichiometric metal oxidants, such as $KMnO_4$ or K_2CrO_4, and metal catalyzed reactions that use secondary oxidants, such as iodosylbenzene, amine-N-oxides, or hypochlorite salts, generate large quantities of waste. Recently, researchers have developed more environmentally friendly oxidation catalysts that use molecular oxygen.(*i*)

Heterometallic complexes are promising catalysts because the organic substrate and oxidizing agent could be activated by different metals within the same complex. Heterometallic catalysts may have advantages over monometallic catalysts if the metals act cooperatively in chemical transformations.(*ii*) Many heterometallic complexes have been prepared, including a few that show catalytic activity.(*iii*)

Our goal is to prepare heterometallic catalysts that will oxidize alcohols and other organic molecules in a selective manner using only molecular oxygen and releasing only water (*Scheme 1*). In the oxidation of alcohols for example, the alcohol O-H unit would add to a metal-oxygen bond. After ß-hydride elimination, the oxidized product and water would be released and the metal complex reoxidized.

Scheme 1

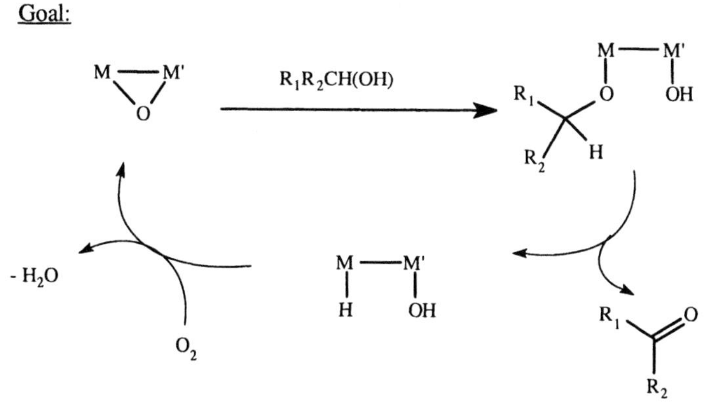

Goal:

Results and Discussion

Anionic osmium(VI) and ruthenium(VI) complexes $[Os(N)(CH_2SiMe_3)_2Cl_2]^-$, $[Os(N)Me_2Cl_2]^-$, $[Os(N)Me(CH_2SiMe_3)Cl_2]^-$, $[Ru(N)Me_2Cl_2]^-$ or $[Os(N)Ph_2Cl_2]^-$ react with aqueous potassium chromate or silver chromate to produce stable organometallic complexes containing a

Scheme 2

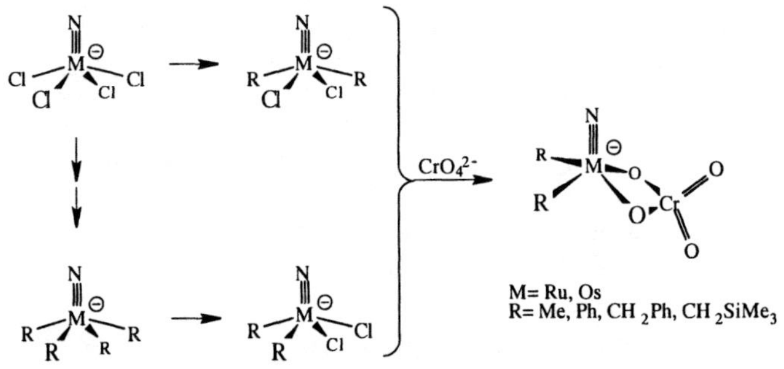

M= Ru, Os
R= Me, Ph, CH$_2$Ph, CH$_2$SiMe$_3$

bidentate chromate group *(Scheme 2)*.(*iv*) We have also prepared related oxyanion complexes with ReO$_4^-$, WO$_4^{2-}$, SO$_4^{2-}$, CO$_3^{2-}$ and tetrathiometallate complexes with MoS$_4^{2-}$, WS$_4^{2-}$.(*v*)

The osmium-chromate complexes are stable in the presence of triphenylphosphine, cyclohexene, carbon monoxide, ethers, ferrocene, and dimethyl sulfide. This is surprising because chromium(VI) oxides are active oxidizing agents. They are capable of oxidizing triaryl- and trialkylphosphines, dialkyl sulfides, alcohols, aldehydes, alkenes, and even some activated hydrocarbons.(*vi*)

The complex [N(*n*-Bu)$_4$][Os(N)(CH$_2$SiMe$_3$)$_2$(μ-O)$_2$CrO$_2$] oxidizes benzylic, primary and secondary alcohols to the corresponding carbonyl compounds *(Table 1)*.(*vii*) In air, these reactions are catalytic. Alcohol oxidation reactions are slow at room temperature and the rate of the reaction depends on the steric bulk of the alcohol. In competition experiments, primary alcohols are always oxidized faster than secondary alcohols. Oxidation of primary alcohols produces only aldehydes. There is no skeletal isomerization with the cyclopropyl-substituted alcohol. With unsaturated alcohols, there is no oxidation of the double bond and no isomerization.

To determine the mechanism of the reaction, we investigated reaction kinetics, characterized intermediate complexes, and examined the effects of isotopic substitution in the alcohol and molecular oxygen. The mechanism of alcohol oxidation with [N(*n*-Bu)$_4$][Os(N)(CH$_2$SiMe$_3$)$_2$(μ-O)$_2$CrO$_2$] proceeds through initial coordination of the alcohol at the osmium center, proton transfer, ß-hydrogen elimination, and activation of molecular oxygen by the Os-Cr bond *(Scheme 3)*. The rate of alcohol oxidation by all of the Os-Cr and Ru-Cr complexes was similar but did depend on both the metal, Os or Ru, and the ligands around that metal. Bulky alkyl groups on the osmium or ruthenium center reduced the rate of the alcohol oxidation reaction. Steric bulk at this metal would impede coordination of alcohol. For complexes with the same alkyl ligands, the Ru-Cr complexes are better catalysts than the Os-Cr analogs.

We have recently prepared chiral analogs of the Os-Cr alcohol oxidation catalyst and are exploring their activity in asymmetric oxidation reactions. We are also examining oxidation reactions catalyzed by osmium and ruthenium complexes of perrhenate and tungstate oxyanions.

Table 1. Catalytic Oxidation of Alcohols by [N(n-Bu)₄][Os(N)(CH₂SiMe₃)₂(μ-O)₂CrO₂]

Substrate	Product
cyclopropyl-CH(OH)-	cyclopropyl-C(=O)-
CH₃(CH₂)₇CH₂-OH	CH₃(CH₂)₇CH=O
2-cyclohexen-1-ol	2-cyclohexen-1-one
geraniol	geranial (citral)
nerol	neral
piperonyl alcohol	piperonal
benzyl alcohol	benzaldehyde
4-methylbenzyl alcohol	4-methylbenzaldehyde
4-methoxybenzyl alcohol	4-methoxybenzaldehyde
4-(trifluoromethyl)benzyl alcohol	4-(trifluoromethyl)benzaldehyde
2-chlorobenzyl alcohol	2-chlorobenzaldehyde

Scheme 3

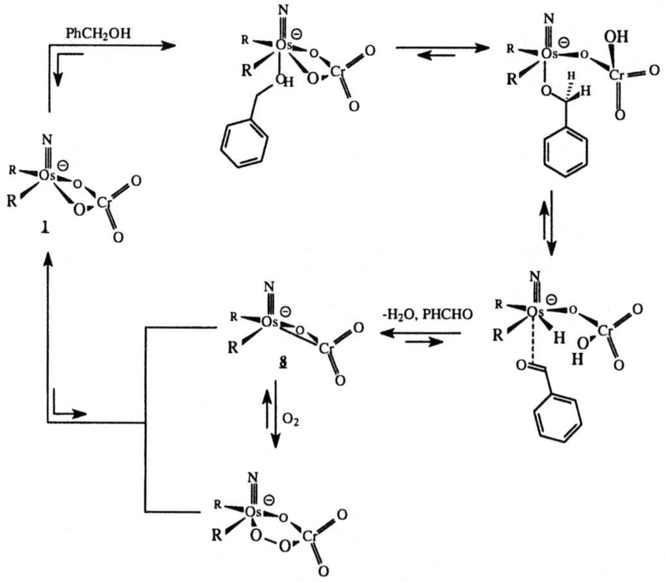

In some instances, a trimetallic complex may be more effective than a bimetallic catalyst for the oxidation of organic molecules, provided that the metals are not coordinatively saturated. We prepared triosmium and triruthenium complexes in which the metals are in a high oxidation state (+6) and have a 16 electron count $\{M(N)R_2\}_3(\mu_3\text{-}S)_2$ (*Scheme 4*).(*viii*) However these homometallic complexes are catalytically inactive.

Scheme 4

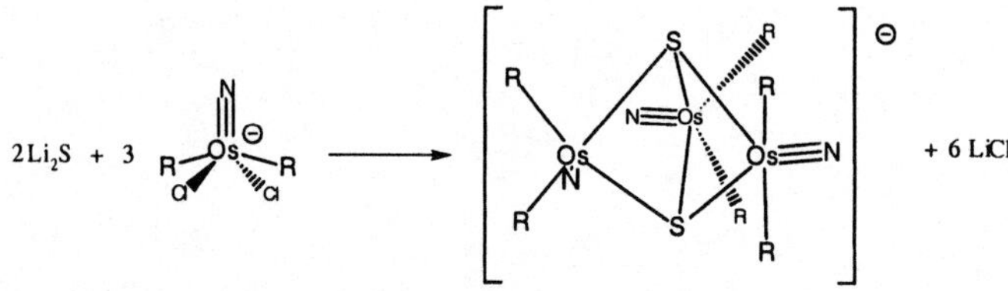

Studies of homogenous Ru-Pt heterometallic complexes in particular may improve our understanding of platinum-containing alloys in catalytic industrial processes and aid in the development of more efficient fuel cells.(*ix*) The addition of ruthenium to platinum catalysts greatly improves their activities for methanol oxidation and the activity is sensitive to structure.(*x*) Although a handful of Ru-Pt heterometallic complexes have been synthesized, only $CpRu(PPh_3)(\mu\text{-}Cl)(\mu\text{-}dppm)PtCl_2$ is an electrocatalyst for methanol oxidation.(*xi*)

There are many strategies for synthesizing heterometallic complexes including bridge-assisted reactions. Sulfido ligands are common bridging groups in metal cluster complexes. Many of the rational syntheses of heteronuclear compounds with bridging sulfido ligands involve displacing a ligand on a metal center with a terminal sulfido ligand on another metal.(*xii*)

New, sulfido-bridged heterobimetallic complexes (dppe)Pt(μ_3-S)$_2${Ru(N)Me$_2$}$_2$, (COD)Pt(μ_3-S)$_2${Ru(N)Me$_2$}$_2$, and (dppe)Pd(μ_3-S)$_2${Ru(N)Me$_2$}$_2$ resulted from the reactions of (dppe)Pt(SSiMe$_3$)$_2$, (COD)Pt(SSiMe$_3$)$_2$, or (dppe)Pd(SSiMe$_3$)$_2$ with [PPh$_4$][Ru(N)Me$_2$Cl$_2$] (*Scheme 5*). The reaction between (dppe)Pt(SSiMe$_3$)$_2$ and [Os(N)(CH$_2$SiMe$_3$)$_2$(NCMe)$_2$][BF$_4$] produced (dppe)Pt(μ_3-S)$_2${Os(N)(CH$_2$SiMe$_3$)$_2$}$_2$.(*xiii*) A similar reaction between (dppe)Ni(SSiMe$_3$)$_2$ and [Ru(N)Me$_2$Cl]$_2$ produced (dppe)Ni(μ_3-S)$_2${Ru(N)Me$_2$}$_2$.(*xiv*) Yields were moderate and ranged from 72 to 53% for the preparation of the trimetallic complexes. In all cases, the driving force for these reactions is the formation of a strong Si-Cl bond and elimination of trimethylsilylchloride. Due to the ease in which the Si-S bond can be cleaved by halides or oxygen nucleophiles, trialkyl- or trialkoxysilanethiolate complexes may be viewed as protected metal sulfides.

Scheme 5

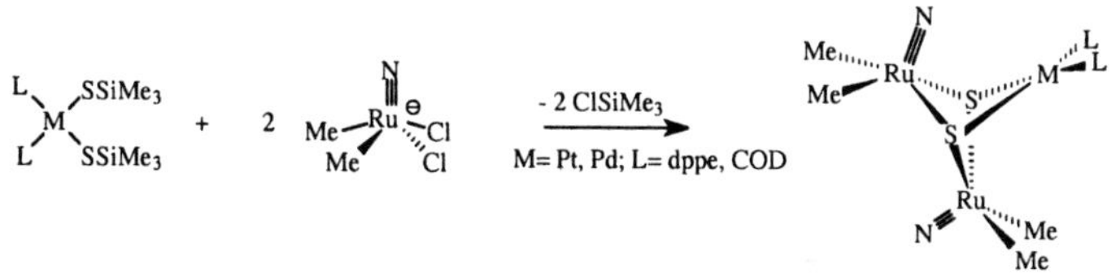

The platinum, palladium, and nickel trimethylsilylsulfide complexes result from the reaction between NaSSiMe$_3$ and the square planar dichloro complexes, (dppe)PtCl$_2$, (COD)PtCl$_2$, (dppe)PdCl$_2$, or (dppe)NiCl$_2$ (*Scheme 6*). The dppe complexes form in high yield (83-87%) but the yield is lower (48%) for the more reactive cyclooctadiene complex.

Scheme 6

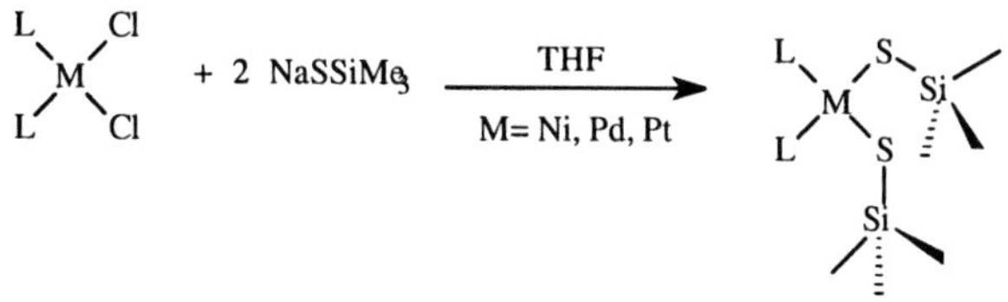

We determined the molecular structures of three of the heterotrimetallic complexes by X-ray crystallography. The structure of (dppe)Pt(μ_3-S)$_2${Ru(N)Me$_2$}$_2$ is shown below. The platinum has a square planar coordination geometry and each ruthenium is in a square pyramidal coordination environment. The three metals and the two nitrido groups are in a plane with one sulfido ligand above and below this plane.

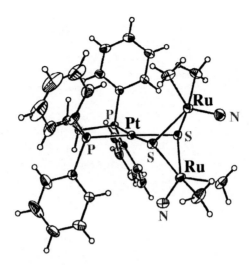

Figure 1. ORTEP Diagram of (dppe)Pt(μ_3-S)$_2${Ru(N)Me$_2$}$_2$

The complexes (dppe)M(μ_3-S)$_2${M'(N)R$_2$}$_2$ are catalysts for the oxidation of alcohols and alkenes with molecular oxygen. The rate of oxidation and the selectivity of the reactions depends on the nature of both M and M'. At 90 degrees, under an atmosphere of O$_2$, the osmium-containing complex (dppe)Pt(μ_3-S)$_2${Os(N)(CH$_2$SiMe$_3$)$_2$}$_2$ is much slower in its oxidation of alcohols than is (dppe)Pt(μ_3-S)$_2${Ru(N)Me$_2$}$_2$. Oxidation of cyclohexene with (dppe)Pt(μ_3-S)$_2${Ru(N)Me$_2$}$_2$ selectively produces cyclohexene oxide when the reaction is conducted in the presence of dimethyl sulfide and molecular oxygen but allylic oxidation products predominate when dimethyl sulfide is absent from the reaction mixture (*Scheme 7*).

Scheme 7

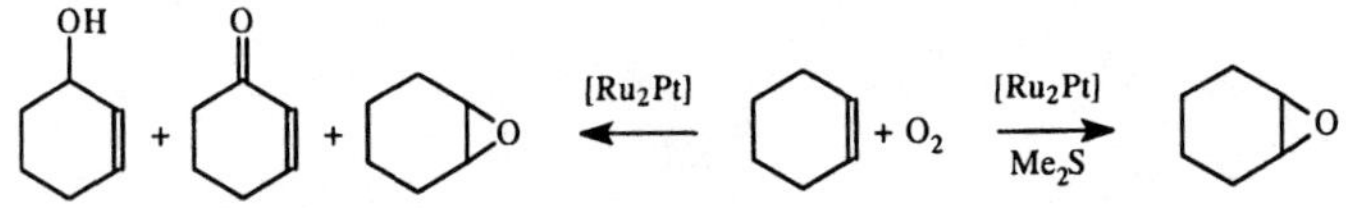

Conclusion

We have prepared a series of sulfido-bridged trimetallic complexes of the form $L_2M(\mu_3\text{-}S)_2\{M'(N)R_2\}_2$ where M is either platinum(II) or palladium(II) and M' is either ruthenium(VI) or osmium(VI). These complexes are soluble in a variety of organic solvents and stable to air and water. We are currently examining the oxidation of hydrocarbon and alcohol substrates with these complexes and molecular oxygen.

Acknowledgements

We gratefully acknowledge the financial support of the Donors of The Petroleum Research Fund, administered by the American Chemical Society (ACS-PRF 36240-AC) and the United States EPA (EPA R 829553). NMR spectra were obtained in the Varian Oxford Instrument Center for Excellence in NMR Laboratory. Funding for this instrumentation was provided in part from the W. M. Keck Foundation, the National Institutes of Health (PHS 1 S10 RR10444-01), and the National Science Foundation (NSF CHE 96-10502). Purchase of the Siemens Platform/CCD diffractometer by the School of Chemical Sciences was supported by National Science Foundation grant CHE 9503145.

References

1. (a) Sheldon, R. A.; Arends, I. W. C. E.; Ten Brink, G.-J.; Dijksman, A. Acc. Chem. Res., *2002*, 35, 774-781. (b) Boring, E.; Geletii, Y. V.; Hill, C. L. J. Am. Chem. Soc., *2001*, 123, 1625-1635. (c) G. T.; Wei, M.; Subramaniam, B.; Busch, D. H. Inorg. Chem., *2001*, 40, 3336-3341.
2. (a) Stephan, D.W. *Coord. Chem. Rev.* **1989**, 95, 41-107. (b) Dobbs, D. A.; Bergman, R. G. *J. Am. Chem. Soc.* **1992**, *114*, 6908-6909. (c) Brunner, H.; Challet, S.; Kubicki, M. M.; Leblanc, J.-C.; Moise, C.; Volpato, F.; Wachter, J. *Organometallics* **1995**, *14*, 6323-6324. (d) Mathur, P.; Hossain, M. M.; Umbarkar, S. B.; Rheingold, A. L.; Liable-Sands, L. M.; Yap, G.P.A. *Organometallics* **1996**, *15*, 1898-1904.
3. (a) Braunstein, P.; Rose´, J. in "Metal Clusters in Chemistry", Braunstein, P., Oro, L. A., Raithby, P. R., Eds.; Wiley-VCH: Weinheim, Germany, in press. (b) Langenbach, H.J.; Keller, E.; Vahrenkamp, H. *J. Organomet. Chem.* **1979**, *171*, 259-271. (c) Hidai, M.; Fufuoka, A.; Koyasu, Y.; Uchida, Y. *J. Chem. Soc., Chem. Commun.* **1984**, 516-517. (d) Casado, M. A.;

Perez-Torrente, J. J.; Ciriano, M. A; Oro, L. A.; Orejon, A.; Claver, C. *Organometallics* **1999**, *18*, 3035-3044.

4. Zhang, N.; Mann, C.; Shapley, P. A. *J. Am. Chem. Soc.* **1988**, *110*, 6591-6592.

5. (a) Zhang, N.; Shapley, P. A. *Inorg. Chem.* **1988**, *27*, 976-977. (b) Shapley, P. A.; Gebeyehu, Z.; Zhang, N.; Wilson, S. R. *Inorg. Chem.* **1993**, *32*, 5646-5651.

6. (a) Piancatelli, G.; Scettri, A.; D'Auria, M. *Synthesis* **1982**, 245-258. (b) Collins, J. C.; Hess, W. W.; Frank, F. J. Tetrahedron Lett. **1968**, 3363-3366.

7 Shapley, P. A,; Zhang, N.; Allen, J. L.; Pool, D. H.; Liang, H.-C. *J. Am. Chem. Soc.*, **2000**, *122*, 1079-1091.

8. Shapley, P. A., Liang, H. C., Dopke, N. C., *Organometallics* **2001**, *20*, 4700-4704.

9. Sinfelt, J. H. *Bimetallic Catalysts. Discoveries, Concepts and Applications;* Wiley: New York, 1983.

10. (a) Iwasita, T.; Hoster, H.; John-Anacker, A.; Lin, W. F.; Vielstich, W. *Langmuir* **2000**, *16*, 522-529. (b) Kua, J.; Goddard, W. A. *J. Am. Chem. Soc.* **1999**, *121*, 10928-10941. (c) Rolison, D. R.; Hagans, P. L.; Swider, K. E.; Long, J. W. *Langmuir* **1999**, *15*, 774-779.

11. Tess, M. E.; Hill, P. L.; Torraca, K. E.; Kerr, M. E.; Abboud, K. A.; McElwee-White, L. *Inorganic Chemistry* **2000**, 39, 3942-3944.

12. (a) Seino, H.; Arai, Y.; Iwata, N.; Nagao, S.; Mizobe, Y.; Hidai, M *Inorg. Chem.* **2001**, *40*, 1677-1682. (b) Ikada, T,; Kuwata, S.; Mizobe, Y.; Hidai, M. *Inorg. Chem.* **1998**, *37*, 5793-5797. (c) Brunner, H.; Challet, S.; Kubicki, M.M.; Leblanc, J.-C.; Moise, C.; Volpato, F.; Wachter, J. *Organometallics* **1995**, *14*, 6323-6324. (b) Massa, M.A.; Rauchfuss, T.B.; Wilson, S.R. *Inorg. Chem.* **1991**, *30*, 4667-4669.

13. Shapley, P. A.; Liang, H. C.; Dopke, N. C. *Organometallic*, **2001**, *20*, 4700-4704.

14. Kuiper, J. L.; Shapley, P. A. submitted to *Organometallics.*

Chapter 23

Catalytic Hydrocarbon Oxidations Involving Coreductants

Joseph E. Remias and Ayusman Sen*

Department of Chemistry, The Pennsylvania State University, University Park, PA 16802

Hydrogen peroxide is a useful oxidant for selective aliphatic and aromatic hydrocarbon oxidations. Unlike dioxygen, which typically does not oxidize these molecules in a selective fashion, a wide body of research already exists reporting selective, catalytic hydrocarbon oxidations using H_2O_2. This, coupled with the fact that H_2O_2 produces only water as a byproduct and has a high percentage of active oxygen, make it an excellent choice for hydrocarbon oxidations. Consequently, great interest exists in using H_2O_2 for selective oxidations. With this has grown interest in generating H_2O_2 from dioxygen and another coreductant for its instantaneous use in organic oxidations. Efficiently doing so would significantly increase a reaction's economic value and its environmental friendliness. In this short review, we focus on the oxidation of alkyl and aromatic hydrocarbons with H_2O_2 generated *in situ* using hydrogen as a coreductant. Special attention is given to the use of palladium in these reactions, and to bimetallic systems that utilize one catalyst for peroxide generation while another performs hydrocarbon oxidations. We also present some of our own data on this topic, particularly on vanadium catalyzed benzene oxidation using H_2O_2 generated *in situ* on a palladium catalyst. Furthermore, we compare our *in situ* results to an analogous system where H_2O_2 is added directly.

Introduction

Coreductants serve an important function in enzymatic oxidations. The monooxygenases such as cytochrome P-450 (*1,2*) and methane monooxygenase (*3*) perform difficult oxidations via reductive activation of dioxygen. In the oxidations, a molecule of water is produced making the maximum efficiency based on oxygen 50%. The ultimate goal of catalysis would be to use 100% of the oxidant, as in "heme" type oxygenases. However, when one considers autoxidation or free radical processes that use 100% of oxygen, they rarely show the necessary selectivity for hydrocarbon oxidations. Consequently, the use of a monooxygenase type catalyst, one involving a coreductant, provides an appealing alternative. A wealth of catalysts and oxidants are available which act in this way and provide more selective reactions. Some examples of oxidants are as follows: hydrogen peroxide, peracids, organic hydroperoxides, inorganic and metallorganic peroxides, sodium hypochlorite, iodosobenzene, and nitrous oxide. Many of these oxidants are too expensive for basic chemical production, the focus here, and find more use in fine chemical production where expensive and even stoichiometric oxidants still play a large role.

Of the oxidants listed above hydrogen peroxide shows the most promise for several reasons. First, H_2O_2 delivers the most active oxygen per molecule, 47%, when compared to those listed above. It is important to note that this is actually ca. 20%, when using H_2O_2 as an aqueous solution. The atom economy of oxidants is important to consider when trying to minimize the environmental impact of a reaction (*4*). It produces only water as a byproduct, a concern when one considers the need for "green" oxidants (ones that produce minimal on non-toxic waste) (*5*). Finally, hydrogen peroxide has proven to be a strong oxidant with good selectivity for many oxidations (*6*).

The use of metal catalysts for the generation of hydrogen peroxide using hydrogen (the coreductant) and oxygen has generated significant interest recently as evidenced by the proliferation of publications. This desire is at least partially fueled by an interest in pursuing "greener" routes than the current anthraquinone process to make H_2O_2 (*7*). Also, generating the oxidant *in situ* maximizes the percent of usable oxygen in the oxidant, no diluent is needed for the oxidant if generated *in situ*. Furthermore, the cost associated with storing and transporting hydrogen peroxide is eliminated. Another advantage is that a bimetallic catalyst system can be envisioned where the H_2O_2 is generated in a slow steady state amount on one catalyst while another performs the oxidation. This should maximize the selectivity and efficiency of the reaction. Finally, the cost of hydrogen peroxide versus that of hydrogen (the only consumable of cost assuming 100% catalyst recyclability and air as the oxidant) is about 5 times more. Thus, a significant cost savings can be realized per mole of oxidant used.

In this short review, we focus on the generation of hydrogen peroxide in an *in situ* fashion for hydrocarbon oxidations. Specifically, we are concerned with the selective oxidation of alkanes and aromatics using H_2O_2 generated on a noble metal catalyst. In all these cases hydrogen will serve as the coreductant either directly or indirectly. The use of bimetallic systems where one catalyst generates the H_2O_2 while another performs hydrocarbon oxidation is given particular attention. We also present some of our recent work in this area looking at the selectivity of *in situ* reactions compared to when H_2O_2 is added directly, and briefly address how the two catalysts in a bimetallic system, as above, might interact.

It is worthwhile to succinctly survey some other oxidations in the presence of coreductants. Barton has developed several Gif systems using iron catalysts in the presence of zinc or iron(0) as a reductant (8). These methods were developed more to model biological monooxygenases than to serve practical industrial functionality. Ascorbic acid has also been used as a reductant to generate hydrogen peroxide (9). There has also been a substantial body of work on alternative organic species to generate hydrogen peroxide in an in situ fashion, (10). Finally, a substantial body of research exists on the epoxidation of olefins using generated H_2O_2 (11-14).

Metal Catalyzed Formation of H_2O_2

Numerous examples exist in the literature detailing the formation of hydrogen peroxide on metal catalysts. During Pd catalyzed aerobic oxidations, it is known that hydrogen peroxide is formed as a byproduct (15). In some instances this peroxide remains in solution. For instance, in the aerobic oxidation of alcohols by a combination of palladium acetate and pyridine, it was suggested that the peroxide may actually compete with the substrate for binding to the metal center and, therefore, its removal by decomposition (e. g., by molecular sieves) promoted substrate oxidation (16). Another notable example is the oxidation of alcohols via Pd(II) complexes producing a significant amount of peroxide, eq. 1 (17). These reactions were carried out in a biphasic mixture to prevent oxidation of the catalyst's ligand; however, this methodology may also have served to prevent peroxide decomposition via the Pd catalyst. The biphasic conditions may partially explain why in other examples no peroxide was observed in alcohol oxidation (see below). The quick separation of the peroxide from the palladium catalyst via phase separation prevented its subsequent metal catalyzed decomposition. Surprisingly, considering the results below, the authors claim that equimolar amounts of ketone and peroxide are formed- a result indicative of no metal catalyzed decomposition of the latter occurring.

$$R_2CHOH + O_2 \longrightarrow R_2CO + H_2O_2 \qquad (1)$$

Additionally, several patents and papers report the palladium-catalyzed generation of hydrogen peroxide in relatively high concentrations in the absence of an oxidizable substrate (see below). These reactions involve the use of dioxygen and either carbon monoxide or dihydrogen as the coreductant. Carbon monoxide used in these conditions is an indirect source of hydrogen through the water gas shift reaction, eq. 2. Both metallic palladium and discrete palladium compounds have been employed as catalysts for the reaction.

$$H_2O + CO \rightleftharpoons H_2 + CO_2 \qquad (2)$$

Leading work showed that hydrogen peroxide could be formed in relatively high concentrations (10-20% by wt.) with a palladium catalyst, hydrogen and oxygen (18). The need for acidic solution and the use of halide promoters, making purification difficult and generating a corrosive environment, have led researchers to look at variations to the above. For example supported heteropoly acids were used to replace inorganic acids (19). Furthermore, others have found that halides can be excluded with certain catalysts but only in organic solvents (20). An interesting study recently performed sheds light on how palladium catalysts form hydrogen peroxide (21). The researchers contend that H_2O_2 formation occurs only on colloidal Pd particles regardless of their source (Pd^{+2} ions or Pd on silica support) at least under acidic aqueous conditions.

In contrast to the above report, discreet Pd^{+2} complexes have been found to catalyze the formation of H_2O_2 using CO as the coreductant (22). Here, a biphasic mixture is again employed to isolate the catalyst from the H_2O_2 formed. It was necessary to carry out the reaction with a large excess of O_2 to prevent reduction to palladium metal. By carefully selecting nitrogen coordinated ligands with appropriate steric and electronic environments, the researchers were able to produce H_2O_2 in concentrations up to 8% by wt.

Though the mechanism of hydrogen peroxide generation on noble metals is still not completely understood, particularly on surfaces in the presence of various promoters and other metals, some conclusions can be made based on discreet Pd and Pt complexes. The first step involves the formation of a metal-peroxo complex. The coordinated peroxide can then be displaced from the metal by the addition of a ligand (e. g., eq. 3) (23). More commonly, in acidic solutions, it is protonated to form hydrogen peroxide (24,25). Finally, it is worthwhile to note that while both Pd and Pt catalysts can effectively form H_2O_2 in the presence of a coreductant, Pt is faster at decomposing the oxidant (26).

$$(R_3P)_2Pt(O_2) + 2\,R_3P + ROH \longrightarrow [(R_3P)_4Pt]^{2+} + HO_2^- + RO^- \qquad (3)$$

Metal Catalyzed Decomposition of H_2O_2

It is pertinent to point out that while H_2O_2 can be formed in reasonable yields, significant decomposition of it once formed is also observed. For example, during olefin oxidation with palladium diacetate and hydrogen peroxide, decomposition of the oxidant was observed (27). This effect was more pronounced in the absence of the substrate. Furthermore, during the aerobic oxidation of alcohols with $Pd(OAc)_2$, it was shown that H_2O_2 in the absence of substrate underwent rapid disproportionation to molecular oxygen and water

(28). Further experiments showed that even while alcohol oxidation was occurring the addition of hydrogen peroxide to the solution resulted in only decomposition of the H_2O_2, eq. 4. No additional oxidation of the alcohol, by the peroxide, was observed, eq. 5.

$$H_2O_2 \longrightarrow H_2O + 0.5\, O_2 \qquad (4)$$

$$H_2O_2 + RCH_2OH \longrightarrow RCHO + 2\, H_2O \qquad (5)$$

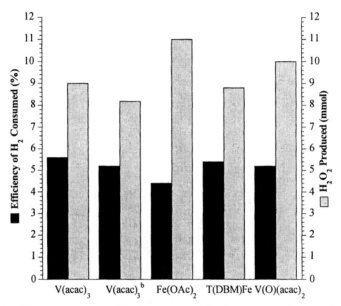

Figure 1: *Efficiency in use of H_2O_2 generated for benzene oxidation. Calculated from total H_2O produced vs. amount of phenol and benzoquinone formed. Ran using 3.2 µmol catalyst, 20.0 mg 5% Pd/Al_2O_3, 2 mL benzene, and 2.5 mL acetic acid, exposed to 100 psi H_2, 1000 psi N_2, and 100 psi O_2 for 2 h at 65°C. Each column is the average of four reactions. Abbreviations: acac = acetylacetonate, OAc = acetate, T(DBM) = tris(dibenzoylmethanato).*
[b] *10.0 mg of 5% Pd/Al_2O_3.*
(Reproduced with permission from J. Mol. Catal. A: Chem. 2003, 203, 179–192. Copyright 2003 Elsevier.)

Figure 1 shows our results on the efficiency with which the dihydrogen (and H_2O_2 formed from it) is used for benzene oxidation with several catalysts. It demonstrates that only about 5.5% of the dihydrogen consumed to make peroxide is used in substrate oxidation. For comparison, under analogous conditions where H_2O_2 was added directly, an oxidant efficiency of 20% was observed. Observation that the efficiency with respect to *in situ* generated H_2O_2 remains constant regardless of the hydroxylation catalyst used is consistent with the palladium catalyst causing almost all of the observed hydrogen peroxide decomposition. Recent work using palladium, gold, and amalgams of the two speculates that the decomposition on Pd may limit the amount of usable H_2O_2 (29).

A plausible solution to decrease the rate of decomposition may involve simply decreasing the percentage of hydrogen in the media. This action would limit the availability of hydrogen for palladium catalyzed H_2O_2 hydrogenation (eq. 6). This has been found effective in other cases (7). However, it seems unlikely that this would completely halt the unwanted reaction and will not stop the disproportionation reaction. Consequently, one must consider the rate of decomposition when seeking out catalysts for the *in situ* generation of hydrogen peroxide.

$$H_2O_2 + H_2 \longrightarrow 2 H_2O \tag{6}$$

H_2O_2 Generation and Hydrocarbon Oxidation at a Single Metal

The use of palladium generated hydrogen peroxide has found use in performing hydrocarbon oxidations, Scheme 1, where Cat = Pd. An elegant example which separated the hydrogen and oxygen, preventing the potential for an explosion, showed that a palladium containing membrane could catalyze the formation of hydrogen peroxide. The formed peroxide then reacted with Pd to oxidize benzene to phenol (30). The method shows high selectivity, 96% for phenol based on benzene consumed; however, anywhere from 5 to 9 moles of H_2O_2 disproportionate for every mole of benzene oxidized. Other work has focused on supporting palladium on highly acidic resins such as Nafion and has successfully generated phenol using H_2O_2 from H_2 and O_2 (31). The use of a heteropolyanion with palladium or platinum catalyst has also been attempted with generated H_2O_2 for benzene oxidation (32).

In the past we have had success in using generated H_2O_2 for hydrocarbon oxidations using palladium. We have shown that supported Pd is effective in oxidizing methane and ethane at mild temperatures in the presence of CO and O_2 (33). In an effort to trap the desired oxidation products at the alcohol stage, trifluoroacetic acid (TFA) was used to generate the corresponding ester. Interestingly, while simply adding TFA did not change the product distribution, the addition of copper salts dramatically increased the selectivity of the reaction with methanol and its TFA ester being virtually the only products (34). The reactivity of the above system in water has been exploited for the deep oxidation (to carbon monoxide, carbon dioxide, and water) of a variety of hazardous organics (35). Several hundred turnovers in 24h can be achieved with numerous substrates including, benzene, phenol, substituted phenols, aliphatic and aromatic halogenated compounds, organophosphorous, and organosulfur compounds.

Scheme 1

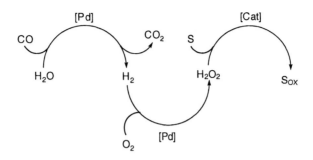

(S = substrate, S_{ox} = oxidized substrate, Cat = Pd or second catalyst)

Other work from our group shows the functionalization of methane to methanol and acetic acid using Rh salts as catalysts (*36,37*). These reactions also employed CO and oxygen with the two products formed arising from competitive attack of a metal alkyl via a nucleophile or insertion of carbon monoxide (Scheme 2). It is presumed that in these reactions CO is performing a dual role: it generates H_2 and thus H_2O_2 via the water gas shift, and it inserts into metal alkyl bonds to form carbonylated products. Indeed, the water gas shift reaction does proceed under similar conditions (*38*), and the oxidation of CO to CO_2 was observed in the absence of methane (*36*).

Scheme 2

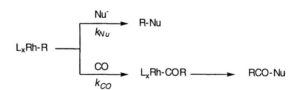

H_2O_2 Generation and Hydrocarbon Oxidation Using Two Different Metals

Utilizing another catalyst for hydrocarbon oxidations once the hydrogen peroxide was generated has also been shown effective, Scheme 1. Miyake has explored the effect of adding metal oxides to silica supported palladium catalysts (*39*). It was found that numerous oxides, vanadium, yttrium, and lanthanum oxides exhibiting the best performance, increased the rate of phenol formation from benzene. Though no details are presented it is hypothesized that the metal oxides may assist in delivering activated oxygen to the palladium catalyst. A more recent paper screens numerous homogeneous metal salts with a heterogeneous palladium catalyst; it is proposed that this reaction occurs via the same pathway as above (*40*). Of course a pathway where hydrogen peroxide forms a catalytically active species from the metal oxide is also a plausible explanation. A comparison of Pd or Pt supported on silica and promoted with vanadium oxides has been performed (*41*). The authors note a significant increase in both yield and selectivity in the use of platinum over palladium; a find contrary to the body of literature on the generation of hydrogen peroxide, where palladium is highly favored (*42,43*). This result may indicate a mechanism not involving the simple generation of H_2O_2 with its immediate use.

The use of Pd-Cu supported on silica has been found to be effective in benzene oxidation via the path shown in Scheme 3 (*44,45*). It is believed that the actual C-H activation occurs by metal promoted formation of hydroxyl radicals attacking the benzene nucleus. The pathway shown in Scheme 3 has the practical advantage of separating the hydrogen and oxygen activation steps- a method eliminating the mixing of these potentially explosive gases. The reaction has been elaborated to the gas phase, which promotes higher selectivities by removing the product phenol from the reaction (*46,47*). Unfortunately, the high temperatures needed for gas phase oxidation also serves to reduce the efficiency of the hydrogen consumed with much of it forming water. Benzene and hexane oxidation have also been attempted with palladium supported on titanium silicates in an attempt to exploit the well known oxidations using TS-1 with generated H_2O_2 (*48*).

Scheme 3

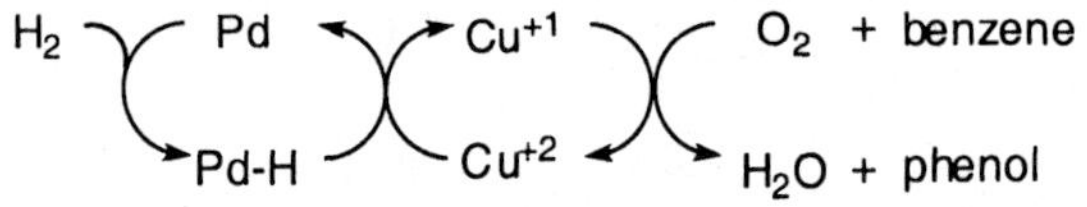

$$H_2 \rightarrow Pd \rightarrow Cu^{+1} \rightarrow O_2 + benzene$$
$$Pd\text{-}H \leftarrow Cu^{+2} \leftarrow H_2O + phenol$$

Beyond benzene oxidation, the oxidation of alkanes using *in situ* generated H_2O_2 has also received attention. The first report of this nature involved the use of a Pd(0) and Fe(II) doped zeolite for the oxidation of alkanes with hydrogen and oxygen (*49*). The catalyst exhibited the remarkable shape selectivity that can be generated with zeolites showing an *n*-octane/ cyclohexane preference of >190. Furthermore, supporting noble metals on TS-1 proved effective in the oxidation of *n*-hexane and *n*-octane in the presence of H_2 with the product being exclusively secondary alcohols and ketones (*10*). The efficiency, based on oxidant used, never exceeded 57%. However, when H_2O_2 was added to the reaction from an external source an efficiency of 86% was obtained even at a significantly higher temperature. Higher temperatures increase the rate of metal catalyzed hydrogen peroxide decomposition. Platinum heteropoly catalysts have been investigated for the oxidation of cyclohexane (*50*). Catalyst redox studies showed that activity is associated with both Pt^0 and Pt^{+2} being present along with a redox couple of Mo^{+5}/Mo^{+6} in the phosphomolybdate. Furthermore, the use gold supported on titanium mesoporous materials have received some attention in the oxidation of propane and isobutanol (*51*). While acetone from propane yield was meager, when compared to propene oxidation, *t*-butanol was formed with reasonable yield and selectivity from isobutane. The $GoAgg^{II}$ system, originally developed by Barton(*8*), has been modified and implemented for the oxidation of cyclohexane using H_2O_2 generated on a Pd surface with an H_2/O_2 mixture (*52*). Based on the typical use of acidic conditions for Pd catalyzed H_2O_2 generation, acetone and acetone/ acetic acid mixtures were employed (instead of the typical basic conditions for the $GoAgg^{II}$ reaction). Under these conditions an increase in the alcohol product of over an order of magnitude was observed compared to the basic conditions. Interestingly, while the alcohol was favored in neutral or acidic conditions, the ketone showed preference under basic conditions. Otsuka has observed a very interesting increase in both methane oxidation and selectivity to methanol when hydrogen is added to certain iron catalysts (*53*). Experiments show that hydrogenation of CO (formed through methane oxidation) is not responsible for the formation of methanol and that similar selectivites are observed when H_2O_2 replaces the H_2/O_2 gas mixture.

Furthermore, we have recently completed work showing the persistence of palladium-generated hydrogen peroxide in solution. The oxidant, though showing marked decomposition, is formed fast enough to interact with another catalyst (e. g., vanadium or iron) and effectively oxidizes aromatic and aliphatic hydrocarbons (*15*). Control experiments showed that very little oxidation occurs in the absence of the second, hydroxylation catalyst (vanadium species) and that $V(acac)_3$ (acac = acetylacetonate) will catalyze the hydroxylation of benzene by hydrogen peroxide, even in the absence of metallic palladium, dihydrogen and dioxygen. Furthermore, no oxidation occurs when the vanadium hydroxylation catalyst is used without some source of hydrogen peroxide (*15*). Thus, metallic palladium is active in the formation of hydrogen peroxide, which in turn generates an active catalytic species from the vanadium complex.

It was of interest what, if any, selectivity advantage could be obtained by employing *in situ* generated H_2O_2 for hydrocarbon oxidations. One can imagine that by carefully adjusting the ratio of the two catalysts (Pd and V) present, better selectivities could be obtained since no excess of H_2O_2 would ever be present during the reaction. We have previously established the oxidations are limited by the amount of hydrogen peroxide formed (*15*). Tables I and II, respectively, show results on the oxidation of benzene comparing the *in situ* reactions to those where H_2O_2 was added directly. It is clear that in all cases a remarkable selectivity advantage can be obtained. For the vanadium-catalyzed oxidations the selectivity difference is around 20% even when the substrate is in huge excess. It is important to note that due to phase separation problems the syringe pumped H_2O_2 reactions employed about one half the amount of oxidant as the *in situ* case (Fig. 1)- a condition that should favor the selectivity for the added H_2O_2 case. In the case of the iron catalyst, the selectivity advantage is nearly 30%.

Table I: Conversion and selectivity to phenol using lower initial benzene concentration *in situ* H_2O_2 generation[a]

Hydroxylation Catalyst	Benzene (μmol)	Phenol (μmol)	Benzoquinone (μmol)	Conversion to Phenol (%)	Selectivity to Phenol (%)[b]
V(acac)$_3$	1400	200(10)	tr.	14	>99
V(acac)$_3$	520	130(10)	tr.	26	>99
FeT(DBM)	1400	59(9)	1.7(0.2)	4.2	97

[a] Performed using benzene as indicated with 3.2 μmol V(acac)$_3$ or FeT(DBM) = tris(dibenzoylmethanato)Fe(III) and 20.0 mg 5% Pd/Al$_2$O$_3$ in 4.5 mL acetic acid exposed to 100 psi H$_2$, 1000 psi N$_2$, and 100 psi O$_2$ for 2 h at 65°C. Data is an average of 2 runs.
[b] Selectivity calculated as mmol phenol/ (mmol phenol + mmol benzoquinone).
SOURCE: Reproduced with permission from *J. Mol. Catal. A: Chem.* **2003**, *203*, 179–192. Copyright 2003 Elsevier.

Table II: Oxidation of benzene using H_2O_2 added via a syringe pump[a]

Hydroxylation Catalyst	Benzene (μmol)	Phenol (μmol)	Benzoquinone (μmol)	Conversion to Phenol (%)	Selectivity (%)[b]	H_2O_2 Efficiency (%)
V(acac)$_3$	22000	600(30)	190(10)	2.7	76	20
V(acac)$_3$	5900	220(60)	41(2)	3.7	84	6.2
V(acac)$_3$	520	47(3)	12(2)	9.0	80	1.4
FeT(DBM)	5900	9.7	tr.	0.16	69	0.37
FeT(DBM)	520	1.7	1	0.33	64	0.075

[a] Performed using 3.2 μmol V(acac)$_3$, or FeT(DBM) = tris(dibenzoylmethanato)Fe(III), benzene as indicated, and acetic acid (making total volume 4.5 mL) for 2 h at 65°C. During this time 0.28 mL, 4.9 mmol of 50% H_2O_2 (diluted to 0.68 mL total with acetic acid) was added dropwise via a syringe pump. Data is an average of 2 runs unless indicated.
[b] Selectivity calculated as mmol phenol/ (mmol phenol + mmol benzoquinone).
SOURCE: Reproduced with permission from *J. Mol. Catal. A: Chem.* **2003**, *203*, 179–192. Copyright 2003 Elsevier.

Another aspect needing consideration in systems where two catalysts operate simultaneously, one generating the H_2O_2 and another performing hydrocarbon oxidation, is whether the H_2O_2 generating catalyst participates in the substrate oxidation. As described above, there are numerous examples where a Pd catalyst generates the oxidant and also uses it for substrate oxidation. In order to truly exploit the available range of catalysts for hydrocarbon oxidations with H_2O_2, it must be ensured that the H_2O_2 generator does not interfere with the catalyst doing the actual substrate oxidation. We have explored this for the vanadium catalyzed oxidation of benzene using palladium generated H_2O_2 (54). The mechanism of benzene oxidation in this system was found to be identical to that where H_2O_2 was added directly. Therefore, the role of the palladium catalyst is to generate H_2O_2 *in situ*, and it does not play a part in the benzene oxidation step. However, this may not be the case under all conditions. In an effort to increase the oxidant efficiency we examined the use of soluble halide salts to retard H_2O_2 decomposition by palladium (18). Rather than the desired effect, a trend was observed showing that the amount of phenol produced decreased in the order $Cl^- > Br^- > I^-$ tracking the coordinative ability of the halide. This result suggests that these species might be showing coordinative inhibition of the vanadium catalyst.

Conclusion

The oxidation of aromatic and aliphatic molecules has clearly been shown possible using *in situ* generated H_2O_2. We have also shown that a significant increase in selectivity can be achieved via the use of *in situ* generated oxidant. However, the reaction is markedly less efficient in its use of oxidant. It is also crucial to address the influence of each catalyst and components present when using bimetallic conditions. The observation that conditions promoting one reaction in the system may inhibit or form alternative products in the second reaction warrants careful scrutiny of how the two systems interact.

One particular concern in reactions generating H_2O_2 *in situ* is the concomitant decomposition or hydrogenation via the catalyst making it. Though using a large excess of oxidant can minimize hydrogenation and using a biphasic solution can slow disproportionation, it seems unlikely that these methods will completely eliminate these competitive reactions. Also, in some situations the use of a biphasic mixture is impossible or impractical, such as in a gas phase reaction. Consequently, there is currently a significant need for catalysts that are less active for the unwanted reactions. It is important to note that the efficiency in the use of generated oxidant is important due to the fact that the economics are significantly impacted by the efficiency with which the coreductant is consumed.

Performing selective hydrocarbon oxidations with H_2O_2 and generating this oxidant *in situ* are currently receiving considerable attention. The current work shows that oxidative hydrocarbon activation can be effectively carried out in this manner. However, there is still a need for H_2O_2 generators showing more efficiency in use of oxidant and a fundamental understanding of how the two parts, H_2O_2 generation and hydrocarbon oxidation, interact. The latter should help in generating oxidation catalysts that work at their maximum rates and through mechanisms unhindered by or assisted by the presence of the H_2O_2 generating catalyst.

Experimental

Caution: Due care must be taken when dealing with gas mixtures under pressure. Special attention must be paid to gas flammability limits (55). All reactions were performed in 300 mL Parr stainless steel autoclaves. Quantitation was typically done using the FID detector excepting water quantitation reactions where the TCD detector was employed. Reactions with benzene used an Alltech EC-5 30 m, ID = 0.32 mm column with 100:1 split ratio and *o*-dichlorobenzene internal standard. Reaction products were confirmed using comparison to known compounds and further confirmed using GC-MS.

3.2 μmol of catalyst and 20.0 mg of 5% Pd/ Al_2O_3 were weighed into ca. 25 mL constricted neck glass liner. To this a small magnetic stir bar was added along with 2 mL of benzene and 2.5 mL of glacial acetic acid. The mixture was sealed in the autoclave and flushed thoroughly with hydrogen. It was then charged with 100 psi H_2, 1000 psi N_2, and 100 psi O_2 in that order and placed in a temperature regulated oil bath for 2 h at 65°C. Internal autoclave temperatures were determined using an autoclave equipped with a temperature sensor. At the end of the reaction, the autoclave was cooled in ice, slowly degassed, and the contents filtered through a 2 μm filter to remove the supported palladium catalyst. After internal standard addition, the liquid was analyzed with GC.

Syringe pump reactions were carried out by weighing 3.2 μmol of catalyst into a 5 mL conical vial. Substrate and solvent were added so that the total volume after hydrogen peroxide addition was 4.5 mL. A solution of acetic acid and 50% hydrogen peroxide (v/v, confirmed by permanganate titration) was made up to deliver 4.9 mmol of hydrogen peroxide in 0.68 mL of solution over 2 h at 65°C using a syringe pump (with a 1 mL syringe entirely of plastic to prevent peroxide decomposition). Reactions were chromatographed after internal standard addition.

Acknowledgement

JER thanks the NCER Star/ EPA fellowship for financial support. This work was funded by a grant from the NSF.

References

1. Ortiz de Montellano, P. R. *Cytochrome P-450 : structure, mechanism, and biochemistry*; Plenum Press: New York, 1986.
2. Que, L.; Ho, R. Y. N. *Chem. Rev.* **1996**, *96*, 2607-2624.
3. Liu, K. E.; Lippard, S. J. *Adv. Inorg. Chem.* **1995**, *42*, 263-289.
4. Sheldon, R. A. *Chem. Ind. (London)* **1997**, 12-15.
5. Sanderson, W. R. *Pure Appl. Chem.* **2000**, *72*, 1289-1304.
6. Strukul, G., Ed. *Catalytic oxidations with hydrogen peroxide as oxidant*; Kluwer: Dordrecht; Boston, 1992.
7. Hancu, D.; Green, J.; Beckman, E. J. *Acc. of Chem. Res.* **2002**, *35*, 757-764.
8. Barton, D. H. R.; Doller, D. *Acc. of Chem. Res.* **1992**, *25*, 504-512.
9. Ishida, M. A.; Masumoto, Y.; Hamada, R.; Nishiyama, S.; Tsuruya, S.; Masai, M. *J. Chem. Soc., Perkin Trans. 2* **1999**, 847-853.
10. Clerici, M. G.; Ingallina, P. *Catal. Today* **1998**, *41*, 351-364.
11. Sinha, A. K.; Seelan, S.; Tsubota, S.; Haruta, M. *Stud. Surf. Sci. Catal.* **2002**, *143*, 167-175.
12. Lu, J. Q.; Luo, M. F.; Lei, H.; Bao, X. H.; Li, C. *J. Catal.* **2002**, *211*, 552-555.
13. Lu, J. Q.; Luo, M. F.; Lei, H.; Li, C. *Appl. Catal., A* **2002**, *237*, 11-19.
14. Jenzer, G.; Mallat, T.; Maciejewski, M.; Eigenmann, F.; Baiker, A. *Appl. Catal., A* **2001**, *208*, 125-133.
15. Remias, J. E.; Sen, A. *J. Mol. Catal., A: Chem.* **2002**, *189*, 22-38.
16. Nishimura, T.; Onoue, T.; Ohe, K.; Uemura, S. *J. Org. Chem.* **1999**, *64*, 6750-6755.
17. Bortolo, R.; Bianchi, D.; D'Aloisio, R.; Querci, C.; Ricci, M. *J. Mol. Catal., A: Chem.* **2000**, *153*, 25-29.
18. Gosser, L. W. U. S. Patent, 4,681,751 (1987).
19. Nagashima, H.; Ishiuchi, Y.; Hiramatsu, Y.; Kawakami, M. U. S. Patent, 5,320,821 (1994).
20. Krishnan, V. V.; Dokoutchaev, A. G.; Thompson, M. E. *J. Catal.* **2000**, *196*, 366-374.
21. Dissanayake, D. P.; Lunsford, J. H. *J. Catal.* **2002**, *206*, 173-176.
22. Bianchi, D.; Bortolo, R.; D'Aloisio, R.; Ricci, M. *Angew. Chem., Int. Ed. Engl.* **1999**, *38*, 706-708.
23. Sen, A.; Halpern, J. *J. Amer. Chem. Soc.* **1977**, *99*, 8337.
24. Stahl, S. S.; Thorman, J. L.; Nelson, R. C.; Kozee, M. A. *J. Amer. Chem. Soc.* **2001**, *123*, 7188-7189.
25. Muto, S.; Tasaka, K.; Kamiya, Y. *Bull. Chem. Soc. Jpn.* **1977**, *50*, 2493.
26. Lin, M.; Sen, A. *J. Amer. Chem. Soc.* **1992**, *114*, 7307-7308.
27. Roussel, M.; Mimoun, H. *J. Org. Chem.* **1980**, *45*, 5387.
28. Steinhoff, B. A.; Fix, S. R.; Stahl, S. S. *J. Amer. Chem. Soc.* **2002**, *124*, 766-767.
29. Landon, P.; Collier, P. J.; Papworth, A. J.; Kiely, C. J.; Hutchings, G. J. *Chem. Commun.* **2002**, 2058-2059.

30. Niwa, S.; Eswaramoorthy, M.; Nair, J.; Raj, A.; Itoh, N.; Shoji, H.; Namba, T.; Mizukami, F. *Science* **2002**, *295*, 105-107.
31. Laufer, W.; Hoelderich, W. F. *Chem. Commun.* **2002**, 1684-1685.
32. Kuznetsova, N. I.; Detusheva, L. G.; Kuznetsova, L. I.; Fedotov, M. A.; Likholobov, V. A. *J. Mol. Catal., A: Chem.* **1996**, *114*, 131-139.
33. Sen, A. *Acc. of Chem. Res.* **1998**, *31*, 550-557.
34. Lin, M. R.; Hogan, T.; Sen, A. *J. Amer. Chem. Soc.* **1997**, *119*, 6048-6053.
35. Pifer, A.; Hogan, T.; Snedeker, B.; Simpson, R.; Lin, M. R.; Shen, C. Y.; Sen, A. *J. Amer. Chem. Soc.* **1999**, *121*, 7485-7492.
36. Chepaikin, E. G.; Boyko, G. N.; Bezruchenko, A. P.; Leshcheva, A. A.; Grigoryan, E. H. *J. Mol. Catal., A: Chem.* **1998**, *129*, 15-18.
37. Lin, M. R.; Hogan, T. E.; Sen, A. *J. Amer. Chem. Soc.* **1996**, *118*, 4574-4580.
38. Baker, E. C.; Hendriksen, D. E.; Eisenberg, R. *J. Am. Chem. Soc.* **1980**, *102*, 1020-1027.
39. Miyake, T.; Hamada, M.; Sasaki, Y.; Oguri, M. *Appl. Catal., A* **1995**, *131*, 33-42.
40. Miyake, T.; Hamada, M.; Niwa, H.; Nishizuka, M.; Oguri, M. *J. Mol. Catal., A: Chem.* **2002**, *178*, 199-204.
41. Ehrich, H.; Berndt, H.; Pohl, M. M.; Jahnisch, K.; Baerns, M. *Appl. Catal., A* **2002**, *230*, 271-280.
42. Izumi, Y.; Hidetaka, M.; Kawahara, S.-I. U. S. Patent, 4,009,252 (1977).
43. Izumi, Y.; Hidetaka, M.; Kawahara, S.-I. U. S. Patent, 4,279,883 (1981).
44. Kunai, A.; Wani, T.; Uehara, Y.; Iwasaki, F.; Kuroda, Y.; Ito, S.; Sasaki, K. *Bull. Chem. Soc. Jpn.* **1989**, *62*, 2613-2617.
45. Kitano, T.; Kuroda, Y.; Itoh, A.; Lifen, J.; Kunai, A.; Sasaki, K. *J. Chem. Soc., Perkin Trans. 2* **1990**, 1991-1995.
46. Kitano, T.; Kuroda, Y.; Mori, M.; Ito, S.; Sasaki, K.; Nitta, M. *J. Chem. Soc., Perkin Trans. 2* **1993**, 981-985.
47. Kitano, T.; Nakai, T.; Nitta, M.; Mori, M.; Ito, S.; Sasaki, K. *Bull. Chem. Soc. Jpn.* **1994**, *67*, 2850-2855.
48. Tatsumi, T.; Yuasa, K.; Tominaga, H. *J. Chem. Soc., Chem. Commun.* **1992**, 1446-1447.
49. Herron, N.; Tolman, C. A. *J. Amer. Chem. Soc.* **1987**, *109*, 2837-2839.
50. Kirillova, N. V.; Kuznetsova, N. I.; Kuznetsova, L. I.; Zaikovskii, V. I.; Koscheev, S. V.; Likholobov, V. A. *Catal. Lett.* **2002**, *84*, 163-168.
51. Kalvachev, Y. A.; Hayashi, T.; Tsubota, S.; Haruta, M. *J. Catal.* **1999**, *186*, 228-233.
52. Kim, S. B.; Jun, K. W.; Kim, S. B.; Lee, K. W. *Chem. Lett.* **1995**, 535-536.
53. Wang, Y.; Otsuka, K. *J. Chem. Soc., Chem. Commun.* **1994**, 2209-2210.
54. Remias, J. E.; Pavlosky, T. A.; Sen, A. *J. Mol. Catal. A: Chem. In Press*.
55. Zabetakis, M. G. *Flammability Charachteistics of Combustible Gases and Vapors: Bulletin 627*; Bureau of Mines: Washington, DC, 1965.

Chapter 24

New Applications of Electrophilic Aromatic C–H Activation with Metal Trifluoroacetates

Vladimir V. Grushin[1,*], David L. Thorn[1,2,*], William J. Marshall[1], and Viacheslav A. Petrov[1]

[1]Central Research and Development, DuPont, Wilmington, DE 19880–0328
[2]Current address: Chemical Science and Technology Division, Los Alamos National Laboratory, Los Alamos, NM 87545

New synthetic applications have been found for high oxidation state trifluoroacetato (TFA) complexes of Rh, Sn, and Ir. Highly selective oxidative carbonylation of arenes has been carried out with "Rh(TFA)$_3$" catalyst under mild conditions (20-65 °C, 1 atm) to produce carboxylic acids via C-H activation. The first intermolecular electrophilic stannylation of arenes has been performed with Sn (IV) TFA species generated from Sn (II) oxide or tetraphenyltin. A new simple method has been developed for exhaustive cyclometalation of fluorinated 2-arylpyridines with Ir (III) TFA complexes to produce tricyclometalated derivatives which are among the most efficient electroluminescent materials known to date.

To the best of our knowledge, the first example of C-H activation with a metal compound is the aromatic electrophilic mercuration reported by Dimroth as early as 1898 (*1*). In a series of his articles published in 1898-1902 (*1, 2*) Dimroth described the preparation of a number of arylmercury acetates from Hg(OAc)$_2$ and various aromatic hydrocarbons (e.g., eq 1).

$$C_6H_6 + Hg(OAc)_2 \longrightarrow C_6H_5-HgOAc \quad (1)$$

The electrophilicity of Hg (II) in such S$_E$Ar mercuration reactions can be enhanced dramatically by replacing the acetate of Hg(OAc)$_2$ with a less nucleophilic anion and by running the reaction in a more ionizing, poorly coordinating medium. At the same time, the counter-anion must be basic enough to successfully deprotonate the Wheland intermediate (σ-complex) to restore aromaticity in the final step of the process (*3*). Trifluoroacetate (TFA) appears to be an excellent counterion to meet both criteria. Thus, mercury trifluoroacetate (Hg(TFA)$_2$) in trifluoroacetic acid (TFAH) mercurates benzene 6.9×10^5 times faster than the classical Hg(OAc)$_2$ – AcOH system (*4*). Furthermore, Kampel et al. have found (*5*) that icosahedral carboranes readily undergo mercuration with Hg(TFA)$_2$ in TFAH under mild conditions, while remaining totally unreactive toward Hg(OAc)$_2$ in AcOH even at elevated temperatures. Deacon and Farquharson (*6*) have demonstrated that up to 5 Hg atoms can be introduced into the benzene ring with Hg(TFA)$_2$, whereas Hg(OAc)$_2$ can afford only the monomercurated product.

We used the above considerations for the development of a series of new applications of metal TFA complexes in synthesis and catalysis. The three sections below describe (i) a new Rh catalytic system for oxidative carbonylation of aromatic hydrocarbons via C-H activation under mild conditions, (ii) the first direct intermolecular stannylation of arenes, and (iii) a simple, novel synthesis of new, highly photo- and electroluminescent tricyclometalated Ir complexes of fluorinated 2-arylpyridines.

Rh-Catalyzed Oxidative Carbonylation of Arenes under Mild Conditions

Finding an efficient catalytic system to directly introduce a small C$_1$ or C$_2$ group (e.g., CH$_3$, CH$_2$OH, CHO, COOH, C$_2$H$_5$, CH=CH$_2$, etc.) in the *para*-position of toluene represents a considerable intellectual and practical challenge.

Such *para*-substituted toluenes could be cleanly oxidized to terephthalic acid, an important monomer for polyester (7). Terephthalic acid is currently produced from *p*-xylene which is significantly more costly than toluene.

Scientifically, upon initial consideration the problem of catalytic introduction of *any* carbon substituent into the *para* (not *meta!*) position of toluene might look like an easy one. Nonetheless, there have been no reports of a catalytic reaction that would furnish a *para*-C-derivative of toluene directly from toluene with high selectivity and a satisfactory catalytic turnover number (TON).

In principle, the Fujiwara reaction (8), also known as oxidative carbonylation of arenes, (eq 2) might provide a solution to the synthesis of toluic acids from toluene and carbon monoxide. However, the *para*-selectivity of the Pd-catalyzed oxidative carbonylation of toluene is in the range of only 40-67%, with low TON (< 10) even under optimized conditions (8).

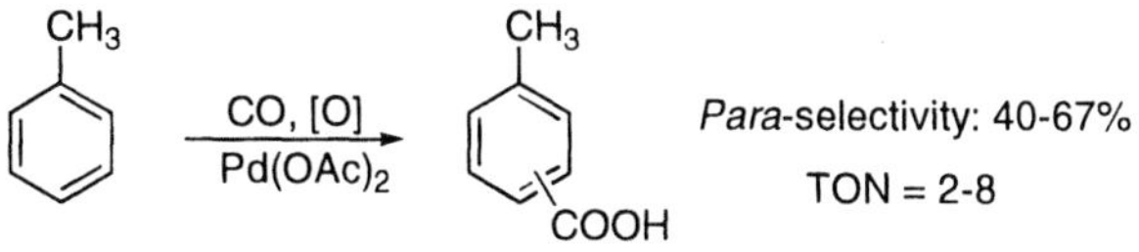

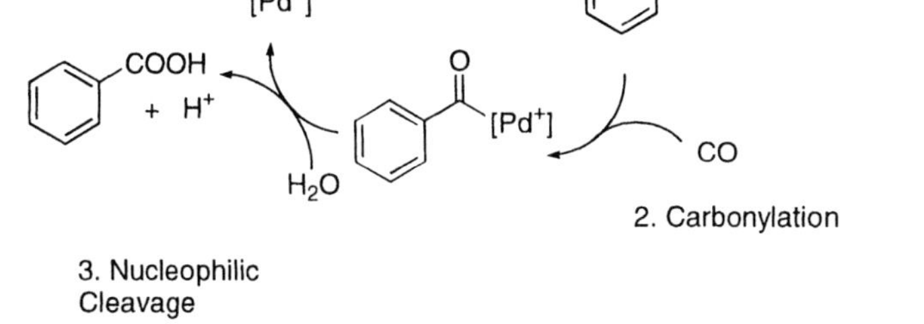

Scheme 1

A widely accepted mechanism for the Fujiwara reaction includes electrophilic attack of Pd^{2+} on the aromatic ring, followed by migratory insertion and nucleophilic cleavage to give a molecule of the acid and Pd (0). The latter must be re-oxidized to the catalytically active Pd (II) which then can commence another catalytic turnover (Scheme 1).

Oxidative carbonylation of toluene has also been performed (9, 10) in the presence of Rh catalysts with higher *para*-selectivities occasionally reaching 90+%. Conditions required for these Rh-catalysed reactions to occur are harsh, up to 270 atm of CO/O_2 at 150 °C. It was not clear if such severe conditions were needed for the C-H activation step or metal re-oxidation.

Keeping in mind the mercury trifluoroacetate chemistry (see Introduction) we intended to develop a rhodium-based catalytic system devoid of strongly coordinating anions. A material formulated as rhodium (III) trifluoroacetate, "$Rh(TFA)_3$", (11) appeared promising but contamination with strongly-binding chloride (12) greatly diminished the electrophilicity of Rh^{3+} and its ability to attack the benzene ring (see above). To eliminate the Cl⁻ contamination problem, we found (12, 13) a new method to prepare genuinely Cl-free "$Rh(TFA)_3$" by exhaustive oxidation of $[Rh(CO)_2(CF_3CO_2)]_n$ with H_2O_2 in TFAH (eq 3).

$$[Rh(CO)_2(CF_3COO)]_n \xrightarrow{\frac{30\% H_2O_2}{CF_3COOH}} \text{"}Rh(TFA)_3\text{"} \quad (3)$$

The product of this oxidation was found (12) to efficiently catalyze H/D exchange between trifluoroacetic acid and benzene-d_6 or toluene-d_8 (Scheme 2).

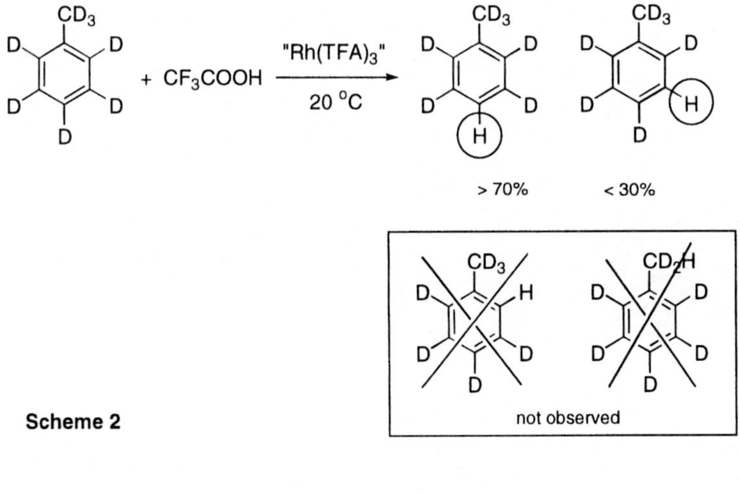

Scheme 2

The expected electrophilic mechanism of this H/D exchange was supported by (i) strong inhibition by chloride, observed upon addition of $[(Ph_3P)_2N]^+$ Cl⁻ ($[PPN]^+$ Cl⁻) and (ii) the positional selectivity observed for the reaction of $CD_3C_6D_5$ with CF_3COOH. Exchange occurred at the *ortho* (major) and *meta* (minor) positions of the toluene molecule (Scheme 2) without involving the *ortho*-position or the CD_3 group. Due to fast rates and multiple exchange we could only roughly estimate the selectivity at > 70% *para* and < 30% *meta* (*12*).

When "$Rh(TFA)_3$" was tested as a catalyst to oxidatively carbonylate toluene in a 1 : 1 (by volume) mixture with trifluoroacetic acid using $K_2S_2O_8$ as an oxidant, the reaction (eq 4) occurred under exceedingly mild conditions, 1 atmosphere of CO and 20-65 °C (*12, 13*). In accord with the results of the H/D exchange experiments, the selectivity for the desired *p*-toluic acid was in the range of 93-98%, the remaining 2-7% being *m*-toluic acid. No carboxylation occurred at the CH_3 group or the *ortho*-positions of the ring. The reaction was catalytic in Rh, with TON = 40-100.

The most remarkable characteristics of the catalytic process (eq 4) are the particularly mild conditions and consistently very high *para*-selectivity. Further studies revealed a number of important features (*12*):

- Under identical conditions, *lower* catalyst concentrations resulted in *higher* turnover numbers (Figure 1).
- The reaction slowed down as CO pressure was increased.
- The catalyst tolerated the reaction products, $KHSO_4$ and toluic acid.
- Small amounts of water did not poison the catalyst, whereas chloride did.
- The *para*-selectivity was consistently in the range of 93-98% whether the reaction was run at 20, 50 or 65 °C.
- The determined kinetic isotope effect of 3 was high, possibly due to a bulky metal electrophile acting under strongly acidic conditions (TFAH).
- Toluene was ca. 2 times more reactive than benzene.
- No reaction was observed in acetic acid.
- No oxidative carbonylation of toluene took place in the presence of C_6Me_6, probably due to the formation of stable $[(C_6Me_6)Rh(CO)_2]^+$ (see below).
- The readily available Rh (II) trifluoroacetate dimer $[Rh_2(TFA)_4]$ exhibited very low catalytic activity, if any.

Considering the above observations, it is believed that the first step of the catalytic process involves electrophilic attack of a Rh (III) TFA complex on the benzene ring to form a Rh aryl which then undergoes carbonylation (Scheme 3). The structure of the reactive Rh species remains unknown but the oxidation state of +3 suggests that the coordination number on the metal is probably 6 in the resting state and 5 for the reactive electrophilic form. Therefore, the active Rh electrophile is expected to be a bulky species, which accounts for the high para-selectivity of the process. Although the Rh center of the electrophile is too electron-deficient to strongly interact with CO, it is reasonable to propose some weak, reversible interactions between the two (Scheme 3). At higher CO concentration the equilibrium would be shifted toward less electrophilic Rh (III) carbonyl species, thus rationalizing slower rates at increased CO pressures.

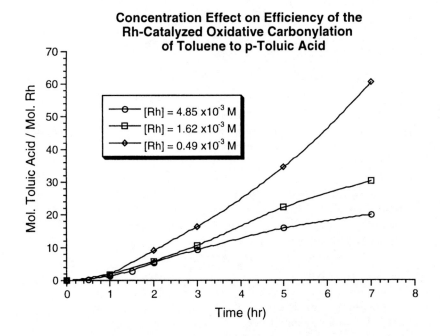

Figure 1. Higher TONs are obtained with lower concentrations of Rh catalyst. (Reproduced from reference 12. Copyright 2001 Wiley-VCH).

In the final step (Scheme 3) an aroyl Rh complex reacts with a suitable nucleophile, such as water or trifluoroacetate anion to give the aromatic acid or its mixed anhydride along with Rh (I). The latter is oxidized back to Rh (III) to close the catalytic loop. As the process occurs, the reaction mixture contains both Rh (I) and Rh (III) species which may comproportionate to produce the stable and catalytically inactive Rh (II) dimer (see above). It is believed that it is the comproportionation side process Rh(I) + Rh(III) → 2Rh(II) that deactivates the catalyst, and is an important factor responsible for more efficient catalysis at lower Rh concentrations (Figure 1).

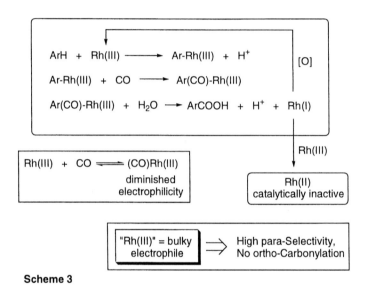

Scheme 3

Interestingly, hexamethylbenzene was found to poison the Rh catalyst, possibly due to the formation of oxidation-resistant [(C$_6$Me$_6$)Rh(CO)$_2$]$^+$. As was demonstrated by a separate experiment (12), strong acids favor the reversible formation of [(η^6-arene)Rh(CO)$_2$]$^+$ from [(CO)$_2$Rh(TFA)]$_n$. The arene Rh (I) complex undergoes very slow oxidation by H$_2$O$_2$ in TFAH at room temperature, possibly via a small equilibrium concentration of [(CO)$_2$Rh(TFA)]$_n$. Hexamethylbenzene is expected to bind to Rh (I) more strongly, thus inhibiting the re-oxidation step.

Using our catalytic system we also synthesized aromatic carboxylic acids from a number of other arenes (Scheme 4). As can be seen from Scheme 4, the C-H activation *ortho* to a methyl group and even two methyl groups appears possible. At the same time, the carboxylation of naphthalene was at least 97% β-selective, indicating that steric effects play an important role in the reaction.

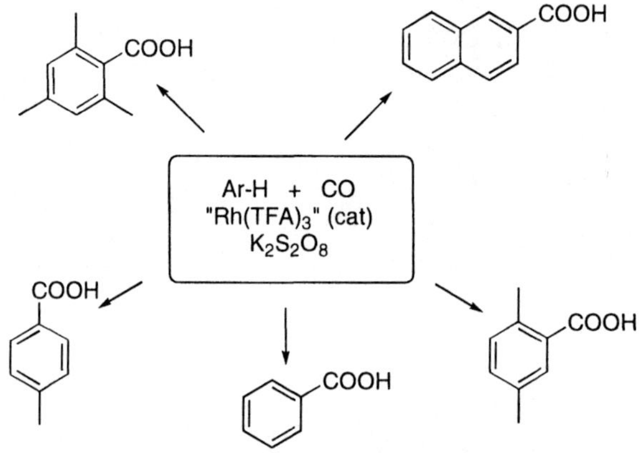

Scheme 4

Electrophilic Stannylation of Arenes

The direct electrophilic metalation of arenes (e.g., eq 1) is clearly the simplest and most economical route to aryl derivatives of the main group elements, which are widely used in synthesis. Besides mercury (II) (*1-6, 14*) (see the Introduction), only two other main group metals, thallium (III) (*15*) and lead (IV) (*16*) form salts which are capable of direct intermolecular electrophilic metalation of aromatics under mild conditions (*17*). Adding tin to this list was desirable because of the importance of organotin compounds. "Tin has a larger number of its organometallic derivatives in commercial use than any other element" (*18*).

Reactivity considerations outlined in the Introduction made us believe that TFA complexes of Sn (IV) should be promising candidates for direct electrophilic aromatic stannylation. First it was found (*19*) that exhaustive dephenylation of Ph$_4$Sn with TFAH resulted in a TFA Sn (IV) species which indeed activated aromatic C-H bonds of benzene, toluene, and *p*-xylene to give aryltin derivatives. We then also developed another route to this electrophilic tin species by oxidizing readily available and inexpensive SnO with 30% hydrogen peroxide in trifluoroacetic acid, followed by treatment with trifluoroacetic anhydride (*19*). Reactions of the thus prepared Sn (IV) electrophilic reagent with arenes afforded organotin compounds which were isolated and fully characterized in the solid state and in solution (Scheme 5).

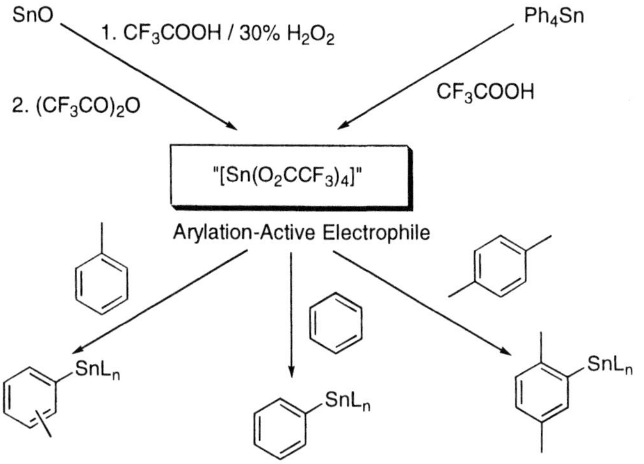

Scheme 5

All attempts to isolate and characterize an analytically pure single compound from the arylation-active material were unsuccessful. It is believed that the electrophilic species may consist of mixtures of $Sn(CF_3COO)_4$ and trifluoroacetoxy-polystannoxanes such as the hypothetical tetramer $Sn_4(\mu^3\text{-}O)_2(CF_3COO)_{12}$ analogous to the aryltin stannylation products (see below). Although the electrophilic stannylation reactions are reversible (^{1}H NMR), the equilibrium can be shifted towards the desired aryltin product by running the reaction with arene under reflux for 5-10 minutes. Thus, the compound $[(Ph)_2Sn_4O_2(CF_3COO)_{10}]$ was the main phenyltin species isolated from the reaction of benzene with SnO-derived "trifluoroacetato-tin (IV)" (Equation 5).

The structure of this organotin cluster was established as $[Sn_4(Ph)_2(\mu^3\text{-}O)_2(\mu\text{-}CF_3COO)_6(\eta^2\text{-}CF_3COO)_2(\eta^1\text{-}CF_3COO)_2]$ by X-ray diffraction (Figure 2) (*19*). The presence of the Ph-Sn bond in solution was confirmed by the observation of Sn-H coupling to the *ortho* protons with $^3J_{Sn-H} = 177$ Hz. A broadened singlet in the ^{19}F NMR spectrum of the sample indicated that the cluster was fluxional at ambient temperature, probably undergoing fast exchange of the TFA ligands. Bromination of the isolated material led to the clean formation of bromobenzene in 91% yield, assuming that

[(Ph)$_2$Sn$_4$O$_2$(CF$_3$COO)$_{10}$] was the only Ph-Sn species present in the bulk product. This assumption, however, is made cautiously as other phenyl tin species may be present in the crude isolated material.

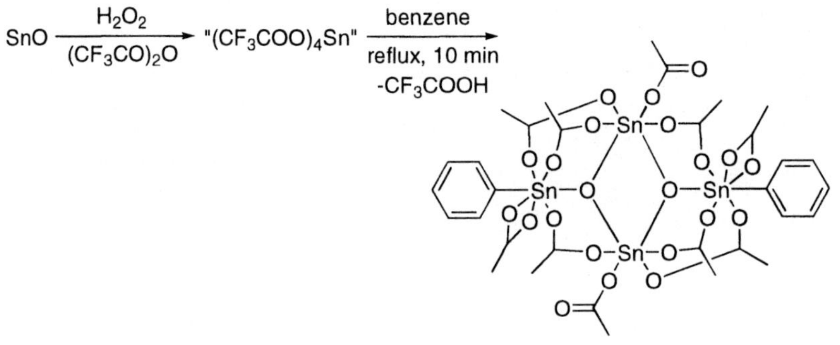

$$SnO \xrightarrow[(CF_3CO)_2O]{H_2O_2} \text{"}(CF_3COO)_4Sn\text{"} \xrightarrow[\text{reflux, 10 min} \atop -CF_3COOH]{\text{benzene}} \quad (5)$$

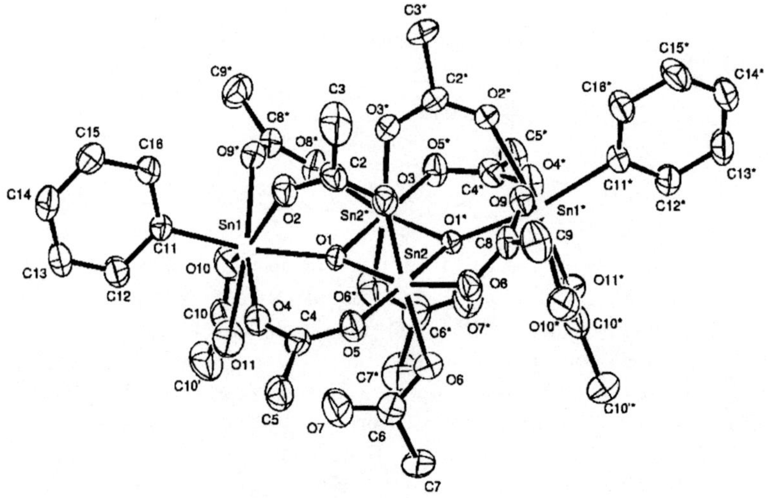

Figure 2. An ORTEP drawing of [Sn$_4$(Ph)$_2$(μ^3-O)$_2$(μ-CF$_3$COO)$_6$(η^2-CF$_3$COO)$_2$(η^1-CF$_3$COO)$_2$] obtained via the direct stannylation of benzene. Fluorine atoms are omitted for clarity. (Reproduced from reference 19. Copyright 2001 Wiley-VCH).

Similarly, direct stannylation of *para*-xylene was carried out (*19*). The structure of the *p*-xylyl Sn complex is very similar to the phenylated compound, but there is also a remarkable difference between the two. In the phenyl derivative (Figure 2), the "organometallic" tin atoms are seven-coordinate, being surrounded by one carbon and six oxygens of the μ-oxo and TFA ligands (**A**; F atoms are omitted). In the xylyl complex however, there are only five oxygens in the first coordination sphere of the arylated tin atom, but the octahedron is distorted, due in part to interaction between the metal and the *ortho*-methyl group (**B**; F atoms are omitted). The non-bridging trifluoroacetato ligand in the phenyl complex is η^2, whereas in the xylyl compound it is η^1.

A   **B**

When toluene was stannylated, a mixture of organotin compounds formed. The positional selectivity of this reaction (^{13}C NMR and bromination studies) was surprising, the *ortho* : *meta* : *para* ratio was ca. 0 : 2 : 1 (*19*). The observed lack of stannylation at the *ortho* positions is likely due to steric bulk of the reactive Sn electrophile, whereas the *meta* to *para* ratio of 2 is uncommon but not unprecedented in S$_E$Ar chemistry, pointing to exceptionally high reactivity of the acting electrophile (*20*).

In concluding this section, tin has now been added to the list of main group metals (Hg, Tl, and Pb) which form compounds capable of directly electrophilically metalating arenes at mild temperatures.

One-Step Synthesis of Highly Luminescent Triply Cyclometalated Organometallic Iridium Complexes of Fluorinated 2-Arylpyridines

It is currently believed that organic light emitting diodes (OLEDs) may dominate the display industry of the future. The tricyclometalated iridium complex of 2-phenylpyridine [Ir(ppy)$_3$] originally described and studied by Watts (*21*), has been recently identified as a green emitter of unprecedented efficiency (*22*). We decided to synthesize fluorinated analogues of [Ir(ppy)$_3$] which we expected to exhibit improved electroluminescent and processing properties, such as volatility and thermal stability.

The reaction of 2-phenylpyridine with $IrCl_3 \cdot nH_2O$ readily produces a non-luminescent doubly cyclometalated Cl-bridged dimer (21, 23, 24). The third cyclometalation which affords the desired luminescent compound [Ir(ppy)$_3$] occurs only sluggishly and usually in low yield under aggressive conditions, even if performed on pre-isolated dicyclometalated intermediate in the presence of silver triflate (24).

Having prepared a series of fluorinated 2-arylpyridine derivatives via Suzuki-Miyaura coupling (25, 26), we were in need of a simple, efficient, and preferably one-step method for their triple cyclometalation with Ir (III). Although mechanistic details of the cyclometalation of 2-phenylpyridine with $IrCl_3$ were unknown, the oxidation state of the metal and some of our preliminary experimental observations suggested an S_EAr-type mechanism for the C-H activation. Therefore, we thought that converting chloro iridium intermediates into their more electrophilic TFA analogues (see above) should facilitate the metalation. Indeed, we were pleased to find (25, 26) that iridium (III) chloride readily reacted with the 2-arylpyridines in the presence of 3 equivalents of $AgOCOCF_3$ to produce the desired tri-cyclometalated complexes in up to 82% yield (eq 6). The reaction is run in excess arylpyridine ligand which is easily recovered and recycled after the reaction.

X = H; Y = H, 2-F, 4-F, 3-CF$_3$;
X = 5-CF$_3$; Y = H, 2-F, 3-F, 4-F, 3-CH$_3$O, 4-CH$_3$O, 3-CF$_3$, 4-CF$_3$O;
X = 3-Cl; Y = 3-CF$_3$;
X = 5-NO$_2$; Y = 3-CF$_3$;

Our recycling technique allowed for the synthesis on a considerable scale, i.e. an excellent green emitter (25-27) shown in Figure 3 was made in the amount of 55 grams (99.98% purity) in 70% yield calculated on the ligand and iridium chloride used. The fluorinated iridium complexes easily sublimed without any sign of decomposition, a critical property for vacuum-deposition technology. Importantly (27), the complexes exhibited very high electroluminescent efficiencies in pure films *without* the host matrix (25-27) which is required (22) for the parent [Ir(ppy)$_3$] complex.

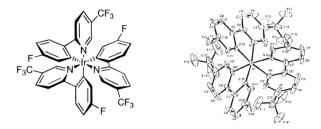

Figure 3. Structural formula and an ORTEP drawing for the most efficient green emitter synthesized via the TFA method (eq 6). (The ORTEP reproduced from reference 25. Copyright 2001 The Royal Chemical Society).

Exploiting the generality of the TFA method, we succeeded in preparing numerous complexes containing different substituents in different positions of the aromatic rings (eq 6). The availability of a large series of the materials led to a better understanding of how molecules of this type should be designed for color tuning via systematic control of the nature and position of the substituents (25), a key issue for the technology of full color displays of the future. Further details of our electroluminescent studies can be found elsewhere (25-27).

Literature Cited

1. Dimroth, O. *Ber. Dtsch. Chem. Ges.* **1898**, *31*, 2154.
2. Dimroth, O. *Ber. Dtsch. Chem. Ges.* **1899**, *32*, 758; **1902**, *35*, 2032; **1902**, *35*, 853.
3. Taylor, R. *Electrophilic Aromatic Substitution*; Wiley: New York, 1990.
4. Brown, H. C.; Wirkkala, R. A. *J. Am. Chem. Soc.* **1966**, *88*, 1447.
5. Bregadze, V. I.; Kampel, V. Ts.; Godovikov, N. N. *J. Organomet. Chem.* **1976**, *112*, 249.
6. Deacon, G. B.; Farquharson, G. J. *J. Organometal. Chem.* **1974**, *67*, C1.
7. Weissermel, K.; Arpe, H.-J. *Industrial Organic Chemistry, 3rd Ed.*; Wiley-VCH: Weinheim, 1997.
8. Taniguchi, Y.; Yamaoka, Y.; Nakata, K.; Takaki, K.; Fujiwara, Y *Chem. Lett.* **1995**, 345. Lu, W.; Yamaoka, Y.; Taniguchi, Y.; Kitamura, T.; Takaki, K.; Fujiwara, Y. *J. Organomet. Chem.* **1999**, *580*, 290. Fujiwara, Y.; Takaki, K.; Taniguchi, Y. *Synlett* **1996**, 591.
9. a) Waller, F. J. Eur. Pat. Appl. EP 32318, 1981; Eur. Pat. Appl. EP 82633, 1983; US Patents 4,356,318; 4,416,801; 4,431,839; 4,463,103, 1982-1983. b) Waller, F. J. *Catal. Rev. Sci. Eng.* **1986**, *28*, 1.
10. Kalinovskii, I. O.; Lescheva, A. A.; Kuteinikova, M. M.; Gel'bshtein, A. I. *J. Gen. Chem. USSR* **1990**, *60*, 108. Kalinovskii, I. O.; Pogorelov, V. V.; Gel'bshtein, A. I.; Akhmetov, N. G. *Russ. J. Gen. Chem.* **2001**, *71*, 1457.

Kalinovskii, I. O.; Gel'bshtein, A. I.; Pogorelov, V. V. *Russ. J. Gen. Chem.* **2001**, *71*, 1463.
11. Gol'dshleger, N. F.; Moravskii, A. P.; Shul'ga, Yu. M. *Izv. Akad. Nauk SSSR, Ser. Khim.* **1991**, 258.
12. Grushin, V. V.; Marshall, W. J.; Thorn, D. L. *Adv. Synth. Catal.* **2001**, *343*, 161.
13. Grushin, V. V.; Thorn, D. L. PCT WO 01/07387 A1, 2001.
14. Makarova, L. G.; Nesmeyanov, A. N. *Organic Compounds of Mercury*; North-Holland: Amsterdam, 1967. Larock, R. C. *Organomercury Compounds in Organic Synthesis*; Springer: Berlin, 1985.
15. Usyatinskii, A. Ya.; Bregadze, V. I. *Usp. Khim.* **1988**, *57*, 1840.
16. Panov, E. M.; Kocheshkov, K. A. *Dokl. Akad. Nauk S.S.S.R.* **1958**, *123* 295. De Vos, D.; Van Barneveld, W. A. A.; Van Beelen, D. C.; Van der Kooi, H. O.; Wolters, J.; Van der Gen, A. *Recl. Trav. Chim. Pays-Bas* **1975**, *94*, 97. Stock, L. M.; Wright, T. L. *J. Org. Chem.* **1980**, *45*, 4645.
17. a) Shilov, A. E.; Shul'pin, G. B. *Activation and Catalytic Reactions of Saturated Hydrocarbons in the Presence of Metal Complexes*; Kluwer: Dordrecht, 2000. b) Intermolecular electrophilic metalation reactions of arenes have also been reported for a few transition metals, e.g., Pt (IV), Au (III), Rh (III), and Pd (II) (*17a*). c) Direct borylation of arenes under prolonged heating has been known for over 40 years: Bujwid, Z. J.; Gerrard, W.; Lappert, M. F. *Chem. & Ind. (London)* **1959**, 1091.
18. Hoch, M. *Appl. Geochem.* **2001**, *16*, 719.
19. Grushin, V. V.; Marshall, W. J.; Thorn, D. L. *Adv. Synth. Catal.* **2001**, *343*, 433. For a review of this work, see: Freemantle, M. *Chem. Eng. News* **2001**, *79*, October 15, 33.
20. Stock, L. M.; Brown, H. C. *Adv. Phys. Org. Chem.* **1963**, *1*, 35.
21. King, K. A.; Spellane, P. J.; Watts, R. J. *J. Am. Chem. Soc.* **1985**, *107*, 1431. Dedeian, K.; Djurovich, P. I.; Garces, F. O.; Carlson, G.; Watts, R. J. *Inorg. Chem.* **1991**, *30*, 1685.
22. Baldo, M. A.; Lamansky, S.; Burrows, P. E.; Thompson, M. E.; Forrest, S. R. *Appl. Phys. Lett.* **1999**, *75*, 4.
23. Sprouse, S.; King, K. A.; Spellane, P. J.; Watts, R. J. *J. Am. Chem. Soc.* **1984**, *106*, 6647.
24. Colombo, M. G.; Brunold, T. C.; Riedener, T.; Guedel, H. U.; Fortsch, M.; Buergi, H.-B. *Inorg. Chem.* **1994**, *33*, 545.
25. Grushin, V. V.; Herron, N.; LeCloux, D. D.; Marshall, W. J.; Petrov, V. A.; Wang, Y. *Chem. Commun.* **2001**, 1494. For a review of this work, see: Freemantle, M. *Chem. Eng. News* **2001**, *79*, August 27, 17.
26. Petrov, V. A.; Wang, Y.; Grushin, V. V. PCT WO 0202714 A2, 2002.
27. Wang, Y.; Herron, N.; Grushin, V. V.; Lecloux, D.; Petrov, V. *Appl. Phys. Lett.*, **2001**, *79*, 449.

Chapter 25

Reengineering of Organic-Based Metal Active Sites for Oxidations and Oxygenations

Sergiu M. Gorun

Department of Chemistry, Brown University, Providence, RI 02912

Structurally abbreviated, functional models of O_2 activating metallo-enzymes based upon metal centers surrounded in three-dimensions exclusively by *aliphatic* C-F bonds are presented. This design is conceptually akin to a real "Teflon-coating" reactor, useful for harsh chemistry. Fluorinated pyrazolyl borates and phthalocyanines, exhibit reversible O_2 binding and unexpected oxidation and oxygenations. The phthalocyanines produce excited state $^1O_2$ and use ground-state O_2 for the oxidation and oxygenation of C-H bonds. Both classes of complexes are sufficiently robust to derivatize external substrates without noticeable decomposition, an important objective of biologically inspired catalysis.

Introduction

Important industrial products include those derived from natural hydrocarbon resources and oxygen. Their transportation and processing, however, is energy inefficient and generates CO_2 and other greenhouse gases. In addition to their usefulness as chemicals, oxygenated hydrocarbons, usually high-energy density liquids, are an advantageous option for transportation of remote gas reserves.

Efficient oxygen incorporation into C-H bonds, especially those of light alkanes, is a thermodynamically possible, yet formidable, unresolved problem. Stated in its most simple form, the reaction: $2CH_4 + O_2 = 2CH_3OH$ is energetically favorable. The direct oxidation of refractory C-H bonds, however, is difficult to stop at initial stages since the C-H bonds of the partly oxygenated products are weaker in comparison with the starting materials. Industrial solutions to this problem have relied primarily upon equipment and reaction conditions design, for example short catalyst-substrate contacts. Nature relies for direct oxygenations mostly on the utilization of pre-reduced oxygen (superoxo, peroxo, oxo groups) and "equipment design", viz. hydrophobic metalloenzymes active sites that favor hydrocarbons over oxygenated products. Heme- and non-heme based exquisite systems perform alkane oxygenations at ambient conditions, thus requiring low energy input. Even methane can be oxygenated at ambient conditions by Cu and Fe non-heme methane monooxygenases (MMOs). Such a reaction has not yet been duplicated in the laboratory. Nature's findings, however, are not directly conducive to optimum catalysis since the pre-reduction of O_2 requires protons and electrons, while enzymes exhibit active-site C-H bonds weaker than those of refractory alkane substrates.

The industrial use of pre-reduced O_2 (H_2O_2) for oxidations may be a viable option in some cases, but H_2O_2 production requires H_2, which is not a natural resource; its production requires energy. Large-scale H_2 production currently uses fossil hydrocarbons, i.e. "stored" solar energy, whose carbon generates CO_2. The ideal pathway to oxygenates is direct oxygenation. Revealing the principles governing biological oxygenation has spawned biomimetic and biologically inspired approaches to catalyst design *(1-3)*, "a great challenge for the synthetic inorganic chemist who is attempting to mimic an enzymatic transformation" by creating "an environment that stabilizes the ligand composition at the metal ion, while offering labile sites for catalysis" *(4)*. Along this line, advances in biomimetic design *(5)* include porphyrin-based catalysts, which, if halogenated, may be close to being industrially viable for free-radical hydroperoxide decompositions. The related phthalocyanines met with limited success either alone or encapsulated in zeolites *(6)* and related materials. Despite advances in organometallic approaches *(3)*, the direct introduction of O_2 into strong C-H bonds, an industrially important goal, remains a challenge *(7)*.

In industry, harsh reactions, including oxidations, are carried out in either glass (inorganic) or Teflon (organic) coated reactors. We propose an analogy between industrial equipment and molecules, with the aim of translating industrial apparati to the molecular level. Formal miniaturization of a glass-coated reactor leads to inorganic lacunar molecules, for example zeolites; the miniaturization of a Teflon-coated reactor, viz. using an organically based material, despite offering a high degree of synthetic, electronic and steric flexibility, is less developed.

This Chapter is concerned with the introduction of new, fluoroorganic-based materials that may illustrate the above analogy, while exhibiting catalytic properties. These materials are inspired by both heme and non-heme metalloenzymes. The focus is on the production of the metal complexes and some proof-of-principle demonstrations of their reactivity and stability under reaction conditions. Particular attention is paid to reactions with dioxygen and derived species. The establishment of reactivity patterns and in-depth mechanistic studies are beyond the scope of this paper.

Conceptually, a catalytic metal center placed within a cavity based on *aliphatic* C-F bonds models a Teflon-coated reactor, Figure 1. These robust complexes should be suitable for the activation of molecules of appropriate shape, size and hydrophobicity, such as O_2 and small hydrocarbons, while possibly excluding reaction products.

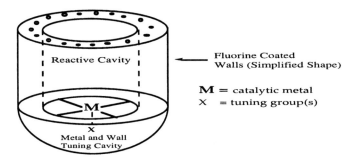

Figure 1. Schematic representation of a "molecular reactor". X is a ligand, or the coordinating group of a tether linked at the other end to a solid support.

Three-Dimensional Fluorinated Ligands: Non-Heme and Heme-like Coordination Environments

Metalloenzymes *(8-11)* that perform oxygenations belong to both heme and non-heme classes. Relevant models include halo tris(pyrazolyl)borates (TPB), porphyrins, unsubstituted (perfluorinated) phthalocyanines (Pc) and fluoroalkyl substituted phthalocyanines (R_fPc), Figure 2a, b, c, and d, respectively.

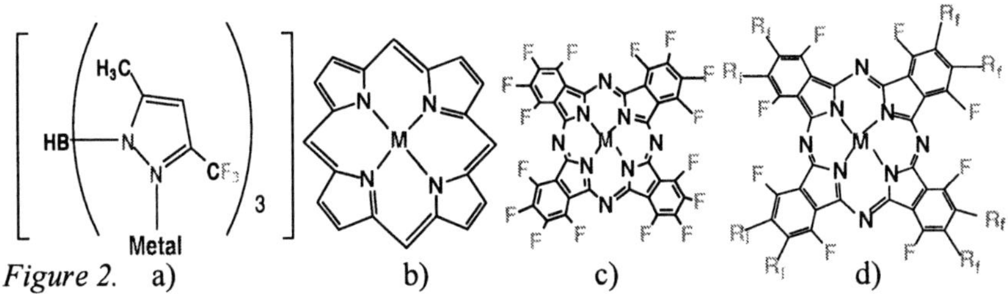

Figure 2. a)   b)   c)   d)

Bis(pyrazolyl)borates, BPBs, have a structure similar to TPBs, but with only two pyrazole rings.

1. Fluorinated pyrazolyl borates. Non-heme type models

The partially fluorinated Tp ligands, Figure 2a, yield several types of first row transition metal complexes *(12-16)*, Figure 3:

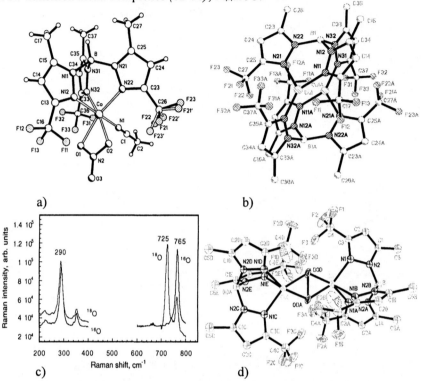

*Figure 3. a) [Co, Mn(Tp^{CF_3,CH_3})(NO$_3$)(MeCN)]   b) [(Cu(I)(Tp^{CF_3,CH_3})$_2$]
c) Resonance Raman and X-ray structure   d) [Cu(II)(Tp^{CF_3,CH_3})$_2$O$_2$].
Adapted from references 13-15.*

These structures reveal (i) the desired F-lined cavity (the CF$_3$ groups point toward the metals), and binding of small, neutral and anionic molecules, (acetonitrile and nitrate, respectively), inside them, Figure 3a; (ii) dinuclear Cu(I) complex (deoxy hemocyanine type), Figure 3b; (iii) a dinuclear μ-η^2:η^2 peroxide complex *(14-15)*, Figure 3d. The peroxo group is located inside a *dinuclear* fluorinated pocket (unlike that of Figure 3a, which is more open) and exhibits the expected 40 cm^{-1} ν$_{O-O}$ shift upon ^{18}O labeling, Figure 3c.

Fluoroalkylation has consequences for stability, O$_2$ binding, and catalysis. Dinuclear [(Cu(I)(Tp^{CF_3,CH_3}))$_2$], Figure 3b, which lacks "intra-dimer" open sites, is in equilibrium with mononuclear [Cu(I)(Tp^{CF_3,CH_3})(solvent)], and both types are stable under argon. The equilibrium is easily monitored by ^{19}F and ^{1}H NMR; no other dinuclear complex is observed. Similar complexes of non-fluorinated ligands are either less stable, or do not dissociate. Importantly, [Cu(I)(Tp^{CF_3,CH_3})(solvent)] binds and reduces O$_2$ in non-coordinating CH$_2$Cl$_2$, toluene etc., forming [Cu(II)(Tp^{CF_3,CH_3})$_2$O$_2$] at ambient conditions. This peroxo complex, stable in solid-state and solution, reversibly loses O$_2$ without ligand decomposition (despite the presence of CH$_3$ groups), being the most stable spectroscopic, structural and functional oxyhemocyanine model to date. A [Cu(II)(Tp^{CF_3,CH_3})(superoxo)] complex is a necessary intermediate. The formation and thermal stability of the peroxide complex, combined with the lack of decomposition of the ligand, proves several points: (i) despite fluorination, the metal center is competent to transfer electrons to O$_2$ forming superoxide and peroxide complexes; (ii) the ligand is stable, the fluorinated pocked sterically surrounds the metal-bonded oxygen species; (iii) the electron-withdrawing CF$_3$ groups, while not preventing Cu(I) oxidation to Cu(II), apparently prevent the oxidation of Cu(II) to Cu(III), as no Cu(III) is observed by X-ray absorption *(15)*. However, a bis μ-OH analogue of the peroxide complex co-crystallizes with it from toluene. From a catalytic perspective, whether the bis OH groups form by H• abstraction by the peroxo complex, remains to be determined.

Mononuclear [Co(Tp^{CF_3,CH_3})(NO$_3$)(MeCN)], Figure 3a, oxygenates cyclohexane to a 1:1 alcohol:ketone mixture more efficiently compared with the analogous complex of the non-fluorinated ligand *(13)*, and without ligand derivatization. [Cu(I)(Tp^{CF_3,CH_3})(solvent)] also exhibits reactivity, in addition to O$_2$ binding. In excess acetone, at 25° C, the solvent is oxygenated aerobically to lactate, Figures 4 and 5, in an unprecedented reaction, and without ligand decomposition (quantitative yield based upon Cu and ligand), *(16)*:

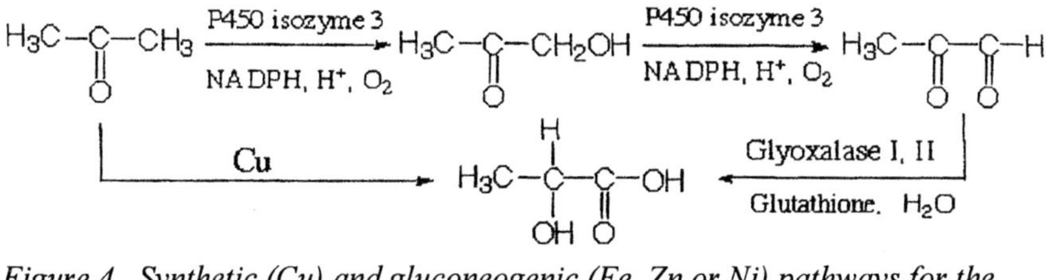

Figure 4. Synthetic (Cu) and gluconeogenic (Fe, Zn or Ni) pathways for the conversion of acetone to lactate. Adapted from reference 16.

In the copper case, acetol is not on the reaction pathway; in fact, acetol is an inhibitor. Methyl glyoxal (MG), on the other hand, is hydrolyzed to lactate, as shown by the incorporation of ^{18}O from H$_2$^{18}O. Thus, acetone is directly oxidized to MG, i.e. skipping the acetol, perhaps via a *cis*-[Tp^{CF_3,CF_3}(acetone)(superoxo)] complex, as mentioned above, intermediate in the formation of the dinuclear peroxo complex. The superoxo group could extract a H• atom from acetone to form the acetonyl radical, CH$_3$COCH$_2$•, consistent with the kinetic isotope effect (KIE) of ~5. O$_2$ capture by the acetonyl radical yields the acetonylperoxy radical, $CH_3COCH_2OO\bullet$. The reaction of a carbon-centered radical with dioxygen is a step in the classical autoxidation chemistry. MG forms by the decay of the acetonylperoxy radical, or by dehydration of the hydroperoxide, CH_3COCH_2OOH, produced by H• abstraction from another acetone molecule. H• abstraction by peroxy radical is another autoxidation step. Considering the lack of product scrambling when a 1:1 mixture of acetone:d_6-acetone is used (only non- and perdeuterated lactate forms), however, a classical solution free-radical chain pathway is probably not operational, again consistent with the inner-sphere mechanism of Figure 5b.

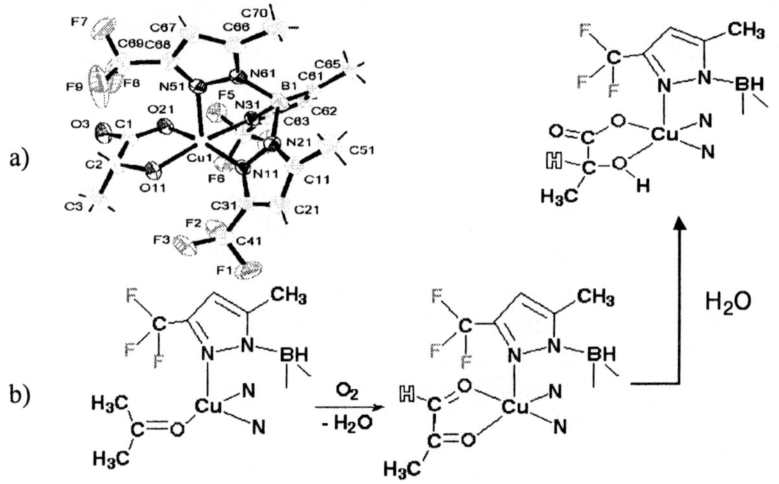

Figure 5. a) X-ray structure of $[Cu(II)(Tp^{CF_3,CH_3})(lactate)]$; the H atoms have been omitted. b) Proposed reaction pathway showing a coordinated methyl glyoxal intermediate and the 1,2 hydrogen (highlighted) determined via deuterium labeling. For $[Cu(I)(Tp^{CF_3,CH_3})(acetone)]$ only one boron-bonded pyrazole ring of is shown in full. Adapted in part from references 16-17.

The effect of increased Cu electron deficiency upon electron transfer and O_2 reactivity was explored briefly *(18)*. $[Cu(I)(Tp^{CF_3,CF_3})(acetone)]$, the first Cu(I) acetone complex, does not react with O_2 at 25° C, unlike $[Cu(I)(Tp^{CF_3,CH_3})(acetone)]$, which could not be isolated due to its reaction with O_2. Electron transfer to O_2 still occurs, but only at subzero temperatures, resulting in the formation of another, stable $\mu-\eta^2:\eta^2$ peroxo group, whose v_{O-O} is similar (5 cm^{-1} difference) with that of the partly fluorinated, $[Cu(II)(Tp^{CF_3,CH_3})_2O_2]$ *(18, 19)*, Figure 3d.

The acetone-to-lactate reaction, a rare example of a complex performing two different functions (oxidation and hydration), occurs without need or loss of reducing equivalents: the water produced in the oxidation step adds back in the second step. The overall functionality is that of a "pseudo-dioxygenase", defined as the formal incorporation of both oxygen atoms of O_2 at the same carbon atom.

The apparently unprecedented direct, oxidative formation of a hydroxy acid from a ketone is stoichiometric thus far, but raises the possibility of derivatizing other ketones, or molecules with metal-coordinating groups such as esters, amines, etc. It is important to recall, in this context, that the proposed acetone oxidation mechanism, while free radical in nature (it includes an autoxidation-type step), takes place at a coordinated substrate and thus, at least in principle, is subject to ligand-induced control. Such control, which can manifest itself in size, shape, enantiomeric or functional group selectivity remains to be explored. The lack of clean acetone oxygenation using non-fluorinated Cu TPBs (and related) complexes, combined with its classical, deep oxidation in solution and gas phase under various conditions, underscores the importance of F substitution

Importantly, *partial, or even complete fluorination does not preclude reduction and activation of O_2*, within the fluorinated pocket; the reducing B-H bond survives oxidation. The discovery of a new oxygenation reaction of acetone at ambient conditions, a solvent used in countless aerobic reactions with countless metal complexes suggests the still unpredictable effect of fluorinated pockets - electron deficient metal combinations.

2. Pefluoroalkyl perfluorinated phthalocyanines. Heme type models.

The isostructural [$MF_{64}Pc(-2)(acetone)_2$] complexes, Figure 2d, (M = Zn, Co) and their dinitrile precursor, **1**, are shown in Figure 6:

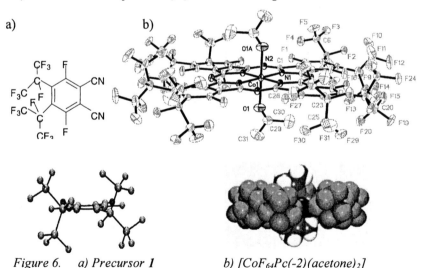

Figure 6. a) Precursor 1   b) [$CoF_{64}Pc(-2)(acetone)_2$]
*   top: scheme   top: X-ray structure*
*   bottom: X-ray structure bottom: space-filling representation*
*   Adapted from references 20, 23 and 24.*

The crystal structure of the perfluoro-(4,5-di-isopropyl) phthalonitrile precursor, **1**, Figure 6a, is viewed along the aromatic ring, with the CN groups pointing toward the back *(20)*. The CF_3 groups of the perfluoroisopropyl substituents are forced above and below the aromatic ring, consistent with the minimum energy conformation of the *ortho* substitution that allows only the tertiary fluorine atoms to point toward each other. This feature, predicted via molecular modeling to generate a cavity around Pc metal centers, is indeed observed in the isomorphous [Zn and $CoF_{64}Pc(-2)(acetone)_2$], Figure 6b, Fe complexes (see below), as well as the non-metallated $F_{64}PcH_2$ *(21)*.

An important potential advantage of the 3D bulkiness, predicted by molecular dynamics simulations *(22)*, Figure 7, is the avoidance of stacking interactions, typical of "normal", 2D (planar) Pcs. Furthermore, the role of the R_f groups in preventing the formation of singly-bridged (oxo) dinuclear complexes was also examined. The underlying assumptions are: (i) an approximate 2.0 Å metal-oxo distance, and (ii) metal location in the Pc ring plane.

The modeling study suggests that site-isolated mononuclear "oxo", and dinuclear, end-on μ-*peroxo* may form, but the thermodynamically stable (and inactive?) μ-*oxo* complexes, are disfavored.

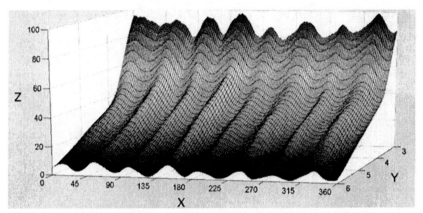

Figure 7. Energy diagram (normalized to 100, z axis) obtained by the 0 to 360 Deg rotation (x axis) of a [$CoF_{64}Pc(-2)$] complex (no axial ligands), cofacially stacked on top of an identical complex at distances between 3 and 6 Å (y axis).

The van der Waals contact boundary occurs ~ 4.1 Å. The expected 8-fold symmetry of repulsions, stronger at shorter distances, is reproduced. Stacking π-π interactions, effective below ~3.5 Å cofacial contact, appear unlikely.

The Zn and Co complexes, the first representatives of 3D perhalo Pcs *(23, 24)* form easily X-ray crystals, a somewhat surprising finding considering the potential disorder of the "greasy" C_xF_y groups and the lack of halogenated MPcs in the Cambridge Crystallographic Database. The perfluoro alkyl groups, while preventing stacking, appear sufficiently rigid for facile crystallizations. This steric feature, consistent with the molecular dynamics fulfills the requirements for a pocket with a well-defined, 3D geometry.

The electronic effects of perfluoroalkyl groups upon both the Pc ligand and MPc complexes are of interest. Let us consider first complexes for which the metal cannot change its oxidation state. Thus, the properties of the complexed R_fPc ligand will be revealed. The $ZnF_{64}Pc$ complex resists photochemical oxidation, but is reduced by hydrazine, or electrochemically in DMF using LiCl as supporting electrolyte to yield $[ZnF_{64}Pc(-2)(Cl)]^-$, and its ring anion radical species, $[ZnF_{64}Pc(-3)(Cl)]^{2-}$, detectable by electrospray MS *(25)*, Figure 8. CBr_4, a sacrificial photooxidant, oxidizes $[ZnF_{64}Pc(-3)(Cl)]^{2-}$ back to $[ZnF_{64}Pc(-2)(Cl)]^-$

A combination of UV–visible and magnetic circular dichroism spectroscopy, mass spectrometry, cyclic and differential pulse voltammetry, and INDO/S and DFT theoretical calculations indicate that perfluoroalkyl substitution red–shifts the Q and $\pi^*-\pi^*$ transitions and narrows the HOMO–LUMO gap, while simultaneously stabilizing the Pc radical anions *(23, 25)*. The reason is that the fluoro-aliphatic substituents, as we have shown previously *(23)*, cannot participate in π backbonding, unlike the aromatic fluoro groups, and thus they are not conjugated with the ring π system. As a result, for photochemically excited $[ZnF_{64}Pc]$, an increase in excited triplet state lifetime, fluorescence quantum yield and $^1O_2$ yield is noticed *(26)* relative to $[ZnF_{16}Pc]$: 131 vs. <<1, 0.39 vs. 0.04, and 0.21±0.03 vs. 0.13±0.01, respectively.

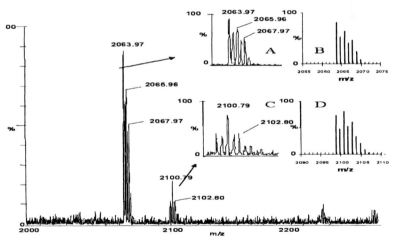

Figure 8. MALDI–MS data for a crystalline sample of $[ZnF_{64}Pc(-2)]$. (A) Observed isotopic pattern for $[ZnF_{64}Pc(-2)]$ at 2101 m/z and (B) predicted isotopic pattern. (C) Observed isotopic pattern for $[(Cl-)ZnF_{64}Pc(-2)]$ at 2101 m/z and (D) predicted isotopic pattern. Adapted from reference 25.

Further details regarding the transient grating methodology used to measure the photophysical properties of fluorinated phthalocyanines are found in Reference 26. Due, at least in part, to enhanced triplet state lifetimes and quantum efficiencies for the formation of $^1O_2$, the [ZnF$_{64}$Pc] complex is more photodynamically active against tumors (23) in comparison with both the parent protio and perfluoro Zn phthalocyanines.

The resistance of the ligand to attack by singlet oxygen is consistent with the observation of lack of bleaching noticed (23) during the photochemical production of tryptophane hydroperoxide. Taken together, this data suggest the stability and suitability of R$_f$Pc ligands for oxidation chemistry.

The [CoF$_{64}$Pc(-2)] complex, in contrast to the Zn complex, exhibits both ligand and metal redox active centers. Since the two complexes are isostructural, the *electronic* R$_f$ influence upon the redox metal may be revealed. Interestingly, the Zn and Co complexes exhibits similar values of their first two reduction potentials (Co in parentheses): -0.47 (-0.40) and -0.96 (-0.94) V vs. SCE, respectively. This data suggest that the LUMO is *ligand* based in both cases. Furthermore, their Q bands are also similar, consistent with the above notion. The one-electron reduced Co species exhibits an ^{19}F NMR spectrum similar with that of the [ZnF$_{64}$Pc(-2)] complex. On the other hand, the 450 nm region is typical of a radical anion only in the Zn complex, while the Co spectrum resembles that of low-spin d^7 Fe(I) phthalocyanine (27), with a similar d$_z^2$ ground state. Thus, it is not clear if the electronic structure of the reduced Co complex is best described as [Co(II)F$_{64}$Pc(-3)]$^-$, or [Co(I)F$_{64}$Pc(-2)]$^-$; perhaps a resonance form would be a more appropriate description. It should be noted that, in contrast, the reduction of Co(II) Pcs yields Co(I), even in the case of perfluorinated Co phthalocyanine, [Co(II)F$_{16}$Pc(-2)] (28), Figure 2c.

[CoF$_{64}$Pc] also exhibits new oxidation reactivity. [CoF$_{64}$Pc(acetone)$_2$] performs unprecedented, environmentally "green" chemistry. Its reaction with both alkyl and aryl phosphines forms ylides (24), equation 1, Figure 9:

$$R'_3P + RCH_3 + 1/2O_2 \rightarrow R'_3P=CHR + H_2O; \quad R' = \text{alkyl, aryl}; R = \text{acetyl} \quad (1)$$

These ylides are currently obtained via Eqs. 2-4 (B = base, M = metal):

$$RCH_3 + X_2 = RCH_2X + HX \text{ (or the equivalent)}, X = Cl, Br \quad (2)$$

$$RCH_2X + R'_3P \rightarrow R'_3P^+\text{---}CH_2RX^- \quad (3)$$

$$R'_3P^+\text{---}CH_2RX^- + B^-M^+ \rightarrow R'_3P=CHR + BH + MX \quad (4)$$

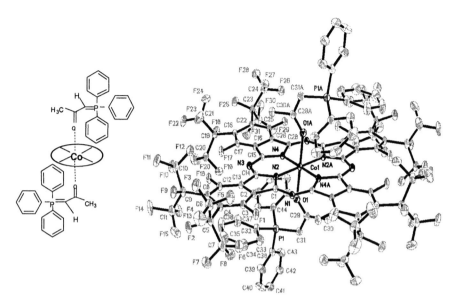

Figure 9. Structural formula and the X-ray structure of [CoF₆₄Pc(Ph₃P=CH-CO-CH₃)₂]. Hydrogen atoms have been omitted. Adapted from reference 23.

In reaction 1, O_2 eliminates two H atoms from the same acetone carbon, to form the C=P double bond. Classically, two halogen atoms perform the same task.

Importantly, reaction 1 is catalytic; the Co ylide complex liberates its ylides in excess acetone to regenerate the starting bis acetone complex, Figure 6b.

Preliminary mechanistic studies revealed that no ylides are produced from acetone and $Ph_3P=O$, in the absence of Ph_3P. Thus, C=P bonds do not form by H_2O elimination from the oxygen of $Ph_3P=O$ and two H atoms of a CH_3 group. The presence of fluorinated alkyl groups is necessary: the 2D perfluoro phthalocyanine cobalt complex, $CoF_{16}Pc$, of similar electronic properties (or not so similar, if the first reduction of $CoF_{64}Pc$ is ligand-based), but with no R_f pocket, is unreactive. Furthermore, the isostructural $[ZnF_{64}Pc(-2)(acetone)_2]$, which exhibits a "pure" ligand-based reduction, is also unreactive. Thus, both redox chemistry and a fluorinated 3D pocket are critical for catalysis. The pocket also imparts some selectivity. Thus, from an electronic point of view, the coupling of the second CH_3 group of acetone with a phosphine is not precluded. On the other hand, as shown in Figure 6b and 9, one of the two CH_3 groups of acetone is half buried inside the fluorinated pocket, thus limiting the coupling to the other, exposed CH_3 group. This result is consistent with the notion that the keto substrate remains rigidly coordinated during the coupling step, as well as with the lack of coupling in solution in the absence of the complex. Moreover, since the role of the pocket is critical, its re-design by using other R_f groups while maintaining the same metal center is likely to modulate the reaction selectivity. This aspect, in addition to the "green" and catalytic features mentioned above, is a significant advantage considering that limited or no selectivity could be expected via the classical procedure, reactions 2-4.

A key question that emerges at this point is whether fluoroalkyl substituted complexes can bind and reduce O_2. For Tp complexes, as shown above, this is the case. In the Pcs case, only the [CoF$_{64}$Pc] complex has been studied. The aerobic oxidation of Ph$_3$P in non-coordinating solvents with [CoF$_{64}$Pc] yields a large amount of Ph$_3$P=O: over 1000 turnovers in 24 hrs. No catalyst decomposition is noticed, and no Ph$_3$P=O forms in the absence of the catalyst. This reaction, considered indicative of oxygen transfer, has been proposed *(29)* in the Co case to occur via a dinuclear peroxo species, but the validity of this mechanism in the case of [CoF$_{64}$Pc] needs to be verified.

The above C-H *oxidation* chemistry is complemented by *oxygenation* chemistry using biologically relevant Fe.

The [FeF$_{64}$Pc(H$_2$O)$_2$]·4acetone complex, Figure 10, was obtained recently *(30)*. Unlike the similar [MF$_{64}$Pc(-2)(acetone)$_2$] (M = Zn, Co) complexes, however, the acetone molecules are not metal-bonded; yet they are still located in the R$_f$ cavity, H-bonded to the coordinated H$_2$O molecules.

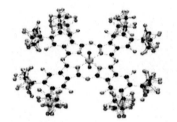

Figure 10. [FeF$_{64}$Pc(H$_2$O)$_2$]·4acetone. X-ray structure and VDW depiction of the non-peripheral F and peripheral i-C$_3$F$_7$ substituents, Fe, H$_2$O and one of the four acetone molecules. The Fe atom lies on the 4/m position; only the axial H$_2$O oxygen atoms are shown. The i-C$_3$F$_7$ groups are disordered across the Pc ring. Each acetone is two-fold disordered, around the pivot C=O carbon. The four acetone molecules are also H-bonded to the 4-fold disordered axial water.

The Fe complexes of related porphyrins (Por) catalyze autoxidations *(31-35)*. However, even fluorinated Por are susceptible to nucleophilic attack and ring distortions *(36, 37)*. Halophthalocyanines have been less studied since they are poorly soluble *(38, 39)*, or unstable *(40)* in the case of partial halogenation. Fluorination of PorFe *(41)*, for example, limits its stability to ~80° C under moderate O_2 pressure. The introduction of *linear* fluoroalkyl and aromatic fluorine substituents via free radical chemistry *(42)* yields ill-defined mixtures of isomers of unknown structures and composition, while not improving their oxidative instability. The formation of inert dinuclear complexes *(43)* represents an additional problem. More complex systems, for example a cyclodextrin "pocket" grafted on a fluorinated porphyrin lengthens the oxidation activity up to 200 turnovers *(44)*. Systems that combine halogenation with encapsulation *(43)*, viz. halogenated Por and Pc in zeolites (Z), [(Por, Pc)Fe@Z], still cannot prevent catalyst deactivation. The most sophisticated design to date *(45)*, a zeolite encapsulated Fe phthalocyanine, imbedded into a membrane (Mem), [PcFe@Z@Mem], oxidizes cyclohexane with organic hydroperoxides, TBHP, (P450 biomimetic model), but not with air.

For Fe complexes, the Haber-Weiss scheme verifies that the decomposition of intermediate ROOH, varies congruently with their Fe(III)/Fe(II) potential *(33, 46, 47)*, consistent with the notion that ROOH reduces Fe(III) slower than it oxidizes Fe(II). Halogenation induces destabilizing ring-distortions, but accelerates the catalyzed autoxidation rate-determining step. Interestingly free *(48)* and encapsulated FePc *(45)*, form catalytically active PcFe(IV)=O. Electronic deficiency does not necessarily preclude the formation of Fe=O: non-peripherally substituted Fe(CF$_3$)$_8$Pc forms with TBHP a ferryl group that oxygenates via oxo group transfer phosphines *(49)*, but not alkanes.

The catalytic effects of R$_f$ pockets/electron poor Fe centers of FeF$_{64}$Pc upon model cyclohexane oxygenation have been examined using both TBHP and air. While steric hindrance disfavors μ-oxo complexes, as suggested by modeling studies, Figure 7, and the location of the H$_2$O oxygen slightly below the VDW edge of the R$_f$ pocket, Figure 10. TBHP coordination and inner-sphere electron transfer, however, are not sterically precluded. The iron complex decomposes TBHP *(30)* with an initial rate of 0.22×10^{-3} M/min, or ~3.9 TBHP/catalyst•sec at 25° C, a turnover rate which is one order of magnitude smaller than the halogenated FePor rate *(46)*. TBHP and FeF$_{64}$Pc oxygenate cyclohexane to a mixture of ~1.0:0.9 cyclohexanone (C$_6$-one):cyclohexanol (C$_6$-ol), with a ~3.0/min initial turnover frequency and a 2.1 kinetic isotope effect. The productive use of TBHP, measured as (C$_6$-one + C$_6$-ol)/TBHP (in moles) reaches 91% after 5 hrs. Addition of fresh TBHP restarts the reaction, which proceeds again with the above parameters. The above data is consistent with a catalyzed autoxidation mechanism.

Unlike [FePc@Z@Mem] *(45)*, [FeF$_{64}$Pc] can oxygenate cyclohexane aerobically, a still challenging reaction" *(50)*, to produces a ~1:2 C$_6$-one:C$_6$-ol mixture, with a turnover frequency (TOF) of 25/hr at 75° C. The catalyst appears to be indefinitely stable below ~100° C, as judged by ^{19}F NMR, with a maximum TOF of 85/hr. While the equatorial R$_f$ substituents render the Fe very electron poor, a *trans* axial donor might induce the oxo transfer reactivity of electron richer, but less stable, Fe Pcs (work in progress).

Conclusions and Outlook

From a catalyst design perspective, Figure 11, the peripheral iso-R_f groups, of which iso-C_3F_7 is the first representative, offer steric protection to residual aromatic F, while making a metal and a pocket available for substrate binding.

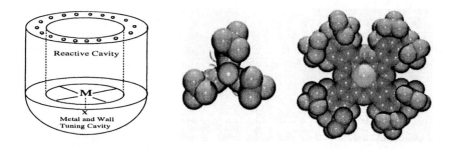

Figure 11. Molecular reactors: 3-fold (TPB) and 4-fold (Pc) reactive cavities.

Reduction and activation of O_2 in the Tp case is not precluded by fluorination; for CoPc, preliminary evidence suggests oxo transfer to R_3P. C-H bond oxidations occur selectively in novel, "green" C-P coupling reactions, while oxygenations proceed efficiently, again without catalyst decomposition. The latter feature is advantageous for drugs that produce $^1O_2$ but resist its action. The *organic* nature of R_fPc ligands offers the opportunity for tuning the size, shape, and chirality of catalytic environments. The R_f groups, "pieces" of super hydrophobic fluorinated solvents, may induce favorable substrates / unfavorable oxygenated products partitions, while favoring solubility in C-H bonds free CO_2. These features, combined with active-site steric restrictions, might, perhaps, help prevent product overoxidation. Additionally, the Pcs could be both tuned and heterogenized via axial anchoring. R_fPc complexes of metals likely to form M=O bonds (Mn, Ru, Re, etc) are currently being explored as well. Last, but not least, a R_f stabilized metallated Pc (radical) anion might activate O_2 for insertion into C-H bonds in a manner complementary to porphyrins and their radical cations.

Acknowledgments

The creative contributions of coworkers and collaborator coauthors, is gratefully acknowledged. Royce and ACS Moissan fellowships, Salomon Foundation, 3M, and Air Products and Chemicals, Inc. are thanked for partial support.

References

1. *Bioinorganic Catalysis;* 2nd edition, Reedijk, J.; Ed., M. Dekker: N. Y., 1999.
2. *Biomimetic Oxidations;* Meunier, B.; Ed., World Scientific Publishing, London, 2000.
3. Arakawa, H.; Aresta, M.; Armor, J. N.; Mark, A.; Beckman, E. J.; Bell, A. T.; Bercaw, J. E.; Creutz, C.; Dinjus, E.; Dixon, D. A.; Domen, K.; DuBois, D. L.; Eckert, J.; Fujita, E.; Gibson, D. H.; Goddard, W. A.; Goodman, D. W.; Kelle, J.; Kubas, G. J.; Kung, H. H.; Lyons, J. E.; Manzer, L. E.; Marks, T. J.; Morokuma, K.; Nicholas, K. M.; Periana, R.; Que, Jr., L.; Rostrup-Nielson, J.; Sachtler, W. M. H.; Schmidt, L. D.; Sen, A.; Somorjai, G. A.; Stair, P. C.; Stults, B. R.; Tumas, W. *Chem. Rev.* **2001**, *101*, 953.
4. Lippard, S. J. *Nature* **2002**, *416*, 587.
5. Breslow, R. *Acc. Chem. Res.* **1997**, *30*, 251.
6. Heron, N. *J. Coord. Chem.* **1988**, *19*, 25.
7. Wolf, D. *Angew. Chem., Int. Ed. Engl.* **1998**, *37*, 3351.
8. Sono, M.; Roach, M. P.; Coulter, E. D.; Dawson, J. H. *Chem. Rev.* **1996**, *96*, 2841.
9. Solomon, E. I.; Sundaram, U. M.; Machonkin, T. E. *Chem. Rev.* **1996**, *96*, 2563.
10. Que, Jr., L.; Ho, R. Y. N. *Chem. Rev.* **1996**, *96*, 2607.
11. Wallar, B. J.; Lipscomb, J. D. *Chem. Rev.* **1996**, *96*, 2625.
12. Gorun, S. M.; Stibrany, R. T. US Patent 5, 627, 164, 1996.
13. Gorun, S. M.; Hu, Z.; Stibrany, R.; Carpenter, G. *Inorg. Chim. Acta* **1999**, *297*, 383.
14. Hu, Z.; Williams, R. D.; Tran, D.; Spiro, T. G.; Gorun, S. M. *J. Am. Chem. Soc.* **2000**, *122*, 3556.
15. Hu, Z.; George, G. N.; Gorun, S. M. *Inorg. Chem.* **2001**, *40*, 4812.
16. Diaconu, D.; Hu, Z.; Gorun, S. M. *J. Am. Chem. Soc.* **2002**, *124*, 1564.
17. Diaconu, D. M.S. Thesis, Brown University, Providence, RI, 2002.
18. Hu, Z. Ph.D. Thesis, Brown University, Providence, RI, 2001.
19. Hu, Z.; Williams R. D.; Tran, D.; Spiro, T. G.; Gorun, S. M. unpublished results.
20. Gorun, S. M.; Bench, B. A.; Carpenter, G.; Beggs, M. W.; Mague, J. T.; Ensley; H. E. *J. Fluor. Chem.* **1998**, *91*, 37.
21. Lee, H.-J.; Brennessel, W. W.; Brucker, W. J.; Lessing, J. A.; Young, Jr., V.; Gorun, S. M. *Chem. Commun.* **2003**, 1576.
22. Sheffler, W.; Gorun, S. M. unpublished results obtained with the program NOMAD: Kalé, L.; Skeel, R.; Bhandarkar, M.; Brunner, R.; Gursoy, A.; Krawetz, N.; Phillips, J.; Shinozaki, A.; Varadarajan, K.; Schulten, K. *J. Comp. Phys.* **1999**, *151*, 283.
23. Bench, B. A.; Beveridge, A.; Sharman, W. M.; Diebold, G. J.; van Lier, J. E.; Gorun, S. M. *Angew. Chem. Int. Ed. Eng.* **2002**, *41*, 748.

24. Bench, B. A.; Brennessel, W. W.; Lee, H.-J.; Gorun, S. M. *Angew. Chem. Int. Ed. Eng.* **2002**, *41*, 751.
25. Keizer, S. P.; Han, W.; Bench, B. A.; Gorun, S. M.; Stillman, M. J. *J. Am. Chem. Soc.* **2003**, *125*, 7067.
26. Beveridge, A.; Bench, B. A.; Gorun, S. M.; Diebold, G. J. *J. Phys. Chem. A* **2003**, *107*, 5138.
27. Lever, A. B. P.; Wilshire, J. P. *Inorg. Chem.* **1978**, *17*, 1145.
28. Cardenas-Jiron, G. I. *J. Phys. Chem. A* **2002**, *106*, 3202.
29. Heinze, K.; Huttner, G.; Zsolnai, L. *Chem. Ber.* **1997**, *130*, 1393.
30. Lee, H.-J.; Brennessel, W. W.; Brucker, W. J.; Lessing, J. A.; Young, Jr., V.; Gorun, S. M., *J. Am. Chem. Soc.*, in revision.
31. Ellis, Jr., P. E.; Lyons, J. E. *Coord. Chem. Rev.* **1990**, *105*, 181.
32. Bartoli, J. F.; Brigoud, P.; Battioni, P.; Mansuy, D. *J. Chem. Soc. Chem. Commun.* **1991**, 440.
33. Grinstaff, M. W.; Hill, M. G.; Labinger, J.; Gray, H. B. *Science* **1994**, *264*, 1311.
34. *Metalloporphyrins in Catalytic Oxidations;* Sheldon, R. A.; Ed., M. Dekker: N. Y., 1994.
35. Dolphin, D.; Traylor, T. G.; Xie, L. Y. *Acc. Chem. Res.* **1997**, *30*, 251.
36. Porhiel, E.; Bondon, A.; Leroy, J. *Eur. J. Inorg. Chem.* **2000**, 1097.
37. Smirnov, V.; Woller, E.; Tatman, D.; DiMagno, S. G. *Inorg. Chem.* **2001**, *40*, 2614.
38. *Phthalocyanines: Properties and Applications;* Leznoff, C. C.; Lever, A. B. P., Eds. VCH: N. Y., 1996.
39. Chen, M. J.; Rathke, J. W. Ref. 38, p. 183.
40. Colonna, S.; Gaggero, N.; Montanari,F.; Pozzi, G.; Quici, S. *Eur. J. Org. Chem.* **2001**, *1*, 181.
41. Moore, K. T.; Horvath, I. T.; Therien, M. J. *Inorg. Chem.* **2000**, *39*, 3125.
42. Horvath, I. T.; Rabai, J. *Science* **1994**, *266*, 72.
43. Balkus, Jr., K. J. Ref. 38, p. 285.
44. Breslow, R. *The Chem. Record* **2000**, *1*, 3, and references therein.
45. Parton, R. F.; Vankelecom, I. F. J.; Casselman, M. J. A.; Bezoukhanova, C. P.; Uytterhoeven, J. B.; Jacobs, P. A. *Nature* **1994**, *370*, 541.
46. Grinstaff, M. W.; Hill, M. G.; Birnbaum E. R.; Schaefer, W. P.; Labinger, J. A.; Gray, H. B. *Inorg. Chem.* **1995**, *34*, 4896.
47. Labinger, J. A. *Catal. Lett.* **1994**, *26*, 95.
48. Monacelli, F. *Inorg. Chim. Acta* **1997**, *254*, 285.
49. Chen, M. J.; Fremgen, D. E.; Rathke, J. W. *J. Porph. Phthal.* **1998**, *2*, 473.
50. Schuchardt, U.; Cardoso, D.; Sercheli, R.; Pereira, R.; da Cruz, R. S.; Guerreiro, M. C.; Mandelli, D.; Spinace, E. V.; Pires, E. L. *Appl. Catal. A-Gen.* **2001**, *211*, 1.

Indexes

Author Index

Ahrendt, Kateri A., 46
Baker, R. Tom, 319
Bercaw, John E., 250, 319
Bergman, Robert G., 46
Betley, Theodore A., 334
Bhalla, Gaurav, 105
Bryant, Jasmine R., 356
Carter, Charles A. G., 319
Chianese, Anthony R., 169
Churchill, David G., 86
Crabtree, Robert H., 169
Czerw, Margaret, 198, 216
Eichele, Klaus, 234
Eisenstein, Odile, 116
Ellman, Jonathan A., 46
Fekl, Ulrich, 283
Goldberg, Karen I., 1, 283
Golden, Jeffrey T., 46
Goldman, Alan S., 1, 198, 216
Gorun, Sergiu M., 407
Grimm, Jost C., 234
Grushin, Vladimir V., 393
Hartwig, John F., 136
Heyduk, Alan F., 250
Iverson, Carl N., 319
Janak, Kevin E., 86
Johansson, Lars, 264
John, Kevin D., 319
Jones, C. J., 105
Jones, William D., 56
Kaska, William C., 234
Klei, Steven R., 46
Krogh-Jespersen, Karsten, 198, 216
Labinger, Jay A., 250, 319
Larsen, Anna S., 356
Legzdins, Peter, 184

Lersch, Martin, 264
Liu, Xiang Yang, 105
Lockwood, Mark, 356
Look, Jennifer L., 283
Mack, Hans-Georg, 234
Maron, Laurent, 116
Marshall, William J., 393
Mayer, Hermann A., 234
Mayer, James M., 356
Milstein, David, 70
Mohammad, Hani A. Y., 234
Norris, Cynthia M., 303
Northcutt, Todd O., 56
Novak, Filip, 234
Pamplin, Craig B., 184
Parkin, Gerard, 86
Periana, Roy A., 105
Perrin, Lionel, 116
Peters, Jonas C., 334
Petrov, Viacheslav A., 393
Pribisko, Melanie A., 319
Remias, Joseph E., 379
Renkema, Kenton B., 198
Rice, Gordon, 356
Rybtchinski, Boris, 70
Sames, Dalibor, 155
Scollard, John D., 319
Scott, Brian L., 319
Sen, Ayusman, 379
Shapley, Patricia A., 370
Speiser, Bernd, 234
Tan, Kian L., 46
Templeton, Joseph L., 303
Thalji, Reema K., 46
Thomas, Christine M., 334
Thomas, J. Christopher, 334

Thorn, David L., 393
Tilley, T. Don, 46
Tilset, Mats, 264
Vetter, Andrew J., 56
Wang, Kun, 356
Wick, Douglas D., 56

Wik, Bror J., 264
Won, Tae-Jin, 356
Wong-Foy, Antek, 105
Yung, Cathleen M., 46
Zhong, H. Annita, 250

Subject Index

A

Abnormal carbene formation, steric crowding at metal center, 171–173
Acceptorless alkane dehydrogenation
  by tri-coordinate "pincer" ligand, mechanism, 218–220
  DFT calculations, mechanism, 216–233
  See also Alkanes
Acetone to lactate conversion, synthetic and gluconeogenic pathways, 411–413
Acetonitrile, associative displacement, methane at platinum, 270–273
Acetylacetonate-iridium(III) complexes, alkane C-H bond activation, 107
η^2-Acetylene intermediates, cyclopentadienyl-M-(NO)-, 192–194
Activation, dihydrogen, methane, and silane by cyclopentadienyl lanthanide complexes, comparison, 129
1,2-Addition, organometallic C-H activation mode, 29–30
Agostic interaction, lanthanide-X bonds, 120–122
Alcohol oxidation by molecular oxygen with heterobimetallic complexes, 372–373, 376
n-Alkane/α-olefin transfer dehydrogenation, free-energy profile, DFT calculations, 210–213
Alkane borylation, mechanisms, 144, 146–149, 151
Alkane C-H bonds
  activation by oxygen-donor iridium complexes, 105–115
  activation reactions, alkane complexes, intermediates, 56–69
  coordination-directed alkyl group activation, 156–166
  giving stable iridium hydrides, intermolecular addition, 11
Alkane carboxylation with carbon monoxide, palladium and/or copper acetates, 8–9
Alkane dehydrogenation, early system, 24–27
  See also Alkanes
Alkane functionalization, Shilov system, 5–7, 252–255
  See also Alkanes
Alkane oxidation, selective, platinum complexes, reaction mechanisms, 283–302
Alkane reaction at terminal position, rhodium complexes, catalyst, 138
Alkane reductive elimination in trispyrazolylborate-rhodium-neopentyl-isocyanide-alkyl complexes, 61–68
Alkane transfer-dehydrogenation, pincer-ligated iridium complex catalyst, 198–215
n-Alkane *versus* cyclooctane, competition experiment, 207
Alkanes
  catalytic functionalization, 139–146
  linear and cyclic, acceptorless dehydrogenation, dissociative pathway, 228–231
  linear, selective activation, terminal C-H bonds, 61–67
  regioselective catalytic borylation at terminal position, 136–154
  See also Acceptorless alkane dehydrogenation; Alkane

dehydrogenation; Alkane functionalization; Shilov system
Alkene oxidation by molecular oxygen with heterobimetallic complexes, 376
Alkenyl-substituted aromatic imines, cyclization, catalyst development, 53
Alkenylated benzimidazoles, cyclization, catalyst development, 53–54
Alkenylation, styrenyl group transfer, 164–165t
1-Alkyl-boronate ester, synthetic yield, 142–143
Alkyl deuteride complexes, rearrangements, 59–61
Alkyl hydride and deuteride reductive elimination studies, 60–68
Alkyl hydride-platinum(IV) complexes, 304–305
Alkylaromatic compounds
  oxidations by manganese-oxo-phenanthroline dimers, 361–363, 367
  oxidations by ruthenium-oxo-polypyridyl complex, 364–367
Alkylidene intermediates, cyclopentadienyl-M-(NO)-, 185–190
Alkylphosphino complexes, group 8 metals, C-H addition, 17
Alkyne insertion into ortho C-H bonds, aromatic ketones, 22
α-Amino acids, regioselectivity trends, 158–159
Arene C-H activation, iridium-ethyl complex, ^{13}C NMR studies, 110–114
Arene C-H bonds, oxidative addition barrier, 312–313
η^2-Arene-platinum(II) intermediates, 311–312
Arenes
  addition, rate-determining step, 13
  borylation, 144–146, 150
  electrophilic stannylation, 400–403
  rhodium-catalyzed oxidation carbonylation, 395–400
Aryl hydride-platinum(IV) complexes, stable, 310
Arylation, t-butyl group, catalytic cycle, 162–163f
Associative coordination, methane at platinum, 270–273
Associative decoordination, toluene from platinum, 273–276

B

Bent-back, square planar, cyclopentadienyl complexes, molecular orbitals, 15
Benzene and tetrahydrofuran ligand self-exchange, 343–344
Benzene at platinum associative coordination, 265, 266, 268
Benzene C-H activation, mechanisms, 346–349
Benzene oxidation, efficiency in use, generated hydrogen peroxide, 383–384
η^2-Benzyne intermediates, cyclopentadienyl-M-(NO)-, 190–192
Bidentate anionic N,N-and N,C-donor ligands, neutral platinum methyl complexes, 321–330
Bis(amino)borate ligand system, preparation, 336–338
Bis(diphenyl-phosphino)ethane complexes. *See* Platinum(IV)-dppe complexes
Bis(o-diisopropylaryl)β-diketiminate trapping ligand, C-H bond activation, 292–294
Bis(phosphine) cyclometalation, metal halide salts, 10–11

Bis(phosphino)borate ligand system, preparation, 336–338

Bis(phosphino)borate platinum(II) system *versus* bis(pyrazolyl) borate platinum(II) system, 349–351, 352

Bis(pyrazolyl)borate ligands, 336

Bis(pyrazolyl)borate platinum(II) system. *See* Bis(phosphino)borate platinum(II) system

Borane reagents, alkane and arene regioselective functionalization, 136–154

Borylation, thermal, catalytic, 139–151

Bridge-assisted reactions, 374–375

Brønsted relation for proton transfer reactions, 359

1,4-Butanediol, production, 250–251

t-Butyl group, catalytic cycle, arylation, 162–163*f*

t-Butyl phosphorus-carbon-phosphorus ligands, reactions
 iridium olefin dimers, 73–74
 rhodium olefin dimers, 72–74

t-Butylaniline substrate, alkenylation, 160

t-Butylethylene, alkane dehydrogenation, 24

t-Butylethylene, cyclooctane
 catalytic reactions *in situ*, 202–204
 pincer-ligated iridium complex, transfer-dehydrogenation, 199
 stoichiometric reactions, 199–202
 transfer dehydrogenation, 204–206

C

C-C and C-H bond activation, steric requirements, 79–82

C-C bond activation, rhodium insertion, 72–74

C-C bond-forming reactions, C-H bond activation use, 52–54

C-C bond *versus* C-H bond oxidation addition in P-C-X ligand systems, 70–85

C-C bonds, direct formation from C-H bonds, 160–162

C-C coupling reactions between aromatic compounds, CO and alkenes, ruthenium catalyst, 53

C-H arene bonds, oxidative addition barrier, 312–313

C-H bond activation
 at platinum(II), 250–263, 319–354
 by methyl-iridium(III)-acetylacetonate-pyridine complex, proposed mechanism, 108*f*
 by platinum(II) species, protonation, Pt(II) alkyls, 19
 cationic platinum(II) diimine complexes, 266–268
 dioxygen oxidation, platinum alkyl complex, 295–298
 electrophilic aromatic, metal trifluoroacetates, 393–406
 hydridotris(3,5-dimethylpyrazolyl)borate platinum(IV) complexes, 303–318
 imidazolium, 171–176
 non-organometallic, 32–34
 overview, 1–43
 oxygen-donor iridium complexes, 105–115
 platinum-catalyzed, trapping ligand use, 288–295
 reactions, alkane complexes, intermediates, 56–69
 reactions, mechanistic investigation, 264–282
 sequential, molybdenum and tungsten complexes, 184–197
 steric requirements, C-C and C-H bonds, 79–82
 synthesis, palladium N-heterocyclic carbenes, 174–176

trigger, boryl ligand, 139–140
versus C-C, cationic phosphorus-carbon-oxygen-rhodium system, 78–82
C-H bond addition, kinetics and thermodynamics, 11
C-H bond oxidation
  addition, double cyclometalation, 234–247
  addition, early examples, 9–11
  addition, silyl groups, 16
  non-organometallic mechanisms, 356–369
  organic-based metal active sites, 410–418
  ruthenium-oxo-polypyridyl complex, 364–368
  See also Oxidative addition
C-H bond reduction, methane elimination, 87–92
C-H bonds
  agostic interaction, 120–122
  functionalization in complex organic synthesis, 155–168
  versus C-C bonds, oxidative addition in P-C-X ligand systems, 70–85
C-O reductive elimination, 285–288
Carbene complexes (N-heterocyclic), rhodium, iridium, palladium, 169–183
Carbon monoxide insertion into C-H bonds, 8–9, 19–21
Carbonyl model complexes, rhodium and platinum, relative CO stretching frequencies, 338–340
Catalysis, fluorinated 3D pocket, 417
Catalysis inhibition, α-olefin binding, 207–209
Catalyst deactivation in alkane dehydrogenation systems, 25
Catalytic cycle, arylation, *t*-butyl group, 162–163*f*
Catalytic functionalization, oxidative addition, C-H bonds, 19–27

Catalytic reactions *in situ*, *t*-butylethylene/cyclooctane, 202–204
Catalytica process, methane oxidation using platinum bipyrimidine catalyst, 285
Cationic phosphorus-carbon-oxygen-rhodium system, C-H versus C-C activation, 78–82
Cationic platinum(II) diimine complexes, C-H activation, 266–268
Chatt, C-H activation by transition metal complex, first example, 3, 9
Chelation effect in C-C bond activation, 78–79
Co-enzyme B_{12}. *See* Vitamin B_{12}
Competition experiment, cyclooctane *versus n*-alkane, 207
Competitive trapping, equilibrating intermediates, 269–270
Competitive trapping experiments, protonation kinetic site, 277–279
Complex organic synthesis, C-H bond functionalization, 155–168
Computational determination, kinetic and equilibrium isotope effects, *ansa*-tungstenocene methyl-hydride complex reactions, 93–100
Coordination-directed alkyl group activation, alkane C-H bonds, 156–166
Copper-catalyzed reduction, metals by dihydrogen, 2
Copper(II) acetate, oxidant in direct arylation, 162–164
Cumene, oxidation by ruthenium-oxo-polypyridyl complex, 364–367
Curtin-Hammet principle, 270, 271, 275, 277
Cyclometalated organometallic complexes iridium-fluorinated 2-arylpyridines, 403–405
Cyclometalation, ligand aryl groups, 10

Cyclooctane/*t*-butylethylene substrate/acceptor couple
  catalytic reactions *in situ*, 202–204
  competition experiment, 207
  pincer-ligated iridium complex, transfer-dehydrogenation, 199
  stoichiometric reactions, 199–202
  transfer dehydrogenation, 204–206
Cyclopentadienyl lanthanide hydride complex
  reaction with dihydrogen, 123–125
  reaction with silane, 127–128
Cyclopentadienyl lanthanide methyl complex
  reaction with dihydrogen, 125–126
  reaction with methane, 127
Cyclopentadienyl lutetium hydride complex, methane activation, 122
Cyclopentadienyl-M-(NO)-η^2-acetylene intermediates, 192–194
Cyclopentadienyl-M-(NO)-alkylidene intermediates, 185–190
Cyclopentadienyl-M-(NO)-η^2-benzyne intermediates, 190–192
Cyclopentadienyl metal complexes, overview, 12–16
Cytochrome P-450, monooxygenase, 32–33

D

Decamethylferrocene, fast-reacting substrate, H/D exchange, 52
Density field theory. See DFT calculations
Deuterated trispyrazolylborate-rhodium-neopentyl-isocyanide-alkyl complexes, kinetics, 62–68
Deuterium. *See* H/D exchange
DFT calculations
  acceptorless alkane dehydrogenation, 216–233
  free-energy profile, *n*-alkane/α-olefin transfer dehydrogenation, 210–213
  thermodynamic values, isotope effects, 94–95
3,5-Di-*t*-butyl catecholboryl substitution, block aromatic C-H bonds, 140
Diborane compounds, photochemical catalytic functionalization, alkanes, 141–142
DigiSim simulation program, 240
Dimroth reaction, 2, 394
Dioxygen in alkane oxidation to alcohol, catalytic cycle, 256–258
Dioxygen oxidation, platinum alkyl complex, C-H bond activation, 295–298
Diphenylketene, hydrophenylation, 21
Diphosphine ligand coordination to rhodium olefin complex, rate determining step, 73
Direct formation, C-C bonds from C-H bonds, 160–162
Direct oxygenations, naturally-occurring processes, 408
Dissociative pathway, alkanes, linear and cyclic, acceptorless dehydrogenation, 228–231
Double cyclometalation, carbon-hydrogen oxidative addition, 234–247
Double-labeling experiment, intramolecular chain walking, 12*f*–13
Doubly cyclometalated iridium(III) complex
  electrochemical studies, 239–242
  preparation, 236–238
  reactivity, 238
  thermodynamical stability, 242–244

E

Early examples, oxidative addition, C-H bonds, 9–11
Electronic structure, lanthanide-X bond, 118–122
Electrophilic activation, benzene, platinum methyl complexes, bidentate anionic N,N-and N,C-donor ligands, 328–330
Electrophilic aromatic C-H activation, metal trifluoroacetates, 393–406
Electrophilic systems, overview, 5–9
Electrophilic *versus* oxidative activation, relationship, 4
Enthalpy surface (calculated), methane, reductive elimination, 93f–94
Enzymatic oxidations, coreductants, 380
Enzymes, omega-hydroxylases, thermal functionalization, alkanes, 151
Equilibrating intermediates, competitive trapping, 269–270
Equilibria, platinum(IV) hydridoalkyl and platinum(II) alkane complexes, 264–282
Equilibrium isotope effects. *See* Kinetic and equilibrium isotope effects

F

Fenton-type chemistry, non-organometallic C-H oxidation, 33
Ferrocene, fast-reacting substrate, H/D exchange, 52
Fischer carbene complex, double C-H activation, 171–172f
Five-coordinate d^6 complexes, computational methods, 17–19
Fluorinated 2-arylpyridines, iridium triply cyclometalated organometallic complexes, 403–405
Fluorinated phthalocyanines, photophysical properties, 414–415
Fluorinated pyrazolyl borates, non-heme type models, 410–413
Fluorine-lined cavity, 410–411
Fluoroalkyl substituted phthalocyanines, 409
Friedel-Crafts reactions, 164, 361–362
Free energy profiles
n-alkane/α-olefin transfer-dehydrogenation cycle, 208–213
Brønsted relation for proton transfer reactions, 359
cyclooctane/t-butylethylene transfer-dehydrogenation cycle, 206
deuterium scrambling, σ-alkane complexes, 62–66
manganese-oxo-phenanthroline dimers, 359–360
propane addition, 225–229
ruthenium-oxo-polypyridyl complex, 359–360
Fujiwara reaction, 395–396

G

Gibbs free energy. *See* Free energy profiles
Gif systems
conversion of alkanes to ketones, overview, 33
iron catalysts in presence of zinc or iron(0) as reductant, 381
Green chemistry
hydrogen peroxide, oxidant, 380
ylide formation, 416–417
Greenhouse gas generation, 407

H

H/D exchange reactions
   alkanes catalyzed with oxygen-donor-iridium complexes, 109
   catalytic, 51–52
   catalyzed by methyl-iridium(III)-acetylacetonate-pyridine complex, proposed mechanism, 108f
   σ-complex intermediates, 88–92
   deuterated trispyrazolylborate-rhodium-neopentyl-isocyanide-alkyl complexes, kinetics, 62–68
Haber-Weiss scheme for iron complexes, 419
Halo tris(pyrazolyl) borates, 409
Heck reaction
   catalysts, palladium CNC pincer complexes, 176
   coupling by palladium-biscarbene compounds, 170
Heterometallic complexes, organic molecules oxidation with molecular oxygen, 370–378
Heterotrimetallic complexes, 375–376
η^4-Hexamethyl benzene complex, rhodium-cyclopentadienyl complexes, octane reaction, 143
Hydridoiridium(III) centers, C-H bond activation, 50–52
Hydridotris(3,5-dimethylpyrazolyl) borate platinum(IV) complexes, 303–318
Hydridotris(3,5-dimethylpyrazolyl) borate trapping ligand, C-H bond activation, 289–292
Hydroborylation, hydrocarbons, 27
Hydrocarbon oxidations
   by palladium-generated hydrogen peroxide, 384–385
   with coreductants, 379–392
   with hydrogen peroxide generation at single metal, 384–385
   with hydrogen peroxide generation at two different metals, 385–389
Hydrocarbons, non-organometallic metal-mediated oxidation, 357–359
Hydrogen atom transfer rates, 357
Hydrogen peroxide generation, 380–389
Hydrosilylation catalysts, rhodium-N-heterocyclic carbenes, 170

I

Imidazole metallation, normal carbene formation, 171–174
Industrial processes, 407–408
Intramolecular chain walking, double-labeling experiment, 12f–13
Ionic nature, lanthanide-X bond, 119–122
Iridium and rhodium complexes, C-H bond activation, 46–55
Iridium catalyst for arene borylation, 145
Iridium complex catalyst, alkane transfer-dehydrogenation, 198–215
Iridium complex with tri-coordinate "pincer" ligand, potential energy profiles, propane addition, 220–225
Iridium complexes, cycloalkane dehydrogenation, 26
Iridium-ethyl complex, arene C-H activation chemistry, ^{13}C NMR studies, 110–114
Iridium olefin dimers reactions, *t*-butyl phosphorus-carbon-phosphorus ligands, 73–74
Iridium oxygen-donor complexes, alkane C-H bond activation, 105–115
Iridium phenyl hydrides, five-coordinate d^8 complexes, stability, 18
Iridium triply cyclometalated organometallic complexes,

fluorinated 2-arylpyridines, 403–405
Iridium(I), *t*-butyl-PCP ligand system, 73, 75–76
Iridium(I)/iridium(III) couple, transfer dehydrogenation catalysts, 218
Iridium(I)-monocarbene complexes, transmetallation, 180–181*f*
Iridium(III) bidentate N-heterocyclic carbene complexes, 178–79
Iridium(III) complex, doubly cyclometalated, 238–244
Iridium(III) intramolecular C-H oxidative addition, 235–238
Iridium(III)/iridium(V) interconversion, 49
Irreversible reductive coupling in secondary alkyl hydride complex, 60
Iron boryl complexes, 140
Iron oxygenation chemistry, biologically relevant, 418–419
Isonitriles, insertion into C-H bonds, 20–21
Isopropyl deuteride complex, metathesis reaction with Cp_2ZrD_2, 59–62
Isostructural neutral and cationic complexes, mechanistic comparisons, 341–343
Isotope effect measurements, isopropyl hydride complex, rearrangement, 59–61
See also Kinetic and equilibrium isotope effects; Kinetic isotope effects

K

KIE. *See* Kinetic isotope effects
Kinetic and equilibrium isotope effects, computational determination, *ansa*-tungstenocene methyl-hydride complex reactions, 93–100
See also Isotope effect measurements; Kinetic isotope effects
Kinetic isotope effects (KIE)
arene C-H bond activation, 313
cyclopentadienyl metal complexes, 13–14
definitions, 94–95
oxidative cleavage, isopropryl deuteride, 59–61
platinum(II) η^2-arene intermediates, 313
reductive coupling, isopropyl deuteride, 59–61
reductive eliminative, methane from *ansa*-tungstenocene methyl-hydride complex, 89–92
See also Isotope effect measurements; Kinetic and equilibrium isotope effects
Kinetic site, protonation, platinum-dimethyl complexes, 277–279

L

Labeling experiment, competition between β-H elimination and C-H bond elimination, 205
Lanthanide center, non-participating 4*f* electrons, 118–119
Lanthanide complexes, electronic structure, H-H, C-H and Si-H bond activation, 116–133
Lanthanide-cyclopentadienyl-hydride complex
reaction with dihydrogen, 123–125
reaction with silane, 127–128
Lanthanide-cyclopentadienyl-methyl complex
reaction with dihydrogen, 125–126
reaction with methane, 127

Lanthanide-X bond, electronic structure, 118–122
Lanthanum, potential energy profile, 124
Light emitting diodes, materials development, 403–405
Lutetium-cyclopentadienyl-hydride complex, methane activation, 122

M

Manganese-oxo-phenanthroline dimers
  oxidation alkylaromatic compounds, 361–363, 367
  oxidation redox potentials, 359–360
Marcus theory, 359–360, 367
Mercat system, organometallic C-H activation mode, 31
Meridional ligands. *See* Pincer compounds
Metal-boryl complexes, stoichiometric studies, 139–141
Metal catalyzed decomposition, hydrogen peroxide, 382–384
Metal catalyzed formation, hydrogen peroxide, 381–382
Metal trifluoroacetates, electrophilic aromatic C-H activation, 393–406
Metallation assisted by weak base, 176–180
Metalloenzymes, heme and non-heme classes, 409
Metalloproteins, histidine residues, possible binding, 180–181
Metalloradicals, organometallic C-H activation mode, 30–31
Metathesis reactions with Cp_2ZrD_2, isopropyl deuteride complex, 59–69
Methane activation, cyclopentadienyl-lutetium complexes, 122
Methane activation reaction, yielding σ-methane complex, 267–268

Methane at platinum, associative displacement by acetonitrile, 270–273
Methane elimination, C-H bond reduction, 87–92
Methane fast-reacting substrate, H/D exchange, 52
Methane monooxygenase, non-organometallic C-H activation, 32–33
Methane oxidation to methanol and methyl trifluoroacetate, 8
Methane oxidation using platinum bipyrimidine catalyst, Catalytica process, 285
Methane oxidative condensation, 9
Methane reductive elimination, equilibrium isotope effects, 96–100
Methane reductive elimination from ansa-tungstenocene methyl-hydride complex, 87–92
Methane to methyl bisulfate conversion, platinum(II) bipyrimidine catalyst/ sulfuric acid, 7
1-Methyl-iridium-acetylacetonate-pyridine complexes, 106–107
Methyl iridium(III) centers, C-H bond activation, 47–50
Methyl-platinum(II) cation generation by anhydrous route, 259
Methylene transfer reaction, 82–83
Model ligand, DFT calculations, 2,6-$C_6H_3(CH_2PMe_2)_2$, 210–211
Molecular reactor, schematic, 409*f*
Molecular oxygen, oxidation organic molecules, catalyzed by heterometallic complexes, 370–378
Molybdenum cyclopentadienyl-NO-η^2-benzyne intermediates, 190–192
Molybdenum-cyclopentadienyl-NO-hydrocarbyl complexes, sequential C-H bond activation, 184–197

Molybdenum cyclopentadienyl-NO-neopentylidene complex, transformations, 188–189
Monooxygenases, non-organometallic C-H activation, 32–33

N

N-heterocyclic carbene complexes
  methane to methyl trifluoroacetate, 8
  rhodium, iridium, and palladium, 169–183
N-heterocyclic carbenes, spectator ions, 170
N,N-ligated platinum complexes, 256–259
N,N,N',N'-tetramethylethylenediame. See Tmeda-platinum(II) complexes
Negative hyperconjugation, electron density delocalization, 121
Neutral phosphorus-carbon-phosphorus and phosphorus-carbon-nitrogen rhodium systems, C-C bond versus C-H bond activation, 75–78
Neutral platinum methyl complexes, bidentate anionic N,N-and N,C-donor ligands, 321–330
Neutral platinum(II) systems, bidentate, uninegative donor ligand, diphenylborate unit, 336–338
Neutral tri-dentate ligands. See Pincer compounds
Non-organometallic C-H activation, 32–34
Non-organometallic mechanisms, C-H bond oxidation, 356–369
Non-organometallic metal-mediated oxidation, hydrocarbons, mechanisms, 357–359
Non-truncated ligand, DFT calculations, 211–213

O

Octane reactions with rhodium-cyclopentadienyl complexes, 143
α-Olefin binding, catalysis inhibition, 207–209
Olefin hydrogenation, platinum catalysts, 4
Olefin insertion into ortho C-H bonds, aromatic ketones, 22
α-Olefin production by dehydrogenation, 138
Open 4f shells, difficulty in computational studies, 117–118
Organic-based metal active sites, re-engineering, 407–422
Organolanthanide silyl complexes, synthesis, 122–123
Organometallic C-H bond activation, (general), 1–43
Organometallic mechanisms, C-H activation versus non-organometallic metal-mediated oxidation, hydrocarbons, 356–357
Osmium-chromate complexes with bidentate chromate group, alcohol oxidation, 371–374
Osmium three-coordinate d^8 species, methane addition in cyclohexane solvent, 17
Oxidation with molecular oxygen, organic molecules, 370–378
Oxidative addition
  barrier, arene C-H bonds, 312–313
  catalytic functionalization, C-H bonds, 19–27
  chemistry, stoichiometry, 12–19
  phosphorus-carbon-X ligand systems, C-C versus C-H bond, 70–85
  See also C-H bond oxidation
Oxygen atom insertion into C-H bonds, 358–359

Oxygen-donor iridium complexes, alkane C-H bond activation, 105–115
Oxygenation chemistry, biologically relevant iron, 418–419

P

Palladium catalysts, silica supported, with metal oxides, hydrocarbon oxidations, 385–389
Palladium catalyzed, hydrogen peroxide formation and decomposition, 381–384
Palladium CCC-pincer complexes, 176–177
Palladium generated hydrogen peroxide in hydrocarbon oxidations, 384–385
Palladium N-heterocyclic carbene complexes, C-H activation, 174–176
Palladium (II) acetate catalyst
  direct arylation, t-butyl group, 162–164
  reaction with imidazolium salt, palladium bis-carbene species, 176
Perfluorinated phthalocyanines, 409
Perfluoro-(4,5-di-isopropyl) phthalonitrile precursor, structure, 413f–414
Perfluoroalkyl perfluorinated phthalocyanines, heme type models, 413–419
Periana's bipyrimidine complex, 256–257, 266, 320
Peroxide reduction by metal in Fenton-type chemistry, 33
Phenydimethylsilanol, phenyl ring donor, 162–164
Phosphine-ligated platinum(IV) alkyl hydride complexes, reductive eliminations, 294–295
Phosphorus-carbon-nitrogen rhodium/cationic phosphorous-carbon-phosphorus rhodium system, 76–77
Phosphorus-carbon-X ligand systems, C-C *versus* C-H bond oxidation addition, 70–85
Photochemical catalytic functionalization of alkanes, 141–142
Pinacolborane complex, rhodium-cyclopentadienyl complexes, octane reactions, 143
Pincer complexes containing (PCP)M fragments, rhodium, iridium catalysts, 26–27
Pincer compounds, iridium(III), double cyclometalation, 234–247
Pincer ligand in iridium complex, mechanism overview, 218–220
Pincer-ligated iridium catalysts, DFT calculations, acceptorless alkane dehydrogenation, 216–233
Pincer-ligated iridium complex catalyst in alkane transfer-dehydrogenation, 198–215
2-Pivaloylpyridine, substrate, arylation and alkenylation reactions, 164
Platinum alkyl complex, C-H bond activation, dioxygen oxidation, 295–298
Platinum associative coordination, methane, 270–273
Platinum associative decoordination, toluene, 273–276
Platinum bipyrimidine catalyst, methane oxidation, Catalytica process, 285
Platinum/carbon and copper(II) chloride, methane oxidation to methanol, 8
Platinum carbonyl model complexes, relative CO stretching frequencies, 338–340

Platinum-catalyzed C-H bond activation, trapping ligand use, 288–295

Platinum complexes, selective alkane oxidation, reaction mechanisms, 283–302

Platinum-dimethyl complexes, kinetic site, protonation, 277–279

Platinum migration, NMR studies, 311–312

Platinum(II) η^2-arene intermediates, 311–313

Platinum(II) bipyrimidine catalyst/sulfuric acid, methane to methyl bisulfate, 7

Platinum(II)-bis(pyrazolyl) borate system, compared to platinum(II)-bis(phosphino)borate system, 349–351, 352

Platinum(II) C-H bond activation, 250–263, 334–354

Platinum(II) catalyst, L-valine conversion to γ-lactone in water, 158–159

Platinum(II) complexes, reaction mechanisms, 255–260

Platinum(II) diimine complexes, cationic, C-H activation, 266–268

Platinum(II) methyl complexes (neutral), bidentate anionic N,N-and N,C-donor ligands, 321–330

Platinum(II) species, C-H activation mechanism, 19

Platinum(IV) alkyl hydride stable complexes, 304–305

Platinum(IV) aryl hydride stable complexes, 310

Platinum(IV)-dppe (bis(diphenylphosphino)ethane complexes, 285–288

Platinum(IV) five-coordinate intermediates, 308–310

Platinum(IV) hydride intermediate, reversible metalation processes, 345–346

Platinum(IV) hydridoalkyl and platinum(II) alkane complexes, equilibrium, 264–282

Platinum(IV) hydridotris(3,5-dimethylpyrazolyl)borate complexes, C-H bond activation, 303–318

Polyolefins, selective borylation, 145–146

Porphyrins, 409

Potential energy profiles, propane addition, iridium complex with tricoordinate "pincer" ligand, 220–225

Primary and secondary isotope effects, methane, reductive elimination, 95–96t

Protonation, kinetic site, platinum-dimethyl complexes, 277–279

Protonation, reductive methane elimination from hydridotris(3,5-dimethylpyrazolyl)borate-platinum(IV) dimethyl hydride, 307–308

Protonolysis, (tmeda)platinum(II) complexes, 255–256

Q

Quantum chemical calculations. *See* DFT calculations

R

Rate-determining step, general, arene C-H addition, 13

Redox properties, platinum methyl complexes (neutral), bidentate anionic N,N-and N,C-donor ligands, 326–327

Reductive coupling, secondary alkyl hydride complex, 59–61

Reductive methane elimination

hydridotris(3,5-dimethylpyrazolyl)borate-platinum(IV) dimethyl hydride, 306–308
tungstenocene methyl-hydride complexes, 87–88
Regioselective catalytic borylation, alkanes, 136–154
Regioselectivity trends, α-amino acids, 158–159
Relative CO stretching frequencies, carbonyl model complexes, rhodium and platinum, 338–340
Reversible metalation processes, platinum(IV), 342–346
Rhazinilam assembly, diethyl segment, selective functionalization, 157–158
Rhodium complexes
  acceptorless dehydrogenation catalysts, 25–26
  terminal position of alkane reaction, 138
Rhodium-cyclopentadienyl complexes, thermal catalytic functionalization, 142–146
Rhodium-nortricyclyl complex, formation mechanism, 179–180
Rhodium(I) and iridium complexes, C-H bond activation, 46–55
Rhodium(I)-carbonyl model complexes, relative CO stretching frequencies, 338–340
Rhodium(I)-catalyzed coupling reactions, heteroaromatic systems, 53–54
Rhodium(I) complexes, C-C versus C-H activation aptitudes, 76–82
Rhodium(I)-monocarbene complexes, transmetallation, 180, 181*f*
Rhodium(I) olefin complex, disphosphine ligand coordination, rate-determining step, 73
Rhodium(III) catalyzed oxidative carbonylation, arenes, 395–400
Rhodium(III) N-heterocyclic carbene complexes, 178–79
Rhodium(III) trifluoroacetate catalyst, oxidative carbonylation, toluene, 396–400
Ring-flip atropisomerism process, 178
Ruthenium boryl complexes, 140
Ruthenium catalysts, C-C coupling reactions between aromatic compounds, CO and alkenes, 53
Ruthenium-catalyzed insertions, "directing" groups, 22–23
Ruthenium-catalyzed "site-directed" addition, C-H bonds to olefins, mechanism, 23–24
Ruthenium complexes with bidentate chromate group, alcohol oxidation, 371–374
Ruthenium-oxo-polypyridyl complex, 359–360, 364–368

S

Sacrificial hydrogen acceptors, homogeneous catalyst systems, 199
Safety alert, gas mixtures under pressure, gas flammability limits, 390
Schiff base transformation, 162–163
Scrambling processes in trispyrazolylborate-rhodium-alkyl deuteride complexes, 62–67
Selectivity, C-H bonds, 11
Shilov system
  alkane functionalization, 5–7, 252–255
  mechanism, 264–266, 283–284
  mild alkane functionalization, 320
  olefin hydrogenation, platinum catalysts, 4
  oxidation, 303–304
  selective alkane functionalization under mild conditions, 5–7
  See also Alkanes

Sigma-bond activation in lanthanide complexes, 122–129
Sigma-bond metathesis, organometallic C-H activation mode, 28–29
Sigma-complex intermediates, *ansa*-tungstenocene methyl-hydride complex reactions, 88–92
Sigma complexes in oxidative addition and reductive elimination, 87
β-Silicon-carbon bonds, agostic interaction in lanthanide complexes, 120–122
Silver(I)-monocarbene complexes, transmetallation, 180
Silyl groups, oxidative addition, C-H bonds, 16
Silylation reaction, lanthanide complexes, 122–123
Single-step carbon-carbon bond activation by metal complex, data, 75–76
Single-turnover studies, alkane and arene borylation, 148–150
Sixteen-electron η^2-allene intermediates, 194–196
Sonogashira reaction, 176
Spectator ions, N-heterocyclic carbenes, 170
Square planar, bent back, cyclopentadienyl complexes, 15
Stacking interactions, 414–415
Stannylation, electrophilic, arenes, 400–403
Steric requirements, C-C and C-H bond activation, 79–82
Stoichiometric alkane borylation, mechanism, 146–148
Stoichiometric oxidative addition chemistry, 12–19
Sulfide exchange, neutral platinum methyl complexes, 329–330
Suzuki-Miyaura coupling, 404
Suzuki reaction, 145, 176

T

Teleocidin B4 core, synthesis, 161–162
Teleocidin natural products, retrosynthetic analysis, 160*f*
Tetrahydrofuran (THF) ligand self-exchange in benzene, 343–344
TFE. *See* 2,2,2-Trifluoroethanol solvent
Thermal catalytic functionalization, alkanes, 142–144
Thermochemistry, non-organometallic metal-mediated C-H bond oxidations, 359–360
Thermolysis, reductive methane elimination from hydridotris (3,5-dimethylpyrazolyl)borate-platinum(IV) dimethyl hydride, 306–307
THF. *See* Tetrahydrofuran
2-Thiomethoxybenzylidene Schiff base, 163
Thorpe-Ingold effect, 244
Three-coordinate d^8 metal centers, other non-cyclopentadienyl-bearing species, 16–19
Three-dimensional fluorinated ligands, 409–419
Tin(IV) trifluoroacetate complexes, electrophilic metalation, arenes, 400–403
Tmeda-platinum(II) complexes, protonolysis, 255–256
Toluene from platinum, associative decoordination, 273–276
Toluenes, oxidation by manganese-oxo-phenanthroline dimers, 361–363, 367
Tri-coordinate "pincer" ligand iridium complex, potential energy profiles, propane addition, 220–225
2,2,2-Trifluoroethanol (TFE) solvent anhydrous methyl-platinum(II)

cation generation, 259
C-H activation, platinum(II) diimine complexes, 266–268
Trispyrazolylborate metal complexes, 12–16
ansa-Tungstenocene methyl-hydride complex, reduction elimination and oxidative addition reactions, 86–104
Transfer dehydrogenation, cyclooctane/*t*-butylethylene, catalytic cycle, 204–206
Transmetallation from silver-carbene complexes, 180
Tungsten boryl complexes, reaction with terminal C-H alkane bonds, 140–141*f*
Tungsten-cyclopentadienyl-NO-η^2-acetylene intermediates, 192–194
Tungsten-cyclopentadienyl-NO-alkylidene complexes, computational mechanistic studies, 189–190
Tungsten-cyclopentadienyl-NO- η^2-allene intermediates, 194–196
Tungsten-cyclopentadienyl-NO-benzylidene complex, transformations, 187–188
Tungsten-cyclopentadienyl-NO-hydrocarbyl complexes, sequential C-H bond activation, 184-197
Tungsten-cyclopentadienyl-NO-neopentylidene complex, transformations, 185–187

U

Unsaturated C-C bonds, insertion into C-H bonds, 21–24

V

L-Valine, conversion to γ-lactone in water, platinum catalyst, 158–159
Vitamin B_{12}, non-organometallic C-H activation, 34

W

Water gas shift reaction, 385
Weak base, role in metallation process, 176–180
Wilkinson's hydrogenation catalyst, 4, 25

Z

Zeolite encapsulated iron phthalocyanine, 419
Zwitterionic resonance, 336